AF763893

BENEFICIAL CO-UTILIZATION OF AGRICULTURAL, MUNICIPAL AND INDUSTRIAL BY-PRODUCTS

BENEFICIAL CO-UTILIZATION OF AGRICULTURAL, MUNICIPAL AND INDUSTRIAL BY-PRODUCTS

Edited by

SALLY BROWN
USDA, ARS, Beltsville, Maryland, U.S.A.

J. SCOTT ANGLE
University of Maryland, College Park, Maryland, U.S.A.

and

LEE JACOBS
Michigan State University, East Lansing, Michigan, U.S.A.

Beltsville Agricultural Research Center

SPRINGER SCIENCE+BUSINESS MEDIA, B.V.

A C.I.P. Catalogue record for this book is available from the Library of Congress.

ISBN 978-0-7923-5189-4 ISBN 978-94-011-5068-2 (eBook)
DOI 10.1007/978-94-011-5068-2

Printed on acid-free paper

All Rights Reserved
© 1998 Springer Science+Business Media Dordrecht
Originally published by Kluwer Academic Publishers in 1998

No part of the material protected by this copyright notice may be reproduced or utilized in any form or by any means, electronic or mechanical, including photocopying, recording or by any information storage and retrieval system, without written permission from the copyright owner.

BENEFICIAL CO-UTILIZATION OF AGRICULTURAL, MUNICIPAL AND INDUSTRIAL BY-PRODUCTS

Proceedings of the Beltsville Symposium XXII held on May 4-8, 1997, at the Beltsville Agricultural Research Center, Beltsville, MD.

Symposium Co-Chairs:

Ronald F. Korcak, Plant Sciences Institute
Patricia D. Millner, Soil Microbial Systems Laboratory
Robert J. Wright, Environmental Chemistry Laboratory

Symposium Committee Chairs:

Program: Rufus L. Chaney, Environmental Chemistry Laboratory
Posters: Lawrence J. Sikora, Soil Microbial Systems Laboratory
Abstracts: Laura L. McConnell, Environmental Chemistry Laboratory
Proceedings: Sally Brown, Environmental Chemistry Laboratory
Budget: Albert E. Herner, Environmental Chemistry Laboratory
Local Arrangements: Tim Badger, Farm Operations Branch; Sally Reynolds, Soil Microbial Systems Laboratory; Charlotte Schomburg, Environmental Chemistry Laboratory; Cecil Tester, Soil Microbial Systems Laboratory; FAR-B Representative: Morton Beroza

Sponsored by:
The Beltsville Agricultural Research Center, Agricultural Research Service, United States Department of Agriculture
with the cooperation of
Friends of Agricultural Research - Beltsville (FAR-B)

Financial Contributions by:

Wheelabrator - BioGro. Inc.
American Coal Ash Association
National Aggregates Association

PREVIOUS BELTSVILLE SYMPOSIA

1976	Virology in Agriculture
1977	Biosystematics in Agriculture
1978	Animal Reproduction
1979	Human Nutrition Research
1980	Biological Control in Crop Production
1981	Strategies of Plant Reproduction
1982	Genetic Engineering: Applications to Agriculture
1983	Agricultural Chemical of the Future
1984	Frontiers of Membrane Research
1985	Biotechnology for Solving Agricultural Problems
1986	Research Instrumentation for the 21st Century
1987	Biomechanisms Regulating Growth and Development: Keys to Progress
1988	Biotic Diversity and Germplasm Preservation - Global Imperatives
1989	The Rhizosphere and Plant Growth
1990	Remote Sensing for Agriculture
1991	Photomorphogenesis in Plants: Emerging Strategies for Crop Improvement
1992	Agricultural Water Quality Priorities
1993	Pest Management: Biologically Based Technologies
1994	Advances in Human Energy Metabolism
1995	Biotechnology's Role in the Genetic Improvement of Farm Animals
1996	Global Genetic Resources - Access, Ownership, and Intellectual Property Rights

Mention of trade names in this publication is solely for the purpose of providing specific information. Mention of a trade name does not constitute a guarantee or warranty of the products by the USDA or an endorsement over other products not mentioned.

FOREWORD

Co-utilization or blending of residuals offers a unique opportunity to develop products with particular characteristics that are able to target specific customer needs. The very notion of deliberateiy blending by-products suggests that the recycling and beneficial reuse industries are taking a quantitative step forward towards developing products rather than simply reusing residuals. At the same time that this step provides unique opportunities, it also presents unique challenges. The science associated with the beneficial use of one product may not apply when that product is mixed with another residual. Blending of materials may alter the chemistry of the components of the mixture. This may offer additional benefits, as in the case of disease suppression in composts, or present unexpected problems, as the use of lime-stabilized biosolids has done in Maryland.

This book consists of the proceedings of the Beltsville Symposium. The organizers of the Symposium attempted to structure a meeting that would outline both the potential benefits of co-utilization as well as concerns. The editors have divided the proceedings into sections that describe the practical basis for co-utilization of residuals as well as the potential benefits. Specific considerations are described. Finally, case studies include descriptions of successful operations and data that detail results of research involving co-utilization materials.

Blending of materials for specific objectives needs to be the focus of any successful co-utilization effort. The scientific implications of the mix need to be determined before a product can be used properly. The proceedings of this Symposium will provide useful information towards this end. There is an important distinction between reuse of residuals that results in no adverse effects and development of products that meet specific needs. The editors of these proceedings hope that this volume helps to move the industry towards the latter goal.

Sally Brown, Editor
Environmental Chemistry Laboratory, NRI
Agricultural Research Service

ACKNOWLEDGEMENTS

The editors would like to thank Dorothy Talmud for her assistance in the final preparations of the proceedings. Without her patience and expertise, the contributions of the participants in the Symposium may have been appreciated only by those who actually attended the meetings.

TABLE OF CONTENTS

SCIENTIFIC AND MANAGERIAL CONSIDERATIONS

SPECIFIC CASE STUDIES

RESEARCH REPORTS

WHY CO-UTILIZATION?

RONALD F. KORCAK
Plant Sciences Institute, Agricultural Research Service, US Department of Agriculture, Beltsville, Maryland 20705, USA
E-mail rkorcak@asrr.arsusda.gov

ABSTRACT

The land application of by-products from agricultural, industrial or municipal sources is certainly not a new phenomena. Wood ashes, manures, crop residues and even some of the first food processing wastes, such as oyster shells, were applied to the land and, dependant upon site specifics, probably showed a beneficial response by the next crop. These positive responses led to agricultural practices which were continued over time. Today, with renewed interest in concepts such as sustainability, biodynamic farming, and natural resource conservation, the practice of applying by-products to land continues.

However, a problem exists. This problem is concentration. Society, has developed technologies for the production of food, fiber, energy and the whole gamut of consumer goods with little regard for the by-products that these technologies produce, or the effects that the processes themselves may have upon the environment. The concentration per se of resources such as manures, coal ashes, waste mineral fines from crushing rock, paper mill sludges, sewage sludges, etc. is further complicated by economics. Even if an individual by-product has some intrinsic value, like the fertilizer value of manure or as a source of trace minerals like waste mineral fines, there is a finite, albeit small, radial area within which these materials can be transported and utilized economically. Thus, the stockpiles continue to get larger and research dollars are spent on better liners for landfilling materials.

The concept of co-utilization is simply the blending, mixing, and/or co-composting of two or more by-products in order to produce a value-added "designer" material which can be beneficially utilized to solve an agricultural problem, remediate soils, and/or fulfill a market niche.

There are numerous examples, many of which will be noted in the presentations throughout the Symposium, where materials that were once called wastes are being co-utilized as value-added products. The goal of this Symposium is then two-fold: first, to identify areas where co-utilization is being practiced and to foster this utilization to better use our natural resource base; and second, to highlight where good science can be practiced in developing co-utilization products that will have a significant impact on our society at large.

S. Brown, J.S. Angle and L. Jacobs (eds.), Beneficial Co-Utilization of Agricultural, Municipal and Industrial By-products, 1–7.
© 1998 *Kluwer Academic Publishers,*

INTRODUCTION

Numerous materials that were once considered "wastes" are now being utilized, either as soil amendments or as components of composts. The amounts of "wastes" generated annually in the United States are large. For example, consider the following list:

- 60 million tons of animal manures (with a fertilizer value of $3.4 billion)
- 90 million tons of coal combustion residues used to generate electricity
- 300 million tons of municipal solid waste (MSW)
- 35 million dry tons of biosolids
- 0.3 million tons of scrap wallboard from residential construction sites
- 0.6 million tons of scrap wood from residential construction sites

Some of the above numbers for manures, biosolids and MSW are familiar to those involved in the arena of "waste" utilization. Inorganic by-products like coal combustion residues consist of a wide gamut of materials, from conventional fly ashes to high alkaline, high gypsum containing fluidized bed combustion residues to flue gas desulfurization residues, which are either predominantly gypsum or calcium sulfite, depending upon the level of forced oxidation. The last two numbers in the above list represent rough estimates for only the single-family home construction industry (Personnel Communication, Peter Yost, National Home Builders Research Center). If multi-unit sites and commercial construction were included, the numbers would be significantly higher.

These numbers set the stage for the magnitude of just a few of the agricultural, municipal and industrial by-products that may be available for the production of co-utilization products.

DISCUSSION

Definition of Co-Utilization

Co-utilization is simply the production of new products from the combination of two or more by-products. The term co-utilization is introduced to place emphasis on expanding the types of by-products utilized in order to broaden the types of products that can be produced. Thus composting, which usually entails a high carbon source (leaves or wood) and a high nitrogen source (manures or biosolids), is considered to produce a co-utilization product. In addition to composts, "designer" or "tailor-made" products would also fit the definition of co-utilization products. These types of products are produced to fulfill a specific agricultural need, e.g., selected crop nutrient requirement, or fulfill a specific need. e.g., achieving a high iron and phosphorous product for use in reducing the bioavailability of environmental lead contamination. Finally, products may be produced from co-utilization without composting, which will be presented later.

Logic of Co-Utilization

The composting or blending of two or more by-products will result in value-added products. Numerous examples of low value materials are scrap wallboard at construction sites, waste mineral fines from crushed stone and aggregate facilities, cotton gin waste, residues from coal combustion plants, etc. As with animal manures and biosolids, these materials are generated at specific sites. Their low value is magnified due to the economics of loading, transporting and, at some point, utilizing these materials. However, combinations of these materials into a co-utilization product may result in higher value which, in turn, increases the maximum distances these combined materials can be economically transported and utilized.

Development of co-utilization products requires that there be an identifiable market need for the product. Products should not be developed just because certain materials are in close proximity or that they are available. Products should be developed to fulfill market needs which can vary from agricultural to forestry to horticultural to remediation product needs.

Beneficial Concerns Related to Co-Utilization Products

In addition to having an identifiable market need for a co-utilization product, the product must be able to provide a beneficial use. The importance of a product providing a beneficial use lies in the concern that, if no beneficial use is obtained, then one is simply disposing of the product. However, beneficial use may take longer to realize for some products. Short-term benefits, such as crop response to applied nutrients, may be easy to document. However, in some instances long-term benefits may be more difficult to document. An example would be the addition of organics as an aid in increasing soil tilth or structure. Such benefits may require annual applications over a number of years before beneficial responses can be documented. Likewise addition of a high-gypsum containing co-utilization product to increase soil water infiltration and alleviate subsoil acidity may not produce beneficial responses for a number of years.

This concern over beneficial use versus disposal has direct effects on regulatory concerns in the utilization of co-utilization products. Many products may be marketed as soil amendments or fertilizers or liming substitutes. However, the documentation required for such labeling of co-utilization products varies from state to state. The best remedy for this variation is to involve appropriate regulatory bodies within states at the onset of product development, testing and evaluation.

Liability concerns of industries providing materials for the formulation of co-utilization products also exists. Comprehensive testing of the individual materials to be used, as well as testing of the final combined products, will greatly reduce the prospect of liability concerns becoming an issue for using co-utilization products.

Cost Factors

Many by-products have low intrinsic value. By definition co-utilization strategies are aimed at producing value-added products. Any activity performed on a by-product, e.g., loading onto trucks, transporting, pelletizing, etc. adds an additional cost. The economics of utilizing these by-products are further complicated by the fact that for many industrial, agricultural or municipal waste streams, the dollars allocated in dealing with the wastes generated are low. Perhaps this is an extension of the "out of sight, out of mind" philosophy of society in general. However, the fact remains that monies are expended to dispose of, or manage the waste, not necessarily to utilize the waste.

This trend may be reversed by a number of ways. Two examples are:

1. Green Energy -- Deregulation of utilities will allow utilities to offer their service to consumers outside their normal market zone, much like the effect deregulation has had on the telephone industry. One of the marketing strategies that may be pursued is Green Energy. The utility would assure the consumer that the electric power provided was produced using "green" technologies, such as flue gas desulfurization, fluidized bed combustion or some other Clean Coal Technology. And, in addition, the utility would assure that the wastes generated would also be utilized, for example as a component of a co-utilization product. However, the cost for this Green Energy service would be a few more cents per kilowatt hour. The additional cost allows the utility to use these new technologies, including co-utilization of the generated residues.
2. Waste Recovery Fund -- This is an old concept which has been discussed by economists over the years. An example of a Waste Recovery Fund would be a two to three cents per pound "assessment" on each pound of chicken broiler produced. The assessment would accumulate in a Waste Recovery Fund with the monies used to better utilize the manures and processing wastes derived in the production of broilers. Such Funds could assist in research on co-utilization products that would have greater value, thereby allowing greater (radial) dispersion of these products to alleviate the concentrating effects of large broiler production farms.

Whether or not strategies such as those described above ever become reality, the fact remains that society at large can find novel ways to deal with wastes as a recognized component in the production or manufacture of most any consumer product or service.

Sustainability and Co-Utilization Products

Production of co-utilization products will have direct impacts on the concept of sustainability, whether one thinks of agricultural sustainability, community sustainability or sustainability on a more global scale. Although the term agricultural sustainability has been met with confusion and frustration in definition, there are four components, that are generally agreed upon, which comprise the term sustainability:

- meeting human needs for food and fiber;
- maintain environmental integrity/quality;
- economically and socially viable; and
- all over a long period of time.

The confusion and frustration enters when one's approach to sustainability is considered. An organic farmer would have a different approach to sustainability compared to a low-input or IPM (integrated pest management) grower.

Independent of one's approach, the production of co-utilization products would directly impact the above four components of sustainable systems. Co-utilization products can reverse the trend in declining soil fertility and enhance soil organic matter and structure, thus assisting in maintaining human needs for food and fiber. Co-utilization products can reduce the potential for surface and groundwater pollution, reduce pesticide utilization, and be a component in nutrient management planning or best management practices, all of which would assist in maintaining environmental integrity and quality. Co-utilization products can converse the natural resource base, provide value-added products, create new industries and associated employment, and assist in handling the wastes produced by all sectors of the population, therein being both economically and socially viable. Thus, production of co-utilization products can be one of the many tools in the arsenal for achieving sustainability.

Integrated Systems and Co-Utilization Products

The vast majority of agricultural research on the utilization of by-products has centered around the application of a solitary by-product and measurement and evaluation of soil and plant responses. This single by-product approach can be considered the classical approach to by-product utilization. However, more recent research reflects the examination of materials following co-application including composts, which can be considered co-utilization products. Two examples will be discussed wherein the classical solitary by-product approach evolved into a co-utilization strategy.

Numerous studies were undertaken in the late 1970's and early 1980's on the potential utilization of a new Clean Coal Technology residue from fluidized combustion of coal. Fluidized bed combustion (FBC) involves the simultaneous combustion of coal and calcitic limestone in a fluid bed, wherein the S that is released from the coal is absorbed on calcium oxide (CaO) particles which result from the conversion of limestone to CaO in the furnace. This technology produces a dry, alkaline residue which is essentially one third each of coal ash, un-reacted CaO and gypsum. This residue was examined as a source of plant Ca and S and as a substitute for limestone, if used as a soil amendment. Numerous crops and soil conditions were studied. The culmination of this research was the publication of a manual for agricultural utilization of the residue (Stout et al., 1988). The manual presents a classical approach to by-product utilization.

Over time, the manual was used in the process of obtaining appropriate permits to apply the FBC residues in New York State as a soil amendment/lime substitute on dairy farms. Logistically, the FBC facility did not have on-site storage for the residues which, in the past, necessitated daily hauling of the residue to a landfill. The permitted utilization of the ash resulted in dairy farmers purchasing the ash as a soil amendment. Associated benefits to the farmer included maintaining soil pH, due to the alkalinity of the ash, and reducing the amounts of fertilizer K and B since the ash was supplying supplemental amounts of both plant nutrients. In fact, since most of the dairy fields were planted to alfalfa, which requires a high level of B for optimum growth, the ash-borne B was especially important. The FBC producer was selling ash is to farmers which normally would have been landfilled, and the farmer was obtaining a beneficial soil amendment at competitive prices compared to agricultural limestone, with the additional benefits of supplying P and B.

Three extraneous factors contributed to the evolution of this classical example of by-product utilization into a co-utilization strategy: (1) Prior research had demonstrated that FBC ash has a composition similar to Portland cement. Thus addition of water to the dry ash produces a cementitious product which remains porous; (2) Dairy barns almost always have a wet, muddy area around the barn yard due to the trampling effects of the herd; and (3) The particular area in New York is prone is to heavy snowfalls during the winter with two or more feet of snow cover the norm, which prohibits land application of ash. The result of these factors was the development of ash "pads" around the dairy barns which would set-up like cement, remain dry due to the porosity of the ash, and could be constructed during the winter, when land application of ash is prohibited. The viability of these pads is the focal point of an ongoing study co-funded by the Department of Energy. The objectives of the study include monitoring water quality over and through the pad as well as examination of a co-utilization strategy. This strategy is the co-mingling of dairy manure, which accumulates on the pad, with ash that is intentionally scraped off when the manure is removed. Thus, a co-utilization product is produced (without the need of composting) which potentially has a number of benefits. These benefits depend upon two different methods of utilizing the removed ash-manure co-utilization product. The farmer could spread the product on fields with a manure spreader. The high calcium in the ash has the potential to reduce the water solubility of manure-derived P which in turn has the potential to reduce surface runoff of P into waterways. Alternatively, the farmer may be able to sell the ash-manure product as a value-added product to other farmers, thus distributing the manure farther away from the farm source. Both of these alternatives are being investigated as value-added co-utilization products. Thus, research performed over more than fifteen years ago has evolved from the classical solitary by-product utilization into a number of co-utilization, value-added strategies.

The second example involves the efforts underway in the home construction industry in exploring ways to recycle new construction wastes. This example again demonstrates the evolution from a classical waste handling approach to a co-utilization strategy. A recent, short-term study funded by the Gypsum Association in collaboration with the National Home Builders Research Center examined the recycling of scrap new construction gypsum wallboard (final report of the study is available from the author). The study looks at the pulverization of scrap wallboard on-site followed by the application of the pulverized material directly on the soil around the site. Measurable benefits of this type of application can be identified for most soils since wallboard is over 95% gypsum. However, some soils will not respond to gypsum application, and and at construction sites for commercial buildings or higher density housing, sufficient land surface may not be available for direct application. In such situations, Missouri home builders have taken scrap wallboard recycling and created an integrated, co-utilization product scheme. The builder transports scrap wallboard and scrap wood from his construction sites to a marshaling center. Here these materials are co-composted with leaves and poultry manure from a nearby farm operation. The resultant co-utilization product is then used to grow turf. The turf, in turn, is used for sodding around the builder's construction sites. A full-cycle approach is achieved using a co-utilization product.

The two examples discussed show the benefit of designing a co-utilization product. In both cases the potential exists that the products formed could be value-added, compared to the materials utilized as solitary soil amendments.

THE FUTURE

Numerous waste streams exist that present opportunities to formulate co-utilization strategies. These may vary from region to region. The possible use of new techniques, such as the Geographical Information System (GIS), to assist in planning, identifying and coordinating the use of wastes in co-utilization product development strategies, may be the next step in fostering these strategies. GIS databases already exist that include transportation networks (road, rail, water), geologic information (underlying rock type), and information about the location and type of quarry operation for the production of aggregates, sand and/or gravel. The potential for expanding GIS databases to include industrial, municipal, and agricultural operations, and the types of associated wastes produced, plus information on soil types, type of farming activity, surface erosion potential, planned development for houses/industries, etc. could prove to be a useful tool in designing co-utilization products. Determinations could be made from the database itself on designing products to combat soil erosion, meeting agricultural needs, providing products to new home owners, re-vegetate disturbed lands, or aid in bioremediation of brown fields. Determinations of costs to transport, produce and distribute co-utilization products could also be made from GIS databases. The potential of using systems like GIS appears limitless.

Linkages between industrial, agricultural and municipal entities in the handling and utilization of wastes by following co-utilization strategies can have significant impact on, not only sustainable agriculture, but also sustainable communities. This Symposium is a first step in fostering these linkages.

REFERENCES

Stout WL, Hern JL, Korcak RF & Carlson CW (1988) Manual for applying fluidized bed combustion residue to agricultural lands. U.S. Department of Agriculture, Agricultural Research Service, ARS-74, 15 pp.

AGRONOMIC BENEFITS OF AGRICULTURAL, MUNICIPAL, AND INDUSTRIAL BY-PRODUCTS AND THEIR CO-UTILIZATION: AN OVERVIEW

MARGIE LYNN STRATTON and JACK E. RECHCIGL
University of Florida, Range Cattle Research and Education Center, Ona, Florida, 33865, USA
E-mail ISCI893@aol.com

ABSTRACT

As landfills close and traditional options for disposal of wastes decrease, many by-product materials are viewed as resources for agronomic uses. Many materials termed wastes are rich sources of nutrients and organic material for use in crop production, improvement in soil physical or chemical properties, or as feed for livestock production. Agricultural, municipal, or industrial by-products may be co-utilized, or combined, so that the materials are more easily land applied, provide more complete nutrition, or enhance the soil conditioning, economic, or environmental value of the individual by-products.

INTRODUCTION

As landfills close and concern for the environment increases, new options are sought for the disposal of the various and numerous wastes and by-products produced. In the United States by-product production has been estimated at more than 1 billion tons annually (Walker et al., 1997). Prior to chemical fertilizer production and large scale farming, farmers in small-scale production commonly recycled nutrients produced on-farm to maintain soil fertility. With industrial growth, commercial fertilizers, and population increases, the tendency arose to rely on chemical fertilizers for food production and to dispose of municipal or industrial wastes rather than to cycle nutrients back into food production. However, with the closure of landfills and the increasing concern for the environment, agricultural, municipal, and industrial by-product materials are being considered for various uses such as nutrient supply, feedstocks, and soil amendments. By-products, which were once wasted by being stockpiled, landfilled, ocean-dumped or otherwise not valued, are now recognized for their agronomic value. Land application of by-products is controlled by several documents such as 40 CFR Part 257: solid waste; 40 CFR Parts 261-268: hazardous wastes; 40 CFR Part 761: polychlorinated biphenyls (PCBs); and 40 CFR Part 503; biosolids, the last of which is being examined for possible application to other land-applied by-products such as composts (Walker et al., 1997).

S. Brown, J.S. Angle and L. Jacobs (eds.), Beneficial Co-Utilization of Agricultural, Municipal and Industrial By-products, 9–34.
© 1998 *Kluwer Academic Publishers,*

Details of these regulations may be found in the Federal Register 40 CFR and are summarized along with a statement of policy encouraging beneficial use of by-products in a chapter by Walker et al. (1997). This chapter presents an overview of by-product utilization and co-utilization in agriculture.

AGRICULTURAL BY-PRODUCTS

For the purposes of this chapter, agricultural by-products are plant or animal residues or wastes that are not collected by a municipality for further processing, and that are not found in industrial production (as are food processing, or tannery, papermill, or textile by-products, for example).

Plant Materials

Traditional on-farm treatment of crop residues often involves harvesting the edible parts of the crop while leaving the remainder of plant material on the plots, either incorporated or as a surface mulch. A cover crop may have been grown in the off season or as a rotation. At a later time, the residues may have been plowed under. Stubble-mulch farming and no-till farming are methods of crop residue management that have been studied with resultant increases in crop productivity, soil fertility, and soil organic matter and are reviewed by Stratton and Rechcigl (1997). Plant material may be composted before it is applied to the soil, and use of plant residues is discussed in sections below and also in a review by Stratton et al. (1995).

Plant materials are valued as a source of nutrients and organic matter, and soil applications have many long-term benefits. In the short term, application of some uncomposted crop residues to soil may initially immobilize N, and this effect has traditionally been overcome by use of leguminous residues, timing of planting after incorporation, or addition of N fertilizer or animal manure. Crop residues and animal manures applied together is perhaps one of the earliest examples of co-utilization of by-products. Co-utilization of by-products in discussed in detail at the end of this chapter.

Crop residues may contain, on a dry weight basis, approximately 1 to 1.5% N, 0.15 to 0.2% P, 1% K, 1% Ca, 0.5% Mg, 0.2% S, 30 mg kg^{-1} Mn, 100 mg kg^{-1} Fe, 30 mg kg^{-1} Zn 5 mg kg^{-1} Cu, 20 mg kg^{-1} B, and about 1 mg kg^{-1} Mo (Mills and Jones, 1996). These values will differ with the crop, the plant part, the season, soil moisture over the growing period, the nutritional history of the crop residues, as well as any other factor which affects plant growth. In general, crop residues contain about half the total plant nutrients in the entire crop (Minnich and Hunt, 1979, Mills and Jones, 1996). Slowly mobile or immobile nutrient concentrations such as S, Fe, Zn, Cu, Mn, Mo, B, and Ca, are essentially all retained in the residue (Mills and Jones, 1996; Minnich and Hunt, 1979).

McCalla et al. (1977) give details of nutrients supplied by different crop residues and different plant parts. Corn (*Zea mays* L.) stover and grain, alfalfa (*Medicago sativa* L.) hay, coastal bermudagrass (*Panicum dactylon* L.) hay, sorghum (*Sorghum vulgare*, Pers.) stover, rice (*Oryza sativa* L.) straw, cabbage (*Brassica capitata* L.) and residues of several other fruit, vegetable, and agronomic crops are reported to supply significant amounts of plant nutrients (McCalla et al., 1977). Table 1 reports amounts of N, P, and K that can be returned to the field by stubble mulch farming (McCalla and Army, 1961).

Table 1. Amounts of N, P, and K, returned to the soil by stubble mulch farming.

Crop	Nitrogen	Phosphorus	Potassium
	--------------------------------kg/Mg------------------------------------		
Corn	9	1.5	8.3
Wheat	7	0.6	9.5
Rye	7	0.6	9.5
Oats	6	0.8	12.5
Alfalfa	24	10	16.5

McCalla and Army, 1961

Nutrient replenishment by crop residues is only one of the important benefits of returning plant material to the soil. The nutrients are released slowly and the nutrient supply to the next crop may be held in the root zone, thus allowing a reserve of nutrients. As well as supplying essential elements, the addition of organic matter in the form of crop residues imparts significant improvements in soil chemical and physical characteristics as reviewed by Stratton et al. (1995). Additional chemical benefits may be an increase in aeration from bulking with plant residues, buffering or changes of pH, immobilization of organic toxins or heavy metals, an increase in cation exchange capacity, or a possible decrease in soluble salts depending on the materials used and the conditions of the soil amended. An increase in certain microbial populations may affect plant nutrient availability or plant disease severity. Physical improvements in residue-amended soils are perhaps the most significant benefits from the addition of organic matter. After crop residues are returned to the soil bulk density and compaction are decreased. Crusting, runoff, and erosion are decreased, and permeability and infiltration are increased. Soil moisture is increased and soil structure is improved.

Plant residues are also a significant source of feed for livestock. Some agricultural by-products used as feed include citrus pulp, corncobs, hulls and husks of various crops, malts and hops, molasses, sisal (*Agave* spp.) pulp, sugarbeet (*Beta vulgaris* L.) culls, sugarbeet tops, sugarcane (*Saccharum officinarium* L.) bagasse, sunflower (*Helianthus* spp.) seed, various straws, tomato (*Lycopersicon esculentum* L.) culls, and wheat (*Triticum* spp.) middlings (Pate and Kunkle, 1989). These by-products can be a significant source of nutrition for livestock (Table 2).

Mixtures of feeds may produce a greater weight gain in cattle than single feeds if properly balanced. For example, cattle feed supplemented with feathermeal from poultry processing resulted in about 0.57 kg of daily weight gain, and molasses-supplemented feed resulted in about 0.45 kg weight gain per day, whereas molasses and feathermeal together (co-utilization) resulted in about 0.79 kg per day gain in cattle (W. Brown, University of Florida, personal communication).

Table 2. Nutritional quality of certain plant residues.

	Wheat Middlings	Molasses	Corn Citrus Pulp	Cotton Soybean Hulls	Fiber Hulls	Seed Hulls
	----------	----------	---------- %(dw)	----------	----------	----------
Crude Protein	18	6	7	12	11	4
Total Digestible Nutrients	80	70	70	80	85	42

National Research Council, 1996

Animal Materials

Animal manures, added to the soil, impart many of the same benefits as plant residues. Organic matter additions, in general, improve soil chemical and physical characteristics. A large portion of the plant nutrients ingested by livestock are excreted and by application are returned to the soil for another season of crops. Significant increases in soil fertility and tilth could be achieved if the great amounts of poultry manure (14 Mg), swine manure (16 Mg), or cattle manure (48 Mg) produced in the United States were applied to the soil (Robert Wright, USDA, personal communication).

Poultry manure is considered the richest of the manures in supplying N. So much of the N is in the ammonium form that care must be taken in its use on crops. Poultry manure is often composted or stored prior to use. These practices are wasteful of nutrients. Manures are rich in P and may contribute to over-enrichment of soil P. Table 3 compares nutrient content of plant residues and manures.

Table 3. Nutrient composition of plant residues and animal manures.

Nutrient	Plant Residues[a]	Cattle Manure[b]	Poultry Manure[b]	Swine Manure[b]
	----------	---------- %(dw)	----------	----------
N	1.0-1.5	1.2 -2.0	1.8 - 4.1	1.2
P	0.15 - 0.2	0.3 - 0.8	1.5 - 3.3	0.4
K	1	1.7	1.5-3.2	0.6
Ca	1	1.9	1.6	0.3
Mg	0.5	0.9	0.4	0.2

[a] Mills and Jones, 1996

[b] Elliott and Swanson, 1976

MUNICIPAL BY-PRODUCTS

Municipal by-products are materials from the greater community. These materials are either collected and processed, or produced by the municipality as a bulk waste, at or by a central facility, often regulated by legislature. These by-products include effluents, sludges, biosolids, and cakes from waste water and sewage treatment, and composts from solid waste disposal.

Biosolids and Sewage Sludges

The term sewage sludge has been applied to the solid human waste collected from wastewater, treated at a central processing plant, and which remains as a residual after the liquid effluent is removed. Now the term biosolids is used. In the United States, in 1990, it was estimated 8.5 million Mg of biosolids were produced, with an estimate 12 million Mg produced annually by the year 2000 (Hue, 1995). According to one estimate, 365,000 kg of dried, anaerobically digested biosolids are produced by 10,000 people in one year (Sabey, 1980). If 3.3% of that was N (11,000 kg of N per year), and if all of the N were available, 10,000 people would produce enough biosolids to fertilize between 40 to 100 ha, depending on the crop and management system (Sabey, 1980). By 1989, 2.34 million dry metric tons of biosolids were land-applied (U.S. EPA, 1993).

With careful application, biosolids can be a good source of nutrients for agronomic use. The liquid effluent portion of sewage treatment may also supply plant nutrients (by dw) 2% N, 1% P, 1.4% K, 2.4% Ca, and 1.7% Mg (Bouwer and Chaney, 1974). Since the 503 Regulations (Table 4) some biosolids are detoxified by removal of heavy metals either at the source or by special processing known as autothermal aerobic digestion or liquid composting (Jewell, 1994). Table 5 gives examples of metal concentrations in biosolids before and after detoxification. The data in Table 5 shows the results of advances that have been made in removal of metals from biosolids. This removal process, along with the prevention of metals from entering the waste stream has done much to improve the quality of biosolids compared to the recent past. Other processes used in biosolids treatment include dewatering or alkali-stabilization. Analysis of these products may be found in Table 6.

Water Treatment Effluents and Solids

Treatment of ground and surface waters for municipal production of clean drinking water results in sediments from the settling processes. Two of the more common sludges are high-alum sludge from treatment of water with aluminum sulfate, and iron humate from the treatment of water with iron salts (Christopher Lind, General Chemical Corporation, personal communication; Rehberg and Smith, 1997). Table 7 shows typical analyses for these by-products. Table 7 clearly indicates the deficiency of macronutrients limiting the use of these by-products as general fertilizers. However high alum sludge has the ability to raise soil pH and bind phosphorus, processes which may be useful in certain soils or with certain crops. Iron humate is a good source of iron and may improve moisture retention in soil (Rehberg and Smith, 1997). These products may be mixed or enhanced

Table 4. Limits for sludge application to cropland as regulated by CFR Title 40, Part 503.

Metal	Ceiling[a] Limit	Cumulative[b] Loading	Pollutant[c] Concentration Limit
	--------(dw)--------		
	mg/kg	kg/ha	mg/kg
As	75	41	41
Cd	85	39	39
Cr	3000	3000	1200
Cu	4300	1500	1500
Pb	840	300	300
Hg	57	17	17
Mo	75	18	18
Ni	420	420	420
Se	100	100	36
Zn	7500	2800	2800

[a] Ceiling Limit- the maximum allowable pollutant concentration in land-applied biosolids
[b] Cumulative loading - the maximum amount of a pollutant that can be land-applied if the pollutant concentration is below the ceiling, but above the pollutant concentration limit.
[c] Pollutant concentration limit - the pollution concentration in biosolids below which sewage sludge can be land-applied without restrictive requirements and management practices.

Table 5. Metal concentrations in biosolids before (B) and after (A) detoxification treatment.

Municipality	Biosolids Metal Content									
	Cd B	Cd A	Zn B	Zn A	Cu B	Cu A	Ni B	Ni A	Pb B	Pb A
	--------mg/kg (dw)--------									
Binghamton	16.1	1.4	2,900	200	749	278	482	60	247	82
Syracuse	27.4	3.1	620	141	871	199	88	21	195	46
Chemung Co.	115	11.9	2,500	169	960	165	43	10	244	178

Jewell, 1994

to use as fertilizer as discussed in the section on co-utilization below.

Municipal Solid Waste Compost

Municipal solid waste (MSW) is the trash, refuse, or garbage produced by each household and put by the curbside to be collected regularly by the municipality. Historically, municipalities incinerated, land-filled, or ocean-dumped this waste, however with escalating disposal costs and risks, MSW is being considered for recycling. It was estimated the U.S. produced 209 Mg of MSW in 1994 (EPA, personal communication). Of this figure 61% was landfilled, 23.6% recycled or composted, and 15.5% was incinerated (EPA, personal communication). Of the 209 Mg MSW, 38.9% was paper and

Table 6. Analysis of alkali-stabilized or dewatered sludge.

Parameter	Alkali-Stabilized[a]	Dewatered[b]
pH	11.9	
EC,ds/m	9.0	
	--------------------%(dw)---------------	
C	12.1	
N	0.89	
P	0.41	2.69
K	1.26	
Mg	1.0	
Ca	25.0	
	---------------- mg/kg (dw)------------	
As	7.6	
Cd	0.83	5.6
Co		10.7
Cr		130
Cu	134	587
Mo	1.38	35.8
Ni	55	
Pb	48	132
Se	1.69	
Zn	186	545

[a] Logan and Harrison, 1995
[b] Richards et al., 1997

Table 7. Analysis of high alum sludge and iron humate.

Nutrient	High Alum Sludge[a]	Iron Humate[b]
	------------------%(dw)------------------	
N	0	0
P	0	1.0-1.5
K	0-0.4	0.5-1.0
Ca	4-7	1.5-2.5
Mg	0-1	0.02
S	0-4	2-3
Fe	0-5	16-20

[a] Christopher Lind, General Chemical Corporation, personal communication
[b] Rehberg and Smith, 1997

paperboard, 28.3% was wood, food, and yard trimmings, and 32.9% was glass, metals, plastic, or other (EPA, personal communication). It is estimated that 67.2% of that material might be easily compostable at municipal waste processing plants which employ separation prior to composting (Stratton et al., 1995). Municipal solid wastes must be processed in some way before considered for agronomic use. Ideally, MSW is collected at roadside, brought to the facility, separated by material (glass, metal, and paper are recycled), and the compostable materials are shredded and composted (Stratton et al., 1995). Analyses of two MSW composts are reported in Table 8. Municipal solid waste compost can be a good source of plant nutrients and organic matter for soil amendment. Some composters combine MSW with biosolids prior to composting, another example of co-utilization.

Table 8. Analysis of municipal solid waste composts.

	Source	
Parameter	Sumterville, FL[a]	Sevierville, TN[b]
pH	7.89	7.1
EC (ds/m)	2.6	2
C/N	20.8	18.6
	--------------------%(dw)--------------------	
N	1.31	1.7
P	0.22	0.65
K	0.18	1.1
Ca	2.92	2.4
Mg	0.18	0.6
	--------------------mg/kg(dw)--------------------	
Zn	1210	800
Cu	413	200
Mn	630	460
Fe	25,650	17,000
B	75	4
Cd	6	6.6
Pb	204	240
Ni	52	56

[a] Stratton and Rechcigl, original data
[b] A. Shiralipour, University of Florida, personal communication

Residential Yard Trimmings

In many municipalities, lawn, tree, and shrub trimmings are collected at curbside and processed at a central facility. The woody by-products are chipped or shredded and used as mulch or added to composts, often as a bulking agent for composted biosolids, an effective co-utilization of these by-products. The material may be composted municipally or sold to a private composter. The mulch is spread on public lands such as highway slopes, parks, and lots, or sometimes sold for homeowner use. The green materials are dried and often composted at a municipal or private facility. Amounts of municipally collected yard trimmings are rapidly changing as more municipalities collect and handle these by-products. Analyses of residential yard trimmings from two sources in Florida are shown in Table 9.

Table 9. Analyses of residential yard wastes.

Parameter	Florida Yard Waste 1	Florida Yard Waste 2
pH	8.02	8.06
Ec (Ds/m)	2.76	2.42
C/n	24.1	24.7
	--------%(dw)--------	
N	0.9	0.67
P	0.11	0.19
K	0.26	0.21
Ca	6.45	4.05
Mg	0.14	0.21
	--------mg/kg(dw)--------	
Zn	251	38
Cu	26	10
Mn	47	56
Fe	2000	3165
B	17	17
Cd	0.5	0.5
Pb	41	9.5
Ni	14	2

Stratton and Rechcigl, original data

INDUSTRIAL BY-PRODUCTS

Energy Production By-Products

Coke is the product of distillation of coal and is still used in some boilers for energy production, or for heating homes. The production process of coke is destructive (involving anaerobic heating) and results in the production of by-product ammonia which historically has been used to make fertilizer (Jones, 1979).

Boiler ash results from combustion, in most cases for energy production. Boiler ash refers to a few different by-products, including base or bottom ash, fly ash, wet scrubber ash, and mixtures of the three, among others. The substrate may be coal, wood, paper, or even some sludges. The composition of the ash is a result of substrate combusted.

Of energy production by-products, those from burning coal are most abundant. Two million tons of coal combustion by-products are produced in the U.S. annually (American Coal Ash Association, personal communication). Of this figure, 58.7% is fly ash, 21.7% flue gas desulfurization product, 16.3% bottom ash, and 3.3% boiler slag (American Coal Ash Association, personal communication). Coal combustion by-products supply plant nutrients, increase soil pH, decrease Al toxicity, enhance root penetration, improve soil structure, reduce bulk density of soil, improve water-holding capacity, and acts as a barrier to weeds (Chang et al., 1989). Composition of some coal combustion by-products is shown in Table 10.

Table 10. Analysis of coal combustion by-products.

Element, Nutrient, Parameter	Bottom Ash[a]	Fly Ash[b]	Fluidized Bed Combustion[c]	Flue Gas Desulfurization[d] Direct Lime	Dual Alkali
pH	8-9.4	12.1			
	---------------------------------mg/kg(dw)--				
P			380-500		
Ca	8-51	0.1-177	240-460		
Fe	27-203	8-289	<1-16		
Mg	5-61	4-32	5-12	24-420	0.1
K	2-35	7-16	<1-8	11-28	320-380
B	--------10-1300-------		95-170	-----------2-530------------	
Cu	--------3.7-349--------		12-19	-----------tr-340------------	
Mn	--------56.7-767--- ---		210-685		
Mo	--------0.84-100-------		0.12-0.28		
S			72-140		
Sulfate			800-4500	-------80,000-84,000-------	
Sulfite			0.9-2.7		
Zn	--------4.0-2300------		12-19		
Na			36-137	-------53,600-55,300-------	
Al			3-20		
As	--------0.5-279-------			-------------tr-52-----------	
Cd	--------0.1-18---------		0.5	-------------tr-180----------	
Cr	--------3.4-437-------		9-23	-------------tr-180----------	
Co	--------4.9-79--------				
Fl	--------0.4-320------			------------0.2-1017---------	
Pb	--------0.4-252------		1.5-7.5	-------------tr-290-----------	
Hg	-------0.005-4.2-----			--------------tr-6------------	
Ni	-------1.8-258--------		13-29		
Se	-------0.08-19--------		0.16-0.58	--------------tr-60-----------	
Ag	-------0.04-8---------				
Sr	-------30-3855-------				
Tl	-------0.10-42--------				
Vn	-------11.9-570------				

[a] U.S. EPA 1988
[b] Page et al., 1979
[c] Stout et al., 1988
[d] Utility Solid Wastes Activities Group, 1982

Gypsum is a by-product of chemical removal of sulfur dioxides from exhaust of coal-burning power production plants. Dry scrubbers or wet scrubbers produce calcium sulfite which, in newer designs, is further oxidized to gypsum (Miller and Sumner, 1977). Flue gas desulfurization gypsum is almost pure gypsum, but may contain 0.2-0.5% Mg if dolomitic lime was used in the scrubber (Miller and Sumner, 1997). In an experimental scrubber near Atlanta, the electrostatic precipitator was turned off allowing fly ash and S to be removed by the scrubber (Miller and Sumner, 1997). The result was a by-product containing about 50% gypsum and 50% fly ash, a mix which is an effective liming and soil conditioning agent and a source of Ca, S, and K. Co-utilization of by-products is encouraged at production in this case. It must be cautioned that this approach caused some wear on the scrubber equipment and is not usually the recommended method for deriving a mix of gypsum and fly ash should such a soil amendment be desired. It should be noted flue gas desulfurization materials may sometimes contain appreciable levels of salts which might be limiting for agricultural use (Miller and Summer, 1997). Land application of gypsum has been reviewed by Shainberg et al. (1989).

One process of flue gas desulfurization produces ammonium sulfate (Wallach, 1997). Development and details of the process are reported by Wallach (1997). The resulting product, named DAKSUL 45 (trademark) has a typical analysis of 21% N and 24% S. Thus far, 16,000 tons of DAKSUL 45 have been produced from a pilot plant (Wallach, 1997). With its high N and S analysis it may well be co-utilized with other by-products to offer a more complete fertilizer.

Wood products which are burned for energy include hardwoods, softwoods, wood chips, bark, logging slash, small trees, pulp from paper mills, and fibrous by-products (Mitchell and Black, 1977). Wood ash is a valuable by-product for liming. Muse and Mitchell (1995) reported an average calcium carbonate equivalent (CCE) of 38% for 19 boiler ash samples. Values for CCE ranged from 0 for a coal ash to 70% for hardwood ash. Lerner and Utzinger (1986) reported CCE values of 88% for black cherry (*Prunus avium* L.) ash to 115% for red oak (*Quercus borealis* Mich.) ash. In Muse and Mitchell (1995) a CCE of 83% was reported for a homeowner's wood burning stove. Most industrial boiler ash has a CCE of 25 to 70% (Mitchell and Black, 1997). Boiler wood ash is considered more effective for neutralizing soil acidity than commercial, ground limestone based on CCE (Muse and Mitchell, 1995). This result is considered a factor of particle size (Mitchell and Black, 1997).

Food Processing By-Products

Food processing wastes can come from plant materials or animal materials. From plant materials come culls, rinds, seeds, pits, pulp, presscakes, marc, malts, hops, and a variety of other by-products from pressing, canning, juicing, pickling, baking, brewing, or other mass food production processes. Often, much water is used in the processes and effluent and sludge result. The sludge may be land-applied as a solid, whereas the effluent may be part of an irrigation program. Analyses of cannery by-products are reported in Table 11, and coffee (*Coffea arabica* L.) processing by-products in Table 12. Food processing by-products may be land-applied (composted or uncomposted) for agricultural use or may be fed to livestock. Some food processing by-products used as feed include tomato culls, sugarcane bagasse, hulls and husks, wheat middlings, malts and hops, seed meals and oils, corncobs, citrus pulp, sisal pulp, paper, and sawdust among others (Owen, 1980). Malting, brewing and distilling of alcoholic beverages yields by-products which have agronomic value for either crops or as animal feed (Table 13).

Food processing of animal products includes such industries as slaughterhouses, packing plants, meat canneries, processed meat plants, dairies, and cheese or yogurt processing plants. In 1995, 11 billion kg of beef, 8 billion kg of pork, and 11 billion kg

Table 11. Nutrients in cannery by-products.

Nutrient	Tomato	Pear	Peach
	%(dw)		
N	1.84	1.37	0.7
P	0.31	0.18	0.12
K	0.012	0.012	0.01
Ca	0.002	0.003	0.001
Mg	0.003	0.001	0.003
S	0.22	0.37	0.065

Wilson and Lemieux, 1980

Table 12. Nutrients in coffee processing by-products.

Nutrient	Coffee Pulp	Coffee Hulls	Dried Coffee Residues[a]
	%(dw)		
N	3.5	1.1	1.8
P	0.1	0.03	0.06
K	2.0		0.04
Ca	0.6	0.16	0.06
Mg	Tr		0.02
	mg/kg(dw)		
Mn	7.2		29-36
Zn	4.6		9-12
Cu	5.7		25-29

[a] Residues from processing instant coffee
Topps, 1980

Table 13. Composition of some brewery and distillery by-products.

Nutrient	Spent Hops	Evaporated Spent Wash
	%(dw)	%(ww)
N	0.6-5.7	2.9-4.5
P (P_2O_5)	0.2-3.4	1.0-1.2
K (K_2O)	0.006-2.6	0.7-1.5
Ca		0.3-0.6
Mg	0.11-0.12	0.4-0.6
	mg/kg(ww)	
Mn		16-69
Zn		52-1012
Cu		192-248

After Berryman, 1970 in Garscadden, 1980; Miller, 1973

of chicken were produced (Bislinghoff, United States Department of Agriculture, personal communication). By-products from food processing of animal matter can be rich sources of nutrients. Effluents may contain from 2 to 5% of the total carcass protein from food processing. Slaughterhouse effluents in one report contained from 10.5 to 11.5 % N (Grant, 1980). Effluents may be land-applied or used as feed. Fats and proteins may be extracted and added to feedmeal. Fats are also removed for gelatin and glue production by boiling and steaming bones from the slaughterhouses, with bonemeal, a valuable phosphorus fertilizer, as a by-product (Jones, 1979).

As a fertilizer, dried blood is a rich source of N, as much as 12% N (Minnich and Hunt, 1979). Slaughterhouse effluent may be used in an irrigation program supplying 64 mg l^{-1} N, 19 mg l^{-1} P, 83 mg l^{-1} K, 220 mg l^{-1} Ca, and 14 mg l^{-1} Mg (Wells and Whitton, 1970). Effluents from meatworks used in such an irrigation study resulted in plants with increased N, P, and K in tissue (Table 14).

Table 14. Composition of white clover and ryegrass grown with or without effluents from meatworks.

	White Clover			Ryegrass		
	N	P	K	N	P	K
	---------------	---------------	---------%(dw)---	---------------	---------------	---------------
Control	3.2	0.21	1.65	2.7	0.2	1.9
Effluent	5.2	0.37	2.65	4.0	0.33	3.3

Wells and Whitton, 1970

Liquid dairy by-products, such as whey, whey permeate, and drug-residue contaminated wastes, are land-applied for disposal (Kelling and Peterson, 1981; Rodenberg, 1991; Sharratt et al., 1959). Liquid dairy wastes can improve soil conditions and are considered good sources for plant nutrients when less than 26,500 l ha^{-1} yr^{-1} of whey are applied (Matzke and Wendorff, 1993). This limit was set due to chlorides from $CaCl_2$ used in cheese processing (Matzke and Wendorff, 1993). Some dairy processing wastes, such as whey, can be fed to livestock.

In 1978, the world marine fish and shellfish harvest amounted to about 70 million Mg yr^{-1} (Swanson et al, 1980). In some processing procedures, 100% of the fish is used; in processing of some types of crabs, 85% of the original material may remain as by-product (Swanson et al, 1980). Seafood processing wastes are a significant source of N, P, and $CaCO_3$ equivalent (Table 15). Seafood processing by-products may be land-applied, fed to livestock, or further processed into pet foods. If used for land application, the by-product may be dewatered, composted, or further processed using procedures such as particle size reduction or amendment addition for easier handling. Seafood processing scraps have been composted with various other by-products (Section VI, this publication). Fertilizer and livestock feed usage of seafood by-products have been recognized for centuries as good management of resources as these wastes are particularly high in protein and nutrients (Swanson et al., 1980). In general, fish meals contain 55-70% protein,

5-10% lipids, 12-33% ash, less than 1% fiber, and ideally about 8% moisture (Brody 1965). Some typical analyses of fish meals are shown in Table 16.

Table 15. Analysis of fish meals for use as feed.

Meal	By-product	Protein	Oil	Ash	Moisture
		----------	----%(dw)----	----------	----------
Herring	Whole Fish and scrap	60-77	8-12	8-11	6-10
Menhaden	Whole fish	50-65	5-15	16-25	5-12
Ocean Perch	Fillet scrap	50-60	8-12	20-24	5-10
Salmon	Cannery scrap	55-65	10-14	12-30	6-10
Tuna	Cannery scrap	60-65	5-10	20-23	7-10

Table 16. Analysis of seafood processing by-products.

Nutrient	Shrimp	Crab	Crayfish
	----------	----%(dw)----	----------
N	5.6	4.4	5.1
P	4.3	3.3	1.2
K	0.22	0.83	0.14
Ca	0.16	0.16	0.18
Mg	0.78	0.92	0.26
CCE	26	42	45

Swanson, Dudley, and Williamson, 1980

Metals Industries By-Products

Steel mills, copper, zinc, and bronze smelters use air-pollution control devices from which micronutrients can be collected as baghouse dusts, electrostatic precipitator sludge, or as scrubber sludges. Since many of these high-metal wastes are declared hazardous under 40 CFR Part 261.3, environmental and legal responsibility concerns have inhibited many growers or fertilizer blenders from utilizing these micronutrient-rich resources. The heavy metals may be removed, but the process is considered too expensive for fertilizer production and use. Much of the material is mixed with a binder such as cement and landfilled (Wyatt, 1997). However metal by-products, if properly formulated and carefully used, could provide plant micronutrients when land-applied.

One common fertilizer and soil conditioner, gypsum, is a by-product of electroplating, metal etching, or mineral processing, especially of titanium ores (Miller and Sumner, 1997). Boron-containing materials include baghouse dusts from glass bead production and slags created by the fluxing of metal recovery smelters. By-product Co is found in off-grade cobalt oxide and cobalt sulfate products. Metallic Cu and some of the Cu in brass slags are not readily bioavailable; however, nitric acid extraction will produce copper nitrate, or heating will produce copper oxide. Copper sulfate is produced from copper scrap. By-product Cu is found in petroleum refining catalysts, Cu recovery operations, smelter effluents streams, furnace slags, scales from Cu wire drawing, Cu plating wastes, and baghouse dust from Cu tube manufacturing. Iron sulfate is produced in waste neutralization from the cleaning of steel coils and sheets prior to galvanization and is a product of the pickling process in steel manufacture (Jones, 1979). Iron sulfate is produced in titanium oxide production, Cu smelting and steel production among other industries. Sources of manganese include baghouse dusts, scrubber sludges, and slags. Zinc is available in many by-products, including plating wastes, baghouse dusts from smelters, boiler ash. Zinc dusts may contain Cd or Pb as a result of the association in parent material rendering the material unsuitable for use on food crops (Wyatt, 1997).

The fertilizer micronutrient industry consumes many by-product materials which would have no other use (Wyatt, 1997). Micronutrients may be supplied in frits, a product which is composed of brown-colored, pulverized, specially compounded glasses that slowly release B, Cu, Fe, Mn, Mo, and Zn, or other waste metals depending upon the industry involved (Jones, 1979). Other metal by-products include oxides, sulfates, nitrates, carbonates, chelates in liquids, suspensions, powders, and granules (Wyatt, 1997). Micronutrients, in small amounts, and in proper proportions may be blended in feeds (Pate and Kunkle, 1989).

Fertilizer Production By-Products

In 1995, the U.S. phosphate fertilizer industry, centered mostly in Florida and North Carolina, produced 40 Mg of phosophogypsum, the by-product resulting from processing 45 Mg of raw mined phosphate rock (Miller and Sumner, 1997). Phosphogypsum is 85-95% gypsum, with 3-17% sand, and 0.1-1.0 % P and F (Alcordo and Rechcigl, 1995; Miller and Sumner, 1997). Typically, phosphogypsum is acidic (pH 4.5-5.5) but not strongly buffered. Phosphogypsum is sometimes mixed with fly ash prior to land application. Potassium chloride can be crystallized from lake brine, leaving a dilute "mother liquor" which can be further cooled in vacuum-type coolers resulting in crystals of crude borax, $Na_2B_4O_7.10H_2O$ as a by-product (Jones, 1979).

Paper and Wood Industries By-Products

Paper products which may be diverted from the waste stream include newsprint, computer paper, cardboard and pasteboard containers, insert advertisements, third class mail, and phonebooks. The composition of these products are given in Table 17. Incorporation of paper or wood products into soil, or use of such products as mulches, affords benefits of organic amendment such as erosion control, reduced bulk density, increased porosity and water infiltration, increased moisture retention, increased CEC, improved soil structure, increased water stable aggregates, increased soil pH, enhanced plant growth, suppressed soil-borne plant diseases, and increased soil enzymatic activity (Edwards, 1997). In drier regions, water conservation efforts can benefit from additions of paper or wood products to the soil. In the southeastern U.S. where hot, humid conditions may rapidly deplete organic matter from the soil, paper and wood products can replace some of the lost organic matter, thereby improving the soil fertility.

Table 17. Analysis of paper products.

Nutrient	Newsprint	Computer Paper	Cardboard Containers	Insert Ads	3rd Class Mail	Phone Books
	--------	--------	--------%(dw)--------	--------	--------	--------
P	0.01	0.01	0.01	0.01	0.01	0.004
K	0.01	0.004	0.008	0.001	0.006	0.009
Ca	0.08	2.86	0.43	0.51	1.0	0.09
Mg	0.01	0.04	0.03	0.02	0.03	0.02
	--------	--------	--------mg/kg(dw)--------	--------	--------	--------
Fe	136	219	298	223	159	57
Mn	31	5	29	28	6	45
B	1.8	1.8	6.6	3.1	15.7	1.0
Zn	51	6	57	151	10	5
Cu	22.5	4.4	16.7	32.3	12.2	5.7

Edwards, 1997

Paper and wood have high C:N ratios (100:1 and greater) and may immobilize N if land-applied (Barker, 1997). Many applicators apply the materials far in advance of crop planting or may compost the material with a high N source such as green plant matter or biosolids (co-utilization of by-products). Paper poses special handling problems in that it is difficult to decompose, may present a physical barrier to water penetration and seedling emergence if layered upon the soil, and if shredded may be so light in weight as to be extremely difficult to apply to land. Pelletization is one method being studied to solve these problems (Edwards, 1997). Metals in paper may pose land application problems, with zinc being the limiting factor in land application of mixed paper wastes in one set of experiments (Goodrich et al., 1988).

In paper production, an estimated 41 kg of fibrous wastes were generated per Mg of finished pulp produced (Millett et al., 1973). Paper manufacturing produces many by-products which may be used as soil amendments. Papermill sludge, wastewater, causticizing residuals, and deinking sludges are a few of the by-products from paper manufacture. Primary sludges are low in plant nutrients and have a high C:N ratio (Camberato et al., 1997). Secondary sludges have higher concentrations of N and P and lower C:N ratios than primary sludges (Table 18) (Camberato et al., 1997). Aluminum concentrations in paper manufacturing sludges were higher than some sewage sludges in one report, however, As, Cd, Co, Hg, Pb, Sn, and Se were lower (Camberato et al., 1997). Paper mill sludge and ash in a set of experiments in Maine were analyzed for metals with results reported in Table 19. In the last decade the paper manufacturing industry has made great efforts to lower the concentration of dioxins and furans in processing, using chlorine dioxide instead of chlorine in the bleaching process thus making land application of paper manufacturing by-products an alternative to land-filling (Camberato et al., 1997).

Table 18. Nutrient analyses of wastewater sludges from paper manufacturing.

Nutrient	Primary Sludges	Secondary Sludges
	--------%(dw)--------	
N	0.27	2.3
P	0.16	0.42
	Primary and Secondary Sludges Combined	
	----%(dw)----	
K	0.22	
Ca	1.4	
Mg	0.15	
S	0.47	
	----mg/kg(dw)----	
Fe	1,540	
Mn	155	
Zn	188	
B	25	
Cu	52	

Camberato et al., 1997

Table 19. Heavy metals concentrations of papermill by-products.

Ash Metal	Sludge	Maximum	Application[a]
	----------mg/kg(dw)----------		
Cd	4.4	2	10
Cr	27	230	1000
Cu	120	33	1000
Pb	59	15	700
Ni	47	31	200
Zn	370	200	2000

[a] maximum Cumulative Application to Food Chain Crops, Maine Department of Environmental Protection
Hatch and Pepin, 1985

Causticizing residuals include lime mud, slaker grits or rejects, and green liquor dregs (Camberato et al., 1997). Lime mud is the residual material from the reaction of CaO with green liquor (Na_2CO_3 and Na_2S solution), yielding $CaCO_3$, NaOH, and Na_2S (Camberato et al., 1997). Most of the NaOH and Na_2S are recovered and reused. The $CaCO_3$ may be removed at this point or may be fired in a kiln to produce CaO. Slaker grits are mainly large, unreactive lime particles and insoluble impurities. Green liquor dregs are the precipitated substances in the green liquor, comprised of C (50% or more) and insoluble metal carbonates, sulfides, sulfates, hydroxides, and silicates (Camberato et al., 1997). Causticizing residuals are used as a liming material. Analyses of lime muds, green liquor dregs, and slaker grits from paper manufacturing are reported in Table 20. Paper manufacturing effluents may also be land-applied, usually in an irrigation program.

Table 20. Analysis of lime muds, green liquor dregs, and slaker grits.

Nutrient Muds	Lime Dregs	Green Liquor Grits	Slaker
	----------------------------	%(dw)	----------------------------
N	0.18	0.27	0.22
P	0.28	0.07	0.08
K	0.04	0.46	0.27
Ca	37.9	19.4	36.4
Mg	0.62	2.27	0.72
S	0.20	1.42	0.12
	----------------------------	mg/kg(dw)	----------------------------
Fe	1,512	8,977	3,300
Mn	310	9,611	192
Zn	255	1,052	488
B	5.9	20.6	7.3
Cu	15.8	240	16.0

Camberato et al., 1997

Newspaper comprising up to 30% of the diet has been fed to dairy cattle (Furr et al., 1974), or sheep (Sherrod and Hanson, 1973) with the general finding that the higher the percentage of paper the lower the digestibility of the feed. Unbleached kraft pulp has been fed to cattle and sheep and goats, with goats showing the best acceptance of the material (Millett et al., 1973). Newsprint and magazine paper are less easily digested than products such as wrapping paper, typing paper, or light cardboard (Dunlap, 1980).

Tannery By-Products

One tannery may produce as much as 12,000 Mg yr^{-1} of tannery wastewater sludges (Mazur and Koc, 1980). Treatment of tannery wastes includes alkalization, followed by coagulation with ferrous sulfate or aluminum sulfate, fermentation and dewatering, often with subsequent stockpiling and composting of the sludges. The metals content of these sludges is dependent upon the process used for tanning. Processes that use Cr produce a by-product considered unsuitable for land application (Mazur and Koc, 1980). Mazur and Koc (1980) reported that tannery sludges with high concentrations of Cr had deleterious effects on plant growth (Table 21). Tannery sludges may contain as much as 88% organic matter (Mazur and Koc, 1980). Mazur and Koc (1980) reported increases in yield of potatoes (*Solanum tuberosum* L.), summer barley (*Hordeum vulgare* L.), horse beans (*Vicia faba* L.), and winter wheat (*Triticum aestivum* L.) with application of tannery sludges. A review article by Hughes (1988) discusses land application of tannery by-products.

Table 21. Analysis of tannery sludge.

Nutrient	Mean	Range
	--------%(dw)--------	
N	3.78	1.97-5.67
P	0.25	0.09-0.49
K	0.09	0.04-0.21
Ca	3.83	1.00-7.51
Mg	0.27	0.01-0.94
S	2.86	1.22-5.45
Fe	0.95	0.001-12.52
Cr	1.29	0.34-2.80

Mazur and Koc, 1980

Textiles and Fibers By-Products

Textile wastes may be land-applied (Overcash et al., 1978). Cotton plant residues separated prior to processing at the textile plant are considered plant residues and are discussed in the beginning of this chapter. The by-products from production are considered here. Cotton waste has been used in potting media (Wang, 1991), and in composting (Shumack et al., 1991; Albanell et al., 1988; Winterlin et al., 1986). It is suggested cotton gin wastes be composted prior to land application may be preferred to eliminate concerns with N immobilization. Some researchers suggest liquifying cotton gin wastes into a fuel oil (White et al., 1996). One intriguing use for cotton gin wastes is as an absorbent in oil spills (Anthony 1994). Wool by-products are typically rich in N and have been land-applied after composting with resulting increases in chickpea (*Cicer arietinum* L.) and wheat growth (Tiwari et al., 1989). The manufacture of acrylic or nylon fibers results in wastewater containing the common fertilizer, ammonium sulfate, as a by-product (Jones, 1973). Wastewater column bottoms from acrylic fiber production contain 2-20% ammonium sulfate by weight, such that a sample of 400g of column bottoms yielded 13.6g ammonium sulfate crystals after precipitation, purification, and drying (Jones, 1973). Wastewater from nylon production contained 89-125 mg l^{-1} PO_4, 104-128 mg l^{-1} SO_4, and 18,000-33,000 mg l^{-1} total N (Jones, 1973) and some organic matter, some of which has traditionally been removed from the wastewater by anaerobic denitrification, algae harvesting, ammonia stripping, ion exchange or carbon absorption (EPA, 1971). Rayon production effluent may contain 2000-2500 mg l^{-1} dissolved Na_2SO_4, 2000 mg l^{-1} $CaSO_4$, 50-100 mg l^{-1} Mg, 0.3 mg l^{-1} Fe, and 1 mg l^{-1} Zn (Rock, 1971). Rayon production requires 19,000 kg of water to produce 454 kg of fiber, 34,000 kg of water to produce 454 kg of acetate fiber (Behmer, 1976). Effluents from fiber production may be land-applied.

Building Industry By-Products

Rock dusts are by-products of the aggregate industry. They are collected in washings of crushed rocks which are cleaned for use in asphalt and cement. As a fertilizer, rock dusts may provide P, K, Ca, Mg, Fe, Mn, and some micronutrients, as well as having a liming effect on soil. Smaller particles of these silicaceous dusts are more effective fertilizers than larger, sand-sized particles (Barker, 1976). Rock dusts have been used as fertilizers since the 1800s (Plummer, 1918), have been researched recently (Hinsinger et al., 1996; Weerasuriya et al., 1993) and are also the subject of mining interests (Korcak, 1996).

Organic synthetic building materials such as countertops contain proteinaceous or amide-type substances which when finely divided as in shavings are sometimes considered as sources of synthetic proteins for feed or fertilizer (Jones, 1979).

Chemical By-Products

Pigment manufacture produces iron oxide waste which may be used to condition soil or as a fertilizer (Behmer, 1976). For the most part, wastes from pigment, paint, or oil industries have poor suitability for land application (Behmer, 1976). However, oily hydrocarbon sludges have been land-applied with decomposition and disposal as the intent (El-Nawawy et al., 1992; Harris, 1976; Loehr et al., 1984).

Antibiotic Fermentation and Drug Production By-Products

Antibiotic fermentation wastes may contain as much as 6% N and 44% organic C as was the case with wastes from Tylosin production (Bewick, 1980). Wastes from penicillin production in this study had about 6% N and 91% organic C (Bewick, 1980). Wastes from Oxytetracyclins production had 2.8% N and 26.5% C, but had a pH of 2.2 which would limit its use as a fertilizer (Bewick, 1980).

Pharmaceutical wastes from vitamin production have been used to grow turf of Kentucky bluegrass (*Poa pratensis* L) or a mix of bluegrass and perennial ryegrass (*Lolium perenne* L.) (May Khabbaz and Martin Petrovic, Cornell University, personal communication). Pharmaceutical wastes or a horse manure comparison treatment produced marketable turf at 8 weeks compared to biosolids which did not produce marketable turf until 12 weeks after seeding. The pH of the pharmaceutical wastes was 7.5, compared to 8.1 for the horse manure, and 5.9 for the biosolids. The fastest germination occurred in the horse manure. In this study, special considerations with the wastes were the depletion of N in the horse manure in 45 days, the high soluble salts of the biosolids, and the need for careful attention to the pharmaceutical wastes as the media tended to be difficult to wet.

CO-UTILIZATION OF BY-PRODUCTS

In many cases, a by-product may not be ideal by itself for land application or for use as a feed. Through co-utilization of by-products more useful agronomic products may result. The benefits of co-utilization may include nutrient balance, improved handling, storage, transportation, and application, reduction of toxins or contaminants, improved moisture content, improved economic value, improved pest and weed control, plant disease control, and various improved soil conditioning effects.

Composting is perhaps the most widely practiced type of by-product co-utilization for agronomic use. The benefits of co-utilization of by-products in composting may include adjusting the C:N ratio, improved bulking and aeration, as well as moisture content, and perhaps a reduction in ammonia volatilization and an enhanced microbial population (Stratton et al., 1995). The term co-composting has been used to designate two or more by-products being composted together, such as biosolids to supply N and wood by-products to supply C. However, many researchers consider the term unnecessary as much of the materials composted are blended prior to the process to achieve optimum composting conditions. Municipal solid waste compost is a good example of co-utilization of by-products in composting as many household wastes may be included in the compost substrate.

Other by-products composted together include pulverized pallets with biosolids and wood ash (BioCycle, 1994), peat with seaweed, crab scrap and fish scrap (Mathur et al., 1986), sawdust with peat, poultry manure, fish offal, and crab processing wastes (Martin et al., 1993), and fish wastes with wood chips (Frederick, 1991). Biosolids have been composted with wood chips, leaves, bark, shredded tires, cereal hulls, straw, or fly ash (Elwell et al., 1994; U.S. EPA, 1984; Pichtel, 1990). Water hyacinth has been composted with gizzard shad, poultry or cattle manure, or scallop offal (Hackett and Thompson, 1991). Cotton gin wastes have been composted with sheep manure (Albanell et al., 1988). These are just a few examples of the many compost mixes researched. Many of the recent advances in co-utilization of by-products have been in the usage of industrial by-products as components of compost. More advances can be expected as different industrial by-products are composted and researched rather than land-filled or otherwise disposed.

Cranberry (*Vaccinum macrocarpon*, Ait.) presscake has been applied to corn and alfalfa fields with rice hulls (Susan Butkewich, Ocean Spray, Inc., personal communication). The presscake contained (on a dry weight basis) 1.25% N, 0.2% P, 0.2% K, 0.06% Ca, 0.07% Mg, and 48.7% C. The C:N ratio was 38:1 in this analysis, and the pH of the material was 3.6. Yield of marketable ears of corn grown with cranberry presscake were equal to those grown with manure, and greater than those grown with compost. Cranberry presscake has been successfully co-utilized with rice hulls or celite in land application experiments. One important finding of the cranberry presscake research is the suppression of weeds (Stephen Herbert, University of Massachusetts, personal communication).

Iron humate from water treatment has been treated to produce an enhanced or value-added product called granulated solubilized dolomitic iron humate, the composition of which is reported in Table 22. Granulated solubilized dolomitic iron humate (GSDIH) might conceivably be applied to land with another product discussed in the section on energy, DAKSUL 45, to supply 21%N and 24%S (dw). Ash may also be land-applied in this co-utilization scheme to provide K, along with bone meal to supply P. Phosphogypsum supplying Ca and S might be applied with the slow release iron humate (SRMAIHP) to more completely supply plant nutrients through co-utilization.

Table 22. Analysis of granulated solubilized dolomitic iron humate (GSDIH), soluble nitrogen iron humate plus minors (SNIHPM), and slow release magnesium ammonium iron humate phosphate (SRMAIHP).

Nutrient	GSDIH	SNIHPM	SRMAIHP
		%(dw)	
Total N		7.4	
Total P_2O_5		17.4	
Available P_2O_5			9.9
K			11.5
Soluble K_2O			3.1
Total Fe	8.0	9.4	9.9
Soluble Fe	1.2	0.1	0.1
Available Fe	4.8	8.9	8.9
S as SO_4=	7.7	13.6	
Mg	5.0	3.3	
Soluble Mg	2.0	2.6	
Ca	7.0	5.2	
Zn		0.2	
Mn		3.0	13.4
Soluble Mn		2.4	0.7
Available Mn			5.9

Rehberg and Smith, 1997

SUMMARY AND FUTURE NEEDS

Co-utilization of by-products may prove so beneficial in the near future as to change social views of resources formally targeted for disposal. Every by-product should be viewed with the intent of finding uses for that by-product rather than disposal. Many by-products have been utilized for years in practices other than those agronomic, for instance, gypsum wallboard in the building industry, or ash added to concrete, however more agronomic uses should be sought.

The economic savings from the use of by-products as fertilizers or feeds may be substantial (Parr and Hornick, 1992; Forste, 1994). For example, the value of cattle manure is estimated at \$23.47 Mg^{-1}, crop residues: \$8.44 Mg^{-1}, biosolids: \$21.40 Mg^{-1}, and MSW: \$3.66 Mg^{-1}, based on values of \$0.30 kg^{-1} for N, \$0.37 kg^{-1} for P, and \$0.20 kg^{-1} for K (Parr and Hornick, 1992). Forste (1994) estimated a savings of \$90.55 in the first year of biosolids land application. The increased perceived value of by-products is expected to provide impetus for co-utilization of by-products in agriculture (Galloway and Walker, 1997).

Environmental issues surrounding the co-utilization, land application, or use as feed, of by-products include food-chain concerns, possible water contamination, and odors, among other perceived concerns. However, the environmental gain from cessation of disposal of by-products in landfills, oceans, or by other wasteful, ecologically unsound methods is expected to outweigh any detrimental effects, provided careful use of by-products is practiced. Future research in this area is expected to increase greatly in the near future.

Future research needs include the evaluation of by-products mixes for land application, feeds, or composting, development of accurate, calibrated analysis of organic content, environmental impact of land-spreading, and economic value of by-products, improved storage and handling of by-products to reduce nutrient loss. Perhaps most importantly, logistical difficulties, such as transportation and application issues, and social acceptance issues which require education of the public, government officials, regulators, and the agricultural community need to be addressed. If the full benefits of co-utilization of high quality by-products can be realized by the agronomic community, by-products may be viewed as the great resources they can be, rather than the difficult disposal problems they were.

REFERENCES

Alcordo IS & Rechcigl JE (1995) Phosphogypsum and other by-product gypsums. In: Rechcigl JE (Ed) Soil Amendments and Environmental Quality (pp 365-425). CRC Lewis Publishers, Boca Raton, FL.

Albanell E, Plaixats J & Cabreró T (1988) Chemical changes during vermicomposting (*Eisenia fetida*) of sheep manure mixed with cotton industrial wastes. Biol. Fert. Soil 6(3):266-269.

Anthony WS (1994) Absorption of oil with cotton products and kenaf. Appl. Eng. Agric. 10(3):357-361.

Barker AV (1976) Using tomato plants to assess the release of potassium from fertilizers. J. Agron. Educ. 5:63-65.

Barker AV (1997) Composition and uses of compost. In: Rechcigl JE & MacKinnon HC (Eds) Agricultural uses of by-products and wastes (pp 140-162). American Chemical Society, Washington, DC.

Behmer DE (1976) Management of chemical wastes. In: Land application of waste materials (pp 216-225). Soil Conservation Society of America, Ankeny, IA.

Bewick MWM (1980) Wastes from the fermentation industries: their use as feedstuffs and fertilizers. In: Bewick MWM (Ed) Handbook of organic waste conversion (pp 165-186). Van Nostrand Reinhold Company, New York.

BioCycle (1994) Pulverized pallets become compost amendment. BioCycle 35(1):40.

Bouwer H & Chaney RL (1974) Land treatment of wastewater. Adv. Agron. 26:133-176.

Brody J (1965) Fishery by-products technology. Thw Avi Publishing Company, Westport, CT.

Camberato JJ, Vance ED & Someshwar AV (1997) Composition and land application of paper manufacturing residuals. In: Rechcigl JE, & MacKinnon HC (Eds) Agricultural Uses of By-Products and Wastes (pp. 185-202). ACS Symposium Series 668, American Chemical Society, Washington, DC.

Chang AC, Page AL, Lund LJ, Warneke JE & Nelson, CO (1989) Municipal sludges and utility ashes in California and their effects on soils. In: Bar-Yosef B, Barrow NJ & Goldshmid (Eds) Inorganic contaminants in the vadose zone (pp125-139). Ecological Studies Vol. 74, Springer-Verlag, Berlin.

Dunlap CE (1980) Feedstuff potential of paper wastes and related cellulose residues. In: Bewick MWM (Ed) Handbook of organic waste conversion (pp 187-208). Van Nostrand Reinhold Company, New York.

Edwards JH (1997) Composition and uses of uncomposted wastepaper and other organics. In: Rechcigl JE & MacKinnon HC (Eds) Agricultural Uses of By-Products and Wastes (pp. 163-184). ACS Symposium Series 668, American Chemical Society, Washington, DC.

El-Nawawy AS, El-Bagouri IH, Abdal M & Khalafawi MS (1992) Biodegradation of oily sludge in Kuwait soil. World J. Microbio. Biotech. 8(6):618-620.

Elliott LF & Swanson NP (1976) Land use of animal wastes. In: Land application of waste materials (pp 80-90). Soil Conservation Society of America, Ankeny, IA.

Elwell DL, Keener HM, Hoitink HAJ, Hansen RC & Hoff J (1994) Pilot and full scale evaluations of leaves as an amendment in sewage sludge composting. Compost Sci. Util 2(2): 55.

Environmental Protection Agency (1971) State of the art of textile waste treatment, Department of Textiles, Clemson University, Clemson, SC. Project 12090ECS 02/71 Water Pollution Control Research Series, United States Government Printing Office, Washington DC.

Environmental Protection Agency (1993) Domestic septage regulatory guidance: A guide to the EPA 503 rule. Project EPA 832-B-92-005 Office of Wastewater Management, United States Government Printing Office, Washington DC.

Frederick L (1991) Turning fishery wastes into saleable compost. BioCycle 32(9):70-71.

Furr AK, Mertens DR, Gutenmann WH, Bache CA & Lisle DJ (1974) Fate of polychlorinated biphenyls, metals and other elements in papers fed to lactating cows. J. Agr. Food Chem. 22:954-959.

Garscadden BA (1980) By-products from malting, brewing, and distilling. In: Bewick MWM (Ed) Handbook of organic waste conversion (pp 135-164). Van Nostrand Reinhold Company, New York.

Goodrich PR, Backus BD, Schroeder DD & Banwart MS (1988) Limiting the land-application of mixed paper based on metal content. Biol. Wastes 24(2):81-94.

Grant RA (1980) Recovery of protein and fat from slaughterhouse effluents by physicochemical means. In: Bewick MWM (Ed) Handbook of organic waste conversion (pp 227-242). Van Nostrand Reinhold Company, New York.

Gray P & Berry DR (1980) The production of feedstuff biomass from liquid organic wastes by fermentation. In: Bewick MWM (Ed) Handbook of organic waste conversion (pp 339-382). Van Nostrand Reinhold Company, New York.

Hackett WC, & Thompson OC (1991) Biofermentation pilot study for water hyacinth and gizzard shad composting. St. Johns River Management Water Management District Spec. Pub. SJ91-SP13.

Hatch CJ & Pepin RG (1985) Recycling mill wastes for agricultural use. Tappi J. 68:70-73.

Hinsinger P, Golland MDA & Gilkes RJ (1996) Silicate rock powder: effect on selected chemical properties of a range of soils from Western Australia and on plant growth as assessed in a glasshouse experiment. Fert. Res. 45:69-79.

Hue NV (1995) Sewage sludge. In: Rechcigl JE (Ed) Soil Amendments and Environmental Quality (pp 199-248). CRC Lewis Publishers, Boca Raton, FL.

Hughes JC, (1988) The disposal of leather tannery wastes by land treatment - a review. Soil Use and Manage. 4(3):107-111.

Jewell WJ (1994) Engineering and cost considerations: Sludge management and land application. In: Clapp CE, Larson WE & Dowdy RH (Eds) Sewage sludge: land utilization and the environment (pp 41-54). SSSA Misc. Publication, ASA, CSSA, SSSA, Madison, WI.

Jones HR (1973) Pollution control in the textile industry, Noyes Data Corporation, Park Ridge, NJ

Jones US (1979) Fertilizers & soil fertility. Reston Publishing Company, Reston, VA.

Korcak RF (1996) Byproduct utilization in agriculture: Past experience and potentials for the future. Mining Engin. 48(8):79-82.

Lerner BR & Utzinger JD (1986) Wood ash as soil liming material. HortScience 21(1):76-78.

Loehr RC, Martin JH Jr. & Neuhauser EF (1984) Disposal of oily wastes by land treatment. Proceedings of the 38th Industrial Waste Conference (pp1-11) Butterworth Publishers, Stoneham, MA.

Logan TJ & Harrison BJ (1995) Physical characteristics of alkaline stabilized sewage sludge (N-Viro Soil) and their effects on soil physical properties. J. Environ. Qual. 24:153-164.

Maine Department of Environmental Protection (1980) Rules for municipal sludge utilization on land. Chapters 565-66. State of Maine.

Martin AM, Evans J, Porter D and Patel TR (1993) Comparative effects of peat and compost employed as bulking agents in composting. Bioresource Tech. 44(1):65.

Mathur SP, Daigle J-Y, Levesque M & Dinal H (1986) The feasibility of preparing high quality compost from fish scrap and peat with seaweeds or crab scrap. Biol. Agric. Hort. 4:27.

Mazur T & Koc J (1980) The fertilizing value of tannery sludges. In: Bewick MWM (Ed) Handbook of organic waste conversion (pp 328-338). Van Nostrand Reinhold Company, New York.

Matzke S & Wendorff WL (1993) Chloride in cheese manufacturing wastes to be landspread on agricultural land. Bioresource Tech. 46(3):251-253.

McCalla TM & Army TJ (1961) Stubble mulch farming. Adv. Agron. 13:125-197.

McCalla TW, Peterson JR & Lue-Hing C (1977) Properties of agricultural and municipal wastes. In: Elliott LF & Stevenson FJ (Eds) Soils for management of organic wastes and wastewaters (pp 11-43). ASA, CSSA, SSSA, Madison, WI.

Miller TB (1973) Malt distillers grain. Symposium of the Agricultural Industries Training Services, England.

Miller WP & Sumner ME (1997) Agricultural and industrial uses of by-product gypsums. In: Rechcigl JE, & MacKinnon, HC (Eds) Agricultural Uses of By-Products and Wastes (pp. 226-239). ACS Symposium Series 668, American Chemical Society, Washington, DC.

Millett MA, Baker AJ, Satter LD, McGovern JN & Dinius RA (1973) Pulp and papermaking residues as feedstuffs for ruminants. J. Anim. Sci. 37:599-607.

Mills HA & Jones JB Jr. (1996) Plant analysis handbook II. Micromacro Publishing, Inc. Athens, GA.

Minnich J and Hunt M (1979) The Rodale guide to composting. Rodale Press, Emmaus, PA.

Mitchell CC & Black ED (1997) Land application of boiler wood ash in the Southeastern United States. In: Rechcigl, JE & MacKinnon, HC (Eds) Agricultural Uses of By-Products and Wastes (pp. 204-225). ACS Symposium Series 668, American Chemical Society, Washington, DC.

Muse JK & Mitchell CC (1995) Paper mill boiler ash and lime by-products as soil liming materials. Agron. J. 87:432-438.

National Research Council (1996) Nutrient requirements of beef cattle, 7th rev. ed. National Academy Press, Washington, DC.

Overcash MR, Klose HC, Rock D, Marshburn R & Pal, D (1978) Pretreatment land application of textile plant wastes. Water 74(178):163-175.

Owen E (1980) Agricultural wastes as feedstuffs. In: Bewick MWM (Ed) Handbook of organic waste conversion (pp 40-71). Van Nostrand Reinhold Company, New York.

Page AL, Elseewi AA & Straughan IR (1979) Physical and chemical properties of fly ash from coal-fired power plants with reference to environmental impacts. Residue Rev. 71:83-120.

Pate FM & Kunkle WE (1989) Molasses-based feeds and their use as supplements for brood cows. Circular S-365, Florida Agricultural Experiment Station, Institute of Food and Agricultural Sciences, University of Florida, Gainesville, FL.

Pichtel JR (1990) Microbial respiration in fly ash/sewage sludge-amended soils. Environ. Pollut. 63:225.

Plummer JK (1918) Availability of potash in some common soil forming minerals. J. Agric. Res. 14:297-315.

Rehberg BE & Smith GL (1997) Iron humate production, processing, properties, and usage. In: Rechcigl JE & MacKinnon HC (Eds) Agricultural Uses of By-Products and Wastes (pp. 268-275). ACS Symposium Series 668, American Chemical Society, Washington, DC.

Richards BK, Peverly JH, Steenhuis TS & Liebowitz BN (1997) Effect of processing mode on trace elements in dewatered sludge products. J. Environ. Qual. 26:782-788.

Rock DM (1971) Zinc precipitation and recovery from viscose rayon waste water. Project 12090ESG 01/71 Environmental Protection Agency Water Pollution Control Research Series, United States Government Printing Office, Washington, DC.

Rodenberg J (1991) Guidance on disposal of contaminated milk. State of Wisconsin Department of Natural Resources File Ref. 3430. Madison, WI.

Sabey BR (1980) The use of sewage sludge as a fertilizer. In: Bewick MWM (Ed) Handbook of organic waste conversion (pp 72-107). Van Nostrand Reinhold Company, New York.

Sanford FP & Lee C (1990) U.S. Fish Reduction Industry Technical Leaflet No. 14. Bureau of Commercial Fisheries, Washington, DC.

Shainberg I, Sumner ME & Miller WP (1989) Use of gypsum on soils: a review. Adv. Soil Sci. 9:1-111.

Sharratt WL, Peterson AE & Calbert HE (1959) Whey as a source of plant nutrients and its effects on the soil. J. Dairy Sci. 42(1):126-131.

Sherrod LB & Hansen KR (1973) Newspaper levels as roughage in ruminant rations. J. Anim. Sci. 36:592-596.

Shumack RL, Eakes DJ, Gilliam CH & Donald JO (1991) Using gin trash in composted soil ingredients. Proceedings of the Beltwide Cotton Conference 1:498-499, National Cotton Council of America, Memphis, TN.

Stout WL, Hern JL, Korcak RF & Carlson CW (1988) Manual for applying fluidized bed combustion waste on the Ca, Mg, S, and Zn levels in red clover, tall fescue, oat, and buckwheat. Agron. J. 71:662-665.

Stratton ML & Rechcigl JE (1997) Organic mulches, wood products, and composts as soil amendments. In: Wallace A (Ed) Handbook of soil conditioners: substances which enhance the physical properties of soil (pp 43-95). Marcel Dekker, New York.

Stratton ML, Barker AV & Rechcigl JE (1995) Compost. In: Rechcigl JE (Ed) Soil Amendments and

Environmental Quality (pp 249-309). CRC Lewis Publishers, Boca Raton, FL.
Swanson GR, Dudley EG & Williamson KJ (1980) The use of fish and shellfish wastes as fertilizers and feedstuffs. In: Bewick MWM (Ed) Handbook of organic waste conversion (pp 281-327). Van Nostrand Reinhold Company, New York.
Tiwari VN, Pathak AN & Lehri LK (1989) Response to differently amended wool-waste composts on yield and uptake of nutrients by crops. Biol. Wastes 28(4):313-318.
Topps JH (1980) Coffee and tea wastes. In: Bewick MWM (Ed) Handbook of organic waste conversion (pp 268-280). Van Nostrand Reinhold Company, New York.
U.S. Environmental Protection Agency (1984) Environmental regulations and technology: use and disposal of municipal wastewater sludge. EPA-625/10-84-003. U.S.EPA, Washington, DC.
U.S. Environmental Protection Agency (1988) Wastes from the combustion of coal by electric utility power plants. EPA/530-SW-88-002. U.S.EPA, Washington, DC.
U.S. Environmental Protection Agency (1993) 40 CFR Part 503. Standards for the use or disposal of sewage sludge. Federal Register 58:9248-9415.
Utility Solid waste Activities Group (1982) Report and technical studies on the disposal and utilization of fossil-fuel combustion by-products. Submitted to the U.S. Environmental Protection Agency, Washington, DC.
Walker JM, Southworth RM & Rubin AB (1997) U.S. Environmental Protection Agency regulations and other stakeholder activities affecting the agricultural use of by-products and wastes. In: Rechcigl JE & MacKinnon HC (Eds) Agricultural Uses of By-Products and Wastes (pp. 28-47). ACS Symposium Series 668, American Chemical Society, Washington, DC.
Wallach DL (1997) Ammonium sulfate fertilizer as by-product in flue gas desulfurization: The Dakota Gasification Company experience. In: Rechcigl JE & MacKinnon HC (Eds) Agricultural Uses of By-Products and Wastes (pp. 240-254). ACS Symposium Series 668, American Chemical Society, Washington, DC.
Wang YT (1991) Evaluation of media consisting of a cotton waste for the production of tropical foliage species. J. Environ. Hort. 9(3):112-115.
Weerasuriya TJ, Pushpakumara S & Cooray PI (1993) Acidulated pegmatic mica: a promising new multi-nutrient mineral fertilizer. Fert. Res. 34:67-77.
Wells N & Whitton JS (1970) The influence of meatworks effluents on soil and plant composition. New Zealand J. Agr. Res. 9:565-575.
White DH, Coates WE & Wolf D (1996) Conversion of cotton plant and cotton gin residues to fuels by the extruder-feeder liquefaction process. Bioresource Tech. 56(1):117-123.
Wilson LL & Lemieux PG (1980) Factory canning and food processing wastes as feedstuffs and fertilizers. In: Bewick MWM (Ed) Handbook of organic waste conversion (pp 243-267). Van Nostrand Reinhold Company, New York.
Winterlin WL, McChesney MM, Schoen SR & Seiber JN (1986) Chemical residues during screening, composting, and soil incorporation of cotton gin waste. J. Environ. Sci. Health Part B Pestic. Food Contam. Agric. Wastes 21(6):507-528.
Wyatt JM (1997) By-product usage in fertilizer micronutrients. In: Rechcigl JE & MacKinnon HC (Eds) Agricultural Uses of By-Products and Wastes (pp. 255-267). ACS Symposium Series 668, American Chemical Society, Washington, DC.

STATUS OF COMPOSTING IN THE UNITED STATES

NORA GOLDSTEIN
BioCycle Magazine, Emmaus, Pennsylvania 18049, USA
E-mail BioCycle@ad.com

ABSTRACT

Composting activity across all categories of organic residuals is steadily increasing in the United States. There are fourteen municipal solid waste composting facilities, 250 biosolids composting sites, 3,260 yard trimmings composting sites, and over 175 food residuals composting projects. Development of composting projects has been driven by several factors over the years: A need for organic soil amendments; restrictions on landfilling yard trimmings; a brief landfill disposal capacity "crisis;" the ability to compete with disposal options in some situations; and a need for flexibility in management of biosolids. The future of composting will be driven by such factors as the need for disposal alternatives; a demand for organic soil amendments; and regulations that stipulate more environmentally sound management of residuals such as manures.

INTRODUCTION

Composting in the United States has come a long way in the past 30 years. A full range of organic residuals -- from municipal wastewater biosolids and yard trimmings to manures and brewery sludge -- are being composted. Technologies and methods have grown in sophistication, as has the knowledge about what it takes to operate a facility without creating a nuisance, and to generate a high quality product.

In the general scheme of waste management alternatives, only a small percentage of residuals from the municipal, agricultural, commercial, industrial and institutional sectors are being composted at this time. Yet the fact that there is a significant level of composting experience in all those sectors lays the groundwork for growth in the future.

This paper will examine composting activity in the municipal, commercial, institutional and industrial sectors. While composting of agricultural residuals on farms is an increasing practice, *BioCycle* does not tabulate the number of projects in that sector. Therefore, no data on on-farm composting will be included.

S. Brown, J.S. Angle and L. Jacobs (eds.), Beneficial Co-Utilization of Agricultural, Municipal and Industrial By-products, 35–44.
© 1998 *Kluwer Academic Publishers,*

BIOSOLIDS COMPOSTING

The first survey of biosolids composting projects was done in 1983. At the time, there were 61 operating facilities and 29 projects in development, for a total of 90. The number of projects has grown steadily over the past 14 years (see Table 1).

Table 1. Biosolids composting project history[a].

Year	Operational	Total
1983	61	90
1985	79	173
1986	89	178
1987	107	197
1988	115	219
1989	119	227
1990	133	255
1991	149	275
1992	159	290
1993	186	321
1994	201	318
1995	228	330
1996	250	338

[a]Source: BioCycle Magazine Annual Biosolids Composting Surveys

As of December, 1996, there were 250 operating biosolids composting plants in the United States and 88 projects in development, according to *BioCycle*'s annual survey of biosolids composting projects (Goldstein and Steuteville, 1996). These facilities compost either raw or digested solids from wastewater treatment plants. Of the 250, 110 utilize the aerated static pile composting method, 72 compost in windrows, 52 have an in-vessel system, eight compost in aerated windrows and eight utilize static piles.

With aerated static pile composting, feedstocks are mixed and put in trapezoidal shaped piles. Blowers are used to aerate in either a positive mode (forcing air up through the pile) or negative mode (pulling air down through the pile). With the windrow method, piles are not aerated with blowers; instead, they are turned mechanically, either with a front-end loader or compost turning equipment. Aerated windrows have both forced aeration and are mechanically turned. Static piles have neither turning nor forced air. There are a variety of in-vessel composting technologies, ranging from agitated bays to silos.

Biosolids are mixed with a bulking agent prior to composting. The bulking agent provides both a carbon source and pile structure. *BioCycle* survey data finds that the most common amendments for aerated static pile composting are wood chips, followed by leaves, grass and brush. In-vessel systems without built-in agitation typically use sawdust and wood chips for amendments, while the agitated bay systems may utilize those materials and/or ground yard trimmings. The most common amendment at windrow facilities are yard trimmings, followed by wood chips. Other amendments utilized in biosolids composting include wood ash (which also helps with controlling odors), newsprint, manure, and peanut and rice hulls. Many facilities also use recycled compost.

The majority of biosolids composting facilities are fairly small to medium in size. The largest sites are a 150 dry ton/day windrow composting facility in Corona, California, a 65 dry ton/day in-vessel facility in Akron, OH, a 60 dry ton/day static pile facility in King County, WA, and a 50 dry ton/day aerated static pile composting plant in Philadelphia, PA.

Biosolids composting facilities typically are successful in marketing or distributing the compost produced. The top paying markets for biosolids compost are nurseries, landscapers and soil blenders. Other end uses include public works projects (e.g. roadway stabilization, landfill cover), application on park land and athletic fields, and agricultural utilization. Many composting plants distribute compost directly to homeowners.

BIOSOLIDS COMPOSTING DRIVERS

There are a number of "drivers" that have contributed to the development of biosolids composting projects in the United States. They revolve around potential difficulties in continuing current practices -- such as landfilling, incineration or in some cases, land application -- to a confidence level to undertake the effort because of the success of other projects.

Research on aerated static pile composting in the early 1970s at the U.S. Department of Agriculture's Agricultural Research Service in Beltsville, MD, helped to stimulate the application of the technology at municipal wastewater treatment plants. While smaller plants may use composting as their primary management option, a number of facilities start a composting project in conjunction with a land application program. Composting provides a backup when fields aren't accessible. For treatment plants in areas where agricultural land within a reasonable hauling distance is being developed, composting not only is a backup, but also is likely to become the primary management method in the future. In other areas, treatment plants that dispose of sludge in landfills may start a composting facility because of the uncertainty of their ability to continue landfill disposal in the future.

In the 1980s, landfill bans on yard trimmings forced many local governments to initiate composting projects to process leaves, brush and grass clippings. In some cases, public works officials joined forces with wastewater treatment plant operators in their towns to create cocomposting projects -- using the yard trimmings as bulking agent for the biosolids. This contributed to the growth of biosolids composting in the late 1980s and early 1990s

Perhaps one of the most exciting and important drivers in the development of biosolids composting plants has been the growing base of knowledge about what it takes to operate a successful facility. For example, many plants were (and still are) plagued by problems with odors. In some cases, they were severe enough to temporarily or permanently cause a site to close. In the early days, little was known about the composition of these odor compounds, the primary sources at a composting site, and technologies that could treat them effectively. Today, there is a great deal of sophistication in this area, and it helps to build confidence in utilizing the composting technology.

Two other drivers -- not just for biosolids composting but for other residuals -- have been the evolution of technologies to handle these materials and demand for compost products. In some municipalities, there is a higher comfort level with composting in a contained vessel or a bay-type system that is in a completely enclosed structure. The availability of these technologies, and the accompanying refinement in controlling odors from these types of systems, have helped to fuel the growth in projects. Research on compost utilization has helped stimulate markets for biosolids compost, especially in the horticultural and landscaping fields. And it is anticipated that demand for these kinds of products will grow in the future. For example, research in Massachusetts with utilization of biosolids compost in a manufactured topsoil shows significant potential for application in landscape architecture projects, an end use that can require vast amounts of finished product.

In terms of the future, growth in the number of biosolids composting projects is expected to continue. Factors contributing to the increase include: 1) A high quality biosolids compost can meet the U.S. Environmental Protection Agency's Class A standards, which give a wastewater treatment plant more flexibility in product distribution and regulatory compliance; 2) Increasing pressure on land application programs due to land development and public acceptance issues. As a result, wastewater treatment plants will need to seek alternatives such as composting; 3) Growing demand for high quality composts; and 4) Continual technology and operational improvements that result in more project successes thus building confidence in composting as a viable management option.

There are several caveats that hamper the development of biosolids composting projects. First, the economics are such that composting can be more costly than other management alternatives, such as land application and landfilling. Second, there is adequate landfill capacity available in many regions, and some treatment plants are taking advantage of that option at this time. As a result, the bottom line with biosolids composting is that we expect to continue seeing steady but not rapid growth in the number of projects in the United States.

YARD TRIMMINGS COMPOSTING

BioCycle began tracking the number of yard trimmings composting sites in the United States in 1989, as part of its annual State of Garbage in America survey. That first year, the survey found 650 projects. In the 1997 State of Garbage survey (which provides data for 1996), there were 3,260 yard trimmings composting sites, a slight drop of about 60 projects from the previous year (Goldstein, 1997).

A majority of the 3,000-plus sites are fairly low technology, smaller operations that are municipally owned and operated. Typically, yard trimmings are composted in windrows. Some of these smaller sites utilize compost turning equipment. Most, however, turn piles with front end loaders. Many simply build windrows, turn them occasionally in the beginning, and then let the piles sit for a number of months, moving material out only when there is a need for more space at the site.

There are some sizable municipal operations that utilize upfront grinding equipment, turners and screens. These sites tend to be managed more intensively because of the higher throughput and thus the need to move finished compost off the site more quickly.

There also is a healthy private sector that owns and operates yard trimmings composting facilities. These sites also tend to be managed more aggressively because the owners rely on income not only from tipping fees but also from product sales. While most of the larger sites also compost in windrows, some have experienced odor problems (particularly from grass clippings) and have started using aerated static piles in order to treat process air and not disturb the piles during active composting.

State bans on the disposal of yard trimmings at landfills and incinerators were the primary driver in the development of yard trimmings composting projects. There are 23 states with disposal bans; several bans only apply to leaves, or leaves and brush. Those 23 states represent 2,510 of the nation's total number of yard trimmings projects, or 77 percent.

BioCycle does not foresee many more states passing bans, thus growth of projects in the future will be driven more by localities trying to divert more green materials from landfills in order to save capacity, or by market demand for composted soil products (and thus the need for more feedstocks). Other possible drivers are the fact that yard trimmings are easy to source separate and thus are accessible for diversion; they are a good fit with biosolids composting; and most states' regulations make it fairly simple to compost yard trimmings, thus there aren't many entry barriers.

In the future, there likely will be some consolidation of yard trimmings projects, as was seen in the 1997 "State of Garbage in America Survey" (a drop of 56). Smaller municipalities may opt to close their sites and contract with a private facility or a larger municipal site in their region. Another trend will be that both public and private sector facilities will start accepting other source separated feedstocks, particularly residuals from institutional, commercial and industrial (ICI) generators. One yard trimmings site in Cedar Rapids, IA, for example, takes paper mill sludge and a pharmaceutical residual. A municipal site in Plano, TX, accepts vegetative trimmings from area grocery stores. A large-scale private site in Seattle, WA, services commercial generators in its region.

MUNICIPAL SOLID WASTE COMPOSTING

Municipal solid waste (MSW) composting has been around in the United States for decades. Projects were started almost 40 years ago, but closed with the advent of cheap landfill space. There was a resurgence in MSW composting in the 1980s due to a number of factors, including closure of substandard landfills in rural areas; rising tipping fees in some regions as well as perceived decrease in landfill capacity; minimal development of waste to energy facilities (due to cost and performance issues); a perceived natural "fit" with the growing interest in recycling; the existence of technologies, primarily European, so that projects didn't have to start from scratch; flow control restrictions that could enable projects to direct MSW to their facilities; and a potential revenue stream from tip fees and product sales.

Solid waste composting in the U.S. emerged on two tracks during the 1980s. The first, the mixed waste approach, involves bringing unsegregated loads of trash (in some cases this includes the recyclables) and doing all separation at the facility, both through upfront processing and/or back end product finishing. The second track, the source separated approach, relies on residents and other generators to separate out recyclables, compostables and trash.

As can be seen in Table 2, interest in MSW composting grew rapidly in the late 1980s and early 1990s. But what also is evident from Table 2 is that the number of operating projects never grew very much. At the peak in 1992, there were 21 operating projects. Today (1997), there are 14 operating facilities, with one in Georgia that is being repaired after two different fires. There are 23 projects in some phase of development, with three most likely to move to an operating status in the next several years (Goldstein, Farrell and Steuteville, 1996).

Table 2. Solid waste composting project history[a].

Year	Operational	Total
1985	1	
1986	1	6
1987	3	18
1988	6	42
1989	7	75
1990	9	89
1991	18	n/a
1992	21	82
1993	17	n/a
1994	17	51
1995	17	44
1996	15	41

[a]Source: BioCycle Magazine Annual MSW Composting Surveys

Six of the 14 operating projects are under 20 tons/day. Five process between 21 and 50 tons/day of MSW, and two are between 51 tons and 100 tons/day. There is only one facility handling more than 100 tons/day (the Sevierville, Tennessee plant, which processes 225 to 250 tons/day of MSW and biosolids). Several large plants did open over the past 10 years (one with a design capacity of 700 tons/day), but have since shut down.

Of the current operating projects, five use rotating drums and either windrows, aerated windrows or aerated static piles for active composting and curing. Three use agitated bed systems and six use only windrows, aerated windrows or aerated static piles. Five of the operating plants receive source separated MSW; the remainder process a mixed waste stream. Twelve of the operating facilities are publicly owned (some of these are privately operated) and two are privately owned and operated. At this time, there are very few vendors in the U.S. selling solid waste composting systems.

MSW COMPOSTING DRIVERS: PRESENT AND FUTURE

Ten years ago, many in the solid waste field felt there would be a landfill crisis in some regions of the country, prompting a surge of interest in alternative management options.

In addition, the federal Subtitle D regulations -- which went into effect in 1994 -- were expected to force the closure of many substandard landfills, again putting pressure on existing disposal capacity.

The expected landfill crisis never really materialized, at least on a national basis. Landfills definitely closed -- from almost 8,000 in 1988 to about 3,000 in 1996. At the same time, however, new state of the art mega-landfills opened, serving disposal needs on a regional (versus a local) basis. When landfills closed in small towns, instead of building small composting facilities, many communities opted instead to build solid waste transfer stations and have waste hauled long distances for disposal. Today, there are more transfer stations than landfills in the United States.

Tipping fees, which did start to rise in many places, never stayed high in most regions. In fact, tipping fees are dropping in the U.S. The national average fell from $34/ton in 1995 to $32/ton in 1996, according to BioCycle's State of Garbage in America survey. And it is not anticipated they will go up significantly any time in the near future.

Solid waste composting projects also were hurt by a 1994 U.S. Supreme Court decision that struck down flow control laws that gave government agencies the ability to direct the waste stream to specific facilities. MSW flow into some composting plants dropped considerably as haulers opted to transport garbage further distances to landfills with lower tipping fees.

Other factors that stymied the development of MSW composting in the U.S. include: generation of odors at some of the larger, higher visibility projects, leading to their failures; inadequate capitalization to fix problems that caused odors and/or put in odor control systems; production of a marginal compost product; and significant skepticism about the technology due to the project failures.

In the future, there will be some development of MSW composting projects, perhaps in areas where it is difficult to implement recycling programs (e.g. major tourist areas). The application of the technology, however, will be very site specific. For example, there may be a few communities that decide to increase diversion by getting households to separate other organics beyond yard trimmings. Many towns, however, have opted to push backyard composting of household organics instead of getting involved in centralized collection.

Experience has shown that composting solid waste on a larger scale requires a lot of capital, as well as deep financial pockets to address problems that arise once the facility starts operating. Projects also need to be able to set tipping fees that are competitive with landfills, which can be difficult when a project needs to make a sizable capital investment in processing (upfront and product finishing) equipment.

FOOD RESIDUALS COMPOSTING

Perhaps the fastest growing segment of the U.S. composting industry is diversion of institutional/commercial/industrial (ICI) organics, primarily food and food processing residuals, including seafood. *BioCycle* began tracking data on this sector in 1995, when there was a total of 58 projects. In 1997, the survey found 214 projects, close to a 270 percent growth rate in two years (Goldstein and Block, 1997).

Of the 214 projects, 176 are fully operational and 22 are pilot or demonstration projects. About 40 percent of the projects are on-site at institutions, e.g. correctional facilities, universities, schools and hospitals. Sixty-four of the projects are commercial sites that service ICI generators. Twenty-two of the projects listed in the 1997 survey are on farms. The types of commercial and industrial generators serviced include grocery stores, restaurants, produce terminals, food and seafood processors, breweries, etc.

The majority of the sites compost feedstocks in windrows; many use yard trimmings as a bulking agent. Feedstocks include pre and postconsumer food residuals (e.g. vegetative trimmings, kitchen prep and plate scrapings, baked goods, meats), out-of-date or off-spec food products and industrial organics such as crab and mussel residuals and brewery sludge. Only a handful of the facilities accept large amounts of food residuals each year. The economics of food residuals composting projects have to be competitive with disposal options because the generators typically deal with private haulers (and thus know current disposal costs).

CONCLUSIONS

Composting serves as both a waste management method and a product manufacturer. As such, a project can generate revenue streams on both the front end (tipping fees) and the back end (product sales). Many companies got into composting mostly based on the upfront revenue from tipping fees, and did not focus a lot of attention on producing a high quality product to maximize sales. But with steady or dropping tipping fees, projects are having to become more market driven and not tip fee driven. Successful companies and operations are those with excellent marketing programs. They have invested in equipment to service their markets, e.g. screens with various sizes to meet different end uses. In short, they know their markets and know how to service them.

There also are exciting developments on the end use side. Composts are being used increasingly not only for their nutrient value and ability to build soil organic matter but also because of their ability to suppress plant diseases. There is an increase in agricultural utilization of compost, and many states are developing procurement programs for compost use on highways and for erosion control. Interesting projects also are developing in the use of compost for bioremediation. In short, while composting will always be around as a waste management option, it is becoming equally (and in some cases more) valuable as a producer of organic soil amendments.

For the most part, major solid waste initiatives that might have a positive impact on the development of composting projects are not expected. There may be some indirect impacts, e.g., from increasing regulation of manure management, which may lead to more composting on farms. But for the foreseeable future, growth in composting may be due more to market demand for compost.

REFERENCES

Goldstein N (1997) The State of Garbage in America. BioCycle 38, No. 4:60-67.

Goldstein N & Block D (1997) Nationwide Inventory of Food Residuals Composting. BioCycle 38, No. 8:46-57.
Goldstein N & Steuteville R (1996) Steady Climb For Biosolids Composting. BioCycle 37, No. 12:68-78.
Goldstein N, Farrell M & Steuteville R (1996) MSW Composting in the United States. BioCycle 37, No. 11:46-53.

NRC COMMITTEE REVIEW OF USING BIOSOLIDS AND EFFLUENTS IN FOOD CROP PRODUCTION

ROBERT K. BASTIAN
Office of Wastewater Management, U.S. Environmental Protection Agency, Washington, D.C. 20460, USA
E-mail bastian.robert@epamail.epa.gov

ABSTRACT

Potential health effects as well as lack of compliance with existing regulatory requirements are frequently raised as concerns when projects are proposed to recycle biosolids on farmland used for the production of food crops for direct human consumption. These concerns were a primary focus of the report, *"Use of Reclaimed Water and Sludge in Food Crop Production,"* issued by the National Research Council of the National Academy of Sciences in early 1996 (NRC, 1996).

The three-year study was undertaken to help answer some of the questions that have been raised about the safety of crops grown in fields where treated municipal wastewater effluent or biosolids have been applied. It provides an independent assessment of the risks associated with these practices and provides recommendations for improving these recycling practices and their acceptance. The committee that conducted the study based its report on existing published literature and discussions with experts in the field. The study reviewed the current state of the practice, public health concerns, existing guidelines and regulations, and implementation issues. While the committee did not conduct a formal risk assessment of possible health effects, it did review the methods and procedures used by EPA in its extensive risk assessment which was the basis of the Part 503 rule.

The National Research Council (NRC) report confirms the basis of Federal policy that properly treated and managed municipal wastewater effluents and biosolids can be safely and effectively used in food crop production, while presenting negligible risk to the crops, consumers, and the environment. To help address the persistent concerns regarding the potential for exposure to contaminants, nuisance problems and adequacy of oversight of programs involving agricultural use of biosolids, the report suggests that POTWs, private processors, distributors, and applicators should not only comply with all regulatory requirements and management practices, but also take extra steps to demonstrate such compliance to various stakeholders (e.g., neighbors, farmers, food processors, and consumers). It recommends full public participation, self-monitoring and reporting programs, and public education campaigns. The study report suggests that it is especially important that these public information efforts be continuing and detailed, as public awareness that safe practices are in fact being followed is necessary if monitoring by state

S. Brown, J.S. Angle and L. Jacobs (eds.), Beneficial Co-Utilization of Agricultural, Municipal and Industrial By-products, 45–54.
© 1998 *Kluwer Academic Publishers,*

or local entities is likely to be minimal.

INTRODUCTION

In March 1996, the National Research Council (NRC), the operating arm of the National Academy of Sciences, released the results of a study by an expert committee that reviewed the current state-of-the-practice, public health concerns, existing guidelines and regulations, and implementation issues associated with the use of reclaimed water and biosolids in the production of food crops. The independent NRC assessment was undertaken to help address the concerns that have been raised by food processors regarding the utilization of these materials in the production of food crops. The study was suggested by the U.S. Environmental Protection Agency (EPA) in 1993 and eventually received funding support from EPA, U.S. Department of Agriculture (USDA), Food and Drug Administration (FDA), U.S. Bureau of Reclamation, National Food Processors Association (NFPA), National Water Research Institute, Water Environment Research Foundation, Association of Metropolitan Sewerage Agencies, Eastern Municipal Water District of California, Metropolitan Water Districts of California, BioGro Technologies, and N-Viro International.

The fourteen-member study panel included experts in the areas of soil and crop science, ecology, wastewater and sludge treatment, risk assessment, toxicology, infectious disease, public health, economics, and environmental law. They represented a range of perspectives from their positions in universities and agricultural extension, state public health agencies, food companies, environmental advocacy groups, and publicly owned treatment works (POTWs). The panel members included:

A.L. Page (Chairman), University of California-Riverside
A. Ayanaba, DelMonte Foods
M.S. Baram, Boston University
G.W. Barrett, University of Georgia-Athens
W.G. Boggess, Oregon State University
A. Chang, University of California-Riverside
R.C. Cooper, BioVir Labs & University of California-Berkeley
R. Dick, Cornell University
S.P. Graef, Western Carolina Regional Sewer Authority
T.E. Long, Washington State Department of Health
C. St.Hilaire, Hershey Foods Corporation
Joann Silverstein, University of Colorado-Boulder
Sara Clark Stuart, Environmental Defense Fund
P.E. Waggoner, Connecticut Agricultural Experiment Station

The primary focus for the NRC study report addresses questions about crop safety and related food processors' concerns. As a part of this effort, the report reviews the current state-of-the-practice (including potential public health concerns), existing guidelines and regulations (including EPA's risk assessment associated with the Part 503 rule), and implementation issues. A summary of the study results published in *Biocycle* in February 1997 by the NRC staff Project Officer and Committee Chairman (Krauss and Page, 1997) was drawn on extensively in preparing this paper.

BACKGROUND

Cropland application of both treated municipal effluents and biosolids represent important management options to POTWs as the volumes of wastewater effluents and biosolids continue to grow as POTWs service increasing populations (Figures 1 & 2). Concerns expressed by the food processing industry over the acceptance of food crops grown on fields amended with biosolids, or irrigated with reclaimed wastewater effluents, have in some areas created a serious constraint to these practices on croplands that currently, or may in the future, be used to grow food crops. When EPA first issued formal regulations addressing land application of biosolids to cropland in 1979, some food processors questioned the safety of food crops grown on biosolids-amended soils and were concerned about potential liability problems. A joint statement of policy and guidance issued by EPA, USDA and FDA in 1981 on the production of fruits and vegetables with biosolids (EPA/USDA/FDA, 1981) did not adequately relieve their concerns. Since NFPA as well as many other groups closely monitored the development of the risk-based Part 503 regulation (EPA, 1993), many interested parties hoped that the issuance of the new regulation would help alleviate the food processors' concerns.

When, in lieu of providing advanced treatment for nutrient removal to protect sensitive receiving waters from excess phosphorus loadings, a growing number of POTWs began to seriously explore marketing treated effluents to farmers growing food crops in the Pacific Northwest, similar concerns were raised by some local food processors over the use of treated wastewater effluents to irrigate food crops. Although extensive use has been made of reclaimed effluents to irrigate food crops in California (Figure 3) and Florida for many years, concerns over the potential for consumer opposition to such practices caused reluctance by some groups to accept planned projects involving the use of reclaimed effluents to irrigate food crops. EPA updated its guidelines for water reuse in 1992 (EPA, 1992), which included recommendations regarding reclaimed water quality used for food crop irrigation.

RESULTS AND DISCUSSION

The observations from experiences in the field where these practices were utilized, as well as the results of extensive research efforts support the findings of the NRC study. Both confirm the basis of Federal policy that properly treated and managed municipal wastewater effluents and biosolids can be safely and effectively used in food crop production, while presenting negligible risk to the crops, consumers and the environment.

Composted and heat dried biosolids have been widely used for many years as organic soil amendments or fertilizer for various uses involving extensive direct human exposure --

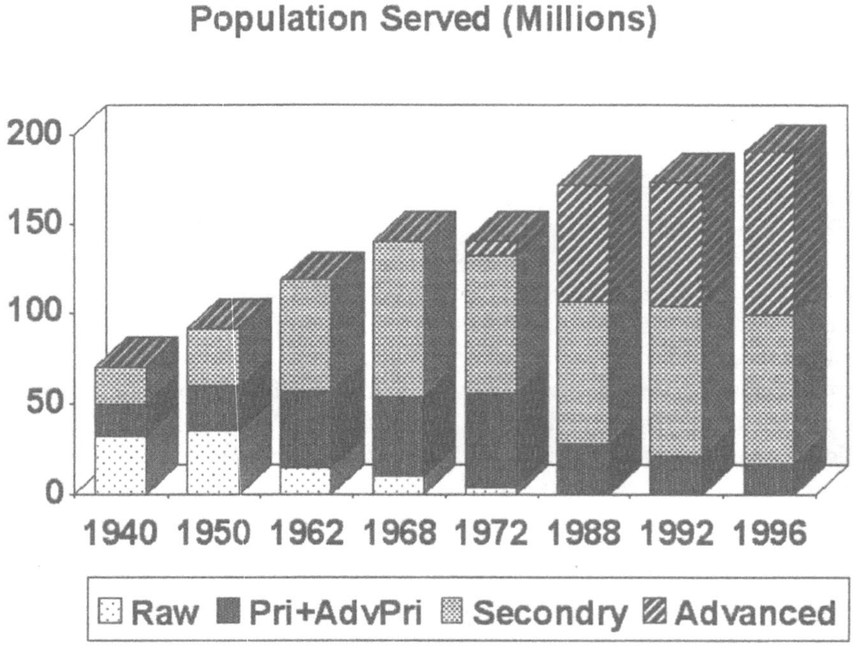

Figure 1. Increasing proportion of the U.S. population served by POTWs.

e.g., turf production and maintenance; landscaping; home gardens and lawns; parks, ball fields and other recreation areas; potting soils and topsoil production, etc. -- including the production of crops (such as citrus) for direct human consumption. These processes meet the Part 503 "Class A" requirements i.e., they reduce pathogens to levels below detection when using testing methods specified in the rule. In spiked samples, these processes have been demonstrated to consistently achieve pathogen reduction rates comparable to pasteurization, as have the other processes listed in the rule as Class A, or Process to Further Reduce Pathogens (PFRP), or processes that have been demonstrated and approved as equivalent to PFRP (e.g., heat treatment, autothermal thermphyllic aerobic digestion, irradiation, high-level alkaline stabilization). Questions regarding the potential for regrowth of fecal coliforms or *Salmonella* spp. in Class A products after processing are also addressed by the Part 503 rule, which requires testing of these products prior to their use or transfer to a user.

Biosolids treated to meet the Part 503 Class B requirements ($\leq 2 \times 10^6$ fecal coliforms) can and are also being applied to land used for the production of food crops (e.g., apples, pears, cherries, sweet corn, small grains, etc.), grazing and turf production, as well as where there will be direct public access. However, the use of Class B treated materials in such areas also requires that specific waiting periods are met prior to allowing general public access, animal grazing, or the harvesting of food crops that touch the ground produced on such sites. These waiting periods allow for natural die-off of the pathogens remaining after Class B processing to occur. The 1996 NRC Report found that the Part

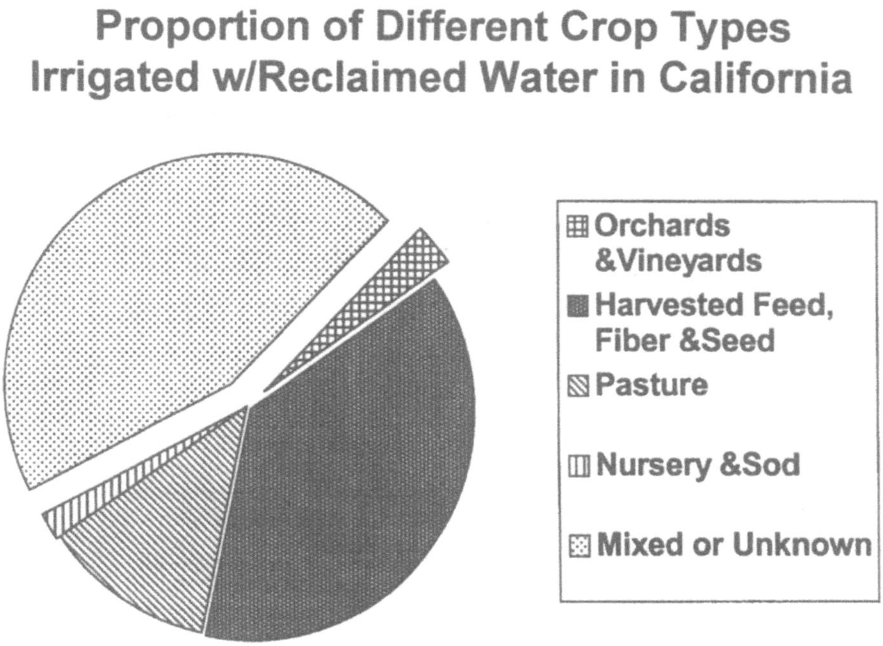

Figure 2. Biosolids use/disposal practices by EPA region.

503 rule appears to be adequate for the protection of the public from the transmission of waste-associated pathogens, with two possible exceptions. The waiting period between the application of Class B biosolids and animal grazing may not be adequate to prevent transmission of beef tapeworm to grazing cattle, and the *Salmonella* test should not be used as a sole indicator of Class A quality because the test, as currently prescribed, lacks adequate sensitivity.

Irrigation of food crops with treated municipal effluents has been effectively and safely practiced in the U.S. on a limited scale. The public has generally accepted the concept of irrigation with reclaimed water as part of larger and more comprehensive water conservation programs that reclaim wastewater for a variety of non-potable uses. When reclaimed water has been used for food crop production, the state standards for wastewater treatment and reuse, along with site restrictions and generally good system reliability, have insured that food crops thus produced do not present a greater risk to the consumer than do crops irrigated from conventional sources. Further, health effects studies addressing the use of reclaimed water to augment potable water supplies indicate that the reclaimed water is as good, if not a better, raw water source than supplies currently used in the study areas.

While limited in number, epidemiological studies conducted in the U.S. on treatment plant workers exposed to municipal wastewater or biosolids, or populations exposed to treated biosolids land application projects, indicate that this exposure was not a significant

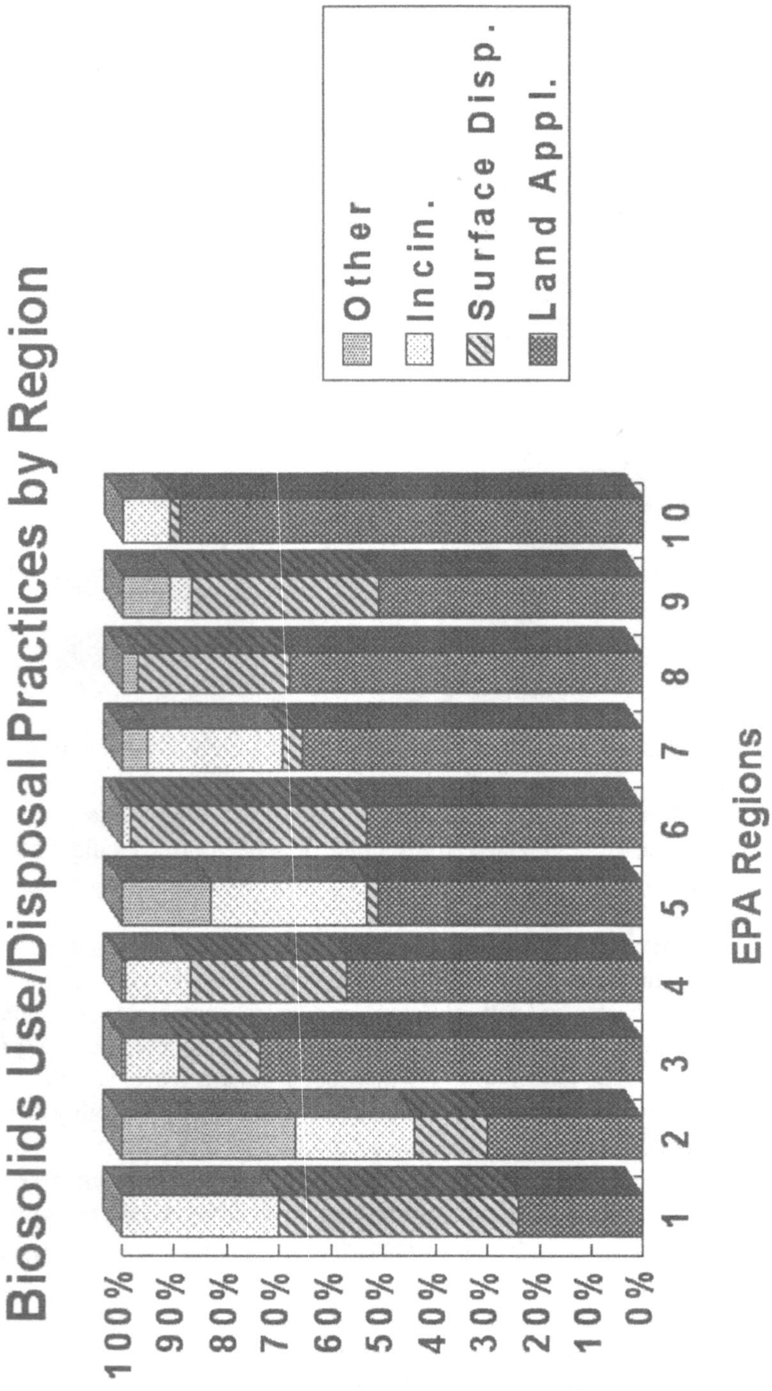

Figure 3. Proportion of different crop types irrigated with reclaimed water in California.

risk factor. Considering the great many sources of infectious disease agents other than the reuse of biosolids, such as prepared foods and person-to-person contact, the potential for additional pathogen exposure from the proper use of biosolids is minimal compared to our everyday exposure to pathogens from these other sources. Community-wide source control and industrial pretreatment programs have resulted in significant reductions of toxic pollutants in wastewater effluents and the biosolids produced by POTWs. Most of the toxic organics and inorganic trace elements that may be present in treated biosolids are either not transferred from soil to plant tissues, or, translocation to edible tissues does not create levels harmful to consumers under normal agricultural conditions. Cadmium is still considered to be the inorganic chemical of greatest concern with regards to human health when treated biosolids are applied to croplands. Research on the bioavailability of toxic organic compounds to plants indicates that the risk to humans consuming food crops grown on soils amended with biosolids is negligible. Toxic organic compounds are typically present at low concentrations and/or are largely not bioavailable to plants.

No human acute or chronic toxicity effects have been reported that resulted from ingestion of food plants grown in soils amended with biosolids. Further, EPA's risk assessment, which involved 14 exposure pathways, suggests that food safety is not likely to be affected by the application of toxic organic or inorganic trace elements in biosolids. The cumulative pollutant loadings determined to be safe, under the Part 503 risk assessment (using the 14 exposure pathways), are significantly lower than the limits that would be allowed by any of the three pathways which assess the production of human crops on biosolids-amended soils. These three pathways are:

Pathway 1: biosolids→soil→plant→human
Pathway 4: biosolids→soil→plant→animal→human
Pathway 5: biosolid→soil→animal→human

This suggests that the "plant-soil barrier" will be an effective protection mechanism for most pollutants, if present in biosolids that are applied to cropland.

Therefore, the Part 503 rule has a sound conceptual basis for protecting public health. In terms of trace elements, biosolids that are applied according to the pollutant loading rates specified in Part 503 should not adversely affect the safety of the nation's food supply. Further, most toxic organic compounds found in biosolids are present at concentrations well below those considered to pose a risk to human health or the environment.

RECOMMENDATIONS

The NRC report makes numerous recommendations regarding the use of reclaimed water and biosolids. These include: 1) the general importance of using agronomic rates and adjusting fertilizer practices to take credits for the nitrogen and phosphorus being applied to avoid potential contamination of water resources; 2) any payment programs designed to promote agricultural use should be carefully structured to avoid incentives to apply at rates exceeding agronomic rates; 3) a more comprehensive survey (using consistent sampling and data-reporting methods) of POTWs is needed to better document whether or not toxic organic compounds are present in biosolids; and 4) Part 503 should be amended to only allow EQ biosolids to be marketed to the general public so that further regulation and management beyond the point of sale or give-away would not be necessary.

The NRC report notes that parties affected by biosolids reuse programs e.g. local communities, crop growers, food processors and consumers - remain concerned about the potential for exposure to contaminants, nuisance problems, liability and adequacy of program management and oversight. To address these concerns, the report recommends that POTWs take extra steps to demonstrate their compliance with all regulatory requirements and their use of stringent management procedures that include such measures as public participation in the decision making process, self-monitoring and reliable reporting, and public education programs.

Reaction of the Food Processors

The NRC report has helped motivate some food processors to begin reexamining their policies regarding the use of treated biosolids and reclaimed water in food crop production. The Northwest Food Processors Association has formed a task force to reconsider their position on the subject. Although the NFPA has not to date modified its official position of not endorsing "the application of sludge to lands that may or will be used in the production of foods for human consumption" (NFPA, 1993), the NFPA does not disagree with the scientific findings of the NRC report which documents the negligible risk to public health of applying properly treated and managed biosolids to food crops. Further, NFPA has described the NRC report as providing a fair presentation of food processor concerns, and that it should be very helpful in the decision-making process. Their reasons for not changing their policy to date have to do with potential liability, marketing and public perception.

CONCLUSIONS

The NRC report provides considerable reassurance that properly treated and managed municipal wastewater effluents and biosolids can be safely and effectively used in food crop production, while presenting negligible risk to crop quality or consumers. This report also notes that public acceptance and implementation issues, rather than scientific information or the health and safety risks from food consumption, may be the critical factors in determining whether reclaimed wastewater effluents and biosolids are used in food crop production. Clearly, the NRC report should prove useful in addressing many of the concerns that continue to be raised about these practices. It also could effectively be used to 1) support the development of tailor-made "designer" biosolids products to meet specialty market uses, 2) encourage use of third party environmental management systems that help ensure compliance with regulatory requirements and assure product quality control, and 3) facilitate the establishment and maintenance of public education and demonstration programs, as well as continue R&D on long-term effects.

Copies of "Use of Reclaimed Water and Sludge in Food Crop Production" are available from the National Academy Press, 2101 Constitution Avenue, N.W., Box 285, Washington, D.C. 20055 (Telephone: 800/624-6242)

REFERENCES

EPA (1992) Guidelines for Water Reuse. EPA 625/R-92/004. Office of Water. Washington, D.C. 247pp.

EPA (1993) Standards for the Use or Disposal of Sewage Sludge; Final Rules. 40 CFR Parts 257, 403 and 503. Federal Register 58(32):9248-9415. February 13, 1993.

EPA/USDA/FDA (1981) Land Application of Municipal Sewage Sludge for the Production of Fruits and Vegetables; A Statement of Federal Policy and Guidance. SW-905. U.S. EPA, Office of Solid Waste. Washington, D.C. 21pp.

Krauss GD & Page AL (1997) Wastewater, Sludge and Food Crops. Biocycle, February 1997:74-82.

NFPA (1993) Statement by the National Food Processors Assoc. on the Use of Municipal Sewage Sludge in the Production of Food for Human Consumption. NFPA. Washington, D.C.

NRC (National Research Council/National Academy of Sciences) (1996) Use of Reclaimed Water and Sludge in Food Crop Production. Water Science & Technology Board/National Research Council. National Academy Press. Washington, D.C. 178pp.

EUROPEAN PERSPECTIVE OF COMPOST CO-UTILIZATION FOR HORTICULTURE

ROBIN SZMIDT
Scottish Agricultural College, Auchincruive, Ayr, Scotland, UK
E-mail szm@au.sac.ac.uk

ABSTRACT

In Europe, legislation demands adherence to specific guidelines covering land-use, effluent management, odour and transportation of wastes. Such factors may demand investment in relatively sophisticated technology including adoption of enclosed systems for composting which are broadly described. However, the market value of composted material and financial pressure on businesses and the public sector means that such investment may not be realistic. The concept of Best Available Techniques Not Entailing Excessive Cost (BATNEEC) is described. If used by planning authorities, it can ensure adoption of the best techniques to meet environmental guidelines while not putting at risk the financial potential of business. A computer 'model' to help optimise investment is described and can be used to predict the cost-benefit of different strategies. In the UK economic factors are heavily influenced by the introduction of tax on landfill operations, and an increase in the quantity of co-composted material is predict. This will have an impact on the market and on the quality of material available to the horticultural industry.

Horticultural industry has readily adopted use of some co-composted materials, particularly in the landscape industries. However, for high value sectors, such as bedding plants and pot plants, adoption has been slow. This should be put in perspective by comparing the enormous volume of research into peat-based growing media which has taken place in past decades with the relatively small investment in research into sustainable growing media such as co-composted materials.

INTRODUCTION

The term co-utilization implies beneficial integration of composted materials with other products or commodities. Specifically, use in horticulture is often considered by composters as a target market for composted products. This pre-supposes that horticulture is one market. However, horticulturists use a wide range of products for which a number of composted materials could be substituted. A lack of understanding of the range of potential horticultural markets has led to failure of some compost businesses through failure to properly target their products. Examples of horticultural uses of composted

S. Brown, J.S. Angle and L. Jacobs (eds.), Beneficial Co-Utilization of Agricultural, Municipal and Industrial By-products, 55–68.
© 1998 *Kluwer Academic Publishers*,

materials include: Pot plant production; bedding plant production; hardy ornamental (tree and shrub) production; soilless growing media; soil amendment; mulching; tree planting; and mushroom cultivation.

None of these uses are wholly novel. With the exception of mushroom cultivation, all of the potential uses of composted materials have primarily relied on alternative materials. Mushroom cultivation is exceptional, because composting and the microbial functions of the compost mix are wholly integrated into the biochemical processes of the crop itself (van Griensven, 1988). By contrast other uses generally demand relatively stable material which will fulfill a purpose over the life of a crop or specific period of growth, such as the establishment season of newly planted trees.

National initiatives such as the UK National Compost Development Association (NCDA) and the Europe-wide Organic Reclamation and Composting Association (ORCA) work to promote composting. While there has been considerable effort to develop technology and to influence guidelines and legislation, less attention has been paid to use of composted material. The principal motivator for the use of compost material has been applied research, such as the work done by the Henry Doubleday Research Association at Ryton, UK, and through the initiatives of the International Society for Horticultural Science (ISHS).

In Europe, peat is the primary material used for the potential uses of composted materials that was previously listed. Peat, an organic medium, is relatively stable and of low plant-nutrient content. The use of peat-based growing media was shown by Bunt (1976) to fulfil many of the needs of horticulturists. More than 30 years of research and development has resulted in optimization of these materials in both physical and nutritional terms, so peat use for horticulture is substantial. In the UK, use of horticultural peat is approximately 2.7 million cubic metres per year ($m^3\ yr^{-1}$). The five major horticultural peat-consuming nations in Europe use c. 17.5 million $m^3\ yr^{-1}$ compared to an estimated 3.75 millon $m^3\ yr^{-1}$ in the USA (Bragg, 1990). A wide range of peat-grades are employed in horticulture and no one substitute is possible (Bragg, 1990). The need for stability may be taken to extremes in uses such as mulching, where plastic field or row covers are now widely employed instead of biodegradable material.

Since the mid 1970s, the European horticultural industry has developed hydroponic systems to use inert material for plant production. Such methods pursue the objectives of controlling the chemical and physical plant root environment (Smith, 1996). The biological growth effects within plant growing media were largely ignored, even in the body of research in centres such as the Glasshouse Crops Research Institute, and has only recently been studied significantly at institutes including the Horticulture Research International (Anon, 1991-96). The objective of using peat or hydroponic systems as a chemically and physically stable substrate, on which are imposed controlled levels of plant nutrients, may be in contrast with the potentially higher biological activity and the potential for degradability of composted materials (Hoitink and Keener, 1993).

The conflict that this perception generates is that industrial practice based on decades of research is not widely applicable to use of composted materials. This does not automatically imply that all composted materials are less good. Most composted materials can grow plants as well as peat-based growing media, and other attributes exist such as disease suppressiveness of composted growing media not known with peat products

(Hoitink et al., 1996a,b) (Schauner, Personal Communication). Commercial horticulturists have yet to see fully integrated research programmes develop these issues to a level where commercial exploitation of composted growing media may be industry-wide. In a number of situations, particularly where peat is not readily available locally, co-utilization of composted material and peat may be recommended (Reis et al., In press, Raviv et al., In press).

To enter into the competitive market, novel composted material must perform as well as peat, be cost effective, show technical advantage such as suppressiveness, and comply with current environmental legislation and expectations.

Sources of Compost Feedstock

The UK and Europe as a whole have a very wide range of potential feedstocks for composting which include agricultural manures and plant waste. Estimated quantities are approximately 2.5 U.S. billion tonnes per annum of organic matter of which food, agricultural and forestry account for approximately 70% (De Bertoldi et al., 1996). Storage and disposal of manures is governed by legislation preventing application directly to land under conditions which may cause pollution (Ministry of Agriculture Fisheries and Food, 1993). Consequently, temporary storage may be necessary. In some cases composting could be used as a volume reduction process potentially to provide feedstock to larger composting ventures. However, many farms are relatively small, and seasonal housing/pasture management may result in an interrupted supply of compostable manure wastes. Similarly seasonal may be the production of straw. Excess straw was traditionally burnt *in-situ,* but in the UK this practice is now restricted (Ministry of Agriculture Fisheries and Food, 1992b) making straw widely available as a compost ingredient.

A number of industrial processes generate waste which could be used as feedstock for composting as well. In many cases, industrial production of biodegradable wastes is intermittent and of relatively small quantities. Material that is highly putrescible should be used as soon as possible. Examples include food wastes (Sinclair, 1996) and bioreactor/biosolid wastes (Hall et al., 1992; Szmidt, In press a; Szmidt and Bryden, 1996; Szmidt and Smith, 1996). In such case co-utilization of wastes (mixing highly putrescible materials with more stable wastes) may have the benefits of improved stability and predictability of compost mixes and improved uniformity of larger volumes in production.

The scientific literature shows required compost parameters, for instance C:N ratio, moisture content and physical structure (Jackson et al., 1992, Lopez-Real and Vere, 1992). However, selecting blends of material which are known to satisfy such criteria may not be easy. Some authors have attempted by tabulating suitable mixes to simplify choices for blending of agricultural by-products for composting (Jeangille, 1991). By contrast there is no easily available 'ready-reckoner' for co-utilization compost recipes. Public information about the wide range of modern industrial by-products is simply insufficient to select blends of material from published lists. In the UK, suppliers must be able on demand to specify the nature and hazard potential of any product or waste, so that compost site operators can be knowledgeable about the characteristics of current materials (Anon, 1974, 1994).

While inert material is not usually appropriate as a compost feedstock, there are some exceptions. Providing there is no negative consequence of incorporation, composting may be a suitable disposal route. Wood chips are resilient and may be considered in this category as they may be little changed during the composting process but can be helpful in regard to air-porosity of material. Other inert material in this category includes sand and grit but, care should be taken as air-porosity may be reduced (Bunt, 1976).

Of particular interest is rockdust fines (less than 200 mesh) the use of which has been advocated by some proponents of organic agriculture and horticulture. Hamaker (1982) proposed that such materials could have a major impact on soil fertility and crop productivity if utilized with compost, to greatly improve soil quality. However, a review of the literature for the use of rockdusts either alone or as a co-utilization component with compost revealed little scientific evidence to support the claims of this technique (Webster, 1995). This literature review was followed by preliminary experimental work which has, for the first time, revealed the potential value of rockdust as a component of compost for improving quality of the end-product, increasing the microbial activity of material, and reducing the odour pollution potential of composted material (Szmidt, In press b).

The use of rockdust demonstrates a paradox in developing improved composted materials. In order to enter material into modern agri-business, product development must be based on robust scientific development and understanding. In many cases quasi-organic materials have a history of unsupported claims and benefits which cannot be proven in practice, or which work under some circumstances but not under others. Until the full reasons for such differences are understood, many such materials will remain on the fringes of commercial horticulture and agri-business.

In addition to calculating the correct physical and chemical balance, further understanding is needed about the pollution potential of industrial wastes (The Scottish Office of Agriculture, Environment, and Fisheries Department, 1997). In addition, compost plant operators should not be complacent regarding the microbial hazards associated with industrial organic by-products (Hoitink and Keener, 1993).

European Legislation

Significant legislation and standards now exist relevant to compost operators and their products. In the European Community (EC), legislation is superior to that of Member States. A number of authors have reviewed the legislative requirements of the EC (Fontaine, 1995, Kent, 1996) and their legislative documentation is readily available. Outside Europe a common misconception is that the EC relates principally to free trade and political aspirations. However, the Maastricht Treaty (Office for Official Publications of the European Communities, 1992) sets out the principles, not only for trade and economic development, but also policy initiatives such as protection of the environment and agriculture, whereas the way in which Member States implement legislation may vary within the EC. Within the Maastricht Treaty (Office for Official Publication of the European Communities, 1992) Title XVI Environment, Article 130R states "Community policy on the environment shall contribute to pursuit of the following objectives:

- Preserving, protecting and improving the quality of the environment;
- Protecting human health;

- Prudent and rational utilisation of natural resources; and
- Promoting measures at an international level to deal with regional or worldwide environmental problems.

Subsidiarity, the concept of derogating responsibility for interpretation and enactment of legislation on a local basis, allows for implementation of legislation in different ways in different areas of the EC. In the case of co-utilization, this can result in different prioritization of related areas such as organic-matter recycling by composting. An example would be setting recycling or waste reduction targets where different Member States may address different industrial sectors to achieve the required reduction. For instance, the Proposed Directive on Landfill of Waste: COM(97)105 Article 5 states that: "By 2010 biodegradable municipal waste going to landfills must be reduced by 25% of the total amount (by weight) of biodegradable municipal waste produced in 1993". Furthermore, Draft Council Resolution COM(96)399 states "the need for promoting waste recovery with a view to reducing quantity of waste and saving natural resources, in particular by re-use, recycling, composting and recovery of energy from waste".

In the UK this is being done by the generation of national and regional strategies such as the Draft National Waste Strategy: Scotland (Unpublished). Implementation of this strategy is encouraged by legal obligation such as the Environment Protection Act (1990) (Anon, 1990b) or by encouragement. An example of the latter is implementation of tax on landfill at different rates for inactive (inert) and other (including biodegradable) material, which in 1997 were £2 ($3) and £7 ($11) per tonne, respectively (HM Customs and Excise, 1996). Of this, tax credit up to 90% of project costs, amounting to no more than 20% of the contributors landfill tax bill may be allocated, not to the Exchequer, but to environmentally related projects, including compost research and development. The broad principles of project eligibility are as follows:

- Reclamation, remediation, restoration or any other operation that facilitates the economic, social or environmental use of land where its use has been prevented or restricted because of previous use.
- Any operation intended to prevent or reduce any potential for pollution or to remedy or mitigate the effects of any pollution on land polluted by a previous activity.
- Research and development, education or collection and dissemination of information about waste management practices. The purpose of which is to encourage the use of more sustainable waste management practices.
- Where it is for the protection of the environment, the provision, maintenance or improvement of a public park or other public amenity in the vicinity of a landfill site.
- Where it is for the protection of the environment, the maintenance, repair or restoration of a building or other structure which is a place of religious worship or of historic or architectural interest.
- The provision of financial, administration and other similar services to enrolled environmental bodies.

In the UK the most significant pollution control and environmental protection legislation is the Environmental Protection Act (1990), which sets out fundamental UK policy, and is enacted through Regulations, as "sub sets" of the parent Act (Anon, 1991a). Regulations may then be changed and implemented with relatively little debate. This process gives an element of flexibility, adapting to political changes or scientific knowledge. In many cases, the Regulations are further supported by Guidance Notes which indicate the way in which the Regulations should be met in order to meet the Requirements of the Act.

Such Guidance Notes are specific to industry sectors (Department of the Environment, 1997a). While such Guidance Notes are not in themselves part of the Law, failure to adhere to such guidance may result in the absence of any "due diligence" defence in the case of a charge under the parent Act. Furthermore, general industry sectors may have access to Codes of Practice which are based on information within Guidance Notes. Such Codes of Practice also may form the basis of "due diligence" strategy. With particular regard to composting and co-utilization in the UK, Codes of Practice exist for the protection of air (Ministry of Agriculture Fisheries and Food, 1992b), water (Ministry of Agriculture Fisheries and Food, 1991) and soil (Ministry of Agriculture Fisheries and Food, 1993). In Scotland a general Code of Practice also covers all three of these elements (The Scottish Office of Agriculture, Environment, and Fisheries Department, 1997).

Within this framework of Law, coupled with policy and acceptable methods of operation, is enshrined the concept of Best Available Techniques Not Entailing Excessive Cost (BATNEEC). This concept facilitates negotiation, particularly where pollution may be caused, which satisfies the requirements of the Codes of Practice, Guidance Notes and ultimately the Act of Law. At the same time, the needs of business to minimise costs, particularly capital, is accepted. In this way, an initial plan for sites such as a compost facility may be negotiated during the early stages of the application process. The chosen design for the facility can conform as high a level of sophistication as possible providing that the requirements imposed by the planning authority are not so onerous as to prevent successful business development. Nonetheless, if an operation is established following a planning procedure that adheres to the BATNEEC principle and ultimately pollutes or causes problems which are covered by the legislation such as the Environmental Protection Act (1990), the parent legislation takes precedence, and the site operators may be challenged in Law. This negotiating stance, to arrive at a facility design and *modus operandi* which meets the requirements of the Law at minimum cost, is in stark contrast to the trend in some other EC Member States. For instance, in The Netherlands and Germany, compost plants are typically approved only on the basis of best available technology, irrespective of location, environmental issues or financial constraints. This essentially results in the closing of facilities in all but a few cases (Lemmes, Personal Communication).

With regard to quality of compost, a number of standards apply in Europe. At the time of this writing, no single guideline exists for compost quality. As an example, Table 1 shows differences in allowable levels of heavy metals and physical criteria in three countries. A proposed pan-European standard for soil improvers and growing media is currently being developed by Comité Européan de Normalisation (CEN) (Gabriels, In press, Verdonck, In press). CEN Standard 223 "Soil Improvers and Growing Media" is unlikely to be published before 1999. Some sections are available as drafts for public comment, covering:

- Labelling requirements
- A safety technical report
- Sampling methodology
- Quantity determination
- Analytical methods

The standard will set a number of criteria which will have to be displayed on each container of product. In the UK the development of this standard is being monitored by British Standards Institute (BSI) Technical Committee AW/20 Topsoil and Other Growing Media who have set up 'shadow' working groups to contribute to the development of the CEN standard (Smith C, Personal Communication). In the absence of a specific standard, guidelines for use of peat, where applicable, tend to be followed (British Standards Institute, 1990). Co-ordination is sought not only in compost quality but how laboratories measure relevant parameters (Willumsen, 1988; Gabriels, 1995). Indications are that there will not be a ready agreement and no deadline is set.

Table 1. Examples of compost standards for metal concentrations and quality criteria.

Elements mg/kg DM	Belgium	The Netherlands	Germany
Zn	300	200	400
Pb	120	65	100
Cr	70	50	100
Cu	90	25	75
Ni	20	20	50
Hg	1	0,2	1
Cd	1,5	0,7	1

Parameters	Unit	Belgium	The Netherlands	Germany
pH_{water}	-	6,5-8,5	8,1	-
Dry Matter (DM)	%	> 55 %	> 65 %	> 55 %
Organic Matter	%	> 16 % DM	> 20 % DM	> 40 % DM (F) > 20 % DM (M)
Impurities	%	< 0,5 %	< 0,5 %	< 0,5 %
Stones	%	< 2 %	< 3 %	< 5%
Maturity	-	$NO_3/NH_4 > 4$	temp. < 40°C	"Rottegrad" 2-3 (F) 4-5 (M)

Process Options

In the UK, horticulture and agriculture are diverse industries with substantial numbers of relatively small farms. In most cases waste management has, until recently, been on a *ad hoc* basis. Waste management planning is now a requirement prior to the awarding of any grants related to farm investment and must be part of any planning application for farm structural development (The Scottish Office of Agriculture, Environment, and Fisheries Department, 1997, c). This means that closer attention is now given to the pollution potential and the market potential for agricultural waste products. Nonetheless, because of the relatively small scale of most operations, on-farm waste management tends to centre on short-term storage of materials prior to disposal off site or by direct application to land. In all but a few cases, animal slurries and manures are direct applied by tractor-drawn spreader, irrigators and rain guns or, less frequently, by sub-surface injection.

Implementation of the landfill tax has meant that farmers are considering importing organic material to carry out on-farm composting followed by application to land. The way in which this may operate is that the farmer charges the supplier a fee which is less than the equivalent landfill tax, providing that the material can be safely applied to land. Because of the inclement weather in most of the UK, a predictable timetable for land application is impossible. This may necessitate on-farm composting as a controlled method of short to medium term storage of biodegradable materials prior to land application. Non-agricultural material may only be applied to land if it can be shown to have either no deleterious effects or, preferably, positive effects in terms of soil fertility. Materials other than biodegradable wastes may fall into this category.

Technology for compost production has been described by Lopez-Real and Vere (1992), Stentiford (1992, 1993) and De Bertoldi et al. (1996). As an underlying principle, the higher the capital investment, the shorter the composting period may be as well as the potential for control of end-product quality. Conversely, low capital investment projects may produce compost which is variable and is produced on a difficult to determine timetable. For horticultural markets, where there is a requirement for composted material having particular attributes, relatively high-investment systems are preferred. For relatively low technology uses or those where the quality of material is less important, low-investment composting processes may be acceptable. Examples are in the production of agricultural-grade or landscape-grade materials. In regard to agricultural-use, compost application to land may be a part of the overall fertilizer strategy. As such, the nutrient content of composted materials should be known before land application, and the balance of nutrients applied as fertiliser, up to the level desired by the farmer or permissible under appropriate regulations.

With respect to landscape use, nutrient content of the compost may not be relevant and the aesthetic appearance and freedom from weed seeds, pests, and disease are more significant. Providing that these basic criteria are met, the production process is not of particular concern to either the composter or the user. In such cases, simple windrow composting may be employed. Windrows may be turned by a fore-end tractor loader with the aeration strategy of the site being as much a part of logistics movement of material to the delivery point, as it is a part of the process control. Self-propelled turners are preferable but may only be cost-effective on a relatively large scale. Quality control should include regular temperature monitoring and controlling the turning of windrows according to temperature profile of the material bulk. Final product grading may or may not be done, unless material is sold to end-user. In some cases, windrow composting is simply done to achieve volume reduction prior to landfilling, or is done to produce material for blending with soils suitable for capping of landfill sites. In such cases grading may be an unnecessary cost. Designs also exist for small scale on-farm compost drums (Malkki and Klemola, In press), although in the UK these are seldom used.

Where material is available in large quantities, high capital investment may be justified to produce uniform compost for sale. A recent European example is in the site at Faenza (I) (Papi et al., 1996) where mixed agro-industrial wastes are co-composted in open-topped bins with a travelling aerator. This runs on rails between neighbouring bins and can be moved between bins using overhead gantries. Bins are aerated from underneath and have fan systems operating independently along the length of the bin, in order to avoid 'hot spots' or cool zones.

Where poor grade material from any of these systems is applied to high value cropping land, problems may result. An example would be the carry-over of pests or disease such as clubroot (*Plasmodiophora brassicae*), a pathogen of brassicas, e.g. cabbages, of worldwide importance. Like a number of plant pathogens, this fungus produces robust resting spores. Such spores and other infective agents may be very difficult to kill during composting (Bollen, 1993, Bollen et al., 1989, Avgelis and Manios, 1992). With regard to yard waste compost (botanical waste), this consideration of plant pests and diseases is significant but does not pose a health risk to operators or product users.

In the case of other materials, particularly with animal wastes and food wastes, such feedstock can be a source of potential human pathogens. For such material it is essential that the composting process should be designed to maximize the pasteurization of the material *en bloc*. This can only be achieved with full containment. Such systems were first developed for the mushroom industry (Vedder, 1978). In order to achieve process control and uniform pasteurization, batch composting was developed whereby material may be loaded to square-section rooms with sub-floor forced ventilation.

Lack of eveness together with compaction of material can contribute to the occurrence of 'hot spots', consequently, facility design must include choice of loading and feedstock transportation and blending. In enclosed systems, ventilation is typically from the pressurized base area through the compost to the space above, from which it can be ducted to recirculate to the sub-floor area and to the compost. Fresh air may be blended with recirculated air and excess exhausted. Because all compost is treated concurrently, pasteurization can be absolute. Furthermore the degree of control is greater than for outdoor or bin systems because the number of control points is generally higher. As exhaust air is focused, odour control for tunnel systems, such as acid-wash or biofilters (Lindberg, 1996), may be more efficient than for other methods, including aerated bins where air from entire buildings may have to be treated (Table 2). Complex interaction means that computer monitoring and control is required and is standard for tunnel composting in the European mushroom industry.

While material known to be hazardous to health is best handled by destructive means such as incineration, other products, particularly agricultural materials, have the potential to become hazardous unless correctly handled. This is particularly important in terms of food poisoning organisms (Hall et al., 1992) such as *Escherisia coli*, particularly strain 0157 (Pennington, 1997), *Salmonella* sp. and *Listeria* (Robinson et al., 1992). At the time of this writing, no evidence of potential carry over of Spongiform encephalopathies through animal wastes exists. In turn no information is available regarding efficacy of

Table 2: Typical control points for tunnel compost systems.

Temperature	Inlet air Exhaust air Recirculated air Compost mass	
Humidity	Recirculated air Inlet air	
Gas profile	Recirculated air	Carbon dioxide Oxygen Ammonia
Odour control	Exhaust air	Ammonia Sulphurous compounds Volatile organic carbons

composting as a part of any future disease minimisation strategy. The situation is complex (Aguzzi and Weissmann, 1996), and a long-term review, research and development strategy to pre-empt such concerns or misinformed judgement is needed.

Process Implementation

Operators must choose from the complex range of process options to generate intended product by selecting from the range of feedstocks locally available. Major decisions on capital investment may be required. Because large compost operations are typically site-specific, readily comparing a number of standard facility designs is not possible. Consideration needs to be given to the decision route by which investment is finally made. A basic structure for such decision making is presented in Figure 1. This may be modified to take account of local issues.

When the basic choice of composting system has been made, further decisions need to be made to determine scale of operation. Because compost production is site and capital specific, traditional gross margin analysis of costings may not apply. A holistic approach is required. External factors, such as transport logistics and seasonality (Szmidt, 1994) may influence costing of compost operations. These should be taken into account in any model. Figure 2 shows a basic structure for a costing model and, as with system options, this may be modified to take into account local issues. Such a structure can then be employed in an interactive spreadsheet-based costing model used as decision support for site design. Such computer-based models can then be used by advisers and consultants to independently verify opinions on land requirements and budget costs that machinery and facility suppliers may offer.

Co-utilization of composted material encompasses a blending of feedstocks to the compost manufacturer and blending of such material with other land-applied material. The relationship between supply and demand lends itself to examination of logistics using Geographic Information Systems (GIS) (Mather, 1994; Boels et al., 1994). This new area of study remains to be fully exploited but may be a useful tool in helping maximise resource use and benefit.

ACKNOWLEDGEMENTS

SAC receives grant in-aid from the UK Scottish Office Agriculture, Environment and Fisheries Department. The author would like to thank Mrs Leslie Gechie for assistance in the preparation of this manuscript and Dr. Ron Korcak for the invitation to prepare this presentation.

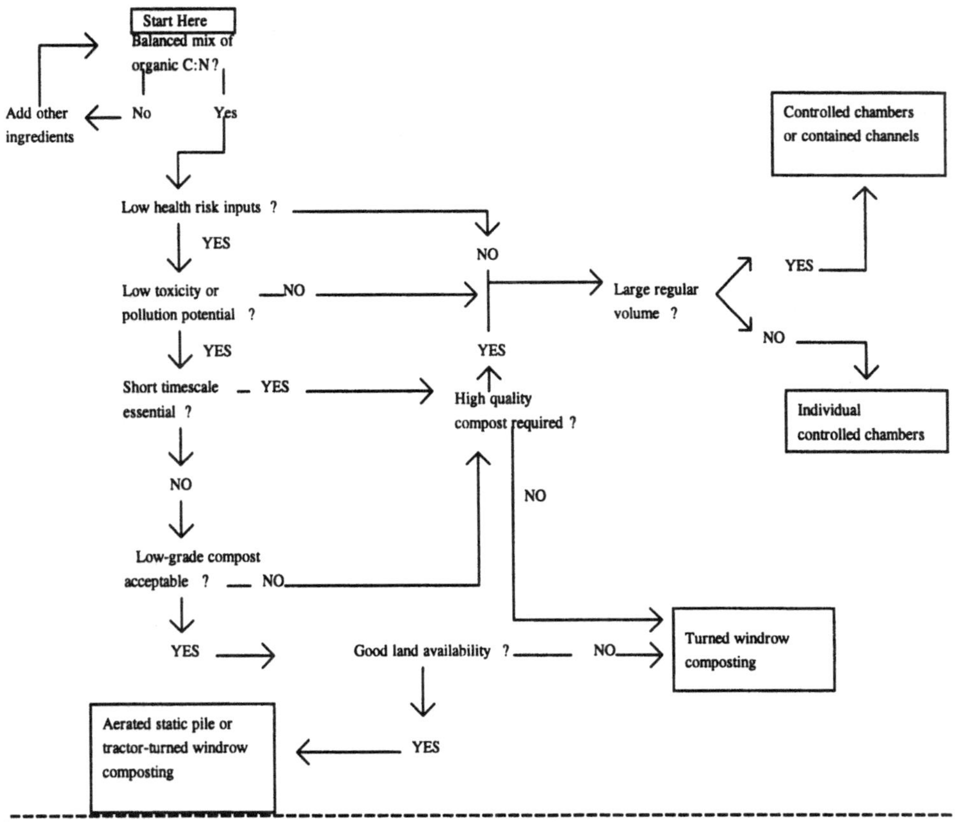

Figure 1. Decision route for deciding which composting options should be considered.

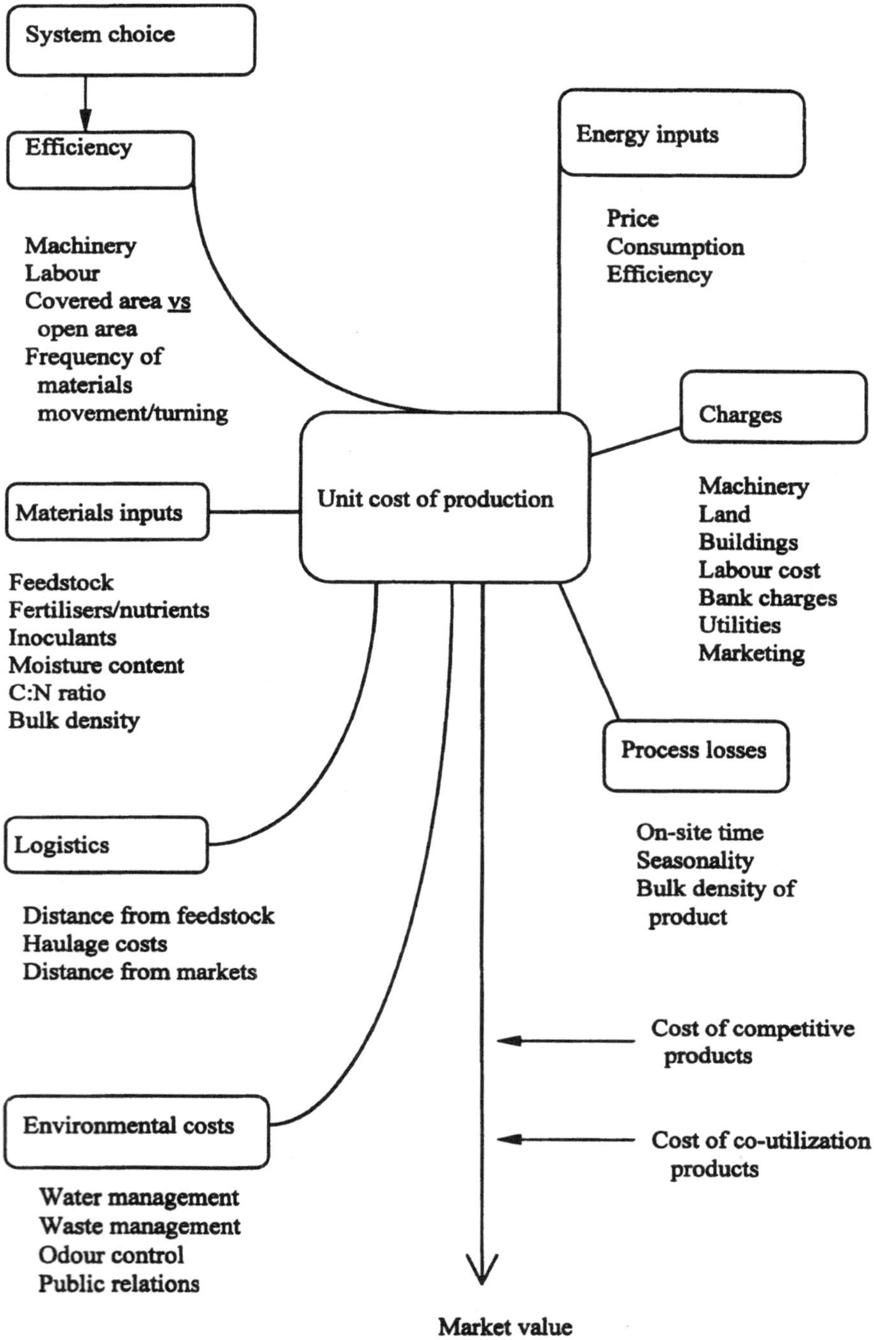

Figure 2. Principal factors in holistic compost costing model.

REFERENCES

Aguzzi A & Weissmann C (1996) Spongiform encephalopathies - a suspicious signature. Nature 383:(6602):666-667.
Anon (1974) Health and Safety at Work Act 1974.
Anon (1990) The Environmental Protection Act 1990 Chapter 43. HMSO, London.
Anon (1991a) Environmental Protection (Prescribed Processes and Substances) Regulations.
Anon (1994) The control of substances hazardous to health regulations 1994 (COSHH 1994).
Avgelis AD & Manios VI (1992) Elimination of cucumber greenmottle mosaic tobamovirus by composting infected cucumber residues. Acta Horticulturae 302:311-314.
Boels D, Bonekamp M & Jonker K (1994) Methodology of choice of location for storage of refuse and waste (Abst.). Rapport Institut. Institut voor Veehouderij en Durgezondheid, Runderweg NL 302, p77.
Bollen GJ, Volker D & Van den Hoek (1989) Inactivation of plant pathogens during small-scale composting of crop residues. Netherlands Journal of Plant Pathology 95 Supp. 1:19-30
Bollen GJ (1993) Factors involved in inactivation of plant pathogens during composting of crop residues. In: Hoitink HAJ & Keener (Eds) Science and Engineering of Compost: Design, Environmental, Microbiological and Utilization Aspects (pp. 301-318). Renaissance Publications, Worthington.
Bragg NC (1990) Peat and its alternatives. HDC, Petersfield.
British Standards Institute (1990) Recommendations for peat for horticultural and landscape use BS4156, London.
Bunt, AC (1976) Modern potting composts. George Allen & Unwin Ltd, London.
De Bertoldi M, Sequi P, Lemmes B & Papi T (1996) The Science of Composting. Blackie Academic & Professional, London.
Department of the Environment (1997a) Secretary of State's Guidance - Production of mushroom substrate PG6/30(97).
Fontaine P (1995) Europe in ten lessons. Office for Official Publications of the European Communities, Luxembourg.
Gabriels R (1995) Standardization of growing media analysis and evaluation: CEN/ISO/ISHS Acta Horticulturae 401:555-557.
Gabriels R (In press) Quality of composts and EC legislation. In: Composting and Use of Composted Material for Horticulture. Acta Horticulturae.
Hall JE, L'Hermite P & Newman PJ (1992) Treatment and use of sewage sludge and liquid agricultural wastes - Review of COST 68/681 programme, 1972-90. Office for Official Publications of the European Communities, Luxembourg.
Hamaker, JD (1982) The survival of civilization. Hamaker-Weaver Publications, Michigan.
HM Customs & Excise (1996) Introduction to landfill tax. Information Note 1/96, Liverpool.
Hoitink HAJ & Keener (1993) Science and engineering of compost: Design, environmental, microbiological and utilization aspects. Renaissance Publications, Worthington.
Hoitink HAJ, Stone AG & Grebus ME (1996a) Suppression of plant diseases by composts. In: De Bertoldi M, Sequi P, Lemmes B & Papi T (Eds) The Science of Composting (pp. 373-381). Blackie Academic & Professional, London.
Hoitink HAJ, Zhang W, Han DY & Dick WA (1996b) Making compost to suppress plant disease. Biocycle 37(4):40-42.
Horticulture Research International (1991-96) Annual reports (various), Wellesbourne.
Jackson DV, Merillot J-M & L'Hermite P (1992) Composting and compost quality assurance criteria. Office for Official Publications of the European Communities, Luxembourg.
Jeangille, P (1991) Substrata for horticulture in subtropical and tropical regions. FAO, Rome.
Kent P (1996) Law of the European Union. Pitman Publishing, London.
Lindberg C (1996) Accelerated composting in tunnels. In: De Bertoldi M, Sequi P, Lemmes B & Papi T (Eds) The Science of Composting (pp. 1205-1206). Blackie Academic & Professional, London.
Lopez-Real J & Vere A (1992) Composting control parameters and compost product characteristics. In: Jackson DV, Merillot J-M & L'Hermite P (Eds) Composting and Compost Quality Assurance Criteria (pp. 131-140). Office for Official Publications of the European Communities, Luxembourg.
Malkki S & Klemola E (In press) Using a drum composter to produce compost from cattle manure. In: Composting and Use of Composted Material for Horticulture. Acta Horticulturae.
Mather PM (1994) Geographical information handling: research and application. John Wiley & Sons Ltd,

London.
Ministry of Agriculture Fisheries and Food (1991) Code of good agricultural practice for the protection of water, London.
Ministry of Agriculture Fisheries and Food (1992) Code of good agricultural practice for the protection of air, London.
Ministry of Agriculture Fisheries and Food (1993) Code of good agricultural practice for the protection of soil, London.
Office for Official Publications of the European Communities (1992) Treaty on European Union, Luxembourg.
Papi T, Marani G, Mannironi R & De Bertoldi M (1996) A new composting plant realized within the THERMIE programme of the European Commission. In: De Bertoldi M, Sequi P, Lemmes B & Papi T (Eds) The Science of Composting (pp. 162-172). Blackie Academic & Professional, London.
Pennington H (1997) The Pennington group report: report on the circumstances leading to the outbreak of infection with E. Coli 0157 in Central Scotland, the implications for food safety and the lessons to be learned. The Stationery Office, Edinburgh.
Raviv M, Reuven R, Krasnovsky A, Freiman L, Medina S & Bar A (In press) Composting as a controlling agent against *Fusarium* wilt of sweet basil. In: Composting and Use of Composted Material for Horticulture. Acta Horticulturae.
Reis M, Martinez FX, Soliva M & Monteiro AA (In press) Composted organic residues as a substrate component for tomato transplant production. In: Composting and Use of Composted Material for Horticulture. Acta Horticulturae.
Robinson J, Naylor PJ & Stentiford E (1992) Development of a method for the isolation of *Listeria monocytogenes* from compost. Acta Horticulturae 302:369-376
Scottish Environment Protection Agency (1997) National waste strategy: Scotland. A blueprint for progress 1997-2001.
The Scottish Office of Agriculture, Environment and Fisheries Department (1997b) Prevention of environmental pollution from agricultural activity, Edinburgh.
Sinclair RG (1996) Managing food residuals through on-site composting. Biocycle 37(1) 34-36.
Smith D (1996) Growing in rockwool. Grower Books, Swanley.
Stentiford EI (1992) The composting process applied to sewage sludge and source separated refuse. In: Jackson DV, Merillot J-M & L'Hermite P (Eds) Composting and Compost Quality Assurance Criteria (pp. 69-80). Office for Official Publications of the European Communities, Luxembourg.
Stentiford EI (1993) Diversity of composting systems. In: Hoitink HAJ & Keener (Eds) Science and Engineering of Compost: Design, Environmental, Microbiological and Utilization Aspects (pp. 95-110). Renaissance Publications, Worthington.
Szmidt RAK (1994) Recycling of spent mushroom substrates by aerobic composting to produce novel horticultural substrates. Compost Science and Utilizaton 2(3):63-72.
Szmidt RAK (In press a) Composting of waste-products of seaweed (*Ascophyllum nodosum*) processing. Compost Science and Utilization.
Szmidt (In press b) Potential for co-utilisation of rockdust and composted material. In: Composting and use of compost for horticulture. Acta Horticulturae.
Szmidt RAK & Bryden G (1996) Composting of bioreactor waste from softwood processing. In: De Bertoldi M, Sequi P, Lemmes B & Papi T (Eds) The Science of Composting (pp. 1357-1360). Blackie Academic & Professional, London.
Szmidt RAK & Smith C (1996) Composted organic wastes for use in horticulture. In: Crop Science, SAC, Edinburgh.
van Griensven LJLD (1988) The cultivation of mushrooms. Darlington, Rustington.
Vedder PJC (1978) Modern mushroom growing. Educaboek VB, Culemborg.
Verdonck O (In press) Compost specifications. In: Composting and Use of Composted Material for Horticulture. Acta Horticulturae.
Webster (1995) Rock dust and mineralised compost (I) report to Scottish Environment Protection Agency, PKDC, Scottish Natural Heritage & SET. SAC, Edinburgh.
Willumsen J (1988) Horticultural substrates and their analysis. Acta Horticulturae 221.

BENEFITS AND DRAWBACKS TO COMPOSTING ORGANIC BY-PRODUCTS

LAWRENCE J. SIKORA
Soil Microbial Systems Laboratory, Agricultural Research Service, US Department of Agriculture, Beltsville, Maryland 20705, USA
E-mail lsikora@asrr.arsusda.gov

ABSTRACT

Composting is a treatment process that requires time, knowledge, experience, equipment and effort. The benefits of establishing a composting process must be balanced against some of the drawbacks of the process and the product. Organic by-products or residuals that are difficult to store, apply to fields uniformly, are unstable or nonuniform are good candidates for composting. Manures, biosolids and food processing residuals are produced daily but often cannot be used on a daily basis and, therefore, must be stored periodically. Composting transforms manures for example to a drier, more uniform and biologically stable product with many uses other than just land application. Composted manures as such have a greater value than untreated manures to the farmer or feedlot owner. Nonuniform materials such as yard trimmings are transformed by degradation and mixing during composting into homogeneous organic mulch. Wet materials such as biosolids become drier as composts and are therefore more easily land applied. By-products that contain human or plant pathogens are safer after the high temperature treatment of composting. Compost products generally have a higher carbon to nitrogen ratio than the original by-product and therefore act as a slow release fertilizer.

Drawbacks of composting by-products are cost for site preparation and equipment, the lengthy treatment period, targeting final use of compost product, and environmental issues such as odors and dust. Some investment in equipment and site preparation is required or recommended. Composting is not a rapid stabilization process and, depending upon technique, could take several weeks to achieve stable compost. Determination of a suitable market for the compost is critical to justify the extra effort in producing compost. Composting is a biological process that can result in significant odor generation if not managed properly. The slow release nature of nutrients in compost requires higher application rates than the original by-product to obtain the same plant response. Higher application rates require more material and more trips across the field than with the original by-product.

Knowing the benefits of composting and composts and its drawbacks provide the generator of by-products a better estimate of cost versus return for starting a composting process. Environmental regulations may govern the ultimate treatment and beneficial use of by-products. The proper selection of by-products to cocompost or use with composts

S. Brown, J.S. Angle and L. Jacobs (eds.), Beneficial Co-Utilization of Agricultural, Municipal and Industrial By-products, 69–77.
© 1998 *Kluwer Academic Publishers*,

can result in value-added products that can be marketed a great distance from their source.

INTRODUCTION

American industries, farms and cities are generating over one billion tons of by-products each year. Several industries have organic by-products that can be a valuable resource when utilized effectively. Simultaneously, agriculture is facing a challenge in developing sustainable farming practices. Sustainable farming practices are characterized as plant and animal production practices that satisfy human food and fiber needs while preserving or enhancing environmental quality and the resource base. A sustainable society, economy or industrial practice may be defined similarly, that is, provide the needs of the nation, but preserve or enhance the environment and resources. To use an agriculture example, confined animal production systems produce 27 million tons of manure solids annually and the nitrogen value of these by-products is equal to the N fertilizer use in the US annually. To substitute these manures for a portion of the inorganic fertilizer would reduce the need for natural gas in fertilizer production and, concomitantly, provide a use for the organic by-product.

Manures cannot be used readily except on the farm because of their physical nature. Treating manures to become more biologically stable and drier material expands the uses for manures. Composting is one treatment process that transforms manures or other organic residues into materials with greater utility and value. Other treatment possibilities for organic by-products are heat drying and alkaline stabilization (Millner et al., 1997). Treatment of organic by-products by composting requires additional effort which must be balanced against the benefits of the process and the final product. An understanding of benefits and drawbacks provides the necessary information for making the correct decision of whether to compost or not.

Material Handling Characteristics of Composts - Benefits and Drawbacks

Composting is a treatment process that reduces the moisture content of organic by-products. Thermophilic temperatures attained in composting masses lead to evaporative cooling which changes water from a liquid state to a vapor (Haug, 1980). With significant air diffusion, forced aeration or agitation, moisture evaporates from the mass. Composting of 30 tons of biosolids by the aerated, static pile method resulted in 19000 liter water loss (Sikora et al., 1981). This volume includes the water generated during the biological breakdown of organic matter. Benefits of a drier organic material are numerous. Compost has lower bulk density, is easier to mix and is more uniform than wet, uncomposted material. Biosolids compost has half the bulk density of uncomposted biosolids (Sikora et al., 1981). Dry material can be screened efficiently to produce additional by-products having different characteristics. Biosolids compost that passed a 1mm screen had an N mineralization rate three times greater than the material that was greater than than 6 mm (Tester et al., 1979). Composts can be screened to recover bulking materials for reuse (On Farm Composting Handbook, 1992)

Composts are more stable or less biologically active than uncomposted material. Therefore composts can be stored with lower potential of odor generation than uncomposted materials. Stabilized and mature composts can be bagged while uncomposted organic materials cannot be bagged without potential heating or odor generation. An exception is mechanically dried organic materials.

Effort that includes compost process training, time to conduct composting and procurement of equipment necessary to successfully transform organic materials to composts is required. Training is available at several compost schools and certification tests for compost operators are available nationwide. Equipment to compost can range from a modest front end loader to mix, build and turn piles to sophisticated mixing machines and compost turners that cost hundreds of thousands of dollars (On Farm Composting Handbook, 1992). Time is required to thoroughly mix organic materials to compost efficiently, to turn piles when needed and to monitor the progress of composting. Therefore, composting of organic material to produce a drier, more stable product requires investment in equipment and time that must be considered when deciding whether to treat organic materials by the composting process.

Slow Release Nutrient Quality of Composts - Benefits and Drawbacks

Organic by-products are amended with a carbon source before composting to achieve a desirable C:N ratio of 30-35 :1. The resulting compost has the majority of its N in the organic form and the rate of release of mineral N from the organic form is generally less than the original organic material. O'Keefe et al. (1986) found that composted biosolids had half the mineralization rate of the uncomposted biosolids. Mattingly (1956) found that the nitrification rates of 18 different composts were 9.3 to 10.6 % of the total nitrogen after 13 weeks. Inorganic N released from composted manures ranged from 11 to 29 % and included inorganic N initially present (two to twelve percent), soluble organic N (one to five percent) and mineralization of insoluble organic N (approximately 10 %) (Hadas and Portnoy, 1994). Low mineralization rates characterize composts as slow release fertilizers that are desirable in several instances. Research has long pursued methods to slow the nitrification step or the transformation of ammonium to nitrate in soils for plants to utilize the N fertilizer before it leaches through the soil profile (Hauck, 1980). By slowing the ammonification step or transformation of organic N to ammonium, composts are accomplishing the same goal. Yakovchenko et al. (1996) determined that organic sources of N (manures or legumes) were more efficiently taken up by crops than commercial fertilizer. An additional benefit of compost amendments is that the organic N that is not mineralized in the year of application is "stored" in the soil and will mineralize in future cropping seasons. The quality of soils under high intensity agricultural practices is declining and the renewal of the organic matter content is one goal in sustainable agricultural practices.

Slow or low mineralizing composts can be a drawback in satisfying the N requirement of crops because large amendments will be necessary. From 40 to 100 Mg ha^{-1} are required to satisfy the N requirement of corn (Hornick et al., 1994). The volume of compost required to provide adequate N may require several application trips across the field. Volumes required to satisfy N needs of crops may not be available when needed and storage of large volumes near the application site may not be possible. Large applications may cause secondary effects such as excessive salt concentrations in soil (Sims et al., 1994, Chaney et al., 1980). If compost applications are based on N to minimize nitrate losses from the soil, an excess of phosphorus is added. Manures, biosolids and their composts have N:P ratios from 3:1 to 1:1 while the N:P requirement for grain crops is eight (White and Collins, 1982). Most importantly, the addition of non-nutrient compost constituents such as heavy metals to soils may result in an unacceptable accumulation that would reduce the value of the land for safe production of food (Leita and DeNobili, 1991). Several countries regulate annual and maximum accumulations of trace metals that result in application rates of composts below that of the N requirement of the crop (DeBertoldi et al., 1990). Blending low amendment rates of composts with fertilizers is suggested as an alternative practice to alleviate problems with high compost application rates (Sikora et al., 1997).

Biodegradation Qualities of Composting Process- Benefit

Composting is an active biological process that degrades organic compounds. High temperatures found in composting result in appearance of thermophilic microorganisms that may be more efficient in biodegradation of certain organics than mesophiles. Forty-seven percent of 2, 4-dichlorophenoxyacetic acid (2, 4-D) was degraded during composting of yard wastes (Michel et al., 1995). Soils contaminated with explosives were remediated through use of compost technology (Kaplan and Kaplan, 1982). Zinc bacitracin was degraded during composting of chicken manure (Vogtmann et al., 1978). Degradation of polyethylene in starch-based polyethylene plastic bags degraded during municipal trash composting (Johnson et al., 1993) High temperatures of composting and the type of plastic bag were more significant factors than microbial action during composting. Several studies report improved degradation of xenobiotics in soils amended with composts (Duah-Yentumi and Kuwatsuka, 1980, Laine and Jorgensen, 1996, Bellinck and Mayandon, 1983). Composting will lead to degradation of availiable C compounds but whether it is more efficient than degradation at ambient temperature is debatable. Extent of degradation of pentachloronitrophenol in a manure-straw mixture by composting and incubations conducted at 25 C were equal (Sikora et al., 1982). Carbaryl added to yard trimmings degraded more rapidly when the mixture was incubated at 25 C than when composted (Racke and Frink, 1989). In general, composting will successively degrade a variety of organic compounds when the guidelines for proper aeration, moisture, and C:N ratio for mixtures are followed.

Odor Generation Potential From Composting Organic Residues - Drawbacks and Benefits

Odor generation from composting operations and their effect on surrounding neighbors is probably the single most cited reason for the closing of composting operations. Inadequate control of composting procedures is most often the main reason for excessive odor. Excessive moisture in piles leading to reduced oxygen content results in anaerobic conditions and production of reduced sulfur compounds, short chained acids and ammonia.

Ammonia volatilization can be reduced with adjustment of C:N ratios of materials prior to composting. In addition to controlling composting conditions, air from compost piles can be scrubbed with odor filter piles. Generally, high organic matter materials such as peat or mulch, or a biologically active material such as screened compost reduce odors in exhaust gases from aerated piles. Chemical scrubbers are also efficient in removing odors.

Standing liquids at compost facilities can result in odors and therefore compost sites designed to direct liquids, either condensate or precipitation, away from compost piles are less likely to be cited for excessive odors. Condensate from the composting of biosolids contained a large percentage of the ammonia lost from the mixture (Sikora and Sowers, 1985).

Manures and biosolids contain odorous compounds and if not processed properly (treated, land applied or heat dried), significant odor generation can result. Composting as a biological process will breakdown odorous compounds in organic materials. Aerobic biological activity that occurs during composting will degrade several of these compounds and, if the compost process is conducted properly, net odor generation from composting organic materials will be significantly less than from organic material with no treatment.

Reduction of Plant and Human Pathogens - Benefit

A significant benefit derived from composting is sustained high temperatures that kill human pathogens in organic by-products. Burge et al. (1981) showed convincingly that 55 C temperatures for three consecutive days significantly reduced the f2 virus (an indicator organism for human pathogens) during the composting of biosolids. United States EPA and most state environmental agencies have used these data to develop guidelines to ensure "further reduction of pathogens" in organic by-products, Some nations require more strict time-temperature relationships than others and guidelines differ whether the process is in static pile or in windrows (Farrell, 1993). As effective as composting is in reducing human pathogen numbers, assuring that composts will not be reinoculated by precompost materials is critical. Regrowth of *Samonella* in finished compost was limited, probably due to because competition for nutrients between saprophytic microorganisms and pathogens (Burge et al., 1987).

High temperatures in composting also reduce plant pathogen populations that may be in the organic by-products. Composting temperatures (> 35 C) are sufficient to severely inhibit the growth and reproduction of plant pathogens (Allen et al., 1983: Cooney and Emerson, 1968). Composting therefore allows spent materials from greenhouses to be reused for growing horticultural crops and straw from grain crops affected by disease to be reused on the same land after composting.

Considerable data suggest that composts reduce plant pathogen populations and disease significantly in pot studies. Millner et al. (1982) demonstrated that biosolids compost reduced the disease severity on some vegetables but had no effect on others (Table 1). Composts differ as to effectiveness of suppression of disease. Ringer et al. (1997) determined that diary manure compost was equal or slightly better than steer manure compost which was equal to or slightly better than poultry manure compost in potting studies that examine suppression of damping-off disease. Compost age did not have an effect on disease suppression, similar to data recorded by Hoitink and Grebus (1994). Craft and Nelson (1996) point to microbial activity of compost and compost amended soils as a possible reason for disease suppresion. Ringer et al. (1997) had less disease suppression when composts were autoclaved. These beneficial properties of composts make them candidates to replace chemicals such as methyl bromide that currently are used to provide protection to plants but are mandated for replacement in the near future.

Table 1. Effect of ten percent biosolids compost amendment on diseases by eight soilborne pathogens as determined in greenhouse pot assays (from Millner et al., 1982).

Pathogen	Host	Disease Severity(%)[a]	
		unamended	compost amended
Aphanomyces euteiches	pea	43	3
Rhizoctonia solani	bean	63	25
Scerotinia minor	lettuce	84	33
Phytophthora capsici	pepper	97	55
Phytium ultimum	pea	39	68
Fusarium solani	pea	33	40

[a] based on root rot severity (0-100%) for *Aphanomyces, Rhizoctonia*, and *Fusarium*, and on diseased plant for *Sclerotinia* and *Pythium*.

Aerosols from Composting Operations - Potential Drawback

Compost operations include mixing and turning of materials that contribute aerosols to the atmosphere (Millner, 1995). The frequency of disturbance or turning of compost piles and the moisture content will effect the emission rate. Contents of aerosols are fugitive dust particles, microbial spores, and gases. Effect of dust particles generated during composting have not been addressed specifically in studies of aerosols. Airborne ammonium nitrate particles contribute to particulate matter less than 10 μM (PM_{10}) from poultry manure application (Moore et al., 1997). Distance microbial spores such as *Aspergillus fumigatus* travel from composting facilities and its effect on nearby residents is an important concern when locating a facility. Data show that dispersion of microbial spores within the boundaries of the site could be substantial and workers at the site need to be screened for sensitivities to allergens. Depending on factors such as meteorological conditions and barriers or buffers around the site, data show that impact on surrounding populations is minimal (Millner et al., 1995). Composting operations will generate aerosols and the extent will depend on frequency of material handling. Static pile composting or passive aeration composting would generate less aerosols than windrow composting. Aerosol generation and dispersion are important, but do not effect the decision to compost. Only the compost technique and location of compost site are affected.

Most composts are highly acceptable products to the public, but when considering composting as a treatment process, careful evaluation of uses or markets for the final product must be included in the decision process. Composts have physical and chemical properties that make them highly suitable amendments for nursery, landscape and home garden markets. However, consideration of competing materials in these markets, distance to markets and necessary quality control during production of final compost should be realized. Number of yard debris compost facilities is increasing and, even though the compost has desirable qualities for landscape markets, wholesale prices seldom are greater than $10 per cubic meter. To obtain a greater return, composts may be blended with other ingredients before or after processing to obtain a value-added product (Korcak, 1997).

CONCLUSIONS

Composting is a treatment process that results in a drier, more biologically stable and easier to apply to land than the original uncomposted material. The effort of composting material must result in equal or better benefits than the original product in order for the process of composting to be worth the effort. Regulations may necessitate using composting as a treatment process. In order for biosolids to be used by the general public, treatment like composting is necessary to further reduce pathogens. If regulations do not require composting of the organic residue, choosing composting is based on economics more than any other factor. Benefits to composting are the physical nature and biological stability of the final product, the slow release of nitrogen fertilizer from composts, and the reduced plant and human pathogen content. The drawbacks are cost of equipment, time and training required to compost efficiently, the high application rate of compost necessary to satisfy the N requirement of crops, the aerosols generated during the composting, and locating a market to sell the final product.

REFERENCES

Allen SJ, Brown JF & Kochman JK (1983) Effects of temperature, dew point, and light on the growth and development of *Alternaria helianthi*. Phytopath. 73:893-895.

Bellinck & Mayandon (1983) Influence dwquilques composts sur la mineralisation du phenmediphame 14C dans un solfrais de prairie et l'mmobilisation des residue 14C. Rev. Ecol. Biol. Sol. 20:291-297.

Burge WD, Enkiri NK & Hussong D (1987) Salmonella regrowth in compsot as influenced by substrate. Microbial Ecol. 14:243-253.

Burge WD, Colacicco D & Cramer WN (1981) Criteria for achieving pathogen inactivation during sewage sludge composting. J. Water Poll. Cont. Fed. 53:1683-1690.

Chaney RL, Munns JB & Cathey HM (1980) Effectiveness of digested sewage sludge compost in supplying nutrients for soilless media. J. Amer. Soc. Hort. 105:485-492.

Cooney DG & Emerson R (1964) *Thermophilic Fungi*. W. H. Freeman and Co., San Francisco. 188 pp.

Craft CM & Nelson EB (1996) Microbial properties of compsots that suppress damping-off and root rot of creeping bentgrass caused by *Pythium graminicola*. Appl Environ. Microbiol. 62:1550-1557.

DeBertoldi M, Civilini M & Corni G (1990) MSW compost standards in the EEC. Biocycle 3, 60.

Duah-Yentumi S & Kuwatsuka S (1980) Effect of organic matter and chemical fertilizers on the degradation of benthiocarb and MCPA herbicides in the soil. Soil Sci. Plant Nutrition. 26:541-549.

Farrell JB (1993) Fecal pathogen control during composting. In: Hoitink HAJ & Keener HM (Eds) Science and Engineering of Composting:Design, Environmental, Microbiological and Utilization Aspects (pp 282-300).

Hadas A & Portnoy R (1994) Nitrogen and carbon mineralization rates of composted manures incubated in soil. J. Environ. Qual. 23:1184-1189.

Hauck RD (1980) Mode of action of nitrification inhibitors. In: Meisinger JJ, Randall GW & Vitosh ML (Eds) Nitrification Inhibitors-Potentials and Limitations. ASA Spec. Publ. 38 (pp. 19-32). American Society of Agronomy, Madison, WI.

Haug RT (1980) Compost Engineering, Principles and Practices. Ann Arbor Science Publishers, Inc, Ann Arbor, MI. 655 pp.

Hornick SB, Sikora LJ, Sterrett SB, Murray JJ, Millner PD, Parr JF, Chaney RL & Willson GB (1984) Utilization of Sewage Sludge Compost as a Soil Conditioner and Fertilizer for Plant Growth. Agricul. Inform. Bull. No. 464. USDA, US Printing Office, Washington, DC. 32 pp.

Johnson KE, Pometto III AL & Nikolov ZL (1993) Degradation of degradable starch-polyethylene plastics in a compost environment. Appl. Environ. Microbiol. 59:1155-1161.

Kaplan DL & Kaplan AM (1982) Thermophilic biotransformation of 2, 4, 6-trinitrotoluene under simulated composting conditions. Appl. Environ. Microbiol. 44:747-760.

Kim KD, Nemec S & Musson G (1997) Effects of composts and soil amendments on soil microflora and Phytophthora root and crown rot of bell peppers. Crop Protection 16: 165-172.

Korcak RF (1998) Why Co-Utilization? Beltsville Symposium XXII on Beneficial Co-Utilization of Agricultural, Municipal, and Industrial By-Products. May 4-8, 1997 (This issue).

Laine MM & Jorgensen KS (1996) Straw compost and bioremediated soil as inocula for the bioremediation of chlorophenol-contaminated soil. Appl. Environ. Microbiol. 62:1507-1513.

Leita L & DeNobili M (1991) Heavy metals in the environment: water soluble fractions of heavy metals during composting of municipal waste. J. Environ. Qual. 20:73-78.

Mattingly GEG (1956) Studies on composts prepared from waste materials. III. Nitrification in soil. J. Sci. Food. Agric. 7:601-605.

Michel FC Jr, Reddy CA & Forney LJ (1995) Microbial degradation and humification of the lawn care pesticide 1, 4-dichlorophenooxyacetic acid during the composting of yard trimmings. Appl. Environ. Microbiol. 61:2566-2571.

Millner PD, Lumsden RD & Lewis JA (1982) Controlling plant disease with sewage sludge compost. Biocycle J of Waste Recycling 23:50-52.

Millner PD, Sikora LJ, Kaufman DD & Simpson ME (1998) Agricultural uses of biosolids and other recyclable municipal residues. In: Wright RJ (Ed) Agricultural Utilization of Municipal, Animal and Industrial Byproducts (pp. 1-47). Government Printing Office, Washington, DC.

Millner PD (1995) Bioaerosols and composting: analysis and health impacts. Biocycle. 36-48-54.

Moore PA Jr, Daniel TC, Sharpley AN & Wood CW (1998) Poultry Manure Management. In: Wright, RJ (Ed) Agricultural Utilization of Municipal, Animal and Industrial By-products (pp. 68-92). Government Printing Office, Washington, DC.

On-Farm Composting Handbook. (1992) Rynk R (Ed) Northeast Regional Agricultural Engineering Service, NRAES-54. Ithaca, NY 186 pp.

O'Keefe BE, Axley J & Meisinger JJ (1986) Evaluation of nitrogen availability indexes for a sludge compost amended soil. J. Environ. Qual. 15:121-128.

Racke KD & Frink CR (1989) Fate of organic contaminants during sewage sludge composting. Bull. Environ. Contam. Toxicol. 42:526-533.

Ringer CE, Millner PD, Teerlinck LM & Lyman BW (1997) Suppression of seedling damping-off disease in potting mix containing animal manure composts. Compost Sci. & Utiliz. 5:6-14.

Sikora LJ, Kaufman DD, Ramirez MA & Willson GB (1982) Degradation of PCP and PCNB in a laboratory composing system. In: Proc of the Eight Ann. Res Symp. Land Disposal of Hazardous Waste (pp. 377-382). EPA-600/9-82-002.

Sikora LJ & Sowers MA (1985) Effect of temperature control on the composting process. J. Environ. Qual. 14: 434-438.

Sikora LJ, Willson GB, Colacicco D & Parr JF (1981) Materials balance in aerated static pile composting. J. Water Pollut. Cont. Fed. 53:1702-1707.

Sikora LJ (1996) Effect of Compost-Fertilizer Blends on Crop Growth. In: deBertoldi M, Sequui P, Lemmes

B & Papi T (Eds) The Science of Composting (pp. 423-429). Blackie Academic & Professional, London. (Book Chapter) .
Sims JT, Visalis BT & Ghodrati M (1994) Effect of coal fly ash and co-composted sewage sludge in emergence and early growth of cover crops. Comm. Soil Sci. Plant Anal. 24:503-512.
Tester CF, Sikora LJ, Taylor JM, Parr JF & Tester CF (1979) Decomposition of sewage sludge compost in soil. III. Carbon, nitrogen, phosphorus transformations in different sized fractions. J. Environ. Qual. 8: 70-82.
Vogtmann HO, Hauser WK, Pfirter HP & Augstburger E (1978) The influence of zinc-bacitracin on the composting process of chicken manure and on plant growth. Schweiz Landwirtsch Monatsh 56:197-207.
White WC & Collins DN (Eds) (1982) The fertilizer handbook. The Fertilizer Institute, Washington, DC, 274p.
Yakovchenko V, Sikora LJ & Kaufman DD (1996) A biologically-based indicator of soil quality. Biol. Fert. Soils 21:245-251.

A NEW PARADIGM TO TAILORING COMPOST STANDARDS

CHARLES HENRY[1] and PEGGY LEONARD[2]
[1]University of Washington College of Forest Resources, Seattle, Washington 98195, USA and [2]King County Department of Natural Resources, Seattle, Washington 98104, USA
E-mail clh@u.washington.edu

INTRODUCTION

Compost... The word sends up images of "earthy richness", "back to nature", and sustainability. ***Contaminants***... This word does just the opposite -- leaving us with images of barren soil, and health and environmental problems. Yet, compounds/elements considered to be contaminants are often naturally occurring, and therefore, compost cannot exist without some concentration of contaminants. As a "compost community" we have attempted to define levels of comfort in the amounts or concentrations of different contaminants in composts by developing compost standards. Yet, acceptable levels of contaminants mean different things to different people.

In Washington State, recycling is reportedly at a higher level than anywhere else in the nation, and very successful programs exist for recycling organic residuals. Last year King County (Seattle area) and their creative partnerships won five national awards for their biosolids management programs. Pierce County (Tacoma area) owns a yard trimmings compost facility, which is considered one of the best in the nation. Biosolids compost has been produced, successfully marketed, and highly accepted in suburban Seattle for over 20 years.

However, no mixed solid waste (MSW) compost facilities exist. In fact two Northwest attempts, with high capital input, have recently failed. This has been due partially to operational difficulties (odors), but also to an aversion to both encourage MSW composting and to the use of MSW compost. The perception of highly dangerous contaminants are associated with some composts. Why does this attitude persist? How do typical approaches to setting compost standards address this attitude? This paper will present an alternate approach to developing compost standards, and important considerations for their development.

ACCEPTABLE LEVELS OF CONTAMINANTS

First, one must define what "acceptable levels" of contaminants means. As a technical

S. Brown, J.S. Angle and L. Jacobs (eds.), Beneficial Co-Utilization of Agricultural, Municipal and Industrial By-products, 79–87.
© 1998 *Kluwer Academic Publishers*,

definition, an element, or compound, isn't a contaminant until it is at a level that causes a detrimental effect. One might even extend that to "contributes" to a detrimental effect, suggesting that cumulative additions of an element/compound could eventually result in a detrimental effect. The term "green" has been applied to composts in which the feedstock is source separated and felt to have a low potential for contamination; yard trimmings and food waste are two examples that are often considered green composts. From a strictly technical viewpoint, we consider any compost which is not "contaminated" to be "green", using the technical definition of a contaminant. This diversity of opinions of what contamination means is reflected in approaches to standards supposedly based on environmental protection. Examples of these are:

- The "zero tolerance" philosophy promoted by some environmental activists, where there is no acceptable level of compounds/contaminants that should be allowed in a compost.
- The community in which recycling is a religion (an "end", rather than a means to an "end"), tending to heavily support the feedstock approach, suggesting that combining waste materials increases contamination, i.e., MSW composting.
- A portion of the regulatory community (especially in Europe and Canada) supports the concept of allowing no compost with concentrations of compounds greater than background soils levels. Others regulators simply adopt the lowest standards they can find -- then add their own factor of safety to "be conservative".
- Industry, combining concerns for the economics of being able to meet standards yet find markets for their product, are interested in formulating standards that encourage acceptance of their product.

However, in the case of most composts, it is difficult to conceive of a compost with high enough concentrations of specific elements/compounds to be classified as contaminated. So, how are standards set on a technical basis that have any relevance to the composts currently being produced? For instance, extensive plant evaluations of a highly metal salts-laden compost concluded that, in most cases, neither application rate or contaminant concentrations caused phytotoxicity, growth reduction, or even increased metal concentrations in plants (Henry et al. 1993). It is our conclusion that, with normal practices, it is hard to chemically contaminate a compost to the extent that it will have detrimental effects on human health and the environment. Thus, we believe tailoring compost standards is not necessarily a technical endeavor.

CONSIDERATIONS FOR COMPOST STANDARDS

There are a wide range of possible uses for compost. Some examples are: agriculture, silviculture, surface mine reclamation, topsoil mix, use in landscaping, retail sales (gardening stores), golf courses, parks, athletic fields, sod production, wholesale container nurseries, and landfill cover. It has always been easy to use analytical parameters (e.g., trace metal concentrations) as the sole driver of compost standards for any of these uses. Yet, even though composts made from different feedstocks may vary in concentrations of certain parameters, there normally will be little or no significance to the environment because the concentrations of all normal composts are far below critical levels (Epstein et al., 1992). Rather than simply technical quality, we see perception and corresponding education about compost quality essential components of setting compost standards, with common considerations to include:

- who the user is
- potential end use/market niche
- what the compost does

As examples of the importance placed on these considerations, three of the major uses of compost in Washington State are described. They include: topsoil manufacture, landscaping, and restoration.

Topsoil Manufacture

Criteria important for topsoil manufacture center around what the compost does for the manufacturer, than what it does for the end user. Manufactured soils normally consist of mined mineral subsoil (sands, silts and clays), with little organic matter or nutrients. Compost additions (often at 10-20% by volume) provide varying amounts of these (in some cases, composts immobilize nutrients rather than provide them), in addition to increasing moisture holding capacity and water drainage characteristics. Perception and education are essential for acceptance of compost as a substitute for peat or sawdust.

In this case the manufactured soil will primarily be used to top dress soil disturbed by construction. A large portion of the product will subsequently be planted with grass, or planted with shrubbery and then covered with a mulch. Since the compost used in this case is mixed with mineral soil, ultimate users of the manufactured soil may often be unaware of compost component. Also, any concentrations of compounds will be greatly diluted, and have little environmental significance. Important criteria for compost standards include compost look (lack of physical contaminants), and maturity (lack of odor and lack of occurrence of mushrooms).

Use in Landscaping

Compost used in landscaping may be as a soil amendment or a mulch. It will be purchased either in bulk or as a bagged product, and used directly by the purchaser. Thus, appearance and smell are primary concerns. First, however, is perception and education as to its use - that the compost has beneficial soil amendment properties, and that it is safe to use (low constituents of concern; metals and pathogens). Lastly, the compost must have consistency; the compost will produce predictable results when used.

Restoration Ecology

Another use of compost is as a soil amendment in restoration ecology projects. Restoration projects have the added attractiveness that they encourage volunteer public participation. Two excellent examples of these are the use of compost in Washington's Mountains to Sound Greenway logging road restoration program along Interstate 90, and the City of Everett's use of compost for constructed wetlands. In both cases, acceptance of compost use was a direct result of education and a change of perception by the public and political and environmental leaders. First, health considerations were addressed, then the beneficial effects were demonstrated. Since the volunteers were in direct contact with the compost, compost appearance and smell were very important. Demonstration projects showed results that were highly visible.

THE NEW PARADIGM TO TAILORING COMPOST STANDARDS

As pointed out in the preceding section, we are finding that the relative importance of analytical parameters has decreased as we learn more about the lack of detrimental effects of normal composts. This awareness is intimately linked to a new paradigm in management of organic residuals. The new management style incorporates a recycling philosophy rather than a disposal mentality and requires new roles for the public, the regulator, the manager, and the scientist. Correspondingly, a "New Work Place" exists.

Development of the New Management Style

The recycling ethic is well established in many parts of the world now. As part of this trend, or on its own merit, all organic residuals management is dramatically changing. Most residuals managers embrace the concept of recycling, either for its intrinsic, economic or environmental benefits; recycling has become part of their vocabulary. Managers wanting a successful recycling program acknowledge that there are new rules of the game. These can be compiled, figuratively, into a "New Book of Residuals Management" that incorporates many new concepts into both the technical and social issues of their use.

The change taking place in organics management in the compost industry seems to be following and learning from that which is occurring in the biosolids industry. This tale of change is reflected in the titles of the popular annual biosolids conferences put on by the Northwest Biosolids Management Association (NBMA) in Washington State. The first two conferences (1989 and 1990) were introductory to the whole concept of "sludge", as a whole host of challenges were facing managers. In 1991 public opposition to beneficial use of biosolids was considerable, and the theme was *In Whose Backyard.* In 1992 managers were paying attention to the public's concerns about contaminants and *Sludge Quality* dominated the conference.

However, biosolids management was about to take a big evolutionary step. New regulations suggested the preferred and potentially the most environmentally sound alternative for biosolids management was reuse. This encouraged the NBMA to become proactive. A certain excitement was born, and pride in programs surfaced. The conference in 1993 was titled *Entering a New Era* -- a theme that now appears right on target. In 1994 education was targeted as a key -- *Conveying the Message.* In 1995 the strength of biosolids management was recognized in *The Biosolids Trilogy: Quality, Collaboration, Perception.* There has been a heralded level of success in biosolids recycling projects in the Northwest, thanks to the public information and collaborative project management techniques advocated by NBMA members. Yet, the organization realizes that the big challenge is now to continue successful, quality programs into the future; 1996's conference theme was *A Bridge to Sustainability*.

The "New Book of Residuals Management" has a distinctive look: cooperation/ collaboration, confidence, acceptance, and opportunities, all based on science and solid technical ground. The major change that has taken place is active **cooperation and collaboration** between all players in residuals management: managers, scientists, regulators, the public and environmental groups. Although they may not always agree on the details of implementation, all of these groups embrace recycling and support reuse as a preferred alternative. There is also **confidence** that recycling has technical merit, economic viability, and environmental compatibility. This confidence is largely based on a significant body of **scientific research**, that has established the credibility of management alternatives. Because of the success of key programs, public **acceptance** has reached an all-time high. All of these factors create new and expanded **opportunities** for recycling projects and partnerships.

To reach this point of success, managers found that traditional roles were not always sufficient; the public has high expectations of managers, scientists, and regulators. The public expects responsiveness to their concerns, thorough research, and regulations that provide long-term protection of the environment and safety for humans and animals. The status quo behavior and assurances from professionals are not adequate for many citizens. These pressures have redefined many of the comfortable traditional roles and thrust us into different approaches as well as consensus making. Our challenge, then, is to embrace these changes, and to readily adapt to new roles for the regulator, the manager, and the scientist. Our new roles and behaviors have created a "New Work Place".

The New Public

In the recent past, neighbors often posted their own signs around proposed biosolids application sites -- signs such as "No Stinking Sludge" or "Sludge Not Lest Ye Be Sludged." Individual citizens as well as environmental groups had a "zero tolerance" for land application of sludge. Many of these proposed sites had to be abandoned, for the citizens were well versed in political action and knew how to create controversy and how to win support from their elected representatives.

The public's involvement in environmental and technical issues continued to grow. As "sludge" became "biosolids" and "disposal and beneficial use" became "recycling", many members of the public began to reassess their positions on this subject. The research and risk assessments used in the development of new federal guidelines were a welcome source of information to those interested in the fate of contaminants. The zero tolerance philosophy eased and productive discussions followed.

Today some citizens' groups actually initiate biosolids programs, realizing that biosolids managers and conservationists can share environmental goals. In the Seattle area, the Mountains to Sound Greenway Trust, a local conservation organization, after much independent investigation, decided that biosolids could be used to maintain thriving commercial forests on the outskirts of the city. The Trust took the lead in developing and negotiating a 50-year agreement between the King County Wastewater Treatment Division, the state Department of Natural Resources, the University of Washington, and the Weyerhaeuser Company. In Toronto, Ontario, the Public Committee for Safe Sewage Treatment has been relentless in investigating and promoting land application of biosolids as an alternative to the city's incinerators.

The New Regulator

The 70s and 80s brought regulators to the forefront on environmental issues. Their task was to clean up the mess -- polluted rivers, smokestacks and hazardous waste dumps! Many regulations were developed under the philosophy that "the more conservative, the better". In some cases it seemed as though regulators were trying to out-do each other by setting lower limits without regard for actual risk assessment. Although those who were regulated by such arbitrary restrictions were frustrated, regulation in general, under the Clean Water Act and others, did much to improve severe pollution problems in lakes and streams.

With certain residuals, such as biosolids, overly restrictive regulations produced little measurable environmental benefit and the marginal improvements did not justify the costs, both economic and net environmental. US EPA was taken to task during review of the proposed federal biosolids regulation, 40 CFR 503 (US EPA 1993) by scientists and managers, for among other things, inappropriately using the results of metal salt studies to determine metal limits in biosolids. EPA listened to the criticism and revised their rule to be consistent with the best available science and risk assessment modeling. This precedent resulted in a remarkable change in EPA's approach towards setting regulations. EPA staff were able to explain regulations by using scientific data that provided the basis for them.

EPA's new science-based approach resulted not only in a sound technical basis for regulations, but in strong support by scientists and managers. Now, the "New Regulator" starts working with their constituents from the start, relying on scientists for technical support. The new regulator is continually challenged to develop sound regulations that are also viable solutions.

The New Manager

Many of today's biosolids programs began 15 to 20 years ago. It was not easy then for a biosolids manager to convince a skeptical public that a sludge application program was a good thing, particularly if it was within sight or smell of passers-by. Usually the manager was in the position of selling the public on a plan that had already been carefully crafted by consultants and technical staff. The public perceived -- correctly -- that the manager had little interest in their input: "This is safe and has been studied extensively. It's the best option and we've already looked at all the alternatives." Not surprisingly, this type of "public information" was the downfall of many technically sound, economical projects.

Our Canadian colleagues in the NBMA refer to the public process as public "consultation." This term is more descriptive of the behavior required from the New Manager. The project manager can usually identify the individuals and the groups who will be protective of the application site. For example, the first biosolids application in state-owned forests in Washington was planned for a multi-purpose forest that was heavily used by hikers and mountain bikers. The forest's Citizen Advisory Group was consulted about planned application. One of the members, fiercely protective of the forest, wanted to review in detail all the planning documents. Because of his life-long familiarity with the forest, he was able to point out a spring immediately downslope of the unit that would make an ideal monitoring site. It was a much better plan than that suggested by the project manager! Throughout the project, the citizen visited the operation, took photographs and offered constructive comments.

Although it may be a difficult paradigm shift, the wise new manager solicits advice and opinions from those outside the organic residuals world. Not only does the biosolids project manager need to consult with interested citizen groups and neighbors, but he needs to build alliances with all those who share his environmental goals. The "New Manager" sustains public confidence by listening and responding to citizens, collaborating and building consensus among the public, his agency, scientists, and regulators.

The New Scientist

In the past, science played an important, but seemingly separate, role in residuals management. Scientists conducted their "statistically designed", "replicated" experiments, but few understood the scientific jargon of the results. Science was not contributing to the development of reasonable regulations or to public understanding of residuals. Scientists were viewed as impartial evaluators in their ivory towers dealing with largely esoteric scientific truths, but speaking an incomprehensible language. Occasionally the relevance of research and presentations was low to meaningless.

Reduction of funding opportunities, and the fact that society now values -- and supports -- only that which has meaning to them, has required research to move towards applied and relevant projects. The public is not only critical of what they spend their money on, but demands that they learn what they paid for.

With the 503's, a "New Scientist" emerged: one that not only investigated questions, but put the answers to use in developing the technical basis for regulations. Additionally, the "New Scientist" was asked to explain and defend these regulations. Today, scientists must know how to talk to the public, how to handle an interview, what "sound bites" are, and occasionally admit they don't know everything. In the "New Book of Residuals Management", scientists are*** needed for their knowledge and- credibility, and to work in conjunction with regulators and managers to develop guidelines. Thus, scientists need to explain and support their work. In most cases this means that scientists become advocates for recycling and application programs because their research reaffirms the benefits and safety of residuals use.

The New Work Place

Similar to a cycle in nature, we find the "New Work Place" to be a pool of people resources with interactions that might resemble Figure 1.

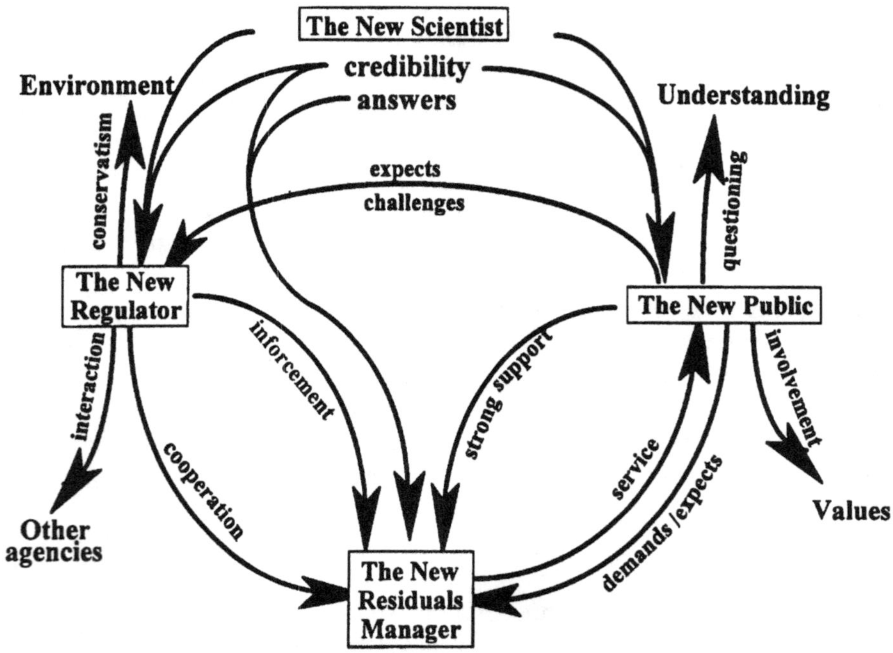

Figure 1. The "New Work Place" conceptually can be a pool of people resources with many interactions, objectives and endpoints.

There are some important dynamics of this workplace as paraphrased from Wheatley (1995). The first is the flow of information. Am I getting the right information at the right time? Are others? What are the barriers to good information flow? The second has to do with relationships in the work place. There are both formal and real relationships. What roles do each play and are they acknowledged? What are the conflicts in the relationships, and how can they be best made effective? The third dynamic aspect is the vision of the work place. Is it a shared vision, and does it give the members a sense of purpose? All of these are pertinent to the dynamics within the "New Work Place".

Wheatley (1995) goes on to describe new science principles that also speak directly to how we approach our new roles in biosolids management. She suggests a shift: 1) From looking at the parts to looking at the whole, 2) to understanding processes, rather than structures, and 3) to understanding that the universe is a web of relationships that are constantly changing and growing. New roles for all of us involved in residuals management have to be taken to heart and matched with a look towards the future.

SUMMARY

Tailoring compost standards to targeted uses encompasses a whole range of emerging philosophies. The first is a definition of what "contaminants" are, and what constitutes acceptable levels of these contaminants. Setting standards based solely on the traditional approaches of easily measured constituents/compounds may ignore the most important considerations.

Perception and corresponding education about compost quality sets the base for developing standards. Standards can then be constructed around the important considerations of:

- who the user is
- potential end use/market niche
- what the compost does

rather than simply contaminants. This approach suggests a new paradigm to residuals management, which builds upon the "New Work Place" and new roles for all the "pools" of people resources. It includes a pathway for establishing or continuing our level of quality by redefining our interactions, objectives and endpoints.

Tailoring standards in the compost and residuals industry will need new approaches. For today's residuals players, success will go to those who are willing to adapt to the rapidly changing rules of play, because, as Wheatley (1995) says, "When things are stable, you're closest to death".

REFERENCES

Epstein E, Chaney RL, Henry CL & Logan TJ (1992) Trace elements in municipal solid waste compost. Biomass and Bioenergy 3:(3/4):227-238.

Henry CL, Wescott HA & Harrison RB (1993) Change in form of trace metals during composting MSW and uptake following field application. Abstr, 2nd International Conference on Metals in Soils, Waters, Plants and Animals. Sept 6-10, 1993. Taipei, Taiwan.

US EPA (1993) 40 CFR 257 and 503: Standards for the use or disposal of sewage sludge; Final rules. Washington, D.C.

Wheatley M (1995) Lessons from the New Workplace. (video recording by CRM Films, Carlsbad, CA).

CARBON AND NITROGEN MINERALIZATION DURING CO-UTILIZATION OF BIOSOLIDS AND COMPOSTS

JOHN T. GILMOUR
Department of Agronomy, University of Arkansas, Fayetteville, Arkansas 72701, USA
E-mail gilmour@comp.uark.edu

ABSTRACT

Increasing the consumer value of organic wastes through co-utilization or mixing is directly related to organic carbon (C) and nitrogen (N) mineralization when the mixture is used as a soil amendment. The objective of this study was to put forward a protocol that can be used to estimate the behavior of a mixture of biosolids and composts in a co-utilization program from the perspective of decomposability and release of plant available nitrogen (PAN) over time. Published literature suggested that the components of such mixtures would decompose in similar manner whether alone or in combination, and that net N immobilization characteristics of components with high C:N ratios could be overcome. Using the proper ratio of the components was essential to the development of a soil amendment which was more valuable than its components. Three mixtures for co-utilization of organic wastes were assessed using the computer simulation model, DECOMPOSITION. Municipal biosolids was mixed with paper mill sludge, leaf litter, and municipal solid waste (MSW) compost. Weather at Baltimore, MD, was selected for these simulations, with mixtures being applied in April. In each case a ratio was found to make a product with good N availability, but each mixture was sufficiently decomposable to limit its ability to increase soil organic matter (OM). The organic wastes most resistant to decomposition, very stable composts, were evaluated for potential to increase soil organic matter. Simulation of very stable compost decomposition showed that they would increase soil organic matter if applied annually for an extended period. These composts were also a good source of PAN after several years of application to soil.

INTRODUCTION

Co-utilization can be defined as "the combination of two or more organic by-products that are either composted, blended, or blended after composting to produce a value-added product which can be beneficially utilized to solve an agricultural problem, remediate soils and/or fulfill a market niche" (R.F. Korcak, Personal Communication, 1997). Increasing the consumer value through co-utilization or mixing organic wastes is directly related to

S. Brown, J.S. Angle and L. Jacobs (eds.), Beneficial Co-Utilization of Agricultural, Municipal and Industrial By-products, 89–112.
© 1998 *Kluwer Academic Publishers*,

organic carbon (C) and nitrogen (N) mineralization, when the mixture is used as a soil amendment.

The decomposability of the mixture determines whether or not long-term increases in soil organic matter can be achieved, while the plant available N (PAN) in the mixture affects its use as a N fertilizer. The key is predicting the relationships among decomposition and net N mineralization, when mixtures of organic wastes are applied to a site with a given set of soil and climatic conditions.

Figure 1 presents a stylized depiction of these relationships. The area labeled *Increased Soil OM* shows that organic wastes resistant to decomposition increase soil organic matter irrespective of C:N ratio. Composts and other organic by-products, such as paper mill sludges which have been stored for extended periods, are examples. As decomposability increases, the potential to markedly increase soil organic matter is lost at the point where the end products of decomposition equal soil organic matter losses (Sikora et al., 1996).

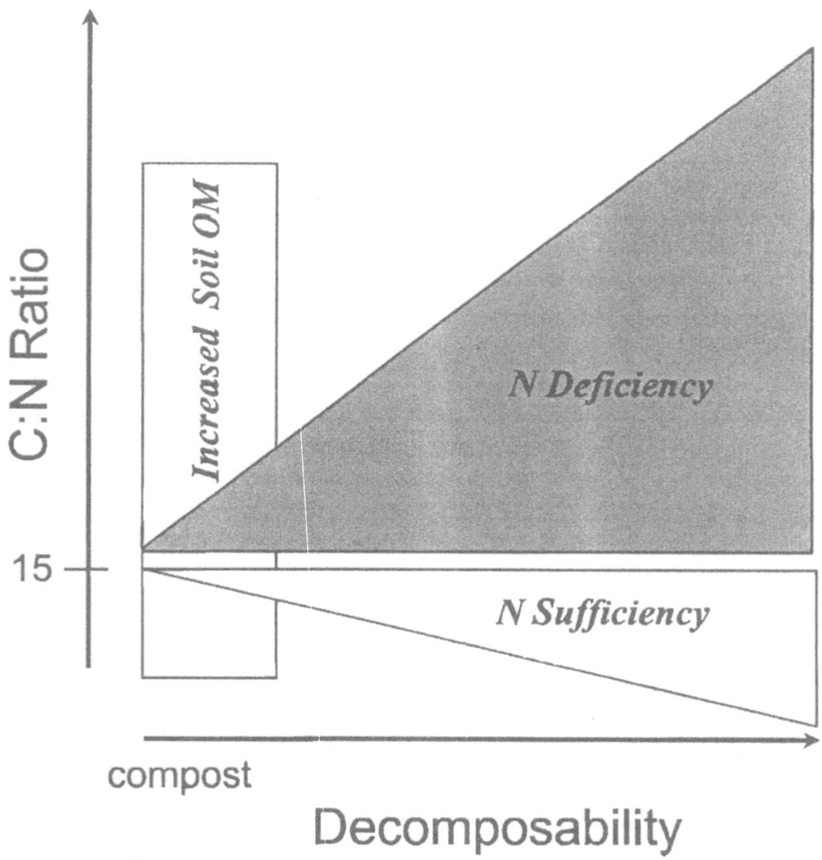

Figure 1. The effects of decomposition rate and C:N ratio on soil N status and soil organic matter levels.

During composting and decomposition (aerobic or anaerobic), the C:N ratio of an organic waste will tend to stabilize near a value of 15 (Blackmer and Green, 1995; Henry and Harrison, 1996; Moorhead et al., 1987). This occurs because net C mineralization generally exceeds net N mineralization at higher C:N ratios, and any inorganic N present in the system will be immobilized (e.g. Zibilske, 1987). Thus, a C:N ratio of about 15 can be used to separate organic by-products into two groups. As organic by-product decomposability increases and the C:N ratio increases above 15, the potential for N deficiency due to N immobilization increases as shown in the area labeled *N Deficiency* in Figure 1. Organic by-products with C:N ratios of about 15 or less can supply sufficient PAN to a crop at higher rates of decomposition and/or lower C:N ratios as shown in the area labeled *N Sufficiency*. Tester et al. (1979) illustrated this concept using differing size fractions of the same composted biosolids (municipal sewage sludge) which had different C:N ratios. Each size fraction had a similar amount of decomposition (7.4 to 8.0%), yet net N mineralization increased by a factor of four as the C:N ratio declined from 18.6 to 10.4

Climate and site factors (soil, crop, management) modify the potential of organic by-products to increase soil organic matter or provide PAN to a crop by altering the decomposition rate of the organic by-products (Gilmour et al., 1977; Sikora et al., 1996). Predicting how an organic by-product mixture will perform must take into account these features and site characteristics, if performance under different conditions is to be estimated.

The objective of this study was to put forward a protocol that can be used to estimate the behavior of a mixture of biosolids and composts in a co-utilization program from the perspective of decomposability and release of PAN over time.

MATERIALS AND METHODS

Laboratory Studies

Three laboratory studies were conducted. The details of each can be found in the Appendix. Study I where paper mill sludge, dairy manure and the combination were evaluated for decomposition and net N mineralization will be discussed below, while Studies II and III are included for analytical and first order kinetic data (Appendix Table 1) that became part of the data set used in computer simulations below.

Description of the Computer Model

The computer model, DECOMPOSITION, is a mechanistic model that uses decomposition kinetics to estimate rates of C and N transfer among pools. A flow diagram of the C and N pathways is presented in Figure 2. The amount of C transferred from one pool to another was estimated by first order kinetics as shown below,

$$C_t = C_{t-1} \times e^{-k \times t} \qquad [1]$$

where C_{t-1} was the amount of C at the start of a day, C_t was the amount of C in a pool at the start of the next day, k was the first order rate constant and t was the time step, which was one day. The difference between C_{t-1} and C_t was the CO_2-C produced.

In the case of the organic waste pool, rapid and slow fraction rate constants were used during the decomposition of the rapid and slow phases, respectively. When sufficient N was present, biomass was created using microbial efficiency, E, described by Eq. 2 below,

$$E = \text{biomass-C}/(\text{biomass-C} + CO_2\text{-C}) \qquad [2]$$

where biomass-C was the new microbial cells in Figure 2 and CO_2-C was estimated using Eq. 1. If sufficient N was not present in the organic waste pool plus the inorganic N pool to make the amount of biomass estimated from Eq. 2, E was decreased and a new, smaller amount of biomass was created. If E = 0, no biomass was created during that day. The initial indigenous biomass pool size was estimated using an equation presented by Insam (1990) which related the ratio of soil biomass to soil organic C, using the ratio of precipitation to evaporation for the location in question. The C:N ratio of both biomass pools (i.e. new microbial cells and indigenous microbial cells) was set to 8 in these calculations (Gilmour et al., 1985).

If the C:N ratio of the organic waste was >15, no organic N in the waste was mineralized or used in the efficiency calculations, and if inorganic N was available, that N was immobilized. For organic wastes with C:N $\leq$ 15, N mineralized was equal to the C decomposed divided by the organic waste C:N ratio. Only decomposition (C loss) occurred for organic wastes with higher C:N ratios which lowered the C:N ratio of these organic wastes as decomposition proceeded. Each time step, E was reset to 0.4.

Biomass created above was distributed among biomass pools as follows (Gilmour et al., 1985): 90% to the new biomass pool, 10% to the indigenous biomass. In turn, 15% of the addition to the new biomass pool was moved to the soil OM I pool which was assigned a C:N ratio of 10 (Gilmour et al., 1985). Organic C in the two biomass pools and the soil OM I pool subsequently decomposed according to Eq. 1.

Several studies have presented rate constant values for decomposition of various biomass pools. Molina et al. (1990) reported a rate constant of 0.33 d^{-1} for a labile biomass pool (56% of total biomass) and 0.04 d^{-1} for a resistant biomass pool in the computer model, NCSOIL. Paul and Van Veen (1978) concluded the value for the biomass pool was 0.01 d^{-1}. In both of these models, the rate constants appear to correspond to the new biomass pool in Figure 1. Gilmour et al. (1985) used a new biomass pool rate constant was 0.07 d^{-1} in an earlier version of DECOMPOSITION. An intermediate value (0.035 d^{-1}) that has provided good simulations of laboratory data by the author was used in this version of the computer model for the new biomass pool.

Ladd et al. (1981) reported that the rate constant for indigenous biomass decomposition was 0.00047 d^{-1} for the period one to four years after addition of a plant residue to soil. Jenkinson and Raynor (1977) presented data (their Table 5) that was used to compute a biomass rate constant of 0.00093 d^{-1} for one to four years after application of several organic soil amendments. The value of Jenkinson and Raynor (1977) was chosen for the indigenous biomass pool.

Initial soil organic matter was equally divided among the soil OM I and soil OM II pools (Paul and Van Veen, 1978). Soil OM II was considered stable for the simulation periods used here. Decomposition of the soil OM I pool was described using the rate

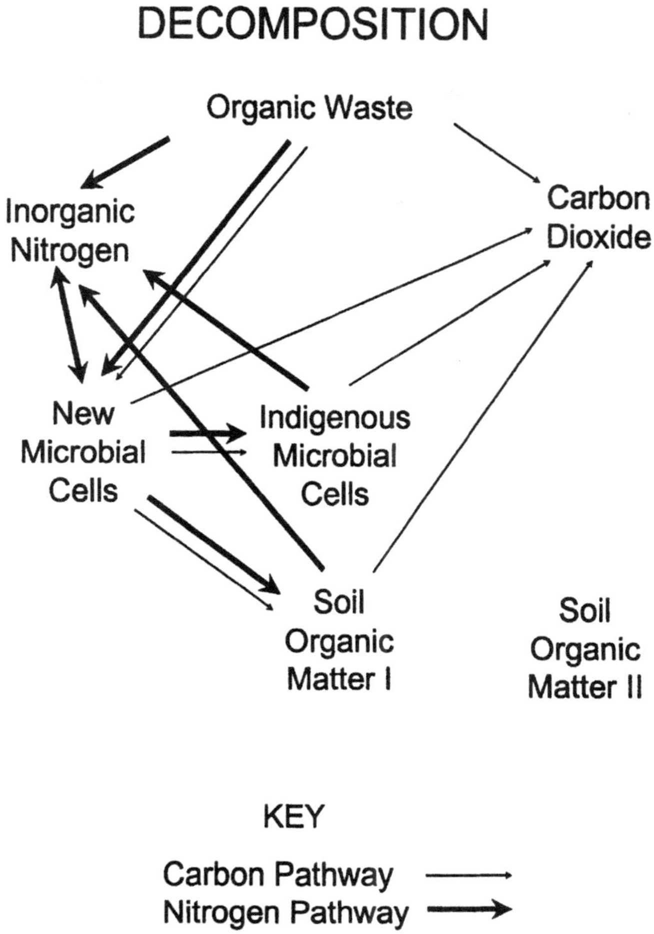

Figure 2. Flow diagram for C and N among various pools used in the computer model DECOMPOSITION.

constant for the physically protected portion of soil OM (0.00020 d^{-1}) presented by Paul and Van Veen (1978). This value is much smaller than the value for the resistant humad pool (0.0065 d^{-1}) used in NCSOIL (Molina et al., 1990).

All rate constants were corrected for land application site temperature and soil moisture status by multiplying rate constants obtained under optimum soil temperature (25°C) and moisture by the factors presented in Figures 3a and 3b. The temperature factor, TF, was computed for temperatures above 10°C using a modified form of the Arrhenius Equation shown below,

$$TF = e^{-A \times (1/T - 1/298)} \qquad [3]$$

where A was estimated for a given Q_{10} and reference temperature (A = 5820°K for Q_{10} = 2 and T_{ref} = 298°K) and T was soil temperature in °K. A linear decline in TF was assumed from 10 to 0°C. Mean monthly air temperature from National Oceanographic and Atmospheric Administration (NOAA) literature for a weather station nearest the land application site was used to estimate soil temperature.

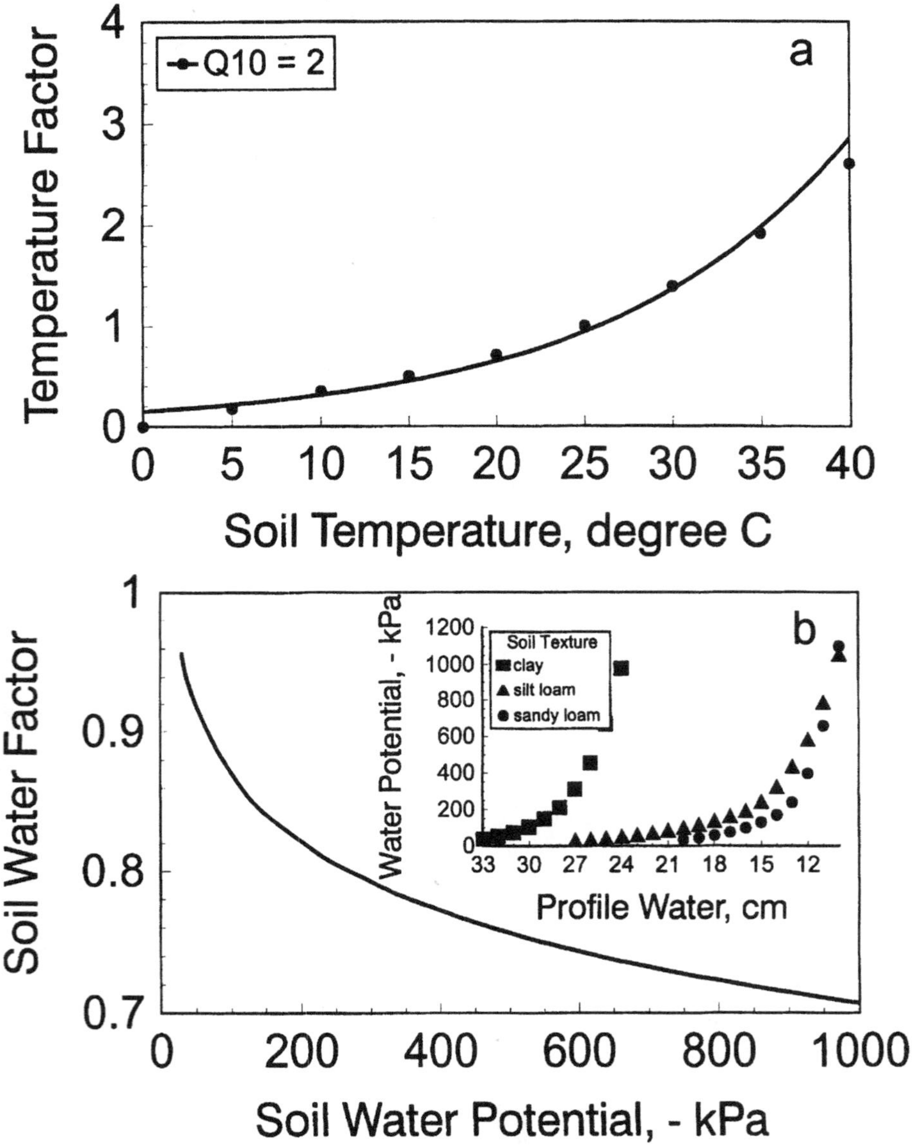

Figure 3. Temperature (a) and soil water (b) correction factor relationships used to adjust first order rate constants for land application site climate.

Soil water moisture correction used a monthly water balance on the 0 to 15 cm soil depth with initial water content set at a water potential of -33 kPa for a given soil texture

(Figure 3b). Monthly rainfall was an addition to this soil moisture level, while 80% of pan evaporation (an estimate of evapotranspiration) was a subtraction. Mean monthly precipitation and pan evaporation from NOAA literature for a weather station nearest the land application site were used in the water balance calculations. When pan evaporation data were absent, monthly pan evaporation was plotted versus monthly temperature to estimate missing values. The soil moisture content was then related to a soil moisture potential using water release curves typical of sandy loam, silt loam or clay soils (see insert in Figure 3b from Hanks and Ashcroft, 1980). Soils were not allowed to become wetter than -33 kPa or drier than -1,000 kPa. The soil water potential was then used to compute the water factor, WF, using an equation presented by Sommers et al. (1980) as shown below,

$$WF = -0.164 \log_{10}(-\psi/1000) + 0.871 \quad [4]$$

where ψ is the soil water potential in kPa. A silt loam texture was used in the simulations described below.

Computer Simulations

Table 1 presents mean analytical data and first order kinetic parameters obtained by summarizing data presented in Appendix Table 1. Each of these organic wastes was considered a candidate for co-utilization. To select components of mixtures, the organic

Table 1. Mean analytical data and first order kinetic parameters used in computer simulations.

Organic Waste	Organic C g kg^{-1}	Organic N g kg^{-1}	C:N	Inorganic N mg kg^{-1}
Paper Mill Sludge	275	7.7	35.7	---
Municipal Biosolid	273	35.6	7.7	22,800
Leaf Litter	451	10.6	42.5	---
Compost -stable	182	12.1	15.0	---
MSW Compost	213	12	17.8	---
Compost - very stable	221	21.8	10.1	---

	Sequential Decomposition Model Parameter[a]		
	k_r d^{-1}	k_s d^{-1}	Rapid Fraction %
Paper Mill Sludge	0.019	0.006	48
Municipal Biosolid	0.025	0.0029	28
Leaf Litter	---	0.0029	---
Compost-stable	---	0.0026	---
MSW Compost	---	---	---
Compost - very stable	---	0.00048	---

[a] k_r and k_s are the rapid and slow fraction rate constants, respectively, while % Rapid Fraction is the amount of organic C undergoing decomposition using first order kinetics and k_r as the rate constant; 100% minus % Rapid Fraction is the amount of organic C undergoing decomposition using first order kinetics and k_s as the rate constant.

wastes in Table 1 were classified relative to decomposability and impact on soil N status. Decomposability followed the order: paper mill sludge ≈ municipal biosolids > leaf litter ≈ stable compost > very stable compost. Relative potential to cause N immobilization considered both decomposability and C:N ratio. The ranking was paper mill sludge > leaf litter > municipal solid waste (MSW) compost. Municipal biosolids, stable composts, and very stable composts had C:N ratios ≤ 15 and were not considered candidates for net N immobilization.

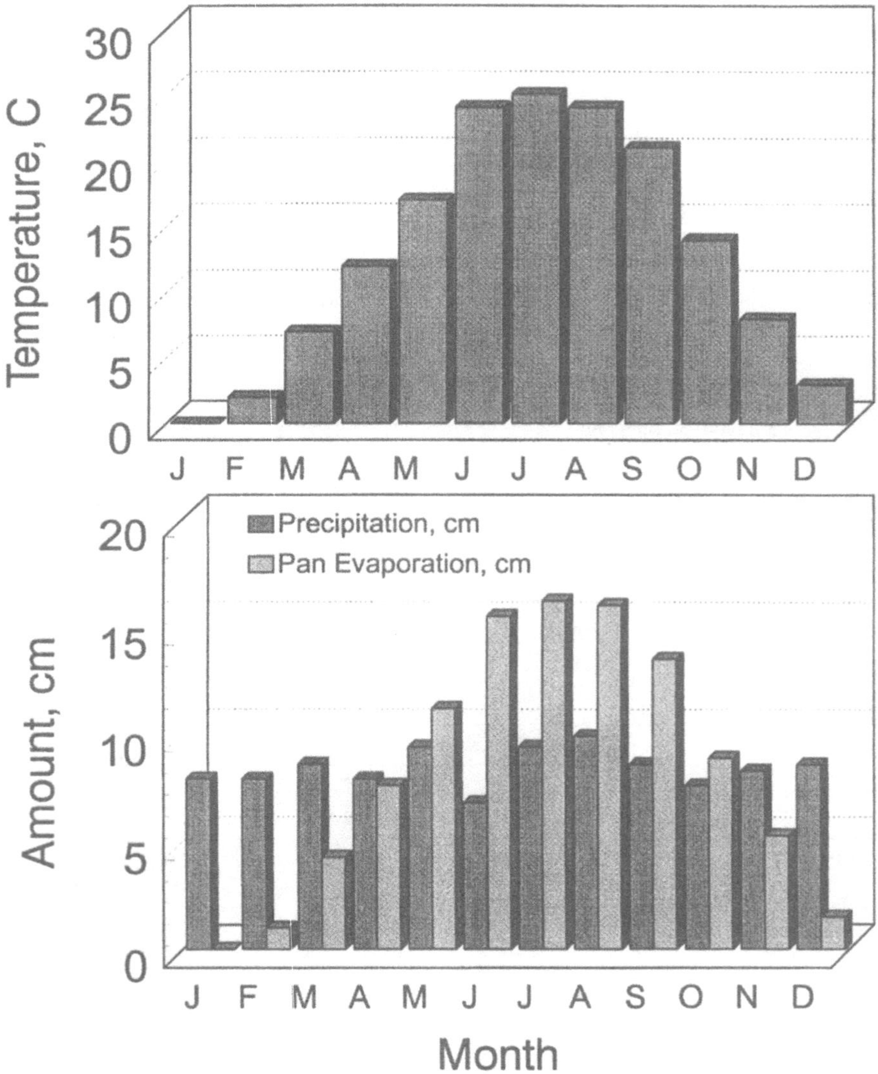

Figure 4. Temperature, precipitation, and pan evaporation data for Baltimore, MD.

Three mixtures were selected to illustrate that co-utilization could reduce or eliminate net N immobilization of one component of the mixture. Paper mill sludge, leaf litter and MSW compost were mixed with municipal biosolids. Weather at Baltimore, MD, was selected for these simulations. Mean monthly air temperature, precipitation and pan evaporation are presented in Figures 4a and 4b, respectively. The mixtures were applied at the first of April and simulations run for 24 months.

The first step was to simulate decomposition of the municipal biosolids at 1 t ha^{-1} to establish the amount of N mineralized each month. The second step was to run simulations using the second component at different application rates until net N immobilization from the second component was compensated for by PAN from the municipal biosolids in any month. That rate established the ratio of the two components. The data presented in Figures 8-10 are the sum of the monthly outputs for the simulations. The first order rate constant was increased 1.5 times for the paper mill sludge rapid fraction based on results from Study I. The kinetic parameters used for the MSW compost were assumed to be the same as for the stable compost. No other adjustments to rate constants were made.

Two additional simulations were made using Baltimore weather: stable compost and very stable compost. The mean annual first order rate constant for these organic wastes was calculated by solving Eq. 1 for k. This k was then used in a spreadsheet program to estimate the accumulation of the organic waste by years of addition where t in Eq. 1 was one year.

RESULTS AND DISCUSSION

Laboratory Studies

The results from Study I are presented in Figure 5. The decomposition of paper mill sludge in the presence and absence of dairy manure, and dairy manure alone, are plotted versus time. Decomposition of paper sludge plus manure was significantly different from the individual treatments for all sample times, while the individual treatments were significantly different from each other at 2.9 d and after 24 d. At 60 days, percent decomposition amounts for the paper mill sludge, dairy manure, and the 50:50 mixture were 58.5, 56.5, and 56.2%, respectively, which were not significantly different.

The first order rate constants for the rapid fractions of the paper mill sludge alone or mixed with dairy manure (assuming dairy manure decomposition was constant) were 0.021 and 0.026 d^{-1}, respectively, while the rapid fractions amounts were 42 and 52% of the total sludge C. These differences were compensated for during decomposition of the slow fractions as slow fraction rate constants were 0.010 and 0.0054 d^{-1}, respectively. In a second part of Study I (data not shown), the addition of 140 kg ha^{-1} N as NH_4NO_3 also did not significantly alter overall decomposition amounts at 60 d. Rate constants for the rapid fraction of the paper mill sludge alone and plus inorganic N were 0.021 and 0.033 d^{-1}, respectively, while the corresponding percentages in the rapid fraction were 42 and 57%. As in the case of the manure addition, the larger rate constant and amount in the rapid fraction in the presence of inorganic N was compensated for by a smaller rate

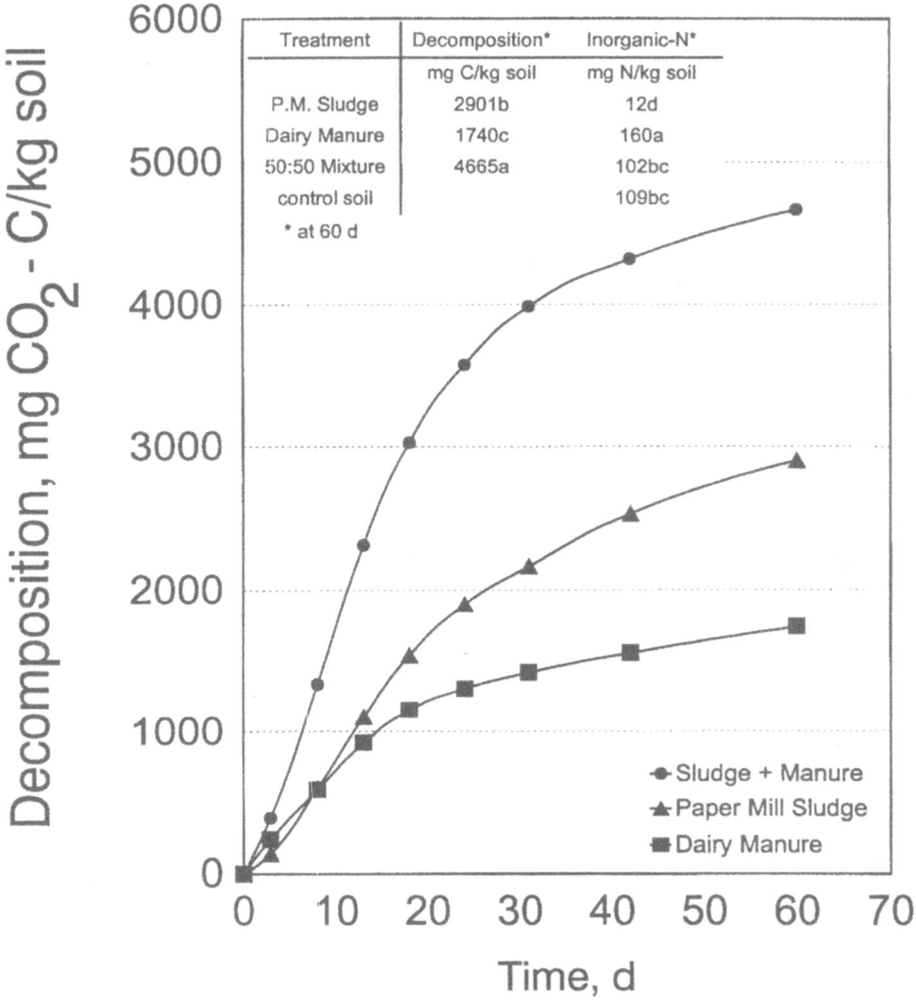

Figure 5. Decomposition and net N mineralizationof paper mill sludge (C:N=49.4) and dairy manure (C:N=12.1) applied at 33.6 t/ha alone and at 67.2 t/ha as a mixture (Study I).

constant during slow fraction decomposition. The values for the slow fraction rate constants were 0.010 and 0.0092 d^{-1}, respectively, a much smaller difference than found in the first part of the study.

At 60 d, total CO_2-C losses were 2901, 1740, and 4665 mg C/kg soil for the sludge, manure and mixture, respectively. The sum of the sludge plus manure was 4641 mg C/kg soil, a value essentially equal to the mix which suggested that decomposition during co-utilization of these wastes in a mix would be similar to the decomposition of the components. Soil inorganic N concentrations at 60 d were significantly different among treatments. The inorganic N levels followed the order: dairy manure > control soil = 50:50 mixture > paper mill sludge. The dairy manure contributed 51 mg kg^{-1} inorganic N above the control soil, while the 50:50 mixture was 7 mg kg^{-1} inorganic N less than the control soil. Paper mill sludge alone was 97 mg kg^{-1} inorganic N smaller than the control soil. Net N immobilization by paper mill sludge was still observed at 60 d and net N mineralization of the dairy manure compensated for this net N immobilization. The observed soil N levels of the mixture would be sufficient for many crops, which illustrates the importance of supplementing an organic waste with a large C:N ratio with one having a small C:N ratio to avoid crop N deficiency.

Figure 6 presents data from Sabey et al. (1975) for decomposition of municipal biosolids alone and in the presence of bark at 22.4 t ha^{-1} over a year period. Biosolids and bark alone each had a decomposition rate of 2.3 mg CO_2-C kg^{-1} soil d^{-1}. The equivalent biosolids/bark mixture exhibited a decomposition rate which was identical to the mean of its components. First order rate constants for the biosolids, bark and mixture were

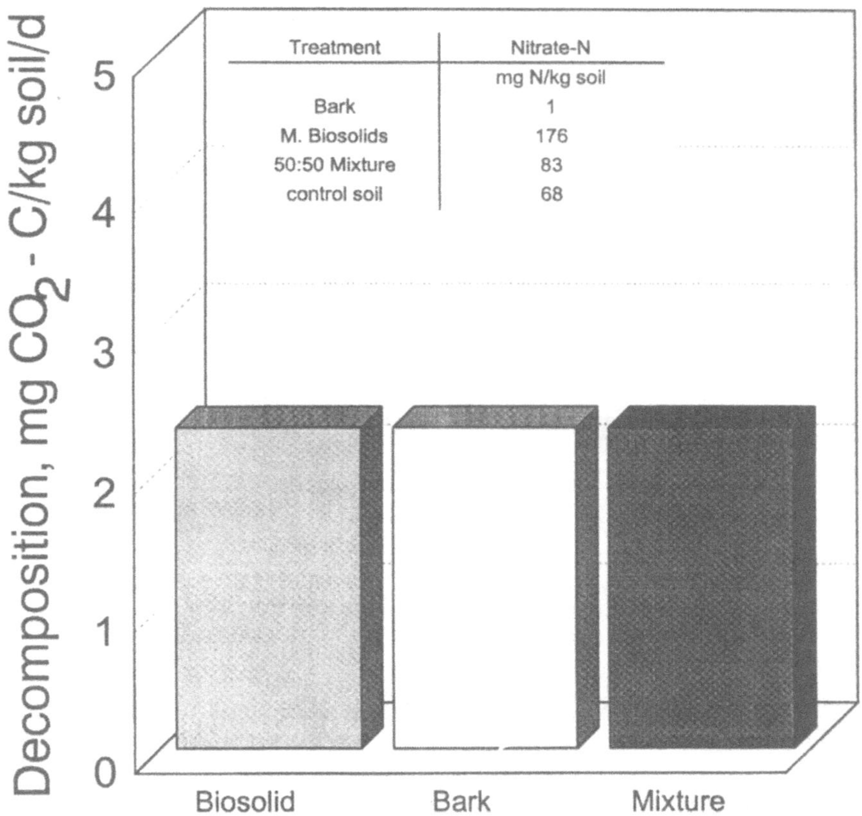

Treatment	Nitrate-N
	mg N/kg soil
Bark	1
M. Biosolids	176
50:50 Mixture	83
control soil	68

Figure 6. Mean decomposition rate and final soil nitrate-N following application of municipal biosolids and bark at 22.4 t/ha alone and as 50:50 mixtures (adapted from Sabey et al., 1975 and Agbim et al., 1977).

0.00078, 0.00071 and 0.00080 d^{-1}, respectively, which suggested that the biosolids and bark were very resistant to decomposition.

In Figure 7, decomposition of wood was 3.1 mg CO_2-C kg^{-1} soil d^{-1} as compared to 2.3 mg CO_2-C kg^{-1} soil d^{-1} for the biosolids/wood mixture and biosolids alone. First order rate constants for the wood and biosolids/wood mixture at 22.4 t ha^{-1} were 0.00084, and 0.00068 d^{-1}, respectively. These rate constants suggested that wood mixed with this biosolids would result in a mixture with slower decomposition kinetics, but the effect appeared to be small.

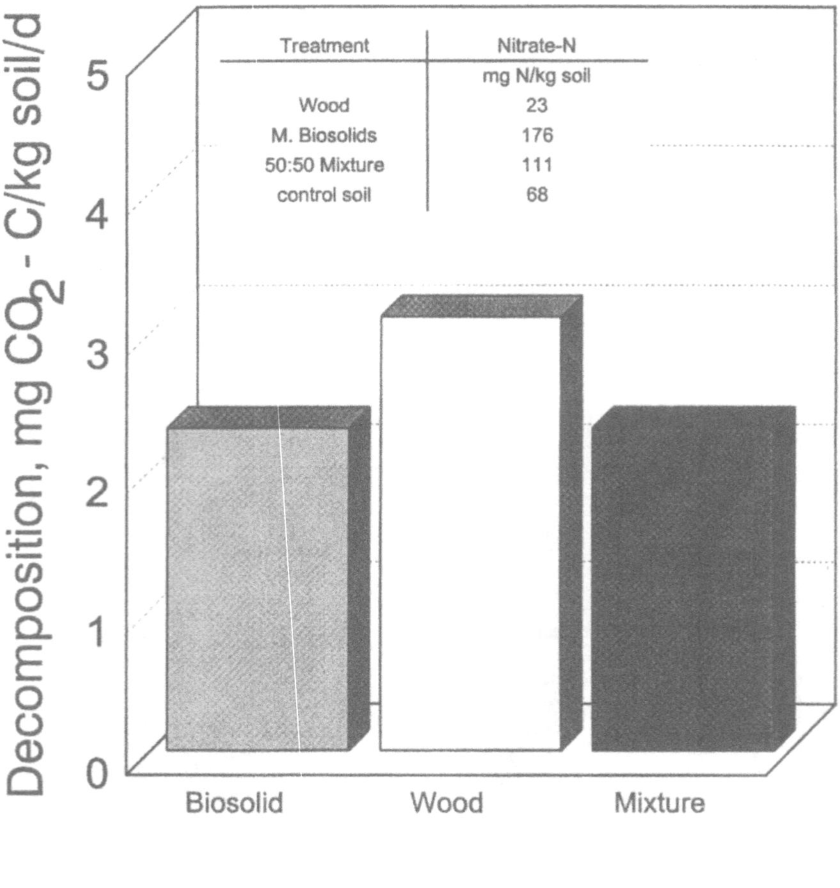

Figure 7. Mean decomposition rate and final soil nitrate-N following application of municipal biosolids and wood at 22.4 t/ha alone and as 50:50 mixtures (adapted from Sabey et al., 1975 and Agbim et al., 1977).

Figures 6 and 7 also present the final NO_3-N concentrations in the soil for the decomposition study described above (Agbim et al., 1977). Biosolids alone resulted in a soil concentration of 176 mg N kg^{-1} soil, while the concentrations in the presence of wood or bark alone were 23 and 1 mg N kg^{-1} soil, respectively. The NO_3-N concentrations in the soil amended with the biosolids/wood or biosolids/bark mixtures were 111 and 83 mg N kg^{-1} soil, respectively. While N immobilization was still apparent, the decomposition of biosolids in each mixture lead to substantial concentrations of PAN which again showed that immobilization and sufficient PAN, for crop production are not mutually exclusive concepts.

Computer Simulations - PAN

The paper mill sludge:municipal biosolids mixture simulation that would result in no net N immobilization for any month was 6.2:1 (w:w), as shown in Figures 8a and 8b. During

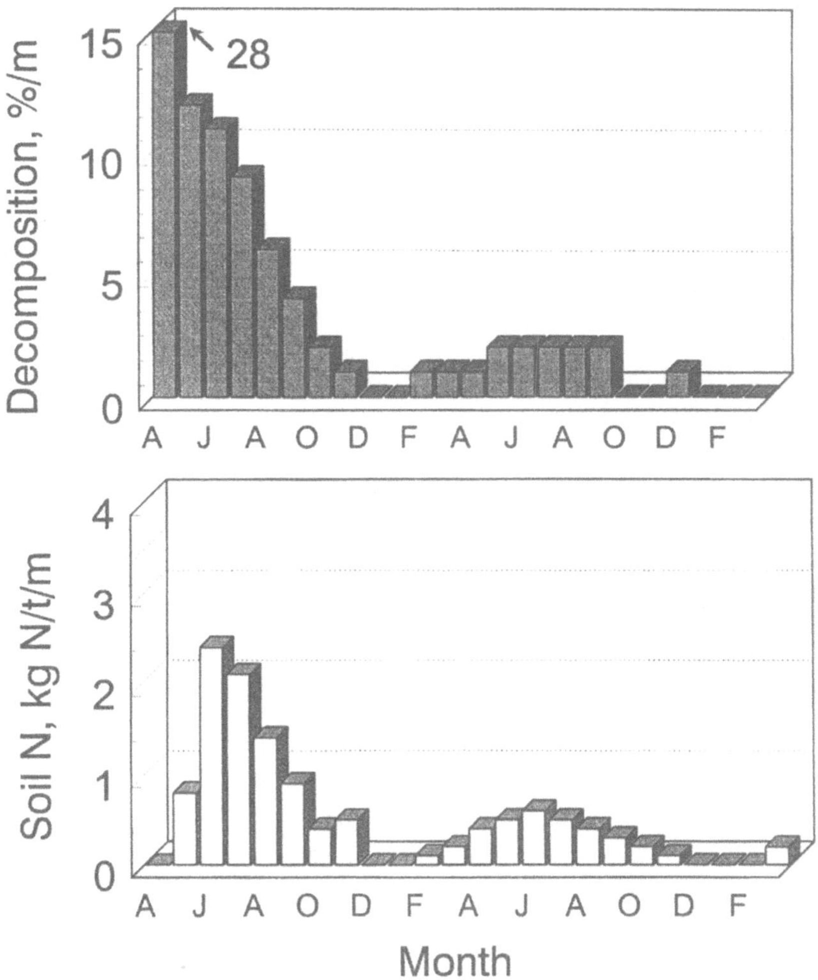

Figure 8. Stimulation results for a 6.2:1 paper mill sludge:municipal biosolids mixture. Soil N is the difference between N mineralization and N immobilization.

the first month after application (April), the change in soil N concentration was predicted to be 0 kg N t^{-1} for the mixture, as the municipal biosolids released PAN in an amount equal to N immobilized by the paper mill sludge. Twenty percent of the municipal biosolids and 29% of the paper mill sludge decomposed during the first month to give an overall amount of 28%. After that time positive PAN values were predicted i.e., soil N (the difference between mineralized and immobilized N) was greater than zero. Honeycutt et al. (1988) reported similar results using a similar paper mill sludge and they also found that N immobilization was a short-lived effect, occurring for about 25 d at 25°C and 33 d at 20°C. In Figure 8a, decomposition slowly declined from May to September, while PAN estimated for each month varied from a high of 2.4 kg N t^{-1} mixture to a low of 0.8 kg N t^{-1} mixture. The total amount of PAN predicted by the simulation model was 7.6 kg N t^{-1} mixture during the first growing season. The second year after application of the paper mill sludge:municipal biosolids mixture, continued decomposition and net N mineralization were predicted by the computer model. Decomposition was 1 to 2 % each month during the April to September period, while monthly PAN was from 0.3 to 0.6 kg N t^{-1} mixture. Total PAN during this period the second year was 2.7 kg N t^{-1} mixture or about one-third the first year value.

At the end of two years, 87% of the paper mill sludge:municipal biosolids mixture had decomposed and total PAN equaled 12.0 kg N t^{-1} mixture. This rate of decomposition suggested that this mixture would be a poor choice for improving soil organic matter, but a good choice as a N source. Eighty-six percent of the PAN was created during the April to September period with the distribution being two-thirds the first season and one-third the second.

Figures 9a and 9b present the computer simulation results for the leaf litter:municipal biosolids mixture. The organic C and N contents of the leaf litter used in the simulation compared well with a recent study conducted by Heckman and Kluchinski (1996) for deciduous trees in New Jersey. The ratio needed to avoid N immobilization was 3:1 leaf litter:municipal biosolids. The amount of decomposition was much smaller in April (8%) than for the paper mill sludge:municpal biosolids mixture. Also, the proportion of municipal biosolids decomposing (20%) was much higher than leaf litter (4%). In consequence, the simulation showed that the amount of PAN created in April was 6.8 kg N t^{-1} mixture as compared to the zero value for the paper mill sludge:municipal biosolids case. During the May to September period, decomposition proceeded at rates of 4 to 8% each month, but PAN created each month was very small (0 to 0.3 kg N t^{-1} mixture) for an overall April to September value of 1.0 kg N t^{-1} mixture, or only 15% of the first month value.

During the April to September period the second year, decomposition ranged from 2 to 4% each month for the leaf litter:municipal biosolids mixture, while PAN ranged from 0.2 to 1.0 kg N t^{-1} mixture. Total PAN during this part of the year was 3.3 kg N t^{-1} mixture, or 42% of the first year value. After two years, total decomposition was 71%, while total PAN was 12.3 kg N t^{-1} mixture. The decomposition amount suggested that the leaf litter:municipal biosolids mixture would also be a poor choice for augmenting soil organic matter. The PAN results showed that applied at the proper rate, the leaf litter:municipal biosolids mixture could provide a large amount of PAN initially with a significant residual PAN value the second year. Ninety percent of the PAN was created

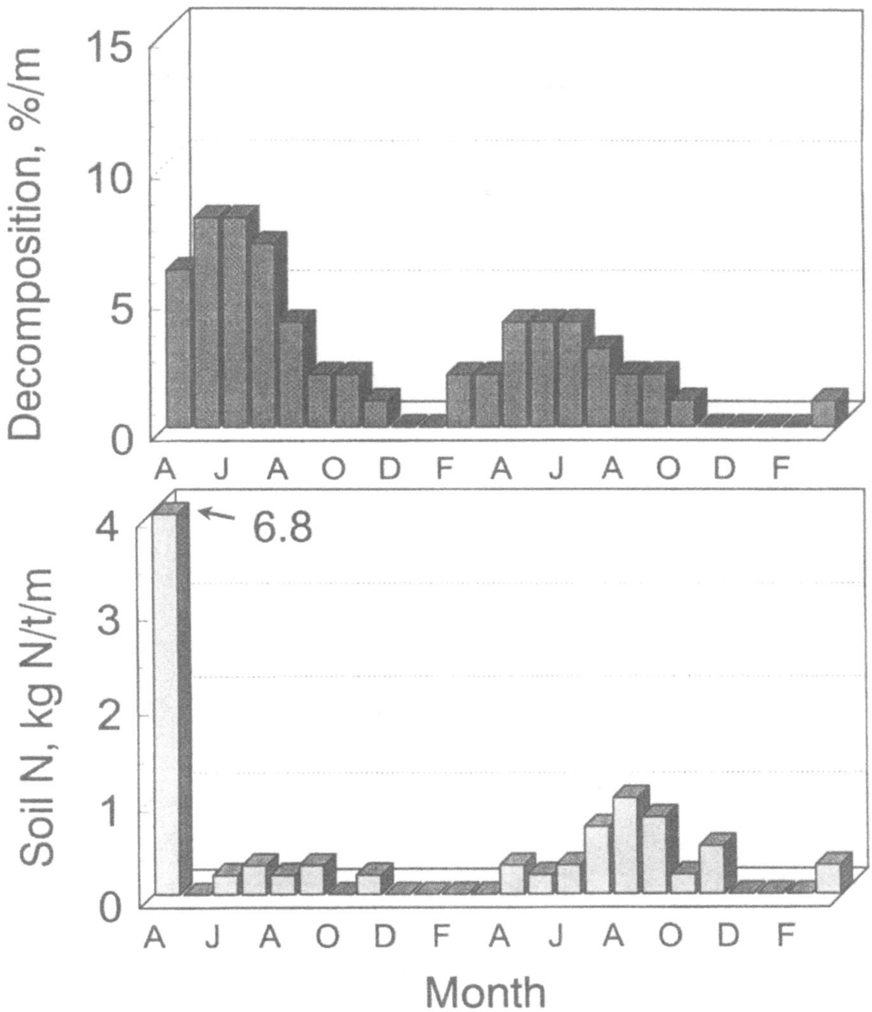

Figure 9. Simulation results for a 3:1 leaf litter:municipal biosolid mixture. Soil N is the difference between N mineralization and N immobilization.

during the April to September period over the two years.

The final mixture was MSW compost and municipal biosolids (Figures 10a and 10b). The ratio that resulted in a prediction of no net N immobilization, i.e., no negative monthly PAN values was 8.2:1. Decomposition of the simulated mixture ranged from 5 to 8% per

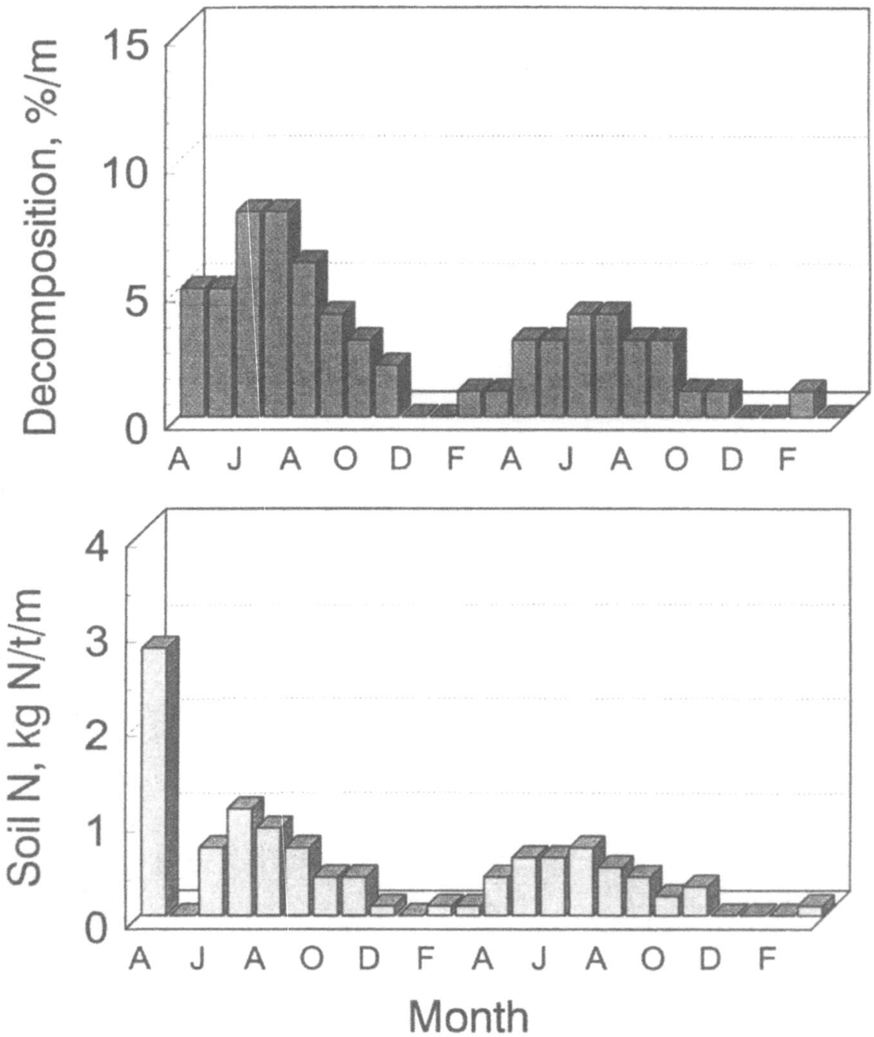

Figure 10. Simulation results for a 8.2:1 MSW compost:municipal biosolid mixture. Soil N is the difference between N mineralization and N immobilization.

month for the April to September period the first year. Monthly PAN followed a similar pattern to that of the leaf litter:municipal biosolids case with two notable exceptions. First, April PAN was much smaller (2.8 versus 6.8 kg N t^{-1} mixture). Second, May to September monthly PAN was much larger (3.4 versus 1.0 kg N t^{-1} mixture). Each of these features was due to the smaller C:N ratio of MSW compost as compared to leaf litter which allowed a smaller proportion of municipal biosolids in the mixture. During the April to September period the second year, the MSW compost:municipal biosolids mixture was predicted to decompose at monthly rates of 3 to 4%. Estimated PAN was 0.4 to 0.7 kg N t^{-1} mixture for a total of 3.2 kg N t^{-1} mixture during the growing season.

Overall, the computer model predicted that 66% of the MSW compost:municipal biosolids mixture would decompose over the two year period with total PAN production of 11.1 kg N t^{-1} mixture. Fifty-six percent of this PAN was available during the first growing season, while 29% was available the second growing season for a total of 85% during the two year period. The amount of predicted decomposition was sufficient to limit the effectiveness of the MSW compost:municipal biosolids mixture as a soil amendment to improve soil organic matter. The mixture does appear to have potential as a N source.

Computer Simulations - Soil Organic Matter and PAN

The mixtures simulated above were sufficiently decomposable to limit their usefulness as soil amendments designed to increase soil organic matter. As a result the organic waste most resistant to decomposition, very stable compost, was evaluated as a potential component of a mixture. Since the very stable compost had a C:N ratio of 10.1 and would not immobilize soil inorganic N, mixtures with other organic wastes were not considered. The various sources of these very stable composts are listed in Appendix Table 1.

Figure 11a presents the results for computer simulation of decomposition following multiple additions of stable and very stable composts. The stable compost was predicted to accumulate at about 2.4 times the annual addition rate and reach a near equilibrium value after 9 to 10 years. The very stable compost predictions showed a much higher accumulation ratio, 8.7 after 20 years, that did not appear to be asymptotic. Hamaker (1966) presented a formula that can be used to estimate the asymptote based on first year decomposition (asymptote = 1/(1-f), where f is the fraction left after one year). That equation yielded asymptotes for the stable and very stable composts of 2.4 and 10, respectively. These results suggest that very stable composts are much better candidates for increasing soil organic matter than stable compost counterparts.

Figure 11b presents the results of the computer simulations for PAN additions that would accrue at the accumulation ratios presented in Figure 11a. Relatively, large amounts of PAN would be available from multiple additions of the stable compost, while much smaller amounts of PAN were predicted for the very stable compost when comparing equal, annual addition rates. The asymptotic value for the stable compost was about 70 kg PAN t^{-1}, while the value for the very stable compost would approach 20 kg PAN t^{-1} using the organic N contents in Table 1. Thus, the stable compost was predicted to be a good N source, while the very stable compost was predicted to be a good source of both soil organic matter and, after several years of application, PAN. The reader is reminded that simulation results shown in Figures 8-11 would only be valid for sites with weather similar to Baltimore, MD.

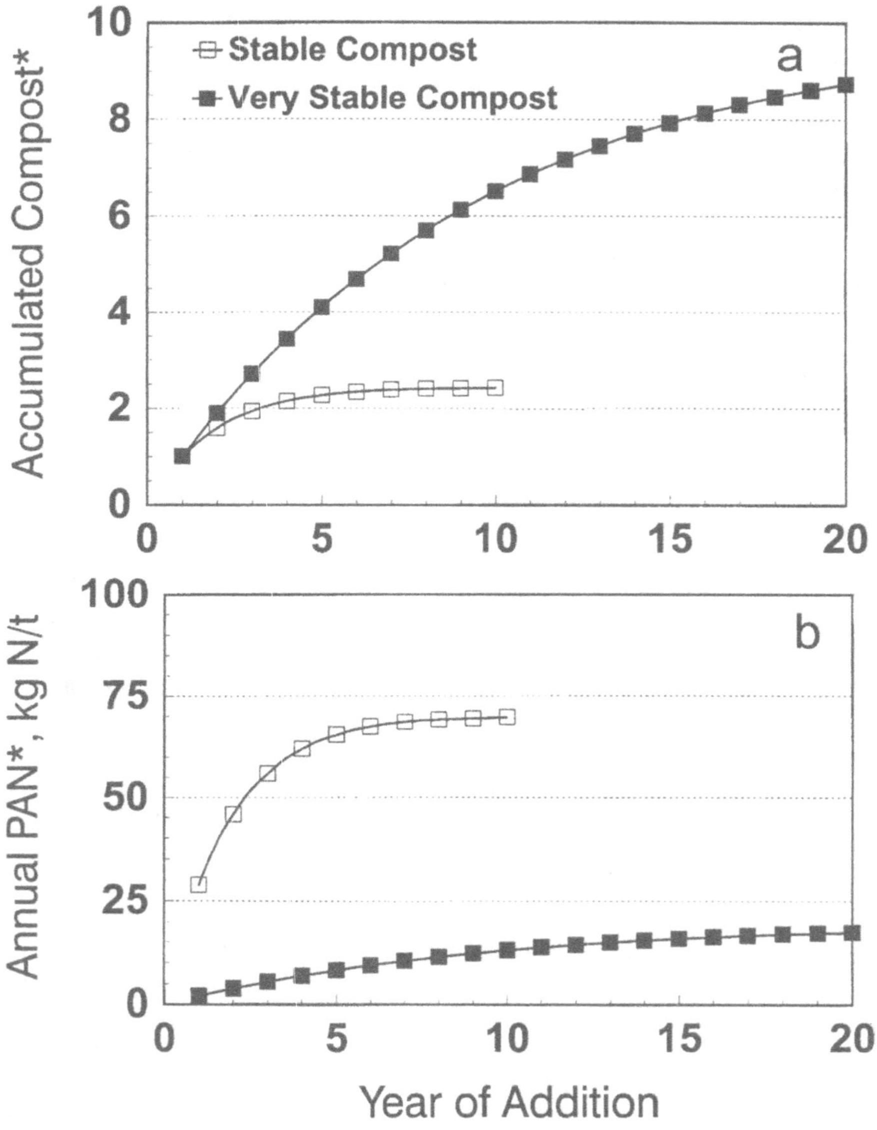

Figure 11. Computer simulation of: a) the amount of accumulated compost and b) the annual plant available N immediately after an annual addition.

RECOMMENDATIONS

When determining whether or not co-utilization will produce a value added product, several steps can be taken to assist in the decision-making process, as follows:

- Determine what properties are desired in the final product:
  - Source of plant available N, or
  - Source of soil organic matter, or
  - Both.
- Evaluate the components of the mixture for:
  - Decomposability and
  - Net N mineralization or immobilization.
- Create mixtures and test using:
  - Computer models, and/or
  - Laboratory and field studies.

Throughout this process, it is important to recognize and estimate variability in the performance of the mixture due to component variability and differences in weather/soil from one location to another.

REFERENCES

Agbim NN, Sabey BR & Markstrom DC (1977) Land application of sewage sludge V. Carbon dioxide production as influenced by sewage sludge and wood waste mixtures. J. Environ. Qual. 6:446-451.

Blackmer AM & Green CJ (1995) Nitrogen turnover by sequential immobilization and mineralization during residue decomposition in soils. Soil Sci. Soc. Amer. J. 59:1052-1058.

Briones MJI & Ineson P (1996) Decomposition of Eucalyptus leaves in litter mixtures. Soil Biol. Biochem. 28:1381-1388.

Clark,MD, Gilmour JT & Gale PM (1988) Computer simulations of decomposition and net N mineralization for land applied papermill sludge. Proceedings 1988 TAPPI Environmental Conference, Charleston, SC.

Cortez J, Denard JM, Bottner P & Monrozier LJ (1996) Decomposition of Mediterranean leaf litters: A microcosm experiment investigating relationships between decomposition rates and litter quality. Soil Biol. Bioch. 28:443-452.

Gilmour CM, Broadbent FE & Beck SM (1977) Chapter 7: Recycling carbon and nitrogen through land disposal of various wastes. In: Elliot LF & Stevenson FM (Eds) Soils for Management of Organic Wastes and Waste Waters. Soil Sci. Soc. Amer.

Gilmour JT, Clark MD & Sigua G. (1985) Estimating net nitrogen mineralization from carbon dioxide evolution. Soil Sci. Soc. Amer. J. 49:1398-1402.

Gilmour JT, Clark MD & Daniel SM (1996a) Predicting long-term decomposition of biosolids with a seven day test. J. Environ. Qual. 25:766-770.

Gilmour JT, Clark MD & Roman F (1996b) Decomposition of biosolids in a disposal site soil. J. Environ. Qual. 25.

Hamaker JW (1966) Mathematical prediction of cumulative levels of pesticides in soil. In: Gould RF (Ed) Organic Pesticides in the Environment. Advances in Chemistry Series 60:122-131.

Hadas A & Portnoy A (1994) Nitrogen and carbon mineralization rates of composted manures incubated in soil. J. Environ. Qual. 23:1184-1189.

Hanks RJ & Ashcroft GL (1980) Applied Soil Physics. p. 44. Springer-Verlag. New York.

He X, Logan TJ & Traina SJ (1995) Physical and chemical characteristics of selected U.S. municipal solid

waste composts. J. Environ. Qual. 24:543-552.
Heckman JR & Kluchinski D (1996) Chemical composition of municipal leaf waste and hand-collected urban leaf litter. J. Environ. Qual. 25:355-362.
Henry CL & Harrison RB (1996) Compost fractions in compost and compost maturity tests. In: Magdoff FR, Tabatabai MA & Hanlon EA Jr. (Eds) Soil Organic Matter: Analysis and Interpretation. Soil Science Society of America Special Publication No. 46:51-67. Madison, WI.
Honeycutt CW, Zibilske LM & ClamphamWM (1988) Heat units for describing carbon mineralization and predicting net N mineralization. Soil Sci. Soc. Amer. J. 52:1346-1350.
Insam H (1990) Are the microbial biomass and basal respiration governed by climatic regime? Soil Biol. Biochem. 22:525-532.
Jenkinson DS & Rayner JH (1977) The turnover of soil organic matter in some of the Rothamsted classical experiments. Soil Sci. 123:298-305.
Keeney DR & Nelson DW (1982) Nitrogen - inorganic forms. In: Page AL et al. (Eds) Methods of soil analysis. Part 2. 2nd ed. Agron. Monograph No. 9:643-698. Am. Soc. Agron. and Soil Sci. Soc. Amer., Madison, WI.
Ladd JN, Oades JM & Amato M (1981) Microbial biomass formed from ^{14}C and ^{15}N-labelled plant material decomposing in soils in the field. Soil Biol. Biochem. 13:119-126.
Molina JAE, Hadas & & Clapp, CE (1990) Computer simulation of nitrogen turnover in soil and priming effect. Soil Biol. Biochem. 22:349-353.
Moorhead KK, Graetz DA & Reddy KR (1987) Decomposition of fresh and anaerobically digested plant biomass in soil. J. Environ. Qual. 16:25-28.
Nelson DW & Sommers LE (1982) Total carbon, organic carbon, and organic matter. In: Page AL et al. (Eds) Methods of soil analysis. Part 2. 2nd ed. Agron. Monograph No. 9:539-557. Am. Soc. Agron. and Soil Sci. Soc. Amer., Madison, WI.
Paul EA & Van Veen JA (1978) The use of tracers to determine the dynamic nature of organic matter. Trans. Int. Congr. Soil Sci. 3:61-102.
Sabey BR, Agbim NN & Markstrom DC (1975) Land application of sewage sludge: III. Nitrate accumulation and wheat growth resulting from addition of sewage sludge and wood wastes to soils. J. Environ. Qual. 4:388-393.
Sikora LJ, Cambardella CA, Yakovchenko V & Doran JW (1996) Assessing soil quality by testing soil organic matter. In: Magdoff FR, Tabatabai MA & Hanlon EA Jr. (Eds) Soil Organic Matter: Analysis and Interpretation. Soil Science Society of America Special Publication No. 46:41-50. Madison, WI.
Sommers LE, Gilmour CM, Wildung RE & Beck SM (1980) Chapter 3: The effect of water potential on decomposition processes in soil. In: Parr JF et al. (Eds) Water Potential Relationships in Soil Microbiology. Soil Sci. Soc. Amer., Madison, WI.
Tester CF, Sikora LJ, Taylor JM & Parr JF (1977) Decomposition of sewage sludge compost in soil: I Carbon and nitrogen transformations. J. Environ. Qual. 6:459-463.
Tester CF, Sikora LJ, Taylor JM & Parr JF (1979) Decomposition of sewage sludge compost in soil III: Carbon, nitrogen and phosphorus transformations in different sized fractions. J. Environ. Qual. 8:79-82.
Zibilske LM (1987) Dynamics of nitrogen and carbon in soil during papermill sludge decomposition. Soil Sci. 143:26-33.
Zibilske LM (1994) Carbon mineralization. In: Mickelson SHd.) Methods of Soil Analysis. Part 2. Microbiological and Biochemical Properties. Soil Sci. Soc. Amer. Book Series No. 5:835-863. Soil Sci. Soc. Amer., Madison, WI.

APPENDIX

Laboratory Studies

Three laboratory studies were conducted to provide information for Appendix Table 1 In each representative samples of the organic waste were collected and stored under refrigeration (4°C) until the experiment was initiated. Surface soil was collected from each land application site. Organic wastes were a primary paper mill sludge (15 t ha^{-1}) and dairy manure (15 t ha^{-1}) alone and in combination (30 t ha^{-1}); composted municipal biosolids (4 t ha^{-1}); and biosolids/wood waste composts (10 t ha^{-1}) for Studies I to III, respectively. Soil series were Onaway sl (Typic Eutroboralf), Maury sil (Typic Paleudalfs), and Willacy l to scl (Udic Argiustoll), respectively. The soil was stored at room temperature. Storage times for organic wastes and soils were less than two weeks. Soil was sieved (~7 mm openings) to remove stones and large pieces of plant material if present.

Percent solids of each organic waste was determined by weight loss during low temperature (45°C) drying. Total N in the dried samples was determined by either Kjeldahl total N (Keeney and Nelson, 1982) or LECO N Determinator. Inorganic N in the dried samples was extracted with 1 or 2 *M* KCl and determined by steam distillation (Keeney and Nelson, 1982). Organic N was the difference between total N and inorganic N for the dried samples. Actual inorganic N for each organic waste was determined on the original moist sample by the KCl extraction/distillation method. Organic C was determined by dry combustion followed by trapping CO_2 in ascarite as outlined by Nelson and Sommers (1982). The results of these analyses are presented in Appendix Table 1.

To initiate a decomposition study, the organic waste was incorporated into 100 g (dry) of the soil and incubated at 25°C under optimum soil moisture (40 to 60% of water holding capacity). Unamended soils served as controls. Evolved CO_2-C was trapped in 1 *M* NaOH which was periodically analyzed by weak acid titration after the addition of 1.5 *M* $BaCl_2$ (Zibilske, 1994). Each treatment was conducted in duplicate or triplicate.

Study II used a flow through system where CO_2 free air was passed over the organic waste/soil mixture and into the NaOH trap, while Studies I and III utilized a static system composed of a 950 ml incubation vessel containing the NaOH trap. The static system was aerated periodically to maintain aerobic conditions. Incubation times were 60, 60 to 63, and 59 d for Studies I, II, and III, respectively. At the end of each incubation, the organic waste/soil mixture and controls were extracted with 1 or 2 *M* KCl and inorganic N determined as before.

Mathematical Description of Decomposition Kinetics

Organic waste decomposition was assumed to follow sequential first order kinetics where one or two fractions, with respect to decomposability (rapid and slow), existed. In sequential decomposition, the rapid fraction was assumed to completely decompose prior to the initiation of slow fraction decomposition. The rapid and slow fractions were described using Appendix Eq. 1, with each being a separate linear segment with a characteristic slope and intercept.

$$\ln(\%C_{remaining}) = -kt + b \qquad [1]$$

where $\%C_{remaining}$ was the percentage of organic waste C remaining in the soil at time t, and k was the first order rate constant. The percentage in the rapid fraction was defined by the intersection of the two linear segments which was determined by equating the linear equations, solving for the time, and reinserting that time into one of the equations. The parameters in Appendix Eq. 1 were estimated using non-linear regression (JMP Version 3.1.5, SAS Institute, Inc.) for Studies I and III as described in detail by Gilmour et al.

(1996b). Since single rate constants were obtained for Study II, linear regression was used to determine k_s (JMP Version 3.1.5, SAS Institute, Inc.). The results of the kinetic analyses are presented in Appendix Table 1.

Several other studies were also evaluated (Agbim et al., 1977; Honeycutt et al., 1988; Tester el al., 1977; Tester et al., 1979; Zibilske, 1987). Percent decomposition was estimated from data presented in each manuscript. The first rate order constant(s) and percentage in the rapid fraction were then calculated using non-linear regression, as described above. For leaf litters (Cortez et al., 1996; Briones and Ineson, 1996), first order rate constants were estimated from mass loss data. In this treatment the fractional mass loss divided by the time period was used to estimate the single, first order rate constant. In all cases where temperature was not 25°C, rate constants were adjusted to 25°C by assuming a Q_{10} of 2.

Appendix Table 1. Characteristics of selected organic wastes.

Source	Organic C	Organic N	C:N	Inorganic N	Sequential Decomposition Model Parameters[a]			Source
					k_r	k_s	Rapid Fraction	
	g kg^{-1}	g kg^{-1}		mg kg^{-1}	d^{-1}	d^{-1}	%	
				Municipal Biosolids				
Several	273	35.6	7.8	22,800	0.025	0.00287	28	Gilmour et al. (1996a)
				Paper Mill Sludges				
Secondary, liquid	292	43.4	6.7	500	0.010	<0.0001	32	Clark et al. (1988)
Combined, 50% primary, 30% secondary, 20% Zimpro	161	9.7	16.6	1,460	0.015	0.0019	40	Clark et al. (1988)
Unknown	346	12	28.8	56	0.032	0.0043	58	Honeycutt et al. (1988)
Primary	331	6.7	49.4	10	0.021	0.010	42	Study I
Combined, 90% primary 10% secondary	262	2.2	119	58	0.014	0.0062	46	Clark et al. (1988)
Kraft Process	383	0.8	479	---	------	0.0010	---	Zibilske (1987)
				Leaf Litter				
Sessile Oak	447	7	63.9	---	------	0.0016	---	Cortez et al. (1996)
Beech	448	7	64.0	---	------	0.0013	---	Cortez et al. (1996)
Sweet Chestnut	458	4	114	---	------	0.0015	---	Cortez et al. (1996)
Holm Oak	452	8	56.5	---	------	0.0013	---	Cortez et al. (1996)
Eucalyptus	---	13	---	---	------	0.0040	---	Briones & Ineson (1996)
Sessile Oak	---	10	---	---	------	0.0031	---	Briones & Ineson (1996)
Common Ash	---	22	---	---	------	0.0064	---	Briones & Ineson (1996)
Silver Birch	---	14	---	---	------	0.0039	---	Briones & Ineson (1996)

			Compost and Other Stabilized Organic Wastes					
Municipal biosolids	190	11.2	17.0	---	0.0066	0.0025	13	Tester et al. (1977)
Municipal biosolids	181	12.3	14.7	---	------	0.0025	---	Tester et al. (1979)
Municipal biosolids	216	11.6	18.6	---	------	0.0025	---	Tester et al. (1979)
Municipal biosolids	140	13.5	10.4	---	------	0.0023	---	Tester et al. (1979)
Municipal biosolids 30 d storage	291	22.6	13.0	400	------	0.00091	---	Study II
Municipal biosolids 60 d storage	284	19.4	14.6	1,100	------	0.00061	---	Study II
Municipal biosolids 1 yr storage	286	19.5	14.7	1,300	------	0.00014	---	Study II
Municipal biosolids 3 yr storage	281	17.7	15.9	<100	------	0.00041	---	Study II
Cattle manure/ poultry litter	80	8	10	---	------	0.00028	---	Hadas and Portnoy (1994)
Cattle manure / poultry litter	165	22	7.5	---	-----	0.00028	---	Hadas and Portnoy (1994)
Cattle manure/ poultry litter	161	26	6.2	---	-----	<0.00001	---	Hadas and Portnoy (1994)
Earthworm humus	202	23	8.8	---	-----	0.00037	---	Hadas and Portnoy (1994)
Municipal Biosolids/ wood waste	237	24.2	9.8	3,700	0.0022	0.00062	3.8	Study III
Municipal Biosolids/ wood waste	221	35.3	6.3	6,800	0.0014	0.00047	3.4	Study III
Municipal solid waste mean values	213	12	17.8	---	-----	------	---	He et al. (1995)
Dairy manure	223	18.4	12.1	1,370	0.024	0.0066	38	Study I

[a] k_r and k_s are the rapid and slow fraction rate constants, respectively obtained directly from the source or computed using decomposition data. % Rapid Fraction is the amount of organic C undergoing decomposition using first order kinetics and k_r as the rate constant. All values adjusted to 25°C ($Q_{10} = 2$) and optimum water content.

SUPPRESSION OF SOIL-BORNE PATHOGENS BY COMPOSTED MUNICIPAL SOLID WASTE

R. COHEN[1], B. CHEFETZ[2] and Y. HADAR[1]
[1]*Dept. of Plant Pathology and Microbiology;* [2]*The Seagram Center for Soil and Water Sciences, Faculty of Agricultural, Food and Environmental Quality Sciences, The Hebrew University, Rehovot, Israel*
E mail hadar@agri.huji.ac.il

ABSTRACT

The objectives of this study were to determine the level of suppressiveness of composted municipal solid waste (CMSW) against soil-borne plant pathogens, and to investigate the mechanism involved in this process. The suppressiveness of CMSW to *Rhizoctonia solani* and *Fusarium oxysporum* f.sp. *vasinfectum* was studied in greenhouse experiments using sandy soil amended with different levels of CMSW. *R. solani* and *F. oxysporum* f.sp. *vasinfectum* disease severity in cotton plants was reduced in the presence of mature compost by 80 and 95%, respectively. After 1.5 months of composting, CMSW still did not suppress *R. solani,* whereas 2 week-old compost already exhibited suppression of *F. oxysporum* f.sp. *vasinfectum.* Reducing the microbial population of CMSW by autoclaving, gamma-irradiation or heat treatment negated *R. solani* suppression. Thus, we suggest that the mechanism involved in the observed suppression is biological. This hypothesis is supported by the fact that the addition of 5% CMSW to heat-treated CMSW resulted in recolonization of antagonistic microorganisms and restored the compost's suppressive ability. On the other hand, heat treatment of CMSW did not eliminate its ability to suppress *F. oxysporum* f.sp. *vasinfectum.* The mechanism governing the suppression of *R. solani is* suggested to involve lysis of fungal hyphae, mostly by the bacterial population, as observed by microscopic examination. Some bacteria isolated from the CMSW showed chitin-degradation ability and the capacity to inhibit *R. solani* growth. One of these isolates, which suppressed *R. solani* in cotton plants in greenhouse experiments, was identified as *Enterobacter agglomerans.* Inoculation of heated compost with *Trichoderma harzianum* suppressed *R. solani* significantly. However, the addition of *T. harzianum* did not improve the suppressive ability of the mature compost, despite the fact that this compost was successfully colonized.

INTRODUCTION

Compost is the stable product of aerobic microbial decomposition of heterogeneous

S. Brown, J.S. Angle and L. Jacobs (eds.), Beneficial Co-Utilization of Agricultural, Municipal and Industrial By-products, 113–130.
© 1998 *Kluwer Academic Publishers,*

organic matter (Hoitink et al., 1993). Various organic wastes such as grape marc, cattle manure, hardwood or pine bark, yard trimmings, municipal solid waste (MSW) and biosolids can be used as raw materials for composting (Bruns et al., 1996; Chef et al., 1983; Hoitink et al., 1993; Mandelbaum et al., 1988). Composting of the organic fraction of MSW can reduce up to 50% of the waste volume reaching landfills and produces a stable product that can be used in agriculture (Chen and Hadar, 1987; Inbar et al., 1993).

The rate of organic matter decomposition and compost maturity affect the compost's ability to suppress soil-borne plant pathogens (Hadar and Mandelbaum, 1992; Inbar et al., 1993). In recent years, compost amendments in agriculture have been investigated as part of an integrated system of biological control due to their ability to suppress soil-borne plant pathogens such as *Pythium aphanidermatum, Rhizoctonia solani, Sclerotium rolfsii, Phytophthora cinnamomi* and *Fusarium oxysporum* f.sp. *vasinfectum* (Chef et al., 1983; Chen et al., 1988; Gera and Wilma, 1996; Hadar and Gorodecki, 1991; Hadar and Mandelbaum, 1986, 1992; Hoitink et al., 1996; Kuter et al., 1988; Mandelbaum and Hadar, 1990)

Suppressive composts can be defined as those which provide an environment in which disease development is reduced, even when the pathogen is introduced in the presence of a susceptible plant (Baker, 1987). Plant-disease suppression is a direct result of the activity of antagonistic microorganisms, which naturally recolonize compost during the cooling phase of the composting process (Hoitink et al., 1993). Thus, compost sterilization generally negates suppression. Although the mechanism of suppression is biological in all cases, different specific mechanisms have been observed in relation to various plant diseases (Hoitink et al., 1993).

Hadar et al. (1992) described the suppression of soil-borne pathogens such as *P. aphanidermatum, R. solani* and *S. rolfsii* by composted grape marc and composted separated cattle manure. The general mechanism of suppression was presumed to be biological in all cases. However, different specific mechanisms were observed. Suppression of *P. aphanidermatum* in composted separated cattle manure was presumed to be due to competition for nutrients, while suppression of *S. rolfsii* in composted grape marc was caused by direct parasitism of germinating sclerotia by antagonistic fungi (Hadar et al., 1992).

Mandelbaum and Hadar (1990) described the effect of carbon-source concentration and microbial activity on the suppression of P. *aphanidermatum* in cucumber plants. Incorporation of a glucose/asparagine mixture (C:N ratio = 10:1) into container media resulted in a rapid increase in microbial respiration rate and enzymatic activity in composted separated cattle-manure medium, but not in peat medium. Hyphae of *P. aphanidermatum* grown on nylon fabric and buried in container media were rapidly lysed in compost medium as compared to peat-based medium (Mandelbaum and Hadar, 1990). Suppression of the Pythium disease creeping bentgrass by brewery sludge and biosolids composts is directly related to the microbial activities in the composts. On the other hand, in poultry-based compost, the suppression is not directly related to the compost microbial activity, and thus microbial activity may not be predictive of suppressiveness of composts (Craft and Nelson, 1996). Zhang et al. (1996) showed that compost (pine or spruce bark)-amended potting mixes suppress Pythium root rot and anthracnose in cucumber foliage. The mechanism involved in this case was suggested to be systemic acquired resistance

(SAR). Peroxidase activity was increased in plants grown in compost suppressive to root rot, as compared to plants grown in a sphagnum peat mix conducive to root rot. The interaction between compost and pathogen appeared critical for rapid activation of SAR-associated gene expression in cucumber plants (Zhang et al., 1996).

Media amended with composted grape marc and composted separated manure were effective in suppressing diseases caused by *S. rolfsii* and *R. solani* (Hadar and Gorodecki, 1991). *S. rolfsii* sclerotia were placed on composted grape marc to isolate mycoparazites of this pathogen. Sclerotial viability decreased from 100% to less than 10% within 40 h. In contrast, viability of sclerotia placed on a conducive peat mix remained close to 100%. *Penicillium* and *Fusarium* species were observed by scanning electron microscopy to colonize the sclerotia. The *Trichoderma* population in the grape marc compost was at a very low level (10^2 cfu/g dry weight) (Hadar and Gorodecki, 1991).

In other composts, specific suppression has been found to be related to a large *Trichoderma* population as well as to the development of antagonistic bacteria. Nelson et al. (1983) identified specific strains of four *Trichoderma* spp. and isolated *Gliocladium virens* as the most effective fungal mycoparazites of *R. solani* present in bark compost. A few of the 230 other fungal species also exhibited activity, but most were ineffective. Kwok et al. (1987) described synergistic interactions between *T. hamatum* and *Flavobacterium balustinum.* Several other bacterial strains, including *Enterobacter, Pseudomonas,* and *Xanthomonas* species, also interacted with the *Trichoderma* isolate in suppressing Rhizoctonia damping-off (Kwok et al., 1987).

The objectives of this study were to examine the ability of composted MSW (CMSW) to suppress the soil-borne pathogens *R. solani* and *F. oxysporum* f.sp. *vasinfectum* in the soil and to investigate mechanisms involved in the suppression.

MATERIALS AND METHODS

Composting of MSW

MSW was collected over a 24-h period from the town of Afula in Israel, and mechanically mixed in a MSW classifying facility. After sieving and eliminating the >5-cm fraction, the raw material was divided into three perforated plastic boxes (about 1 m^3 each) and composted for 187 days. The compost was turned, mixed, and sampled after removal of glass, plastics and metals by hand-picking at 14, 30, 45, 75, 120 and 160 days. Samples (10 L each) were stored at 4°C. Smaller subsamples were dried at 65°C, then ground and sieved (Chefetz et al., 1996).

Mature (6-month-old) CMSW was taken from Amnir Afula Ltd., Israel.

Fungal Isolation

Rhizoctonia solani Kuhn was isolated from field-grown diseased cotton seedlings on a potato dextrose agar (PDA) medium. *Fusarium oxysporum* f.sp. *vasinfectum* (Ask) and Hans. was isolated from field-grown diseased cotton seedlings (Sivan and Chet, 1986). An isolate of *Trichoderma harzianum* (Y) was isolated from the rhizosphere of Fusariumwilted cotton seedlings on a medium selective for *Trichoderma* spp. (TSM) (Elad et al., 1981).

Greenhouse Experiments

Ten cotton seeds *(Gossypium barbardense* L. Pima- 177) were planted at a depth of 1 cm in polypropylene pots containing 500 ml infested soil. Sandy loamy soil (82.3% sand, 2.3% silt, 15.0% clay, 0.4% organic matter, pH 7.2) was mixed with various amounts of CMSW. All experiments were conducted under controlled conditions in the greenhouse (27°C, 14-h illumination). In greenhouse bioassays, each replicate consisted of a pot containing 10 seeds, six pots per treatment. Each experiment was performed twice, and the data points presented are means of diseased plants.

Suppressiveness to *R. solani* damping-off was determined using cotton plants. Soil inoculum of *R. solani* was produced in Ko and Hora's (1971) chopped potato-soil mixture. The amount of inoculum added to the soil resulted in 40-70% diseased plants in the control. After 14 days, the percentage of diseased plants and the disease index (0--no disease, 5--dead plant) were determined.

Suppressiveness to *F. oxysporum* f.sp. *vasinfectum* damping-off was determined with cotton plants. Soil inoculation of *F. oxysporum* f.sp. *vasinfectum* was prepared as described by Sivan and Chet (1986). The amount of inoculum added to the soil gave 40-70% diseased plants in the control. After 21 days, the percentage of diseased plants was determined. The percentage of surviving seedlings was calculated 21 days post-planting.

Antagonistic bacteria isolated from the compost were used in the biocontrol assay. The bacteria were cultured for 48 h in nutrient broth (Difco) as shake cultures (50 ml in a 250-ml flask, 120 rpm, 30°C), washed by centrifugation and resuspended in water to population levels of approximately 10^8 cfu/ml. These cell suspensions were added to a sandy soil medium.

Inoculum of *T. harzianum* was prepared on bran-peat medium as described by Elad et al. (1981). This preparation was added to plastic bags containing 3 L of soil and incubated for 3 days. The population of *T. harzianum* added to the soil reached a level of 10^4 cfu/g soil (dry weight).

Isolation and Identification of Bacterial Antagonists

Bacteria were isolated from nylon meshes (described in 'lysis' section) on nutrient agar (Difco) medium and identified by analysis of their fatty-acid profiles by Microbial Identification System (MIDI DOS System Version 4.15, Dept. of Plant Protection, Ministry of Agriculture, Israel).

Sterilization and Heat Treatments

Four different treatments were used to reduce the microbial population in the compost: 1) 2 L of medium (soil+10% CMSW, 50% moisture) were incubated in an oven at 60°C for 5 days (HC); 2) 3 L of medium (soil+10% CMSW) stored in polypropylene bags were autoclaved at 121°C for 1 h on three consecutive days (AC); 3) autoclaved compost was washed with five volumes of sterilized distilled water to remove the soluble compounds (AWC); 4) CMSW samples (1 L per polypropylene bag) were irradiated with 5 Mrad/h (Sorvan Inc., Nahal-Soreq, Israel) (GRC).

Microbial Activity

Microbial activity in the different media (described in "sterilization and heat treatments") was determined by three different methods: 1) CO_2 evolution--3 g samples (in triplicate) of wet media were placed in 25-ml glass bottles equipped with a gas-tight rubber cap. The bottles were flushed with fresh sterile air (0.035% CO_2 before sealing), then incubated at 30°C. CO_2 levels were determined by gas chromatograph (GC model 580, Gow-Mac Instrument Co., Bound Brook, NJ); 2) hydrolysis of fluorescein diacetate (FDA) was performed by the method of Shnürer and Rosswall (1982), with slight modifications: FDA (Sigma Chemical Co.) was dissolved in acetone (analytical grade) and stored as a stock solution (2 mg ml^{-1}) at -20°C. Media samples (1 g, wet weight) were dispersed in 100 ml of sterile buffer (60 mM potassium phosphate pH 7.6) and then amended with FDA to a final concentration of 10 μg FDA $^{ml-1.}$ The suspension was incubated in a 250-ml Erlemeyer flask at 30°C in a rotary shaker (120 rpm). The amount of fluorescein was measured as absorbance at 494 nm using an HP 8451A diode-array spectrophotometer (Hewlett Packard, Palo Alto, CA). FDA hydrolysis was measured as fluorescein units released from FDA within 2 h after adding FDA to the media suspension; 3) bacterial and fungal counts--viable bacteria were counted using a dilution-plating technique (three replicates) on compost or soil extracts (500 g of compost or soil autoclaved in 1 L of deionized water for 30 min. then sieved through cheesecloth and centrifuged 5 min at 6000 g; 100 ml of supernatant was amended with 1 g of glucose, 0.5 g of K_2HPO_4, 20 g of agar, to a final volume of 1 L with deionized water, and autoclaved). Total fungal cfu's were determined by dilution-plating on Martin's rose Bengal medium (Martin, 1950).

The population density of *T. harzianum* was determined by dilution-plating on selective medium according to Elad et al. (1981).

Lysis of Hyphae of *R. salami*

A modification of the nylon-fabric technique developed for *Pythium* by Lumsden (1980) and Mandelbaum and Hadar (1990) was used to support and recover *R. solani* hyphae. Nylon monofilament fabric (pore size of 100 μm; Swiss Silk Bolting Co. Ltd., Zurich), was cut into $5mm^2$ pieces and autoclaved (121°C, 20 min). Ten nylon pieces were placed on Petri dishes containing PDA and inoculated with a *R. solani* disk in the center. After 48 h of incubation, the nylon pieces containing *R. solani* mycelium were buried in plastic-cylinder vessels (5 cm diameter, 5 cm height) with nylon fabric covering the bottom. The vessels were incubated at 30°C for different times, then the nylon fabrics were taken for microscopic observation.

Statistical Analysis

Data analysis was performed by ANOVA. All significant differences were at P=0.05.

RESULTS

Effect of CMSW on Cotton-Plant Development

The effect of increasing levels of mature (6-months-old) CMSW in sandy soil on cotton-plant growth was studied. Addition of 5% CMSW to the soil (equivalent to about 150 tons per hectare, calculated for a depth of 25 cm) did not affect emergence rate relative to the control (100% soil). However, soil amended with 20, 40 and 80% CMSW exhibited 2-, 10- and 11-day delays in time to 50% emergence, respectively (Figure 1).

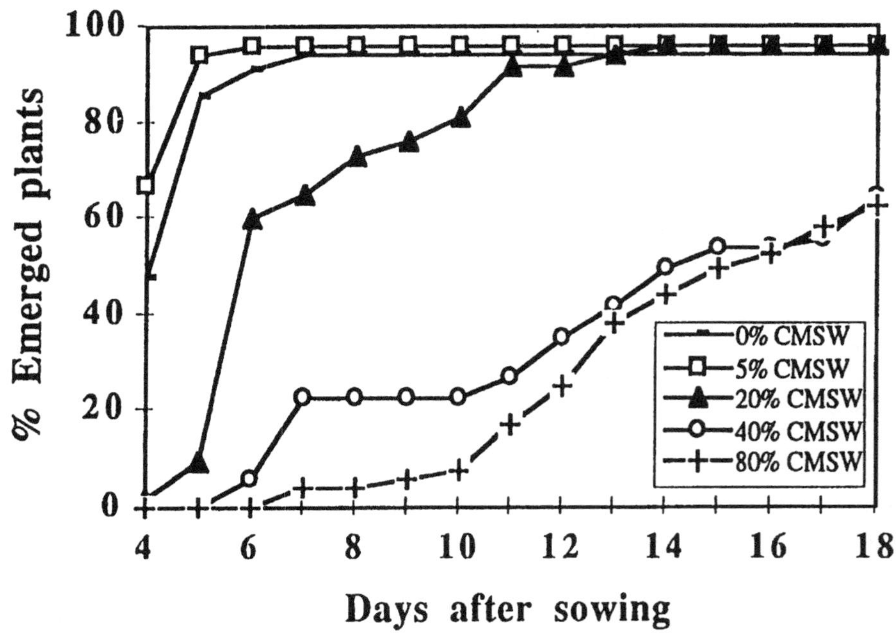

Figure 1. Effect of percent CMSW amended to sandy soil on cotton-plant emergence. cotton plants were grown for 18 days. Percent plant emergence was determined daily.

Figure 2 shows the effect of CMSW soil amendments on plant growth. The addition of 2 and 5% CMSW enhanced plant growth relative to the control. However, addition of 10% or higher resulted in significantly reduced cotton-plant height. The same trend was recorded with cotton-plant dry weight.

Effect of Compost Age and Level on Suppression of Soil-borne Diseases

Small-scale composting was performed to study the effect of compost age on its suppression of soil-borne diseases. CMSW samples taken at various stages of composting were added to soil (5%, v/v) inoculated with *R. solani* or *F. oxysporum* f.sp. *vasinfectum*. *R. solani* was only suppressed by CMSW 2.5-month or older (Figure 3). A different suppression trend was observed when *F. oxysporum* f.sp. *vasinfectum* was inoculated into

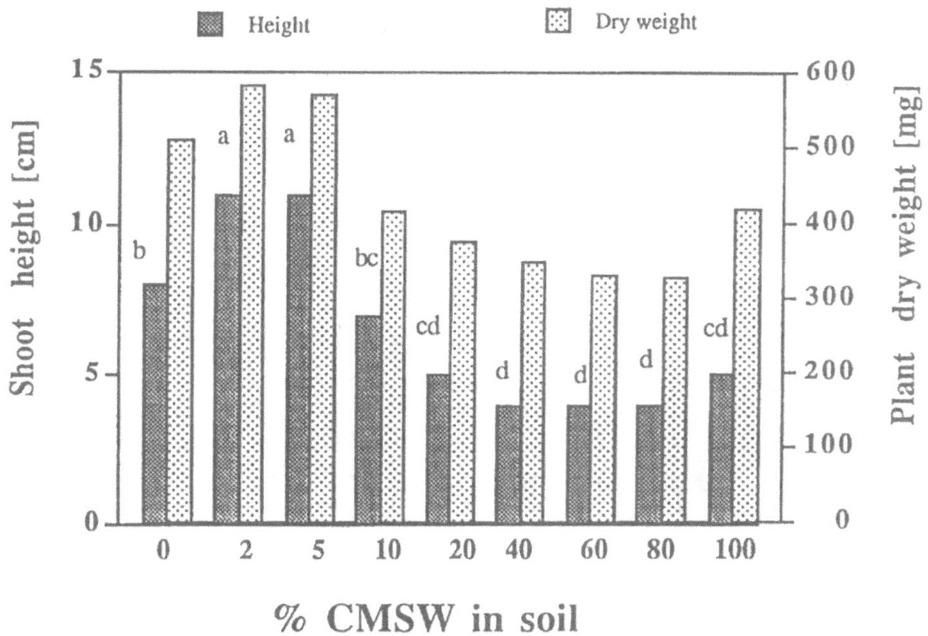

Figure 2. Effect of percent CMSW amended to sandy soil on cotton-plant height and weight. Different letters heading the columns indicate significant differences between treatments (P=0.05).

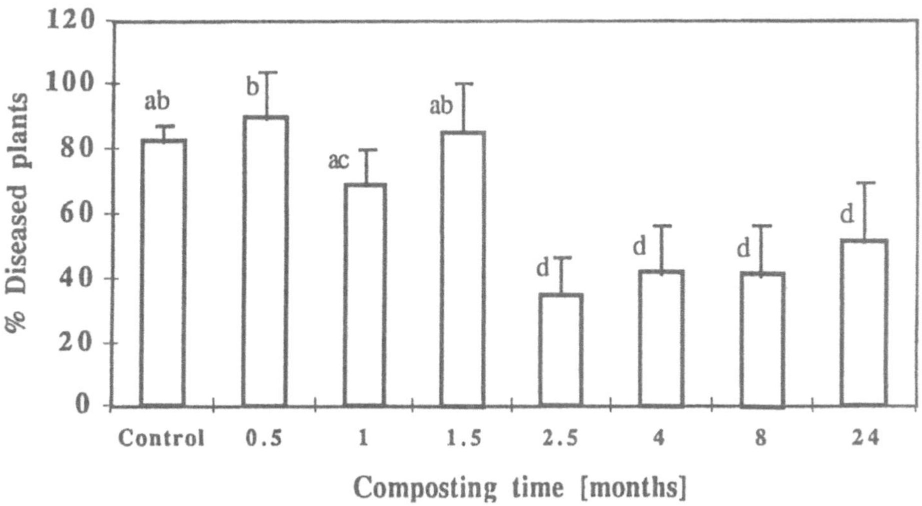

Figure 3. Effect of MSW composting time on the suppression of *R. solani* in cotton plants. CMSW samples taken at various stages of composting were added to soil inoculated with *R. solani*. Percent of diseased plants was determined 14 days after sowing. Control--sandy soil. Different letters heading the columns indicate significant differences between treatments (P=0.05).

soil amended with CMSW at different stage of composting (Figure 4). The level of disease, due to Fusarium was reduced in the presence of CMSW samples regardless of their composting age.

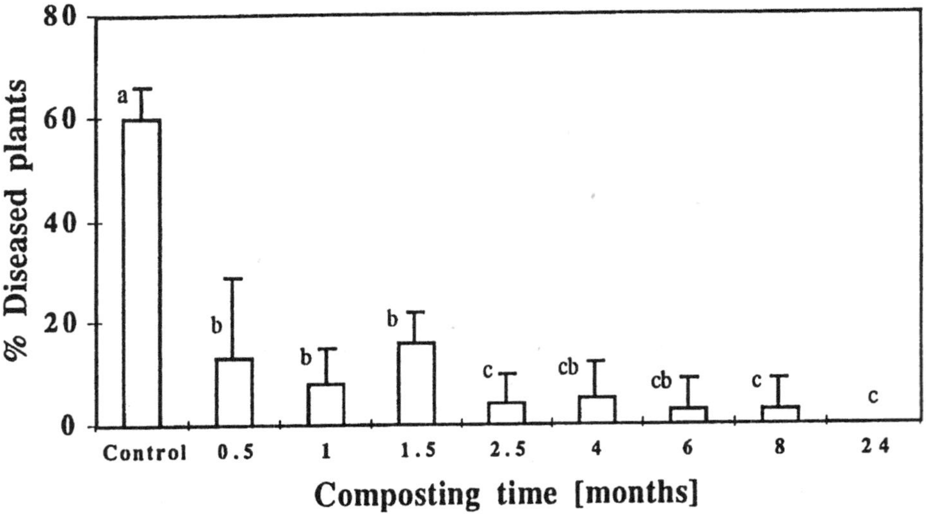

Figure 4. Effect of MSW composting time on the suppression of *F. oxysporum* f.sp. vasinfectum in cotton plants. CMSW samples taken at various stages of composting were added to soil inoculated with *F. oxysporum* f.sp. *vasinfectum*. Percentage of diseased plants was determined 21 days after sowing. Control-- sandy soil. Different letters heading the columns indicate significant differences between treatments (P=0.05).

The level of mature (6-month-old) compost required to suppress *R. solani* and *F.* oxysporum f.sp. *vasinfectum* was also studied under greenhouse conditions (Figs. 5, 6). *R. solani* disease incidence was reduced from 70% to less than 20% at all CMSW levels tested (2, 5, 10, 20 and 40% CMSW in soil). A similar trend was observed for disease severity (Figure 5B).

F. oxysporum f.sp. *vasinfectum* was tested over two growth cycles. During the first growth cycle, 93 and 97% disease reduction were observed with 5 and 10% CMSW, respectively (Figure 6). During the second growth cycle, disease incidence increased significantly in the 5% CMSW treatment, whereas this increase was not significant in the 10% CMSW treatment.

Mechanism of *R. solani* Suppression by CMSW

Three of the treatments used to reduce the microbial population in the CMSW negated suppression (Figure 7). Disease severity following autoclaved, autoclaved-washed and gammairradiation treatments were 92, 90 and 86%, respectively, as compared to the soil (control) and mature CMSW, which resulted in 86 and 15% disease severity, respectively.

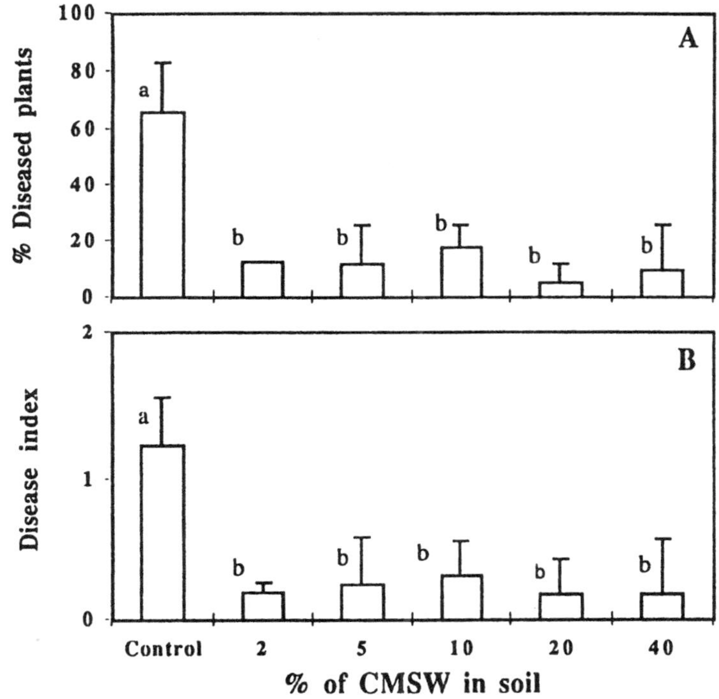

Figure 5. Suppression of *R. solani* in cotton plants by CMSW. a) Percentage of diseased plants, determined 14 days after sowing; b) disease index: disease severity ranked from 0-5, with 0--healthy plants, 5--dead plants. Control-- sandy soil. Different letters heading the columns indicate significant differences between treatments (P=0.05).

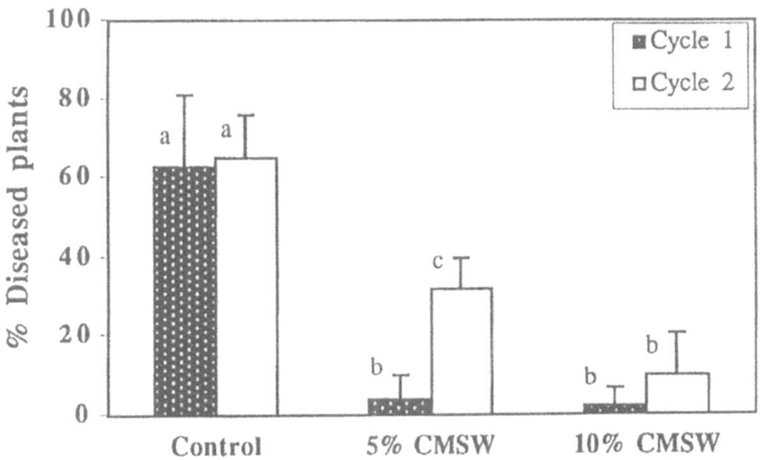

Figure 6. Suppression of *F. oxysporum* f.sp. *vasinfectum* in cotton plants by CMSW during two growth cycles. Percentage of diseased plants was determined 21 days after sowing. Control-- sandy soil. Different letters heading the columns indicate significant differences between treatments (P=0.05).

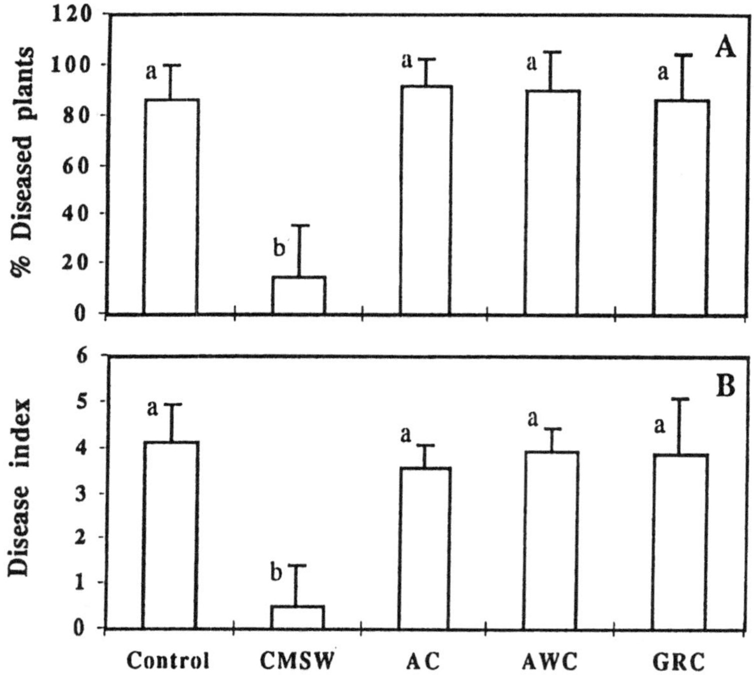

Figure 7. Effect of autoclaving and gamma-irradiating CMSW on suppression of *R. solani* in cotton plants. Control-- sandy soil; AC-- autoclaving of CMSW three times for 1 h each, at 121 °C; AWC-- autoclaved CMSW washed with five volumes of sterilized distilled water; GRC- gamma-irradiation (5 Mrad/h) of CMSW. A) Percentage of diseased plants determined 14 days after sowing; B) disease index: disease severity ranked from 0-5, with 0--healthy plants, 5--dead plants. Different letters heading the columns indicate significant differences between treatments (P=0.05).

Additional evidence for the involvement of the microbial population in the suppression of *R. solani* was provided by adding 5% (v/v) mature CMSW to the heat-treated CMSW. The latter also negated the compost's ability to suppress the pathogen. Adding 5% (v/v) mature CMSW to the heat-treated CMSW restored suppression to the level observed with mature CMSW (Figure 8). The incubation of heat-treated CMSW with 5% mature CMSW for 5 days resulted in a damping-off level of 43%, as compared to 95% with heat-treated CMSW alone.

Under the same conditions, we followed bacterial and fungal populations, as well as total microbial activity, as measured by FDA hydrolysis and respiration. The results presented in Table 1 demonstrate the differences between the CMSW treatments and between the fungal and bacterial populations. Heat treatment of CMSW resulted in reduced bacterial and fungal populations, from 10^8 to 10^3 cfu/g, and from 10^5 to less that 10^2 cfu/g, respectively. The inoculation of heat-treated CMSW with 5% mature CMSW restored bacterial and fungal populations to 10^8 and 10^3 cfu/g, respectively. The same treatment also restored the suppressiveness of CMSW against *R. solani*.

These observations are in agreement with the FDA hydrolysis results. The amount of

FDA hydrolyzed was 1.6 and 1.03, μg flourescein/g in the mature and heat-treated CMSW, respectively. Restoration of the CMSW microbial population after inoculation of the heat-treated CMSW with 5% mature CMSW resulted in increased FDA hydrolysis (Table 1).

Table 1. Microbial population and activity in CMSW.

	Mature CMSW raw CMSW	CMSW-heated	CMSW-heated +
Fungi (cfu/g	1.5 x 10^5	< 10^2	0.6 x 10^3
Bacteria (cfu/g)	4.5 x 10^8	1.4 x 10^3	1.5 x 10^8
FDA (μg flourescein/g)	1.6a[a]	1.03b	1.9a
CO_2 emission (μl/g DM)	7.1a	4.4b	4.84c

[a] Different letters indicate significant differences between treatments (P=0.05).

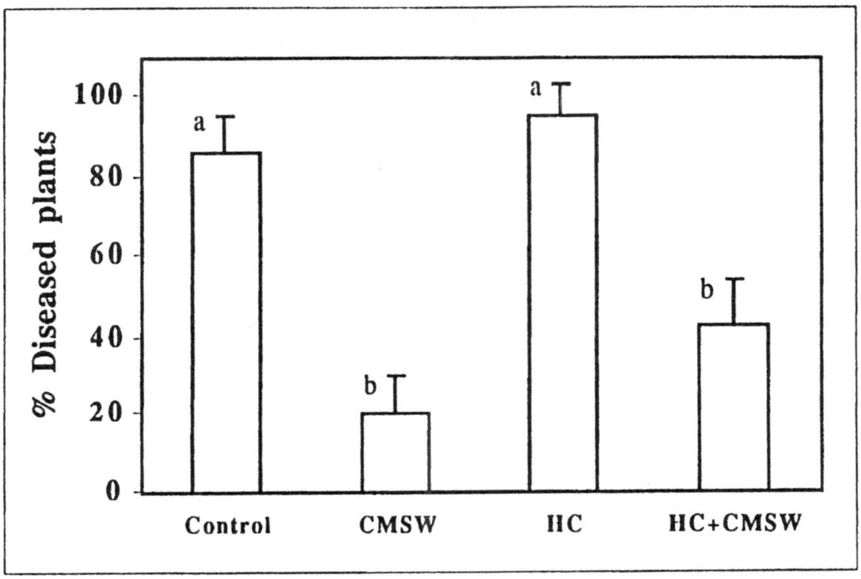

Figure 8. Effect of adding mature CMSW to heat-treated CMSW on its suppression of *R. solani* in cotton plants. Control-- sandy soil; HC-- heat-treated CMSW (5 days at 60°C); HC+CMSW-- heat-treated CMSW inoculated with 5% mature CMSW. Percentage of diseased plants was determined 14 days after sowing. Different letters heading the columns indicate significant differences between treatments (P=0.05).

The same trend was observed by measuring CO_2 emission in the different treatments after 72 h of incubation. However respiration rates after the inoculation of heat-treated CMSW with 5% mature CMSW did not reach the same level as the mature CMSW alone.

Microscopic observations were performed to visualize microbial interactions involved in the suppression of *R. solani* by CMSW. The fungus was grown on a 100-μm nylon mesh and then transferred to media containing soil or soil with 10% CMSW. After 1 h (as observed by light microscopy x400), there were many bacteria attached along the fungal hyphae in the CMSW treatment (data not shown). In the control treatment (soil), only a few bacteria were observed, and these did not enclose the hyphae. *R. solani* hyphae incubated in mature CMSW for 12 h changed their original shape and after 24 h complete lysis had occurred.

Role of Antagonistic Microorganisms in Suppression

Bacteria were isolated from CMSW meshes to determine antagonistic activity against *R. solani.* Some of the bacteria exhibited chitinolytic activity when grown on colloidal chitin (Rodriguez-Kabana et al., 1983) as the sole carbon source. Colonies capable of hydrolyzing the colloidal chitin, as evidenced by clearing of the medium around the colony, were subcultured. These strains also exhibited growth inhibition of *R. solani* as tested in dual cultures (data not shown). Eight strains showing inhibitory effects in Petri-dish assays were tested under greenhouse conditions for their ability to reduce *R. solani* disease severity. Upon addition to the soil, some strains caused increased disease severity (B1-B3), while others (B6-B8) did not have a significant effect, relative to controls. Two of the isolates (B4 and B5) suppressed *R. solani* by about 50% (Figure 9). Strain B5 was identified by fatty-acid profile analysis as *Enterobacter agglomerans,* a strain which is known as an efficient antagonistic bacterium (Chernin et al., 1995).

T. harzianum (Y), a known *R. solani* antagonist (Sivan and Chet, 1986), was added to CMSW to improve its suppressiveness against *R. solani.* We were not able to isolate *Trichoderma* from mature CMSW using TSM. Soil, mature CMSW (5% in soil) and heated CMSW were inoculated with *T. harzianum* (Figure 10). The antagonistic fungus successfully colonized all three media, reaching a level of 10^6 cfu/g soil. *T. harzianum* added to the soil (control) suppressed *R. solani* by 56%. Similar levels of suppression were obtained in the media containing CMSW or CMSW inoculated with *T. harzianum.* Heated CMSW showed a high level (100%) of disease severity due to the biological vacuum created after the heat treatment; this biological vacuum was successfully filled by *T. harzianum,* resulting in a high level of suppression, similar to that obtained with the other treatments.

DISCUSSION

The use of suppressive compost provides effective biological control of plant pathogens. Thus, the composting of a wide variety of waste materials and by products is an economically and ecologically sound alternative to fungicides and fumigation with methyl-bromide (Chef et al., 1997). The aim of this study was to investigate the

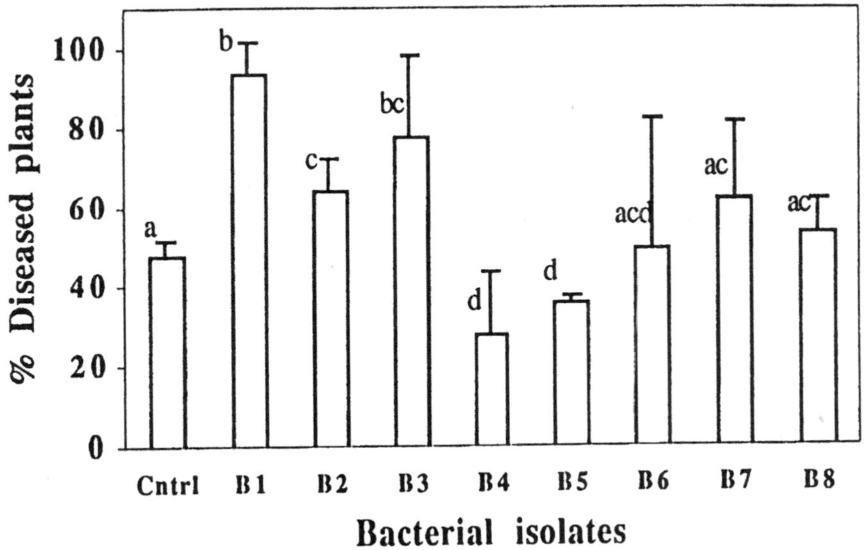

Figure 9: Effect of bacteria isolated from CMSW on suppression of R. solani in cotton plants. Percentage of diseased plants was determined 14 days after sowing. Control-- sandy soil. Different letters heading the columns indicate significant differences between treatments (P=0.05).

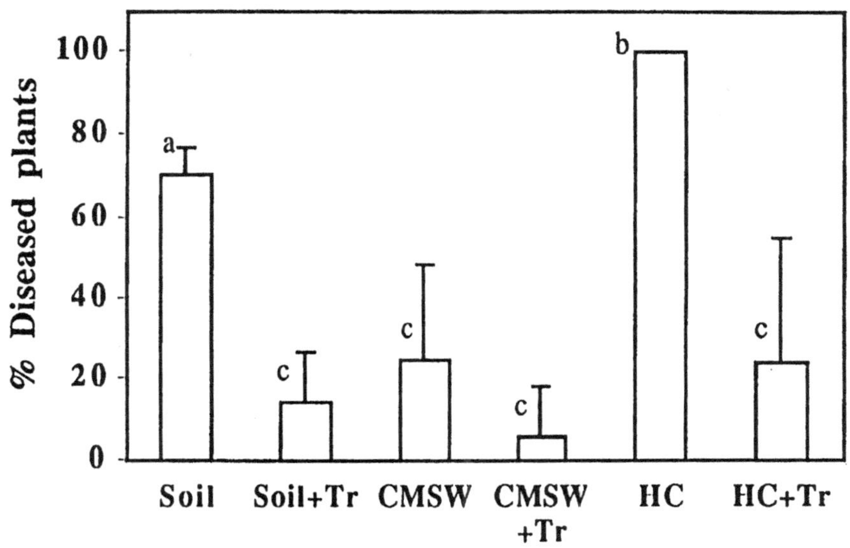

Figure 10: Effect of adding *Trichoderma harzianum* (104 cfu/g soil) to CMSW on suppression of R. solani in cotton plants. Soil+Tr-- soil inoculated with *T. harzianum*; CMSW+Tr-- CMSW inoculated with *T. harzianum*; HC-- heat-treated CMSW (5 days at 60°C); HC+Tr-- heat-treated CMSW inoculated with *T. harzianum*. Percentage of diseased plants was determined 14 days after sowing. Different letters heading the columns indicate significant differences between treatments (P=0.05).

suppressive ability of CMSW against soil-borne pathogens. This compost was studied by Cohen et al. (1996), who described its suppression of P. *aphanidermatum* in container media (containing CMSW:peat:vermiculite 1:1:1 v/v/v) and sandy soil (containing 20% CMSW); CMSW reduced disease incidence in plants by approximately 50% and 70%, respectively.

In this study, CMSW suppressed the soil-borne pathogens *R. solani* and *F. oxysporum* f.sp. *vasinfectum,* when added to soils at levels as low as 2 and 5%, respectively. These levels are equivalent to 60 and 150 tons per hectare, respectively, which are commercially acceptable level for nonedible crops such as cotton. In cotton plants, CMSW (2% in the soil) reduced damping-off caused by *R. solani* by about 80% (Figure 5). Wilt disease caused by *F. oxysporum* f.sp. *vasinfectum* was reduced by 98% by using 5% CMSW. In this case a 10% application of CMSW was more effective in the second growth cycle than 5% (Figure 6). A similar trend was reported by Serra-Wittling et al. (1996), who studied suppression of Fusarium wilt of flax by a loamy soil amended with CMSW. The addition of compost increased the suppressiveness of the control soil proportionally to the application rate (10, 20 and 30%).

The level of CMSW in the soil affected plant growth, as reflected by emergence rate, height and weight of cotton plants (Figs. 1, 2). CMSW at a level of 2-5% in the soil enhanced plant growth with no delay in emergence. This is in contrast to higher levels of CMSW which delayed plant emergence and above 10%, reduced plant height and weight. The negative effect of CMSW application was due to compost salinity: this effect was much more pronounced when CMSW was used in container media vs. its application in the f~eld (Hoitink et al., 1993).

The age of the compost, and the level of maturity and decomposition of the organic matter within it affect its ability to suppress soil-borne plant pathogens (Hoitink et al., 1993; Inbar et al., 1993). Indeed, in this study MSW which had been composted for only 1.5 months did not suppress *R. solani;* suppression only occured after 2.5 months of composting (Figure 3). On the other hand, in the case of *F. oxysporum* f.sp. *vasinfectum,* 80% disease reduction was obtained with 0.5-, 1- and 1.5-month-old compost; moreover, 93% suppression was found with compost after 2.5 or more months of composting (Figure 4). As reported by Chefetz et al. (1996), CMSW after 2.5 months has a C:N ratio of 12; thus, it is concluded that the compost was stable and the decomposition rate very low. Since the temperature of the 0.5-month-old compost pile is around 70°C (Chefetz et al., 1996), it may be assumed that thermophilic microorganisms are involved in the suppression of *F. oxysporum* f.sp. *vasinfectum. Bacillus* spp. are known to survive at such temperatures, and to be antagonistic to soil-borne pathogens (Hoitink et al., 1993). Involvement of thermostable microorganisms in suppression of soil-borne pathogens by compost has also been reported by Mandelbaum et al. (1988), who described suppression of *Pythium* in compost (separated manure and grape marc). This suppression was improved by heating (55°C for 2 h), and microorganisms in the composted separated manure were suggested to be more thermostable than the peat microflora (Mandelbaum et al., 1988). Suppression of *F. oxysporum* f.sp. *vasinfectum* could also result from the presence of chemical inhibitors in the CMSW, a phenomenon which merits further investigation.

Several treatments were used to study the mechanism of suppression by CMSW. In the experimental system with *R. solani,* autoclaving, gamma-irradiation and heat treatment all negated suppression (Figs. 7,8). Thus, the mechanism involved in *R. solani* suppression is most likely related to the microbial population in the compost. Further evidence of the involvement of microorganisms in this phenomenon was provided when 5% CMSW was added to the heat-treated CMSW (Figure 8), causing the recolonization of antagonistic microorganisms and restoration of the compost's suppressive ability. Similar results have been reported by Mandelbaum et al. (1988), who showed that suppression of *P. ahanidermatum* damping off in cucumbers by composted separated manures and composted grape marc was negated by heat treatment and could be restored by incorporating small amounts of medium back into the autoclaved soil.

The microbial population and its activity in the different treatments were investigated to better understand their role. Microbial activity before and after reinoculation of heat-treated CMSW with mature, untreated compost was compared (Table 1). Bacterial and fungal counts, FDA hydrolysis and CO_2 emission were determined. Microbial activity was reduced after heat treatment; this reduction, as well as suppressiveness, were restored after reinoculation of the heat-treated CMSW with mature CMSW. The bacterial population was restored to the same level after reinoculation, as compared to the fungal population, which was increased only ten fold after the inoculation of heated CMSW. It may be concluded that bacteria were more involved in the suppression of *R. solani* than fungi. This conclusion is supported by the microscopic observations, in which bacteria were seen in close relation to lysing hyphae of *R. solani,* while no fungal interactions were observed. Similar involvement of bacteria in suppression has been reported by Mandelbaum et al. (1988), who studied the effect of heat treatment on the suppression of *P. aphanidermatum* by compost and peat. Heated compost (55°C for 2 h) remained suppressive even after two growth cycles. The fungal population is highly sensitive to heat treatment. In both composted separated manure and peat, heat treatment decreased the fungal counts to less than 1%. Since the same heat treatment did not negate the suppressive effect, it seems that saprophytic fungi are not as important as bacteria in controlling Pythium damping off (Mandelbaum et al., 1988).

In the experimental system with *F. oxysporum* f.sp. *vasinfectum,* heat treatment (5 days at 60°C) did not eliminate the suppressive effect, suggesting that either a thermophilic population or chemical components were involved in this suppression. This phenomenon was also observed by Serra-Wittling et al. (1996): heat-treated compost was as effective as control compost in inducing suppressiveness in control soil. The authors suggested that the microflora of both the soil and the compost are involved in the suppression, acting mainly via competitor for nutrients and space with the pathogen population. These results contradict those reported by Trillas-Gay et al. (1986), who found that heating destroys the suppressive effect of composted hardwood bark on *F. oxysporum* f.sp. *conglutinans.*

In the case of *R. solani,* the suppressive mechanism is presumed to be lysis of fungal hyphae. Microscopic examination of *R. solani* grown on nylon meshes in this work and incubated in CMSW-amended soil suggests bacterial involvement. Some of these bacteria were isolated from the compost meshes. Eight isolates showed the ability to degrade chitin and to inhibit *R. solani* growth in Petri dishes. Two of these isolates suppressed *R. solani* damping-off in cotton plants under greenhouse conditions (Figure 9). One of them (B5) was identified by fatty-acid-profile analysis as *E. agglomerans,* a strain which is known as an efficient antagonistic bacterium (Chernin et al., 1995); the other could not be identified by this method. Kwok et al. (1987) isolated eight species of bacterial antagonists, including *Enterobacter agglomerans,* in container media amended with composted hardwood-tree bark; combinations of some of these strains of bacterial antagonists with *T. hamatum* were consistently more effective than the fungal isolate alone. In this study, *T. harzianum* was added to mature or heat-treated composts. The fungus colonized the heated compost and significantly restored its suppression of *R. solani* (Figure 10). In contrast, *T. harzianum* did not significantly improve the suppressive effect of the mature compost alone, despite the fact that it colonized it successfully. *T. harzianum* was not found in mature CMSW, in contrast to composted hardwood bark which is naturally colonized with *T. hamatum* (Kwok et al., 1987). Nelson et al. (1983) demonstrated the role of antagonistic fungi in suppressing *R. solani* in container media amended with composted hardwood bark, and emphasized the role of *Trichoderma.*

The application of disease-suppressive compost in agriculture has the additional benefit of making use of organic-by-product composts. CMSW provided effective biological control of plant pathogens and may therefore serve as an alternative to fungicides and methyl bromide fumigation. It is essential to use mature compost colonized by antagonistic microorganisms to obtain suppressiveness and to enhance plant growth. Therefore, it is crucial to define parameters of compost maturity and quality to avoid possible damage to crops. Understanding the microbial mechanisms involved in the suppression of soil-borne pathogens could contribute to the quality improvement of compost via its enrichment with beneficial microorganisms (Hoitink et al., 1993; Kwok et al., 1987). The key to large-scale utilization of disease-suppressive compost is the development of reproducible products with defined, consistent properties.

ACKNOWLEDGEMENTS

This research was supported in part by Deutshe Forschungsgemeinschaft (DFG), and Binational Agricultural Research and Development Fund (BARD).

REFERENCES

Baker KF (1987) Evolving concepts of biological control of plant pathogens. Annul Rev. Phytopathol. 25:67-85.

Bruns C, Ahlers S, Gatinger A, SchŸler C, Vogtman H & Wolf G (1996) The suppressive effects of composted separately collected organic waste and yard waste compost on two important soilborne plant pathogens. In: de Bertoldi M, Sequi P, Lemmes B and Papi T (Eds) The Science of Composting (pp. 1094-1095). Blackie A & P. London.

Chef DG, Hoitink HAJ & Madden LV (1983) Effects of organic components in container media on suppression of Fusarium wilt of chrysanthemum and flax. Phytopathology 73:279-281.

Chefetz B, Hatcher PG, Hadar Y & Chen Y (1996) Chemical and biological characterization of organic matter during composting of municipal solid waste. J. Environ. Qual. 25:776-785.

Chen W, Hoitink HAJ, Schmitthenner AF & Tuovinen OH (1988) The role of microbial activity in suppression of damping-off caused by *Pythium ultimum.* Phytopathology 78:314-322.

Chen Y & Hadar Y (1987) Composting and use of agricultural wastes in container media. In: Ferranti MP, L'Hermite P & Zucconi F (Eds) Composts: Production Quality and Use (pp. 71-77). Elsevier Science, London.

Chernin L, Ismailov Z, Haran S & Chet I (1995) Chitinolytic *Enterobacter agglomerans* antagonistic to fungal plant pathogens. Appl. Environ. Microbiol. 61:1720-1726.

Chet I, Inbar J & Hadar Y (1997) Fungal antagonists and mycoparasites. In: Esser K & Lemke PA (Eds) The Mycota-Environmental and Microbial Relationships (pp. 165-184). Springer, New York.

Cohen R, Bar-Lev N, Chefetz B, Chen Y & Hadar Y (1996) Suppression of soil-borne pathogens by composted municipal solid waste. Phytoparasitica 24 (2):145.

Craft CM & Nelson EB (1996) Microbial properties of composts that suppress damping-off and root rot of creeping bentgrass caused by *Pythium graminicola.* Appl. Environ. Microbiol. 62:1550-1557.

Elad Y, Chet I & Katan J (1981) A selective medium for improving quantitative isolation of *Trichoderma* spp. from soil. Phytoparasitica 9:56-67.

Gera JVO & Wilma JMVG (1996) Possibilities for biological control of Pythium root rot in ornamental bulb culture with composted organic household waste. In: de Bertoldi M, Sequi P, Lemmes B & Papi T (Eds) The Science of Composting (pp. 1374-1375). Blackie A & P. London.

Hadar Y & Gorodecki B (1991) Suppression of germination of sclerotia of *Sclerotium rolfsii* in compost. Soil Biol. Biochem. 23:303-306.

Hadar Y & Mandelbaum R (1986) Suppression of *Pythium aphanidermatum* damping-off in container media containing composted liquorice roots. Crop Prot. 5:88-92.

Hadar Y & Mandelbaum R (1992) Suppressive compost for biocontrol of soilborne pathogens. Phytoparasitica 20:113-116.

Hadar Y, Mandelbaum R & Gorodecki B (1992) Biological control of soilborne pathogens by suppressive compost. In: Tjamos EC, Papavizas GC & Cook RJ (Eds) Biological Control of Plant Diseases (pp. 79-83). Plenum Press, New York.

Hoitink HAJ, Boehm MJ & Hadar Y (1993) Mechanisms of suppression of soilborne plant pathogens in compost-amended substrates. In: Hoitink HAJ & Keener HM (Eds) Science and Engineering of Composting: Design, Environmental, Microbiological and Utilization Aspects (pp. 601-622). Renaissance Publications, Worthington, OH.

Hoitink HAJ, Stone AG & Grebus ME (1996) Suppression of plant disease by composts. In: de Bertoldi M, Sequi P, Lemmes B & Papi T (Eds) The Science of Composting (pp. 373-381). Blackie A & P. London.

Inbar Y. Hadar Y & Chen Y (1993) Recycling of cattle manure: the composting process and characterization of maturity. J. Environ. Qual. 22:857-863.

Ko W & Hora HF (1971) A selective media for the quantitative determination of *Rhizoctonia solani* in soil. Phytopathology 61:707-720.

Kuter GA, Hoitink HAJ & Chen W (1988) Effect of municipal sludge compost curing time on suppression of Pythium and Rhizoctonia diseases of ornamental plants. Plant Dis. 72:731-756.

Kwok OCH, Fahy PC, Hoitink HAJ & Kuter GA (1987) Interactions between bacteria and *Trichoderma harzianum* in suppression of Rhizoctonia damping-off in bark compost media. Phytopathology 77:1206-1212.

Lumsden RD (1980) A nylon fabric technique for studying the ecology of *Pythium aphanidermatum* and other fungi in soil. Phytopathology 71:282-285.

Mandelbaum R & Hadar Y (1990) Effect of available carbon source on microbial activity and suppression of *Pythium aphanidermatum* in composted and peat container media. Phytopathology 80:794-804.

Mandelbaum R, Hadar Y & Chen Y (1988) Composting of agricultural wastes for their use as container media: effect of heat treatment on suppression of *Pythium aphanidermatum* and microbial activities in substrates containing compost. Biological Wastes 26:261-274.

Martin SB (1950) Use of acid rose bengal and streptomycin in the plate method for estimation of soil fungi. Soil Sci. 69:215-232.

Nelson EB, Kuter GA & Hoitink HAJ (1983) Effect of fungal antagonist and compost age on suppression of *Rhizoctonia solani* damping-off in container media amended with composted hardwood bark. Phytopathology 73:1457-1462.

Rodriguez-Kabana R, Godoy G, Morgan-Jones G & Shelby RA (1983) The determination of soil chitinase activity: condition for assay and ecological studies. Plant Soil 75:95-106.

Serra-Wittling C, Houot S & Alabouvette C (1996) Increased soil suppressiveness to Fusarium wilt of flax after addition of municipal solid waste compost. Soil Biol. Biochem. 28(9):1207-1214.

Shnürer J & Rosswall T (1982) Fluorescein diacetate hydrolysis as a measure of total microbial activity in soil and litter. Appl. Environ. Microbiol. 43:1256-1261.

Sivan A & Chet I (1986) Biological control of *Fusariurn* spp. in cotton, wheat and muskmelon by *Trichoderma*

harzianum. J. Phytopathology 116:39-47.
Trillas-Gay MI, Hoitink HAJ & Madden LV (1986) Nature suppression of Fusarium wilt of radish in container media amended with composted hardwood bark. Plant Dis. 70(11):1023-1027.
Zhang W, Dick WA & Hoitink HAJ (1996) Compost induced systemic acquired resistance in cucumber to Pythium root rot and anthracnose. Phytopathology 86:1066-1070

USING COMPOST IN THE ORNAMENTAL HORTICULTURE INDUSTRY

FRANCIS R. GOUIN
Professor Emmeritus, Dept of Natural Resource Sciences and Landscape Architecture, University of Maryland, College Park, Maryland 20742, USA
E-mail -R-GOUIN@umail.umd.edu(fg7)

INTRODUCTION

For decades, the ornamental horticulture industry has relied on peatmoss and commercial fertilizers for amending soils and potting media and for establishing and maintaining landscapes. In the last few decades, however, milled pine bark, pine bark mulch and composted hardwood barks, and other organic amendments have become available. Also, under many uses, chemical slow-release, organic, and water soluble fertilizers have replaced agricultural grade fertilizers.

Until the mid 50's the most common inorganic amendment used in reducing bulk density and increasing the aeration of soils and potting media was washed sand. With the introduction of vermiculite (expanded mica), expanded shale, and perlite (expanded volcanic ash), came the introduction of light-weight and soil-less potting media. This media was made by blending the inorganic components with peatmoss and/or milled composted barks. Although many nurserymen and growers of greenhouse crops were initially reluctant to accept soil-less potting media, they are now widely used and commercially available in bag and bulk.

The ornamental horticulture industry is highly dependent on soil amendments and fertilizers to grow and maintain their crops, gardens and landscapes. Amendments are important because horticultural industries grow and maintain a wide variety of plant species requiring a range of cultural practises. Nurserymen use a variety of cultural techniques including container culture, raised beds, and manufactured root balls, and must often grow and maintain plants under adverse soil conditions. Home gardening, the most popular national past time, has also created a demand for soil amendments and specialized fertilizers. Therefore, the ornamental horticultural industries as well as home gardeners are important markets for quality compost.

POTENTIAL USES FOR ORGANIC AMENDMENTS

Container Culture

Nearly 60% of all ornamental plants and vegetable transplants marketed are grown in

S. Brown, J.S. Angle and L. Jacobs (eds.), Beneficial Co-Utilization of Agricultural, Municipal and Industrial By-products, 131–138.
© 1998 *Kluwer Academic Publishers,*

containers. Fifty to eighty percent of potting media in those containers is organic in nature while the remaining twenty to fifty percent is inorganic. Soil is seldom used because it is too dense, must be sterilized and lacks uniformity.

Growers can produce 20,000 to 100,000 container plants per acre, depending on container size and plant species. Based on this information one can appreciate the organic amendment needs of this industry.

Field and Bed Culture

Nurserymen who market their plants as balled and burlap (B & B), balled and potted (B & P) or in manufactured root balls are also potential users of soil amendments. When harvesting plants with the ball of soil that surrounds roots, 200 to 250 tons of top soil per acre are removed with each harvest. Even when harvesting plants bare-rooted, some soil is removed. When manufacturing balls, nurserymen utilize up to 20 liters of rooting medium per plant.

Experience and research have demonstrated that soil productivity in most nurseries cannot be maintained using conventional crop rotation practices. To maintain soil productivity, heavy and repeated applications of animal manures and/or compost immediately following harvesting and continuing for one to two years before planting the next crop is required. However, the demand for ornamentals often exceeds supply and many nursery operators choose to neglect good soil management practices and follow one crop directly with another. This results in declining growth rates due to a loss of nutrients and organic matter.

Landscape Contracting and Home Gardening

Landscape contractors and home gardeners are also heavy users of organic and inorganic amendments to establish and maintain landscapes. Organic amendments are used to improve existing top-soil or to create top-soil by amending existing sub-soil. Although top-soil may be purchased, the quality is often questionable, it may be very costly, and it may not always be compatible with the existing sub-soil. Mulches are also used extensively in maintaining landscapes. In addition to their aesthetic value, mulches provide weed control, maintain more uniform soil temperatures, and reduce irrigation needs.

Mulch

The careful selection and use of mulch adds the finishing touch to any landscape. Landscape contractors, landscape maintenance and garden centers are the primary outlets for mulches. The factors that distinguish quality mulch are aesthetic appearance, weed control properties, stability, compatibility with species, consistency, and availability. Dark colored organic mulches are preferred because of the contrast they provide, giving the ground a rich organic appearance. In addition, organic mulches are generally preferred because they are capable of supplying some of the nutrient needs of the ornamentals.

Ornamental Horticulture - A Unique Agricultural Industry

The nursery and greenhouse growing industries are the only agricultural industries that require that products sold must be capable of continued growth. This means that the majority of plants are sold either growing in containers or with soil surrounding the roots. Thus the demand for organic amendments to replenish soils that are being depleted or components to formulate potting media is great.

Ornamentals are high value plants and include a wide and diversified selection of species. They often require years to grow to marketable size and/or to maturity, and can also require specialized propagation practices. The market value of such crops may vary from $5,000 to $35,000 or more per acre while the annual value of greenhouse crops may vary from $300,000 to $600,000 per acre.

Ornamentals include a highly diversified number of plant species. Species vary from those that may grow under desert conditions to others that are native to forest, bog or aquatic environments. Being able to grow such a diversified group of plants is often dependent on modifying soils and/or potting media to satisfy the particular needs of each species. It is well known that soils or potting media high in organic matter are better able to supply nutrients over a wider range of pHs than mineral soils. These features explain the value of organic amendments for modifying root media environments to satisfy the needs of a range of plant species.

A BRIEF HISTORY OF CONTAINER MEDIA AMENDMENTS

Until the mid 60's, most of the peatmoss used in the USA ornamental horticultural industries was imported from Germany, Poland and Ireland. Since the mid 60's, most of the peatmoss used in the U.S.A. has been imported from Canada. The change in source of peatmoss necessitated a change in cultural practices. European peatmoss tends to have a pH of 5.0 and above and is coarse textured. Canadian peatmoss tends to have a pH between 3.0 and 4.0 and is fine textured. Neither have sufficient nutrients to support plant growth.

Until the mid 80's the availability of Canadian peatmoss was determined by spring weather conditions, often resulting in shortages and fluctuating prices. Improved harvesting technology combined with improved packaging and storage, and improved transportation has resulted in year-round availability and stable prices.

The introduction of milled pine bark and composted hardwood bark in the late 70's provided an alternate source of organic amendments. However, the use of these materials required changes in cultural programs. Barks tend to be more porous, retain less water and have a lower initial C.E.C. than peatmoss. This meant that plants had to be watered and fertilized more often to maximize growth. However, it is important to note that both peatmoss and bark do not contribute substantially to the nutrient needs of plants.

TESTING THE SUITABILITY OF COMPOST PRODUCTS

Ornamentals as Test Plants

Ornamentals are ideal crops for testing and utilizing compost amendments made from different feed stocks since plants and plant parts are generally not grown for human consumption. Even if the compost feedstocks used contained toxins that could be harmful to man, using the compost made from such feedstocks for growing and/or maintaining ornamentals, is unlikely to have a negative impact. Transfer of toxins from soil to plant to human is unlikely as uptake of toxins is minimal and plants are primarily used for decorative purposes.

Also there should be little concern about using contaminated compost for growing and marketing plants in containers. Most container grown plants are either transplanted into landscapes or discarded in landfills. Thus there is little chance of contaminating land with compost that may contain questionable minerals and/or compounds.

Evaluating Compost in Horticulture

Identifying and evaluating compost for growing and maintaining horticultural plants is based on the availability, uniformity, consistency, chemical and physical composition, stability and cost of the material. All factors are equally important and the product that can best satisfy all of these needs is most likely to be widely used.

Many ornamental horticultural plants are grown on tight schedules. Timing with regards to delivery of compost that is stable and consistent at all times is critical. Any variation in the chemical and physical properties of compost that causes compost to fall below a minimum standard could be disastrous to growers. Success in producing a marketable crop of ornamentals is dependent on rapid establishment and optimum growth.

Compost for Formulating Potting Media

Much effort has been placed on maximizing the use of compost in formulating potting media for the production of nursery and greenhouse crops (Link, et al.,1983; Purman and Gouin, 1992; Gouin, 1993). Particle size and particle size distribution, maturity, pH and soluble salt concentrations are of immediate concern when compost is to be used for the production of plants in containers. Compost particles larger than 1.25 cm in potting media may interfere with the mechanics of potting a broad number of plant species and contributes to the media shrinkage problem. Maintaining media volume is very important especially for plants that are to be grown in containers for several months. A high concentration of large particles also reduces the water holding capacity of the potting media. However, the distribution of particle size, does not appear to be critical when normal screening practices are followed (Wootton et al.,1981; Marcotrigiano et al., 1985).

The level of compost maturity should be such that microorganisms in the compost do not compete with the plant roots for available N, unless provisions are made to supply sufficient additional N to satisfy the needs of both (Vega-Sanchez et al.,1987). Using immature compost in formulating potting media without supplying additional fertilizer can result in stunted growth and even death of plants (Brown, 1995). Microorganisms appear to have a greater ability to absorb and fix nutrients than do plant roots.

The pH of the compost amended potting medium must either be easily adjusted or compatible with the species to be grown. The soluble salt concentration should also not exceed the tolerable level of the species. Compost made from lime de-watered biosolids, when used as a media amendment may cause nutritional problems and stunted growth because it raises pH above 7.0 (Ku, 1996). Furthermore, the pH of the potting media containing lime- dewatered biosolids cannot be economically and efficiently lowered to satisfy the growing needs of such species as: *Kalmia* L. *Rhododendron* L., *Pieris* D., *Leucothoe* D., *Vaccinium* L. *Quercus* L. etc.

There are few ornamentals, other than ornamental grasses that can tolerate an electrical conductivity in excess of 2.5 mmhos/cm (saturated paste extract). Therefore, the amount of compost that can be used in formulating a potting medium will be dependent on its soluble salt concentration and the exchange capacity of other amendments. Because of this, most compost made from biosolids or animal manures generally cannot be used in excess of 1/3 by volume in formulating potting media for most nursery and greenhouse plants (Gouin, 1985). Excess salt effects can also be alleviated by leaching compost materials before use. It is possible to use higher levels of compost made from yard trimmings and/or municipal solid waste without leaching. These generally contain lower levels of soluble salts (Purman and Gouin, 1992).

For the production of bedding plants and vegetable transplants, compost levels in excess of 25% by volume may limit the number of species that can be grown. Some species of bedding plants are more sensitive to soluble salts than others. Such species as *Coleus Lour.* and *Salvia* L. are extremely sensitive to soluble salts while other species such as *Tagetes* L. and *Zinnia* L. are more tolerant (Jacques, 1992). Since most growers produce many species of bedding plants it is important to develop one potting medium that can be used to efficiently grow a broad number of species. In general, the electrical conductivity of the potting medium for bedding plants should not exceed 2.0 mmhos/cm as measured by saturated paste. Most vegetable transplants appear to be tolerant to electrical conductivity of 2.5 mmhos/cm (Falahi-Ardakani et al., 1988).

At these compost concentrations, potting media are capable of supplying all of the nutrient needs of the plants for the first 2 to 3 weeks of growth. After this period the plants must be fertilized regularly (Falahi-Ardakani et al., 1987). However, because many composts contain an abundant supply of macro and trace elements and compost containing biosolids tend to have high levels of P, quality plants can be produced using one-third to one-half the regular amount of fertilizers and without having to amend with trace elements and high levels of P (Korcak, et al., 1979; Falahi-Ardakani et al., 1987 b; Falahi-Ardakani et al., 1988). This can result in substantial savings in fertilizers.

Studies currently in progress, indicate that the use of compost in potting media can be further maximized by blending compost made from yard trimmings with compost made from biosolids or chicken manure (personal communication Catherine Ku, J. C. Bouwkamp, Univ. of Maryland, Dept. of Natural Resource Science and Landscape Architecture). By using blends of these composts at 1-1 or 2-1 by volume (Designer compost), potting media may contain 50% to 75% compost by volume and 50% to 25% peatmoss and perlite or vermiculite. The use of such blends in formulating potting mixes has resulted in a 50% reduction in the use of water soluble fertilizers in the production of potted chrysanthemums and/or poinsettia (research in progress). In addition because of early top growth suppression, without sacrificing axillary shoot development, the use of Designer compost, in formulating potting media, may eliminates the need to use chemical growth regulators to control plant height.

Although no formal studies have been conducted on the effects of compost on the shelf-life of potted and bedding plants, observations by the author and by commercial growers has indicated improved shelf-life and quality. Poinsettias grown in compost amended media have been observed to maintain bracts and leaves through April and early May under home conditions while receiving only water while similar plants growing in commercial soilless mixes defoliated by early February. Growers operating garden centers have observed bedding plants growing in compost amended media maintained dark green foliage color for three to four weeks after leaving greenhouse conditions. Similar plants growing in commercial soilless mixes may begin to exhibit symptoms of chlorosis within 1 to 2 weeks after leaving greenhouse conditions. This response can be attributed to the improved exchange capacity of the potting media and to the mineralization of the compost.

Compost for Growing Forest Seedlings

Soils in forest nurseries are rapidly depleted of organic matter and nutrients because in harvesting, top-soil is removed with the harvest of each crop and ground must remain in continuous production in most instances. Frequently soils used for forest seedling production are generally sandy and not likely to contain high levels of organic matter.

Incorporating 224 T/ha of compost made from biosolids and woodchips into soil just prior to fall seeding of *Liriodendron tulipifera* L., *Cornus florida* L., *Juglands nigr*a L., and *Fraxinus pennsylvanica* Marsh. not only resulted in the largest number of seedlings produced but also in seedlings with greater stem length. Increasing compost concentration to 448 T/ha significantly reduced the seedling population do to elevated soluble salt levels but produced taller seedlings of those that survived (Gouin and Walker, 1977). This study also demonstrated that it is important to use only compost that had been screened through 1.25 cm mesh. The incorporation of unscreened compost resulted in stunted top growth of all species and chlorotic looking plants (most likely do to N stress).

In addition to satisfying the nutrient needs of trees through their first growing season, the use of compost as a soil amendment resulted in improved winter survival regardless of the amount of compost used. All plant species growing in the check treatment suffered winter die-back while all plants growing in the compost amended soils did not exhibit any symptoms of winter die-back.

The germination of coniferous seed of *Pinus strobus* L. and *Picea abies* L. was significantly reduced when seeded in soils that had been amended with biosolids compost prior to seeding. The poor germination was attributed to elevated levels of soluble salts present at the time of germination. However, mulching the seedlings after they had grown 5 to 8 cm tall with 224 T/ha of biosolids compost produced the greatest number of quality seedlings. Mulching the seedlings with screened biosolids compost not only provided weed control, but also appeared to have provided the seedlings with sufficient nutrients for normal growth and development (Gouin, 1977).

Compost for Establishing and Maintaining Landscapes

Based on the information obtained in forest seedling production, compost being made from biosolids, animal manures, crab chum and/or yard trimmings are being used extensively in establishing and in maintaining landscapes (Gouin. 1993) Regardless of soil test results, the amount of compost used for amending soils in landscape should not exceed 3 m^3 per 100m^2 to minimize compost abuse under normal soil conditions. When used at the maximum level, there is no need to apply any commercial fertilizers through the first year of growth even under the worst soil conditions. Furthermore, in the transplanting of trees and shrubs it is recommended to amend backfill soils with 1/3 by volume of compost. The use of compost in the backfill appears to promote early establishment and nourishment through the first year of growth.

In recommending compost to the landscape contracting industry, it is important to set maximum allowable levels as a means of protecting one's reputation. In this industry there is a tendency to multiply recommended levels by 2X and 3X until there has been a bad experience. The 3 m^3 per 100 m^2 recommendation is sufficient to establish a landscape even on the worst possible soil and supplemental adjustments can always be made at a later date.

Unscreened compost is also being used as a mulch at the at the rate of 1 to 2 m^3 per 100 m^2. In addition to providing good weed control, the compost appears to supply sufficient nutrients to sustain normal growth and development.

CONCLUSIONS

There is a wide variety of potential uses for compost products in the horticulture industry. To be accepted, these products have to be a viable alternative to the peat based products currently in use. Product consistency, availability, and cost are critical factors to assure product acceptance. Evaluating the ability of composts for cultivation of ornamentals is an excellent method to determine the potential for use of compost products. Initial studies have shown both positive and negative effects of compost addition to potting media. Negative effects are related primarily to high EC and pH of some materials. Positive effects include savings on micro and macro nutrient fertilizers, increased shelf life of plants, and more vigorous plants. Initial studies testing the potential for blending compost products to maximize the positive responses and minimize negative effects are very encouraging. Ornamental horticulture represents an enormous potential market for compost. Quality products are the way to guarantee acceptance.

REFERENCES

Brown, K. (1995) Composting municipal solid waste (MSW) for horticultural uses. M.S. thesis. University of Maryland, College Park, MD.

Daniel J, Gouin FR & Konjoian PS (1992) Nitrate Fertilization and root media effects on growth and shelf-life of Honeycomb marigolds and Novette Red impatiens. J. of Plant Nutrition 15:569-578.

Falahi-Ardakani F, Bouwkamp JC, Gouin FR & Chaney RL (1987) Growth response and mineral uptake of vegetable transplants grown in a composted sewage sludge amended medium. Part I. Nutrient supplying power of the medium as measured by tissue analysis. J. Environ. Hort. 5:107-112.

Falahi-Ardakani F, Bouwkamp JC, Gouin FR & Chaney RL (1987b) Growth response and minéral uptake of vegetable transplants grown in a composted sewage sludge amended medium. Part II. As influenced by time of application of N and K. J. Environ. Hort. 5:112-116.

Falahi-Ardakani F, Bouwkamp JC, Gouin FR & Chaney RL (1988) Growth response and mineral uptake of lettuce and tomato transplanted grown in media amended with composted sewage sludge as influenced by fertilizer and amendments. J. Environ. Hort. 6:130-132.

Gouin FR & Walker JM (1977) Deciduous tree seedlings 12:45-47.

Gouin FR (1977) Conifer tree seedling response to nursery soil amended with composted sewage sludge. HortScience 12:341-342.

Gouin FR (1985) Growth of chrysanthemums in containers of media amended with composted municipal sewage sludge. J. Environ. Hort. 3:53-55.

Gouin FR (1993) Utilization of sewage sludge compost in Horticulture. HortTechnology 3:161-163.

Korcak RF, Gouin FR & Fanning DS (1979) Metal content of plants and soils in a tree nursery treated with composted sludge. J. Environ. Qual. 8:63- 68.

Ku, CSM, Bouwkamp JC, & Gouin FR (1996) Effect of compost sources and fertigation frequency on growth of potted poinsettia. HortScience 31:584(abstr.)

Link CB, Gouin FR & Shanks JB (1983) Composted sewage sludge and soil amendment for greenhouse and woody ornamental crops. Acta Horticulture 133:199-204.

Marcotrigiano M, Gouin FR & Link CB (1985) Growth of foliage plants in composted raw sewage sludge and perlite media. J. Environ. Hort. 3:98-101.

Purman J, & Gouin FR (1992) Growth response of potted plants and transplants in media from processed garbage and sewage sludge. J. Environ. Hort. 10-52-54.

Vega-Sanchez F, Gouin FR & Willson GB (1987) The effects of curing time on physical and chemical properties of composted sewage sludge and on the growth of selected bedding plants. J. Environ. Hort. 5:66-77.

Wootton FD, Gouin FR & Stark FC (1981) Composted digested sludge as a medium for growing annuals. J. Amer. Soc. Hort. Sci. 106:46-49.

POTENTIAL FOR UTILIZING COAL COMBUSTION RESIDUES IN CO-UTILIZATION PRODUCTS

K. D. RITCHEY, M. A. ELRASHIDI, R. B. CLARK and V.C. BALIGAR
Appalachian Soil and Water Conservation Research Laboratory, Agricultural Research Service, US Department of Agriculture, Beaver, West Virginia 25813, USA
E-mail dritchey@asrr.arsusda.gov

ABSTRACT

Approximately 90 millions tons of coal combustion by-products are produced annually in the United States, consisting of 54 million tons of fly ash (solid particles removed from smoke), 15 million tons of bottom ash, 3 million tons of boiler slag, and 20 million tons of flue gas desulfurization (FGD) material. Markets for boiler slag are well developed, but only 33% of bottom ash, 25% of fly ash and 8% of FGD residues are beneficially used.

Numerous opportunities for agricultural use of these materials have been identified. Fly ash (a very fine, relatively inert, dry powder consisting mostly of Fe, Al, Ca, Si, and O) provides a means of reducing water contents of wet mixtures, and can also provide B and other micronutrients to plants. Fly ash is currently being used to improve the texture and water-holding capacity of potting mixtures and artificial soils. Class C fly ash (produced from burning coal from Western US) can have a calcium carbonate equivalency of up to 50% and may serve as a substitute for aglime. Mixtures of fly ash and sewage sludge produced an effective medium for growth of apple trees in northern West Virginia, and mixtures of fly ash and municipal compost produced an acceptable soil substitute for nurseries in Ohio.

FGD materials contain mainly calcium sulfate or calcium sulfite, as well as some unreacted alkaline sorbent. These products can be used as substitutes for aglime, as an insolubilizer for phosphorous from animal manure in environmentally sensitive watersheds where excess P occurs in run-off, as a source of Ca and S in potting mixtures, as an anti-sodicity agent for displacing Na from certain degraded mine lands being treated with sawmill by-products, and to improve subsoil root growth enhancement properties of surface-applied amendments. Fluidized bed combustion materials, which contain both ash and FGD components, can be used for similar purposes.

S. Brown, J.S. Angle and L. Jacobs (eds.), Beneficial Co-Utilization of Agricultural, Municipal and Industrial By-products, 139–147.
© 1998 *Kluwer Academic Publishers,*

INTRODUCTION

Useful by-products from the combustion of coal have been available for a long time. Recent legislation emphasizing cleaner air has resulted in additional types and amounts of by-products. Data collected by the American Coal Ash Association (Brendel et al., 1997) showed that approximately 90 millions tons of coal combustion by-products were produced in the United States in 1995, consisting of 54 million tons of fly ash, 15 million tons of bottom ash, 3 million tons of boiler slag, and 20 million tons of flue gas desulfurization (FGD) material. Markets for boiler slag are well developed, but only 33% of bottom ash, 25% of fly ash and 8% of FGD residues produced annually are being used beneficially. Descriptions of some coal combustion by-products and their uses are given in Table 1.

Table 1. Some coal combustion by-products and their uses.

By-Product	Properties	Uses
Boiler slag	insoluble gravel-sized glassy particles	roofing granules, blasting grit
Bottom ash	insoluble grit, 0.1 to 10mm diameter.	structural fill, aggregate, traction enhancement, soil texture improvement
Fluidized bed combustion residue	highly alkaline mixture of bottom ash, calcium carbonate, calcium oxide, calcium sulfate, calcium sulfite and unburned coal	substitute for aglime, B and K source in agriculture, use in low-strength concrete formation, stabilization of wet organic sludges
Fly ash (Class F)	neutral or acid silt- to clay-sized glassy spheres containing Al, Fe, Si and O with some trace elements	texture improvement, substitute for inorganic particles in sewage sludge-based soils, ingredient for cements
Fly ash (Class C)	higher Ca and Mg content than Class F	Class F uses, and as aglime substitute
Flue gas desulfurization residue high in calcium sulfite	high moisture, thixotropic, toothpaste-like slurry	mixing with fly ash and calcium oxide forms low-strength cement
Flue gas desulfurization residue high in calcium sulfate	moist, easily handled, relatively pure gypsum	wallboard manufacture, cement set retarder, agricultural Ca and S source

Numerous summaries and reviews of individual use of coal combustion by-products (CCBP) have been prepared (Alcordo and Rechcigl, 1995; Bilski et al., 1995; Carlson and Adriano, 1993; Korcak, 1995; Ritchey et al., 1995a; Stout et al., 1988; Shainberg et al., 1989; Terman, 1978). In this paper, we specifically examine the potentials that CCBP materials offer for co-utilization in mixtures with other materials, and discuss examples of their agronomic use.

ORIGIN AND NATURE OF CCBP MATERIALS

Coal-burning plants vary in design, and designations of by-products and their mixtures are not completely standardized. Because of this, it is important to obtain accurate information on the origin and chemical composition of any particular by-product. The following descriptions (Korcak, 1995) are given as a rough guide to some of the broader classes of CCBP materials.

Boiler Slag

Boiler slag is formed by water-quenching of molten material collected from the base of so-called wet bottom boilers. It forms black glassy particles that look like crushed obsidian. Boiler slag is in high demand for use as blasting grit, granules for roof shingles, and for spreading on icy roadways. Because boiler slag is very marketable, little emphasis has been given to develop agricultural uses of this material.

Bottom Ash

Bottom ash is also collected from the base of boilers, but is generally more heterogeneous than boiler slag. Bottom ash from stoker-fed boilers where larger pieces of coal are burned may contain up to 30% unburned coal. Stoker-fed boilers are generally older and are often found at industrial and institutional locations which produce their own steam or electricity. Boilers which burn pulverized coal tend to produce bottom ash that is more uniform than stoker-fed bottom ash. Bottom ash is used for structural fill, as aggregate for concrete block manufacture, and for traction enhancement on roadways.

Fluidized Bed Combustion Residues

In certain types of boilers, pulverized coal is fed into closed or open combustion chambers where burning coal particles are kept in suspension to facilitate combustion. Simultaneously, powdered limestone is injected into the combustion bed. At the high temperatures of the chamber, calcium carbonate in the limestone is converted to calcium oxide, which combines with sulfur dioxide formed by oxidation of sulfur present in the coal. The solid products of this reaction are calcium sulfate and calcium sulfite. These products, together with bottom ash and unreacted calcium carbonate and calcium oxide, are termed fluidized bed combustion (FBC) residues.

FBC materials are usually generated in the eastern US where higher-sulfur coal is commonly burned.

Flue Gas Desulfurization Residues

Many different processes are used to remove sulfur dioxide from the gas stream after it leaves the combustion chamber (for a fuller discussion, see Korcak (1995)). By-products resulting from these processes are called FGD residues. Some producers also classify FBC materials as FGD by-products.

The most prevalent flue gas desulfurization techniques involve passing flue gas through a spray of calcium hydroxide or calcium carbonate slurry. The by-product consists of mixtures of calcium sulfite and sulfate, depending on the amount of oxidation involved. In some cases, calcium sulfite is oxidized to calcium sulfate in subsequent operations. These by-products usually contain considerable amounts of water, which in the case of calcium sulfite, is difficult to remove.

In the so-called dry FGD processes, by-products contain less water and larger amounts of unreacted calcium hydroxide or calcium carbonate. Fly ash concentration in these products may also be higher than in the wet FGD by-products. Some power plants mix high calcium sulfite residue with fly ash and a small percentage of calcium oxide to form a cementitious mass that will solidify to facilitate disposal in dedicated landfills.

FGD residues are usually produced in the eastern US. The high calcium sulfite materials come predominantly from Ohio River Valley electric power plants.

Fly Ash

Fly ash consists of small smoke particles removed from flue gas by cyclone separators, electrostatic precipitators or bag houses (fabric filters). Separation of fly ash from flue gas before the flue gas is scrubbed to remove sulfur dioxide results in purer FGD calcium sulfate by-products, which can be sold for wallboard manufacture. However, in some dry FGD systems, desulfurization processes occur before fly ash removal. This increases fly ash contents of FGD residues and reduces the amount of fly ash produced by the separator.

Class C fly ash usually has higher Ca concentrations than Class F fly ash, and can neutralize up to 50% as much acidity as an equal weight of calcium carbonate. Class C fly ash usually comes from burning coal and lignite from the western part of the US. Class F fly ash is prevalent in the Eastern US. Both types of ash are extensively used in production of cement, concrete, and grout. The value of these residues for cement manufacture decreases if they have high levels of unburned carbon (measured by weight loss on ignition).

Commercial Value

The prices that power plants obtain for CCBP materials depend on type and quality of the by-product and distance between plant and user. In some situations, high-quality cement-grade fly ash can be sold for $40 per ton. In other cases, wallboard grade calcium sulfate is sold for $1 per ton. Some of the materials presently have negative value, because power producers pay for on-site storage in landfills or lagoons or off-site disposal.

CHARACTERISTICS OF BY-PRODUCTS AND EXAMPLES OF THEIR USE IN MIXTURES

Bottom Ash

Sell et al. (1989) describe bottom ash as a coarse gritty material the size of sand or fine gravel (0.1 to 10 mm diameter particles), consisting mostly of fused silica, suitable for mixing with clayey soil to improve texture, tilth, and aeration.

Bearce et al. (1997) evaluated use of various proportions of bottom ash together with peat-vermiculite potting mixtures for use as growth media for ornamental plants, principally as a means to reduce costs. The ash was sieved to exclude particles larger than 2 mm. Up to 50% could be included without inducing excessive levels of Ca or soluble salts. Bottom ash also serves as a source of nutrients when included in growing media.

Fluidized Bed Combustion Residues

Fluidized bed combustion residues consist of bottom ash, calcium carbonate, calcium hydroxide, calcium oxide (if fresh), calcium sulfate, calcium sulfite, and possibly some unburned coal. The alkalinity of FBC makes it a valuable substitute for agricultural limestone. It can also serve as a source of B, an important nutrient for plants (especially legumes such as alfalfa). However, if B concentration in the material is high, leaching may be required to reduce concentrations below potentially phytotoxic levels.

Because of its ability to absorb moisture and to increase pH to 12-12.5, fresh FBC material has been successfully used for stabilizing organic-rich industrial biosolid filter cake resulting from microbial production of pharmaceuticals, which in the absence of treatment begins to putrefy after 4-5 days (Jody Tishmack, personal communication). The stabilized filter cake can then be mixed with yard waste and composted for use as a synthetic top soil.

The process of pasteurizing municipal waste water treatment plant sewage sludge as developed by the N-Viro Corporation involves treatment with lime kiln dust or cement kiln dust. Trials were conducted to determine if dry FGD or FBC materials could be substituted for the dust (Beeghly et al., 1992). The FGD material consisted of a lime injection multistage burner (LIMB) by-product resulting from a sulfur scrubbing process where calcium oxide is injected into the flue gas. The FBC used in the test was obtained from a circulating bed combustor. Both LIMB by-product (no longer being produced) and FBC residue were effective in meeting the requirements for producing a soil substitute. Free calcium oxide in the by-products increased pH of the sludge to over 12 for a minimum of 7 days. The heat of hydration, released as anhydrite (anhydrous calcium sulfate) and calcium oxide reacted with moisture in the sludge, maintained temperatures at 52-62 °C for 12 hours. Both of these steps are important for pasteurization of municipal sewage sludge. The hydration reaction also dried the sludge. The final soil amendment material was judged to be superior to that produced using lime kiln dust or cement kiln dust in terms of granularity, odor control, and general appearance.

Bhumbla et al. (1997) recently reported use of FBC residue in successful production of apple trees in West Virginia. A "technogenic soil" (consisting of FBC residue, top soil and sewage sludge) was placed on top of mine-land soil specifically contoured for slope and aspect to more closely meet the requirements for fruit tree production than the natural sites usually available to fruit growers in the state. Approximately 1-meter thick layers of various mixtures consisting of FBC material (up to three-fourths of the total volume), stockpiled top soil, and sewage sludge were prepared and placed over the soil remaining after surface mining. The FBC residue sufficiently increased soil pH to eliminate symptoms of manganese toxicity that appeared in the limed and fertilized topsoil treatment.

A mixture of FBC and fly ash was used to form thick caps for preventing weed growth under apple trees (Korcak, 1997). Caps of this nature allowed downward movement of water, but retarded evaporation losses (Ritchey et al., 1996).

Fly Ash

Fly ash is a fine powder consisting of Al, Fe, Si and O, with variable Ca and Mg contents, chemically bound into a glassy material. It can contain from 0.6 to 3.2 % K and some S (Terman, 1978). Small amounts of many plant nutrients and trace elements, such as B, Se, Cd, Mo, and As, are also present.

The particles making up fly ash are rounded, sometimes hollow, spheres 0.001 mm to 0.1 mm in diameter. The largest particles are equivalent to very fine sand particles in size, and the smallest to clay particles. The material was effective in improving texture of many mixtures. Increased air-filled porosity, decreased bulk density, and improved moisture retention capacity were attributed to fly ash incorporation in West Virginia (Bhumbla et al., 1993).

Class C fly ash generally has higher Ca content than Class F fly ash. The neutralizing ability of some Class C fly ashes is equivalent to up to 50% of their weight in calcium carbonate. This makes them useful in mixtures where alkalinity is desired.

Fly ash, particularly material with high unburned carbon content which is unsuitable for cement manufacture, has been used for mixing with municipal sewage sludge to form technogenic soils for mine land reclamation. In this capacity, fly ash can substitute for inorganic constituents of soil, while sewage sludge provides benefits of adequate levels of organic matter including improved N and P availability. Organic matter can also help provide physical stability to fly ash, which due to the spherical shape of its particles is subject to erosion. When fly ash makes up a major portion of the final material it is important to be aware of possible deleterious effects such as excessive concentrations of soluble salts, B, or other potentially toxic trace elements, and reductions in availability of P and N (Carlson and Adriano, 1993).

To overcome cementing effects that some fly ashes have when mixed with soil, Cawood (1967, cited in Terman (1978)), suggested co-application with organic materials such as crop residues, sewage sludge, or peat. The ability of fly ash to aid in formation of cement-like materials has been utilized in mixtures with FBC materials to construct agricultural pond liners, loafing pens for dairy and beef cows, and water-permeable weed barrier caps (Korcak, 1997).

Fly ash has also been used as a sand substitute in production of compost-based synthetic soil to provide water-holding capacity and improve drainage characteristics (Franciosi, 1997). Other components of the synthetic soil are composted cotton gin stems and hulls, and composted hardwood fibers. Potting mixes formed by combining fly-ash based compost with pine bark performed as satisfactorily as the commercial comparison standards.

Flue Gas Desulfurization Residue

One important use for calcium sulfate produced by scrubbing flue gas is as feedstock for manufacture of wallboard. Because specifications for this material are stringent, FGD gypsum tends to be relatively pure, the most common other components being calcium sulfite and calcium carbonate. The presence of calcium carbonate is usually beneficial for agronomic applications, but calcium sulfite can have some disadvantages. Direct incorporation of large amounts of calcium sulfite (30 g/kg, equivalent to 67 Mg/ha) into soil at pH 5.5 or less can generate sulfur dioxide gas, which can be toxic to germinating plants (Ritchey et al., 1995b). However, the toxicity is short-lived in soil since calcium sulfite is relatively insoluble, and rapidly oxidized to calcium sulfate (Ritchey et al. 1994). When surface-applied as dry pellets, FGD material has been shown to be an effective calcium source for peanuts (Fisher and Franciosi, 1997).

Stehouwer and Mafi (1997) reported on the use of a fixated FGD residue consisting primarily of calcium sulfite that had been treated with fly ash and 2-3% calcium oxide. The fixated FGD had a calcium carbonate equivalent of 15%. The FGD was mixed with yard compost prepared from leaves, brush and grass clippings, and incorporated to a depth of 51 cm in hyper-acidic coal refuse. The FGD contributed sufficient alkalinity to allow successful establishment of vegetation on the refuse material.

Calcium sulfate is often used to prepare growth media for mushroom production, and sometimes utilized as an ingredient in general compost preparation. Franciosi (1997) reported using a rate of 1.7 pounds calcium sulfate per cubic yard. The solubility of calcium sulfate is about 2 g/L (Weast, 1978) which provides a saturated solution of approximately 15 mM calcium, a solution Ca level higher than that usually found in limed soil. Crops such as sweet potatoes, peanuts, roses, blueberries, and Irish potatoes which are grown at low pH or need higher amounts of soluble Ca for production of underground fruit or tubers can benefit from high solution Ca concentrations. Calcium sulfate is also sufficiently soluble to leach below the zone of incorporation and reduce many effects of subsurface acidity.

Schuman (1995) reported benefits of addition of gypsum to bentonite mine spoil lands in northeastern Wyoming that had previously received Ponderosa pine wood residues (woodchips, bark, and sawdust) as an aid in overcoming sodicity problems. The wood residue was important to improve physical condition of spoils by enhancing water infiltration and vegetation growth. Posterior gypsum addition facilitated displacement and leaching of Na for long-term mine land reclamation.

Calcium sulfate is generally useful as a source of the important plant nutrients Ca and S (Franciosi, 1997). This makes FGD materials an excellent source of easily available Ca, while avoiding problems of excess salinity.

The inclusion of CCBP gypsum with mixtures of rock phosphate and cellulose added to acid soil reduced rock phosphate dissolution and soil level of labile P, presumably by shifting the equilibrium reaction towards insoluble Ca-P compounds (He et al., 1996). The reduction of P solubility is being investigated as a means to reduce P contamination of runoff water which results in export of P from farms to streams and estuaries eventually polluting areas such as the Chesapeake Bay and the Gulf of Mexico (Anderson et al., 1995; Stout et al., 1997). Amendment use of gypsum has been demonstrated to reduce soil P losses from manure-enriched fields in Florida, possibly by reduction of mineralization rates (DL Anderson, personal communication).

In conclusion, we have seen that there are a number of CCBP materials with widely varying properties. The potential for beneficial use of CCBP materials can be increased by judiciously combining them with other by-products to create or accentuate positive attributes and diminish deleterious ones.

REFERENCES

Alcordo IS & Rechcigl JE (1995) Phosphogypsum and other by-product gypsums. In: Rechcigl JE (Ed) Soil Amendments and Environmental Quality (pp. 365-425). Lewis Publishers, Boca Raton, FL.

Anderson DL, Tuovinen OH, Faber A & Ostrokowski I (1995) Use of soil amendments to reduce soluble phosphorus in dairy soils J. Env. Eng. 5:229-246.

Bearce BC, Myers S, Burch M, Engstrom B & Smutna L (1997) Coal bottom ash as a growing medium for poinsettia, Easter lily and peperomia. In: Proceedings 12th International Symposium on Coal Combustion By-Product (CCB) Management and Use (Vol 1. pp. 4-1 to 4-22). Electric Power Research Institute. Palo Alto, CA.

Beeghly JH, Grathwol K & Demuth J (1992) Limb ash mixed with sewage sludge to produce a soil amendment. In: Proceedings Ninth Annual Pittsburgh Coal Conference, (pp. 205-211), Pittsburgh, PA.

Bhumbla DK, Bearce B, Singh RM, Keefer RF, Hamrick R & Bayl D (1997) Using coal combustion residues to create technogenic soils on mined land for fruit tree production. In: Proceedings: 12th International Symposium on Coal Combustion By-Product (CCB) Management and Use (Vol. 3, pp. 102-1 to 102-16). American Coal Ash Association, Alexandia, VA.

Bhumbla DK, Keefer RF & Singh RN (1993) Ameliorative effect of fly ashes on acid mine soil properties. In: Proceedings 10th International Ash Use Symposium (Vol 3. pp. 86-1 to 86-31.) American Coal Ash Association, Washington, DC.

Bilski JJ, Alva AK & Sajwan KS (1995) Fly ash. In: Rechcigl JE (Ed) Soil Amendments and Environmental Quality (pp. 327-363) Lewis Publishers, Boca Raton, FL.

Brendel GF, DiGiola, Jr. AM & Tyson SS (1997) Developing an effective CCB marketing program. In: Proc. 12th Int. Symp. on Management and Use of Coal Combustion Byproducts (CCBs): Innovation for a Sustainable Future. (Vol. 3. pp. 94-1 to 94-23). American Coal Ash Association. Alexandria, VA.

Carlson CL & Adriano DC (1993) Environmental impacts of coal combustion residues J. Environ. Qual. 22:227-247.

Cawood MJ (1967) The concretion of pulverized fuel ash in relation to plant growth. Ph.D. thesis. University of Leeds, England.

Fisher BC & Franciosi FF (1997) Pelletization technology and product marketing for FGD residue for agricultural applications In: Proceedings: 12th International Symposium on Coal Combustion By-Product

(CCB) Management and Use (Vol. 2. pp. 59-1 to 59-7). Electric Power Research Institute, Palo Alto, CA.

Franciosi F (1997) Production and marketing of potting and landscape soils using coal combustion by-products. In: Proceedings: 12th International Symposium on Coal Combustion By-Product (CCB) Management and Use (Vol. 1. pp. 1-1 to 1-8). Electric Power Research Institute, Palo Alto, CA.

He ZL, Baligar VC, Martens DC, Ritchey KD & Kemper WD (1996) Factors affecting phosphate rock dissolution in acid soil amended with liming materials and cellulose. Soil Sci. Soc. Am. J. 60:1596-1601.

Korcak R (1997) FBC residue as an apple orchard surface mulch: final results. In: Proceedings: 12th International Symposium on Coal Combustion By-Product (CCB) Management and Use (Vol. 1. pp. 3-1 to 3-15). Electric Power Research Institute, Palo Alto, CA.

Korcak RF (1995) Utilization of coal combustion by-products in agriculture and horticulture. In: Karlen DL, Wright RJ & Kemper WD (Eds) Agricultural Utilization of Urban and Industrial By-products (pp. 107-130) Spec. Publ. No. 58. Am. Soc. Agron., Madison, WI.

Ritchey KD, Kinraide TB, Wendell RR, Clark RB & Baligar VC (1994) Strategies for overcoming phytotoxic effects of calcium sulfite applied to agricultural soils. In: Chiang S-H (Ed), 11th Annual International Pittsburgh Coal Conference Proceedings (Vol. 1, pp. 457-462) September 12-16, 1994, University of Pittsburgh, Pittsburgh, PA.

Ritchey KD, Feldhake CM, Clark RB & de Sousa DMG (1995a) Improved water and nutrient uptake from subsurface layers of gypsum- amended soils. In: Karlen DL, Wright RJ & Kemper WD (Eds) Agricultural Utilization of Urban and Industrial By-products Spec. Publ. No. 58 (pp.157-181). Am. Soc. Agron., Madison, WI.

Ritchey KD, Kinraide TB & Wendell RR (1995b) Interactions of calcium sulfite with soils and plants. Plant and Soil 173:329-335.

Ritchey, KD, Korcak RF, Feldhake CM, Baligar VC & Clark RB (1996) Calcium sulfate or coal combustion by-product spread on the soil surface to reduce evaporation, mitigate subsoil acidity and improve plant growth. Plant and Soil 182:209-219.

Schuman GE (1995) Revegetating bentonite mine spoils with sawmill by-products and gypsum. In: Karlen DL, Wright RJ & Kemper WD (Eds) Agricultural Utilization of Urban and Industrial By-products Spec. Publ. No. 58. (pp. 261-274) Am. Soc. Agron., Madison, WI.

Sell N, McIntosh T, Severance C & Peterson A (1989) The agronomic landspreading of coal bottom ash: using a regulated solid waste as a resource. Resources, Conservation and Recycling 2:119-129.

Shainber I, Sumner ME, Miller WP, Farina MPW, Pavan MA & Fey MV (1989) Use of gypsum on soils: A review. Adv. Soil Sci. 13:1-111.

Stehouwer RC & Mafi S (1997) Efficacy and transport of wet FGD scrubber sludge and compost amendments for revegetation of acidic coal refuse. In: Proceedings: 12th International Symposium on Coal Combustion By-Product (CCB) Management and Use (Vol. 1. pp. 6-1 to 6-11). Electric Power Research Institute, Palo Alto, CA.

Stout WL, Sharpley AN & Pionke HB (1997) Reducing soil phosphorus availability with by-products from power-generation plants. In: Tunney H (Ed) Phosphorus Loss to Water from Agriculture. CAB International Press, Cambridge, England (in press).

Stout WL, Hern JL, Korcak RF & Carlson CW (1988) Manual for applying fluidized bed combustion residue to agricultural lands. ARS-74, United States Dept. of Agriculture-Agricultural Research Service. Government Printing Office, Washington, DC.

Terman GL (1978) Solid wastes from coal fired power plants: use or disposal on agricultural lands. Bull. Y-129. Tennessee Valley Authority, Muscle Shoals, AL.

Weast RC (1978) Handbook of Chemistry and Physics. CRC Press, Boca Raton, FL.

COSTS AND ENVIRONMENTAL IMPACTS OF CO-UTILIZATION OF BY-PRODUCTS IN NURSERY OPERATIONS

CONRAD A. SKIMINA
Monrovia Nurseries, Azusa, California 91702, USA
E-mail d_merhaut@lightside.com

ABSTRACT

Historically, nurseries have been co-utilizing organic wastes such as sawdust, barks, mushroom composts and manure litters for a long time. Materials used now are sawdust, barks, poultry and stable litters, jute fiber, coconut coir, rice hulls, nut shells, yard wastes and biosolids composts. Some manufacturers are re-using poly plastics from film and containers in making durable, plastic boards and laminates. With many states banning yard wastes from landfills, nurseries are utilizing this source of organic matter. Use of yard waste and/or in house organic wastes has two potential benefits for any nursery operation: the cost savings associated with producing their own organics, and potential reduction in the use of fungicides due to the suppressive capabilities of compost. A container nursery will generate about 114 $m^3ha^{-1}yr^{-1}$ (60 yd^3A^{-1}) of waste. After consideration of all operating costs in grinding its own green waste, a nursery can save approximately \$10.24 per m^3 (\$7.83 per yd^3) of waste. This would require grinding 4883 m^3 (6386 yd^3) of nursery waste to recover the cost of a \$50,000 grinder, but the machine would stand idle 95% of the time. Hindrances to utilization of these wastes are 1) cost of machinery, 2) maintenance of equipment, 3) lack of knowledge and 4) availability of an abundance of yard wastes from municipalities. In an experiment to find alternatives to methyl bromide, composts containing nursery wastes were equal to or better than growing media fumigated with methyl bromide. The heat of composting eliminated the pathogens and weed seeds, and the resulting "suppressive" compost provided growth stimulation and biocontrol of diseases. In this experiment, 54 out of 55 treatments were equal to or better than methyl bromide treatments. Properly prepared composts eliminate composts the use of fumigants and reduced the need of some fungicides.

BY-PRODUCT UTILIZATION

The nursery industry has been recycling wastes such as sawdusts, barks, mushroom composts, biosolids composts and various locally available organics for many years. Prior to 1960, metal food containers were cleaned, painted and re-used as plant growing containers until plastics began to replace these containers. Plastics in turn, created a

S. Brown, J.S. Angle and L. Jacobs (eds.), Beneficial Co-Utilization of Agricultural, Municipal and Industrial By-products, 149–155.
© 1998 *Kluwer Academic Publishers,*

disposal problem. Recently, with the scarcity of landfills and the enactment of yard waste diversion regulations in many states, the use of recycled green waste is increasing tremendously. Some nurseries are realizing that reuse of yard wastes as well as their own wastes can result in substantial savings. Current organic co-utilization materials used in growing media are sawdust, stable and poultry litters, barks, rice hulls, nut shells, nursery wastes, yard wastes, paper, cardboard, worm castings, biosolids composts and coconut coir. Non-biological origin materials such as poly film and container melt down are recycled into the manufacture of plastic board or panels. Some plastic formulators are re-utilizing the waste poly by washing, re-processing and blending with virgin materials. This material is then sold to the manufacturers of containers.

COSTS ASSOCIATED WITH BY-PRODUCT UTILIZATION

In California, the current diversion law states that counties must divert 25% of their waste stream from landfills by January 1, 1995, and 50% by the year 2000. (California State Government, 1990) Yard wastes constitute a large portion of this volume and diversion of these wastes is being actively encouraged. Now with almost 50% of the states banning or restricting the disposal of yard wastes, nurseries are being forced to look at ways to utilize this resource. The average tipping (disposal) fee for the nation is \$34.00 per U.S. ton or approximately \$8.11 per m^3 (\$6.20 per yd^3) (Steuteville, 1996).

If a nursery elects to re-utilize its own waste it can 1) save the cost of the disposal fee, 2) save on the purchase of organics and 3) save by reducing the usage of pesticides. Potential savings could be as high as \$36.56/m^3 (\$27.96/yd^3) of raw, ground nursery waste. (Table 1)

Table 1. Potential savings for a nursery co-utilizing its own by-products.

	Ground Waste	
	per m^3	per yd^3 (U.S. \$)
Value of organic matter	\$10.46	\$8.00
Cost to grind waste[a]	-(2.86)	-(2.18)
Disposal fee savings[b]	28.96	22.14
Potential savings	\$36.56	\$27.96

[a] 28% remaining in volume after grinding. 0.80 divided by 0.28, per m^3.
[b] 28% remaining in volume after grinding, assuming loose waste is being disposed. 8.11 divided by 0.28, per m^3.

If the cost of processing loose waste and the value of the organic matter is known, the cost recovery of processing equipment can be determined by the formula:

$$\frac{\text{Cost of Grinder}}{(\text{VOM} \times \text{VR}) - (\text{CTP}) + (\text{DFS})} = \text{Number of } m^3\ (yd^3) \text{ which must be processed to recover the cost of equipment}$$

Where: VOM = value of organic matter per m^3 (yd^3)
VR = volume of original m^3 (yd^3) remaining after grinding
CTP = cost to process waste per m^3 (yd^3) of loose waste
DFS = disposal fee savings per m^3 (yd^3)

Taking into consideration that it costs, 1) \$0.80 to grind a m^3 (\$0.61 per yd^3) of loose waste, 2) \$2.86 to produce a m^3 of ground waste (\$2.18 per yd^3) and that 3) the value of the organic matter is about \$10.46 per m^3 (\$8.00 per yd^3) and that 4) there is a savings of \$8.11 per m^3 (\$6.20 per yd^3) in disposal fees, it would require grinding 4853 m^3 (6386 yd^3) of waste to recover the cost of \$50,000 for a grinding machine. Since a container nursery will generate approximately 113 m^3 of waste per hectare per year (60 yd^3 per acre), it would take a 43 ha (106 acre) nursery to recover the capital cost in one year, a 21.5 ha (53 acre) nursery 2 years and a 11 ha (27 acre) nursery 4 years. With such a grinder, it would take the 43 ha nursery only 15 days to grind all of its waste of one year. In contrast, an 11 ha nursery would grind all of its annual waste in 4 days! Indeed, there is a problem here, since for the 11 ha nursery, the machine would stand idle 99% of the time! Manufacturers will have to come up with a lower cost machine of lower capacity that a smaller nursery could utilize more efficiently. An alternative would be for several nurseries to form a cooperative to handle all wastes jointly. Another alternative is for a nursery to accept yard wastes from a local municipality, charge them a tipping fee, which in turn will help them recover costs more quickly. Some nurseries are now doing this. The result is that organic matter does not require outlays of capital. Where this is not feasible, the smaller nursery can take advantage of the low price of large volumes of ground yard wastes which are becoming more available.

ENVIRONMENTAL IMPACTS

One great advantage of composting green wastes, is the ability of the process to generate heat. This results in self pasteurization and elimination or reduction in the use of fumigants. Our research demonstrates that the heat of composting i.e., more that 3 weeks of temperatures greater than 55°C (131°F) is superior to some fumigants in the kill of weed seeds and pathogens (Table 2). In one test, it was found that even after 8 weeks, temperatures hovered at 55°C. In fact, our last extensive, "alternatives to methyl bromide" experiment demonstrated that appropriate manipulation of biologicals and heat of composting eliminates or reduces the need of chemical fungicides. In this experiment, 54 out of 55 treatments using compost products were equal to or better than methyl bromide fumigation (Table 3). One probable reason for this is that methyl bromide killed off many of the beneficial organisms. Use of compost alone or innoculated with beneficial microorganisms resulted in greater growth of 7 ornamental species than the use of methyl bromide (Table 4). In addition, use of compost products in combination with alternative inorganic treatment proved to be a highly effective alternative to methyl bromide treatment (Table 5). Because yard wastes, nursery wastes, barks and litters generate a great amount of heat of composting and are a good food base for microorganisms, sustained bio-control of diseases is feasible.

Table 2. Percent mean seed kill of 9 genera by alternative chemical/physical treatment[a].

Treatment	% Seed Kill
Solarization	67
Solarization + metham sodium, low rate	99
Solarization + Enzone, low rate	70
Solarization + Dazomet, low rate	100
Metham sodium, high rate	95
Ca cyanamide	55
Dazomet, high rate	100
Steam, 80 to 84°C for 12 min.	90
Composting	100

[a] Capsella, Cardamine, Cyperus, Malva, Poa, Portulaca, Senecio, Sisymbrium, Tribulus.
See Table 5 for treatment rates.

Another benefit associated with recycling nursery waste is the recycling of nutrients. Nursery waste, by its nature, has a high nutrient content, generally due to luxury fertilization (Table 6). Consequently, the C:N ratio often is 23:1.

Recycling of plastics used in the nursery business has an inherent problem of attached soil or sand. Cleaning the plastic in order to recycle it into new containers is costly. We investigated co-utilization by grinding polyethylene containers and using the ground plastic in growing media. This idea was short lived because in certain cases grinding such plastic, produced filamentous forms. Another problem is that polyethylene has a specific gravity of 0.963. This means that it may stay suspended in water or float, which causes problems with a water recycling system such as that used at Monrovia. Recently, container manufacturers have been forced to accept used waste containers to maintain favorable customer relations. They in turn supply this waste plastic to re-processors for cleaning and re-utilization.

BENEFITS TO THE ENVIRONMENT FROM CO-UTILIZATION OF WASTES ARE:

1. Diversion of wastes from landfills
   a) nursery waste is recycled into soil media
2. Conservation of mined resources (e.g., peat, petroleum)
   a) ample supplies of organic matter reduces the need for peat
   b) petroleum conserved by the reduction in manufacture of plastic
3. Reduction or elimination of fumigants
   a) a 0.36 kg of methyl bromide elimination per m^3 of compost
4. Reduced use of fungicides, herbicides
   a) suppressive composts reduce the need for drenching with soil fungicides
   b) the heat of composting kills weed seeds

5. Re-use of nutrients
   a) nutrients in nursery wastes are recycled into soil media
6. Good public image
   a) besides beautifying our surroundings with plants, the public sees the nursery industry as environmentally responsible

Table 3. Percent relative plant growth by interaction of inorganic x organic treatments[a]. Means of 7 species

Treatment Inorganic x Organic	% Relative Growth	Treatment Inorganic x Organic	% Relative Growth
19	110	47	113
20	107	48	108
21	121	49	115
22	114	50	114
23	106	51	120
24	112	52	109
25	109	53	107
26	112	54	116
27	107	55	115
28	108	56	111
29	109	57	111
30	112	58	108
31	102	59	109
32	110	60	109
33	115	61	109
34	112	62	107
35	114	63	109
36	113	64	109
37	113	65	109
38	109	66	107
39	116	67	109
40	121	68	106
41	107	69	109
42	115	70	109
43	110	71	97
44	112	72	112
45	114	73	105
46	106	74 methyl bromide[b]	99
		75 methyl bromide[b]	100

[a] See Table 5 for 11 inorganic treatments, Table 4 for 5 organic treatments
[b] Treatment 74: inorganic fraction only; Treatment 75: entire medium

Table 4. Percent relative growth of 7 species of ornamentals by composting treatment[a]

Aerated Compost Indigenous	Aerated Compost *Trichoderma*	Aerated Compost *Penicillium*	Aerated Compost *Gliocladium*	Aerated Compost *Trichoderma* *Pseudomonas*[b]	Methyl Bromide Fumigated
113	114	107	111	108	100

[a] Means of 7 plants of each species in 2 replicated blocks. Organisms added at appropriate compost temperature.
[b] *Pseudmonas* added to roots, not compost

Table 5. Percent relative growth of 7 species of ornamentals by alternative chemical/physical treatments of the inorganic fraction.

Inorganic Fraction Treatment		Organic Fraction Treatments[a]	% Relative Growth
1.	Solarization[b]	19 through 23	113
2.	Solar + metham sodium, low rate[c]	24 through 28	114
3.	Solar + Ca cyanamide, low rate	29 through 33	111
4.	Solar + Enzone, 0.61 kg/m^3	34 through 38	117
5.	Solar + Dazomet, low rate	39 through 43	114
6.	Solar + (proprietary)	44 through 48	118
7.	Metham sodium, 0.46 L/m^3	49 through 53	113
8.	Ca cyanamide, 1.48 kg/m^3	54 through 58	116
9.	Enzone, 1.23 L/m^3	59 through 63	111
10.	Dazomet, 0.19 kg/m^3	64 through 68	112
11.	Steam, 82°C, 12 min.	69 through 73	107
12.	Methyl bromide, 0.33 kg/m^3	none (indigenous)	99
13.	Methyl bromide, 0.33 kg/m^3	methyl bromide	100

[a] See Table 4 for the organic fraction treatments. Each array consists of 5 organic treatments. The medium consists of 20% inorganic, 80% organic
[b] 55 days, 20 cm depth, 0.1 mm poly
[c] Low rates are approximately 25% of the label rate

Table 6. Chemical composition of ground nursery waste[a].

		Kg/m^3					
pH	DS m^{-1}	N	P	K	Ca	Mg	Fe
5.7	6.00	2.60	0.196	1.01	1.96	0.59	0.88
Mn	Zn	Cu	Na	B	Bd[b]		
0.047	0.013	0.005	0.071	0.005	200		

[a] Extraction with 6N HCL, 16 hrs
[b] Bulk density, dry wt

REFERENCES

California State Government (1990) AB 939: 1095, AB 1820: 145, AB 3992: 145.

Steuteville R (1996) The state of garbage in America, Biocycle 37(4): 54-61.

Quarles W & Grossman J (1995) Alternatives to methyl bromide in nurseries - Disease suppressive media. The IPM Practitioner 17:1-13.

TEAM APPROACH TO RESIDUALS UTILIZATION

BILL SEEKINS
Maine Department of Agriculture, Maine Compost Team, Augusta, Maine 04333, USA
E-mail bill.seekins@state.me.us

ABSTRACT

The Maine Compost Team is an interagency group focusing on the advancement of composting as a practical and environmentally sound method of by-product utilization. The group is a self-directed team, whose members work for the different agencies that have an interest in the utilization of organic residuals. The Team is comprised of representatives from three state agencies and the state university, all working toward the goal of improved utilization of residuals through composting. The Team members are:

Dick Verville, University of Maine Cooperative Extension.
Bill Seekins, Maine Department of Agriculture, Food & Rural Resources.
Mark King, Maine Department of Environmental Protection.
Geoff Hill, Maine State Planning Office

The Team has undertaken a variety of projects that have increased the visibility and adoption of composting in Maine. Although these projects have themselves been successful, it is the Team approach to conducting these projects that is considered noteworthy and the focus of this presentation. The Team has adopted mission and vision statements, set out a number of objectives for its activities, and has basic operating principles and criteria for accepting projects that help guide its work. The Team's work was recognized in the August 1996 issue of Biocycle magazine.

THE MAINE COMPOST TEAM

The Maine Compost Team is an interagency group focusing on the advancement of composting as a practical and environmentally sound method of by-product utilization. The Team tries to take an active approach to addressing composting issues in Maine, rather than waiting for problems to develop and react to them. Team members include:

1. Dick Verville (University of Maine Cooperative Extension), the Team leader and the one responsible for coordination of the various players, making the necessary links to pull a project together, developing the educational component of most projects, and providing public relations; he has many contacts in both the public

S. Brown, J.S. Angle and L. Jacobs (eds.), Beneficial Co-Utilization of Agricultural, Municipal and Industrial By-products, 157–162.
© 1998 *Kluwer Academic Publishers,*

sector and in the agricultural community.

2) Bill Seekins (Maine Department of Agriculture, Food & Rural Resources) provides the technical background necessary for successful composting; responsible for recipe development, system evaluation and design, process trouble-shooting, and writing technical reports; has connections with agricultural businesses and agencies, as well as a knowledge of the compostable feedstocks within the state, based on his "Usable Waste Products for the Farm" survey and reports.

3. Mark King (Maine Department of Environmental Protection) a regulator in the sludge (biosolids) and residuals program provides the Team with knowledge of rules that govern composting in Maine; helps with site and layout facilities, system options and troubleshooting, and advises participants about steps necessary to meet requirements; greatest contribution may be the boundless energy he provides in seeing that whatever needs to be done, gets done. He provides links to the sludge (biosolids) compost facilities and to several industries.

4. Geoff Hill (Maine State Planning Office) a recycling specialist with the State's recycling program, which has a goal of recycling 50% of Maine's solid waste; has worked with many municipalities in establishing both home composting and municipal leaf composting; provides the Team with links to communities and regional recycling programs that have interest in cooperating on compost projects.

In addition to the regular Team members, most of the Team's projects involve other players. These include county field staff for the USDA Natural Resources Conservation Service (NRCS), farmers, municipalities and businesses. Recently, the University of Maine Waste Management Specialist has been involved with several of the Team's projects. These other players change depending on the nature of the project being undertaken and the geographic area of the state in which it is located.

Several benefits exist by using a team approach to work with the composting industry. The Team provides a 'one stop shop' for answers to composting questions that may be addressed to any one of the agencies. Because of their different backgrounds, Team members can provide information on all aspects of establishing a compost operation, such as siting and regulatory issues, system design, recipe development, pile management, troubleshooting, business planning, marketing, finding feedstocks, linking with other businesses or municipalities, and other issues. Working closely together on a number of projects has also resulted in a significant amount of cross training among Team members.

When a controversial or unusual question is raised, the Team is able to look at it from the viewpoint of each agency and formulate an answer to address all concerns, thus avoiding potential conflicting advice from different agencies. Since each agency is represented on the Team, each can share in Team successes. Finally, a team synergy exists, i.e., the Team is greater than the sum of the individuals of which it is comprised. The interaction and brainstorming among individuals with very different backgrounds and skills results in ideas and solutions that would never have been developed by any one of them individually.

MAINE COMPOST TEAM HISTORY

The Team had its genesis as a result of the Waterville-Winslow Leaf Compost Project in 1990. After that project, the Team was approached by the Lincoln County Recycling Program, which handled the recycling activities for 26 towns in coastal Maine. This resulted in three demonstration projects, all of which are still ongoing. Throughout the Lincoln County Projects, the Team worked on an Ad Hoc basis, coming together as requests for assistance were received. In 1991, the Team realized that there was a growing demand for the type of assistance they were able to provide and decided to adopt a name and establish themselves as an ongoing Team. In 1995, the Team became more formalized by adopting a mission statement with a vision, goals and objectives. The Team has conducted more than a dozen projects since its inception.

MISSION

The mission of the Maine Compost Team is to expand the use of composting to a wide variety of organic materials in many industries using multiple technologies in a cost-effective and environmentally-sound manner.

OBJECTIVES

In order to fulfill its mission, the Team has set the following objectives:

1. Conduct projects that introduce composting to new sectors of the Maine economy and demonstrate its feasibility as a by-product management method.
2. Educate others about the demonstration projects through:
   - written technical reports;
   - videos illustrating how a project was done;
   - handouts describing the project;
   - industry field days at the demonstration sites;
   - tours of the facility once it is established as an ongoing operation (after the demonstration project is completed); and
   - presentations at public workshops, conferences, and symposiums.
3. Identify the appropriate composting technology to fit each particular situation.
4. Establish a compost management system that can continue to be successful after the Team is no longer involved.
5. Hand off the management and follow-up activities to the organization (either public or private) that will continue it in the future.
6. Be ahead of the industry in identifying the next step needed for the development of composting in Maine.

TEAM OPERATING PRINCIPLES

The Team operates under several basic principles when undertaking a demonstration project. These are:

- provide Sound Technical Information;
- focus on making the best use of existing resources;
- commitment to keeping operations low cost;
- operations are not artificially subsidized, i.e., demonstration costs must reflect actual costs of operation; and
- farmers are seen as key players in many projects involving municipalities and businesses

PROJECT CRITERIA

Over the last six years, Team members have been approached by many individuals and organizations wishing to undertake a composting project. Some are provided assistance by Team members on an individual basis, and some are brought to the Team for consideration as a Team project. Because of the limited amount of time that each member of the Maine Compost Team can give to Team projects, the Team has had to be selective in accepting projects. The criteria used in making the decision to accept a project include:

1. No higher and better use is available for the feedstock being proposed as the primary compostable material for the project. For example, if a food waste could be used as an animal feed and an appropriate livestock operation is nearby, the Team would attempt to link the two rather than set up a compost operation.

2. The project must have an educational component. The principles involved with the project must agree to share the knowledge gained by working with the Team with other members of their industry. This may mean having field days or site tours, producing a video describing the project, and/or distributing a report detailing the project.

3. The project must utilize a new feedstock, a new technology, or a new approach to using an existing technology not previously demonstrated by the Compost Team. The intent of the Team is not to assist with the start up of every new compost operation, but to demonstrate the feasibility of composting to an industry and provide the necessary documentation so it can be repeated by other members of the industry. Or

4. The project helps move the composting industry in Maine in a critical new direction. Two of the projects undertaken by the Team fall into this category - the Compost Maturity Test Project and the University of Maine Cooperative Extension Compost Operators School. Both of these projects (discussed later) result from the recognition that some compost management issues must be addressed on a statewide basis rather than on an industry by industry basis.

MAJOR COMPOST RELATED PROJECTS

The focus of the Maine Compost Team activities has been to promote and improve composting in all sectors of the Maine economy where use of organic wastes can be enhanced. A description of major compost projects selected by the Team follow.

Maine Wild Blueberry Compost Project. The Team worked with Maine Wild Blueberry Inc. to test the compostability of the residuals from their blueberry processing facility. In that project, the Team developed a methodology and recipe to compost its waste berries and process waste from its blueberry freezer plant. The blend of ingredients included woodash, sawdust, and fish.

The Compost Maturity Test Project. In 1995, Woods End Laboratory of Mount Vernon, ME, approached the Compost Team with a request to conduct a field study of their new 'Solvita' Compost Maturity Test. The test promised to be a quick, low cost approach for compost system operators to test their compost to determine if it was ready to market and for what purposes. In addition, the test offered the potential to be used as a monitoring and pile management tool. The test could be conducted in four hours, compared to a minimum of five days for tests that must be performed at a laboratory, and the cost of the test would be about $14 each compared to $30 to 50 for a lab maturity test.

The Compost Team visited over 40 compost operations in Maine. At each site, the Team collected samples of compost which were then subdivided into two portions. One portion was sent to the lab for testing using standard laboratory techniques (Dewar's Flask test, O_2 uptake and CO_2 generation), while the other was tested using the 'Solvita' test kits. The results of the study were published in the August 1996 issue of Biocycle magazine. Based on these results, the State of Washington adopted a requirement that all compost products in their state will be tested using the 'Solvita' test kit prior to marketing. Several other states are considering similar requirements.

UMCE Compost Operators School.

The Maine Compost Team initiated the University of Maine Cooperative Extension Compost Operators School to meet the need for training and certification of compost facility operators and managers. The first session of the school was June 23 through 27, 1997 at the University of Maine Witter Research Center in Orono, ME. Additional sessions are planned for 1998. The school offers classroom and field training in all aspects of compost facility operation, from introductions to basic biology and composting systems to site and pile management to business planning and marketing. Interest in the school has extended well beyond the boundaries of Maine.

OTHER TEAM PROJECTS

Ricker Farm Project - linked Bates College with farm; feedstocks are leaves, hay, silage, dairy manure, food waste using a turned windrow system.

EMR Crab Compost Project - feedstocks are crab and lobster shells and sawdust with windrows turned by a front-end loader.

State of Maine Cafeteria/Hallowell Leaf Compost Project - linked City of Hallowell, the Hallowell Pre-release Center (Prison) and a cafeteria; feedstocks were leaves, chipped brush and food waste using a bin compost system.

Apple Pomace Compost Project - linked Town of Monmouth, Chick Orchards and Dorothy Egg Farm; feedstocks are apple pomace, chicken manure, leaves and wood ash using a turned windrow system.

J&L Urchin Compost Project - linked sea urchin processor with small farm; feedstocks were sea urchins and horse bedding using a passively aerated static pile system.

Windham Correctional Facility Compost Project - linked correctional facility with two communities, a major supermarket chain and a breeding facility for goats; feedstocks are leaves, goat bedding and food waste using a turned windrow system.

Coast of Maine Compost - linked a compost marketing firm (Coast of Maine) with several seafood and blueberry composters; provided technical assistance in blending, finishing and labeling the product for sale.

1996 Northeast Biocycle Conference - worked closely with conference organizers and with composters in Maine to focus the conference on activities in Maine; set up tours of several significant compost related activities in the state.

FUTURE PLANS

The Maine Compost Team will continue to evaluate the state of composting in Maine. The Team has already initiated some contacts and drafted a proposal to address marketing for small, farm based composters in the next two years.

CO-UTILIZATION OF BY-PRODUCTS FOR CREATION OF SYNTHETIC SOIL

L. DARRELL NORTON[1], ROBERT ALTIERI[2] and CLIFF JOHNSTON[2]
[1]*National Soil Erosion Research Laboratory, Agricultural Research Service, US Department of Agriculture, West Lafayette, Indiana 47907, USA and* [2]*Department of Agronomy, Purdue University, West Lafayette, Indiana 47907, USA*
E-mail norton@ecn.purdue.edu

ABSTRACT

Agriculture has been practiced in the USA for hundreds of years. The general trend has been to produce foodstuffs and send them to towns and villages where most of the population lives. As a largely agrarian society, concentration of wastes was not a big problem because much of the wastes produced were applied back to the land where it came. With the population shift from farms to large cities the waste streams became larger and more concentrated. Disposal of these wastes was generally in landfills. Today with the increased cost of landfilling, less landfill space and regulations restricting what can be filled, land application of many waste streams is becoming more economically desirable. Also given the fact that many of these waste streams contain beneficial organic materials and nutrients that came from the soil to begin with, it may be beneficial to amend the soil with them to improve soil organic carbon content, nutrient status and control erosion. The objective of this study was to develop a methodology to create a synthetic soil from waste materials. We studied two waste streams from a coal-fired power plant and a pharmaceutical operation in order to develop a co-blending technology. The method consisted of mixing the hot ashes with wet ORIS. Different combinations of mixtures were studied to determine the optimum ratio of wastes to provide a soil like material. By combining the organic rich industrial sludge (ORIS) with fly and bottom ash from a fluidized bed combustor we were able to reduce the adverse properties of both materials and create a soil like material with favorable properties. The ash provided a structural backbone for the ORIS and the ORIS served as a proton donor to reduce the alkalinity of the ashes. The impact of this research is that two waste materials with adverse properties can be combined such that they form a beneficial material with soil like properties that can be used as a synthetic soil for soil amending to improve organic matter content and potentially erosion control.

S. Brown, J.S. Angle and L. Jacobs (eds.), Beneficial Co-Utilization of Agricultural, Municipal and Industrial By-products, 163–174.
© 1998 *Kluwer Academic Publishers,*

INTRODUCTION

The Clear Air Act (CAA) of 1976 as amended in 1990 mandates that emissions from coal burning facilities reduce the amount of SO_2 delivered to the atmosphere. This is accomplished by burning low-sulfur coal and using Clean Coal Technologies (CCT). Among these, newer plants often choose Fluidized Bed Combustion (FBC) technique due to the lower costs. In the FBC system the combustion air passes through a bed of lime, coal and ash particles, enveloping them in the fluidizing gas and presenting an increased surface area for combustion. During the combustion process, the injected, finely ground $CaCO_3$ burns to CaO which reacts with the SO_2 converting it directly to $CaSO_4$ (anhydrite) which is falls to the bottom of the bed as the particles become larger and is removed. A large mass of bed waste materials accumulate from this process. Approximately one-third of a ton of wastes (fly ash plus anhydrite) are generated for each ton of coal-burned. The types of products produced include fly and bottom ash (mainly anhydrite). The former is generally very fine sand or silt sized and capable of being carried by the flue gases from the combustion and are typically removed from stack gases by electrostatic precipitators (Heat Recovery Ash) or particle filters [Baghouse Fly Ash (BFA)]. The latter is the residue waste removed from the bottom of the fire box [Fluidized Bed Combustion Bottom Ash (FBCBA)].

Coal combustion by-products have proven to be an effective soil amendment for neutralizing soil acidity, due to the presence of considerable amount of calcium sulfate and calcium oxide (Page et al., 1979; Adriano et al., 1980). Norton et al. (1993) found that FBCBA decreased soil surface sealing, increased infiltration and reduced water erosion by releasing electrolytes to the low electrolyte rainwater. Coal combustion by-products have been used to improve the water-holding capacity of coarse-textured soils (Chang et al., 1977; Campbell et al., 1983). These waste materials have been evaluated for potential agricultural land application (Stout et al., 1988). However, fly ashes are treated as a toxic industrial waste because of high concentrations of trace elements, such as B, Se, Al, As, Cr, Cu, Zn in some cases (Andren et al., 1975; Adriano et al., 1980; Elseewi and Page, 1984; Aitken and Bel, 1985; Bennet et al., 1985). Therefore, they can have harmful effects on plants growth or could increase plant tissue retention of trace elements to levels potentially toxic to animals feeding on those plants (especially Mo).

Recycling organic rich municipal or industrial sludge through the soil plant system has also been widely recognized as a promising method of disposal (Page et al., 1987). Although attention must be given to determine the sludge loading rates in order to avoid reaching harmful levels of enrichment of some trace elements in the soil solution that are potentially toxic to plants (Baham and Sposito, 1994).

It was hypothesized that BFA and FBCBA could be co-blended with Organic Rich Industrial Sludge (ORIS) in order to optimize the benefits obtained from the organic matter (OM) and nutrient content of ORIS and to minimize the environment and food chain pollution potential of the ashes. ORIS is the byproduct of a pharmaceutical fermentation product which has a high water content and must be refrigerated at 98 percent water content prior to land disposal by injection. There is a finite amount of storage and when land application is not possible due to frozen soil of other problems, costly shutdowns in production occur. The ORIS can be de-watered, but it has little structural integrity upon putrefying. This causes problems with storage, handling and disposal. The co-blending of the ashes with the ORIS can provide a structural "backbone" for the ORIS giving it better structural integrity. In addition, it was hypothesized that potential harmful heavy metals in the coal combustion by-products and the nutrient elements present in ORIS and may be chemically bound into the organic matrix through a combination of chelation, precipitation and sorption phenomenon. Thus, the potentially of their leaching loses to surface or groundwater would be reduced (Gu et al., 1994; Hugg et al., 1993 and Jardine et al., 1989). Also, ORIS can increase the microbial organism population that contributes to stabilize the enzymatic pool of such co-blended materials (Garu et al., 1991). Also, the addition of ORIS to the ashes could partially buffer the alkalinity of the ashes since the organic acids serve as a proton donor. A reduction in alkalinity would permit the material to be used as a soil amendment to increase OM status of soils. This increase in organic mater would cause soil to have a lower soil erodibilty (Wischmeier et al., 1971) and reduce the environmental threat of erosion.

Although some information is present in literature on the chemical and properties of these two different waste materials (Roy et al., 1984; Mattigod et al., 1990; Eary et al., 1990; Campbell et al., 1978; Schaumberg et al., 1980), no information is available concerning co-blended FBCBA, BFA and ORIS mixtures.

The main objective of this study was to develop a co-blending method that would produce a material that would have favorable properties and have good handling qualities that could be used as synthetic soil or soil amendment to reclaim land that had lost topsoil from disturbance or erosion.

MATERIALS AND METHODS

This study deals with an innovative co-blending strategy of these waste materials, with hot ash (~300 °C) being immediately mixed with ORIS, therefore, utilizing the excess heat to dry, and sterilize, the ORIS. For this purpose, different amounts of ORIS were mixed with hot ash in order to develop an appropriate co-blending methodology that produced a soil like material having desirable OM content, pH value, water holding capacity, nutrient retention. The material needed also to have a water content that would make it easy to be spread onto soil to improve and re-vegetate it.

In this study, we focused on the application of Fourier Transformed Infrared Spectroscopy (FTIR) to study the OM of the blended materials and X-ray Diffraction (XRD) to characterize their mineralogical composition. The first step was to determine the OM content of the blended materials. In addition, batch equilibrium leaching tests and pH titrations were performed to evaluate the stability of the organic coatings on the ash and the neutralization of the alkalinity of the ashes by the organic acids of the ORIS.

The main approach taken to produce the mixtures and the methods of characterization used are presented in Figure 1. In order to create the mixtures, stored samples of FBCBA (ground) and BFA were both collected from the Purdue University Power Plant in West Lafayette, IN. This plant is a coal burning power plant equipped with an atmospheric

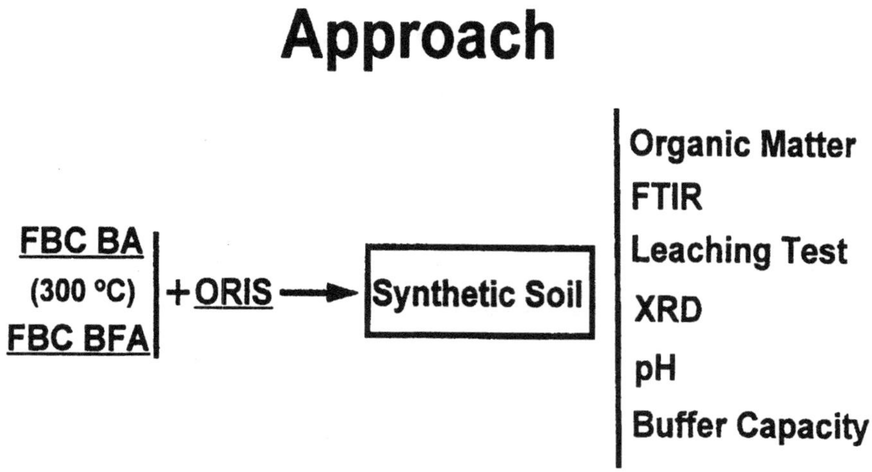

Figure 1. Schematic of the approach taken to produce and characterize mixtures of ORIS and ash.

circulating fluidized bed combustor. Aerobically digested ORIS (12% solids) was collected from Eli Lilly's Tippecanoe Laboratory Facility, Lafayette, IN. This facility produces ORIS as a result of a pharmaceutical fermentation process. These were the starter materials for creating the mixtures in this study.

Aliquots of ash samples were heated to 300°C for 20 minutes and immediately mixed with different amounts of ORIS [1:1, 1:1.5, 1:2, 1:2.5, 1:3 (w/v)] in order to obtain substrates having desirable OM content.

The mineralogical composition of the ashes was determined by X-ray diffraction analysis, using Co $K\alpha$ radiation and a Philips PW 1216/90 wide-range goniometer. The nature of the OM of the blended materials was studied by FTIR, using a Perkin Elmer 1600 Spectrophotometer and preparing KBr pellets. Moisture content of the mixtures were determined gravimetrically. The OM content was determined using a Leco Carbon Hydrogen Nitrogen analyzer [(CHN-600) Leco Corporation St. Joseph, MI] using standard procedures. The pH titrations were performed on batch solutions of the mixtures using a Mettler DL25 Titrator. The FBCBA suspensions were prepared by dissolving 0.1 g of material in 40 ml of de-ionized water, while the BFA suspensions, by dissolving 0.35 g in the same amount of distilled water. The different weights were used because of the less intrinsic alkalinity of the BFA.

In order to evaluate the stability of the OM on the ashes, the co-blended materials were submitted to a washing procedure with de-ionized water, by shaking overnight 1:50 (w/v) suspensions in N_2 atmosphere (to avoid pH changing due to the CO_2 atmospheric dissolution). After centrifugation, the resulting supernatants were discarded, while the solid phases were air-dried and measured for the carbon content.

RESULTS AND DISCUSSION

The FTIR proved useful technique to identify the presence (Figures 2, 3 and 4) and, indirectly, quantify the amount of OM on the ashes by calculating the integrated area of the main peak due to the OM [ν(C-H) of aliphatic groups (2980-2820 cm^{-1})]. Figure 2 is the FTIR spectra of pure ORIS containing 80.2% organic matter and shows the strong peak at this wavelength. Figure 3 shows the FTIR spectra of the various mixtures of ORIS with BFA with differing OM contents while the same is shown for ORIS+FBCBA in Figure 4. Figure 5 shows the strong positive relationship between the OM content and the integrated peak area of the blended materials.

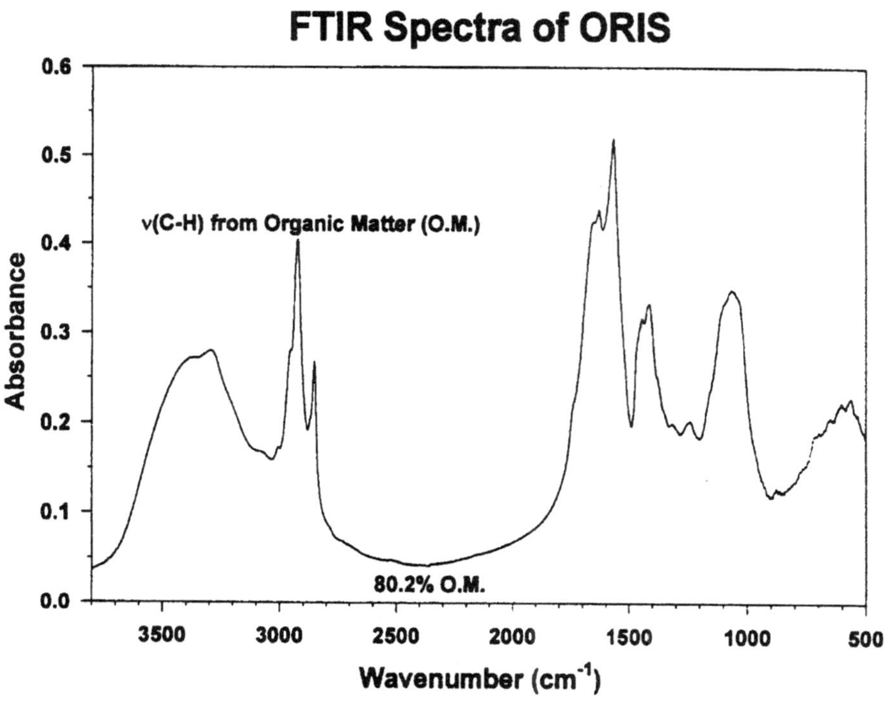

Figure 2. FTIR spectra of ORIS exhibiting strong C-H peak.

The XRD patterns revealed that the FBCBA contained mainly anhydrite ($CaSO_4$), whereas hematite ($Fe2O_3$) and quartz (SiO_2) were the dominant components in BFA (Figure 6). The FBCBA also contained lime, portlandite, calcite and quartz in minor amounts. The lime, portlandite and calcite were due to the inefficiencies of the scrubbing of the fluidized bed combustor. In the mixtures, the loading of OM onto the ashes did not change their mineralogical composition, except for the increased hydration of lime into portlandite in the FBCBA mixtures (data not shown). The mixtures of ORIS+BFA did not change the mineralogical composition, only the organic matter content.

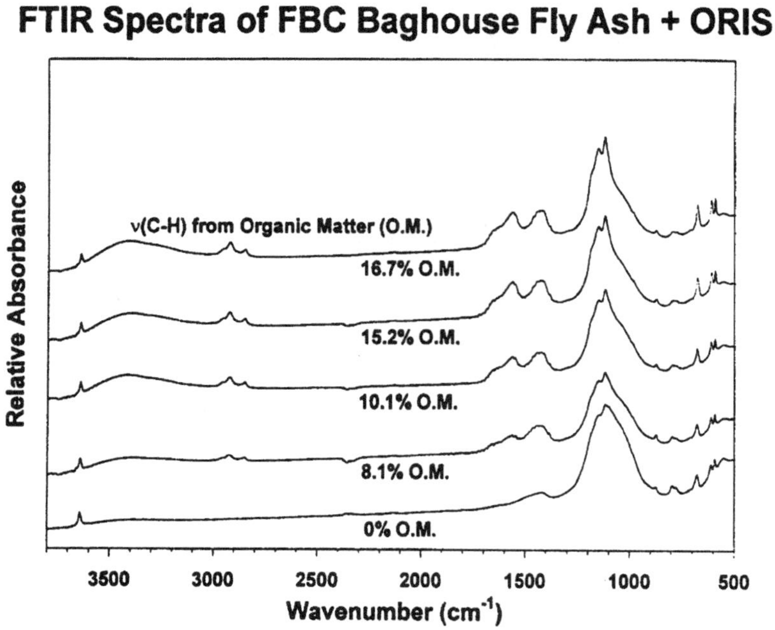

Figure 3. FTIR spectra of the ORIS and BFA mixtures.

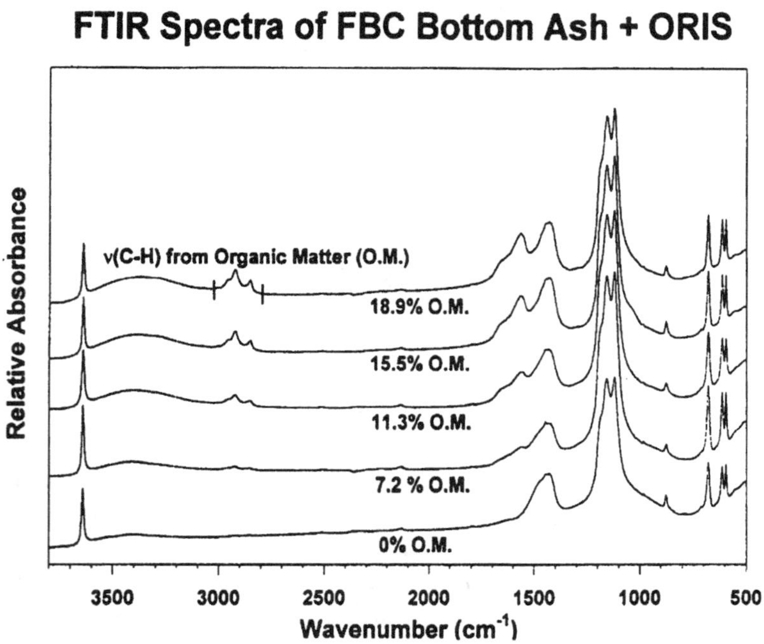

Figure 4. FTIR spectra of the ORIS and FBCBA mixtures.

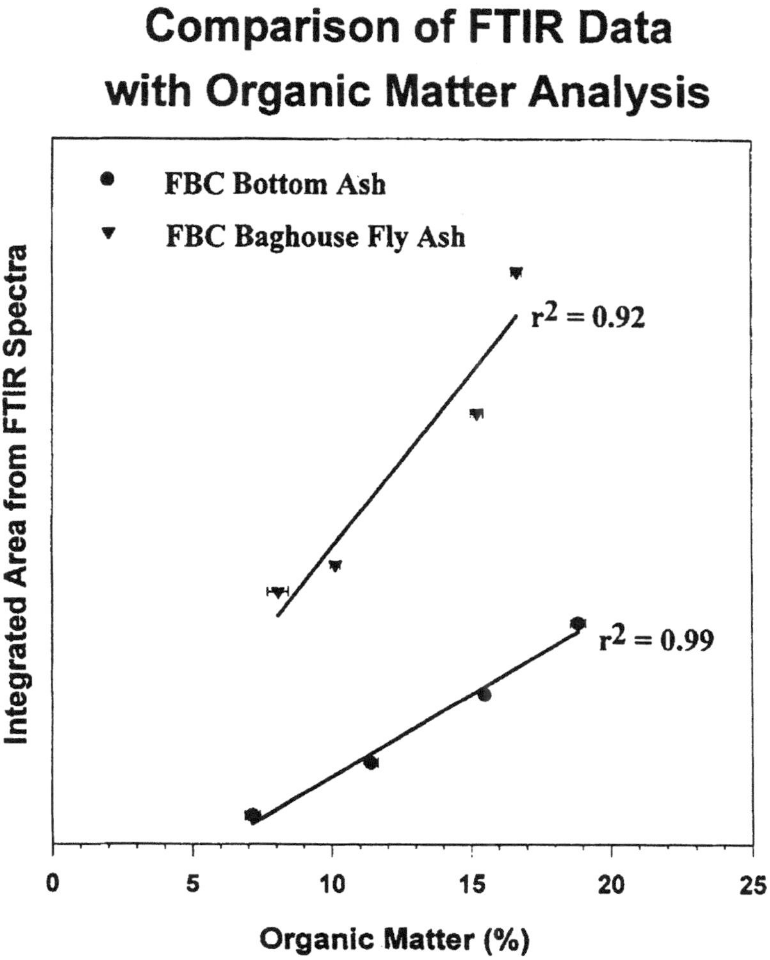

Figure 5. Relation between integrated C-H peak area andorganic matter content of the ORIS/Ash mixtures.

The OM content of the FBCBA+ORIS mixtures ranged from 7.1% to 22.8% before washing and from 6.4% to 25.2% after washing. This greater content after washing meant that some of the soluble components of FBCBA, such as lime or anhydrite, were lost in the washing. This is not necessarily bad since the release of electrolytes from FBCBA has been shown to prevent soil surface crusting, reduce clay dispersion, increase infiltration and reduce water erosion (Norton et al., 1994). The OM content of the BFA+ORIS mixings ranged from 8% to 21.4% before washing and from 8.5% to 16.4% after washing. The decrease in OM at the higher content indicated that some of the organic mater was soluble. Because of the less soluble mineralogy of BFA the organic matter content did not increase with washing and therefore the mixtures would not be a good electrolyte source to prevent erosion. The leaching tests did demonstrate that little of the OM was soluble (except for the

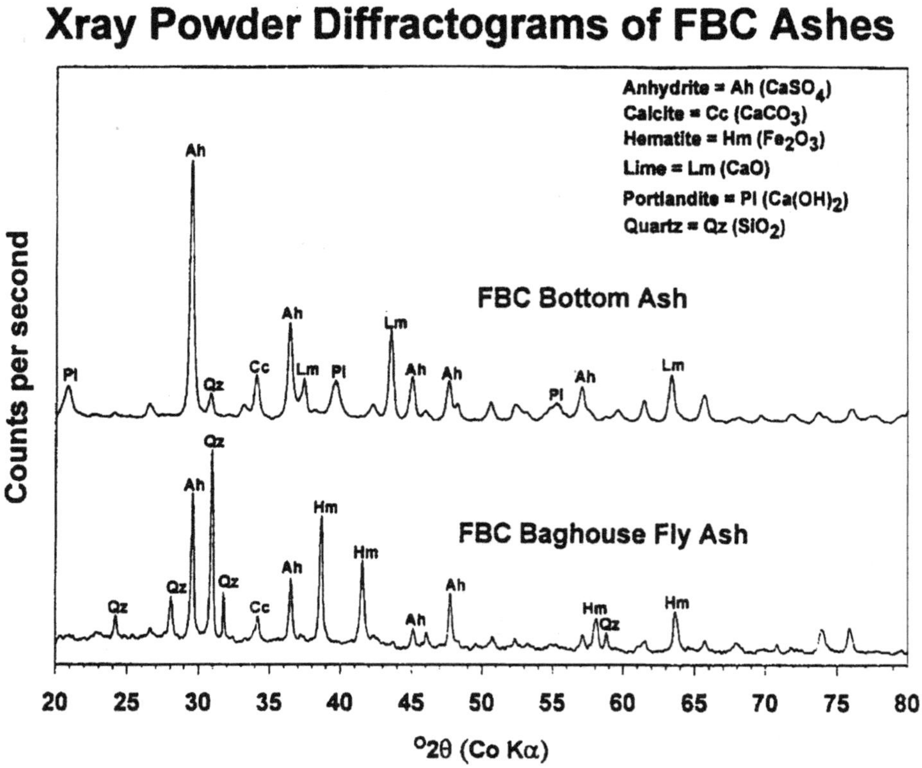

Figure 6. X-ray powder diffractograms of the BFA and FBCBA

BFA) and that it remained intimately associated with the ashes and and that the mixtures were stable (Figure 7). Figure 7 also demonstrates that mixtures with BFA+ORIS and less stable than those with FBCBA+ORIS because of the lesser slope of the line relating the OM content before versus that after washing. A completely stable material would have a line with a slope of one.

The most significant findings were those from the pH titrations of the mixtures (Figures 8 and 9). The titrations showed that the FBCBA had an alkalinity 3.7 times greater than the BFA. This was due to the greater amount of lime and portlandite as compared to BFA (Figure 4). The addition of the ORIS to FBCBA significantly reduced the buffering capacity of the ashes, in particular by 1.7 for the FBCBA with 18.9% of OM and by 3.4 for the BFA with 16.7% of OM. These values were calculated, for each sample, by measuring the amount of 0.05 M HCl needed to titrate to pH=7, starting from their natural initial pH (Figures 8 and 9). The reduction of the initial pH was also significant because in addition to reducing the buffering capacity of the ashes the pH of the mixtures were below 11. This is the value that causes the FBCBA to be classified as a toxic waste in Indiana and many .

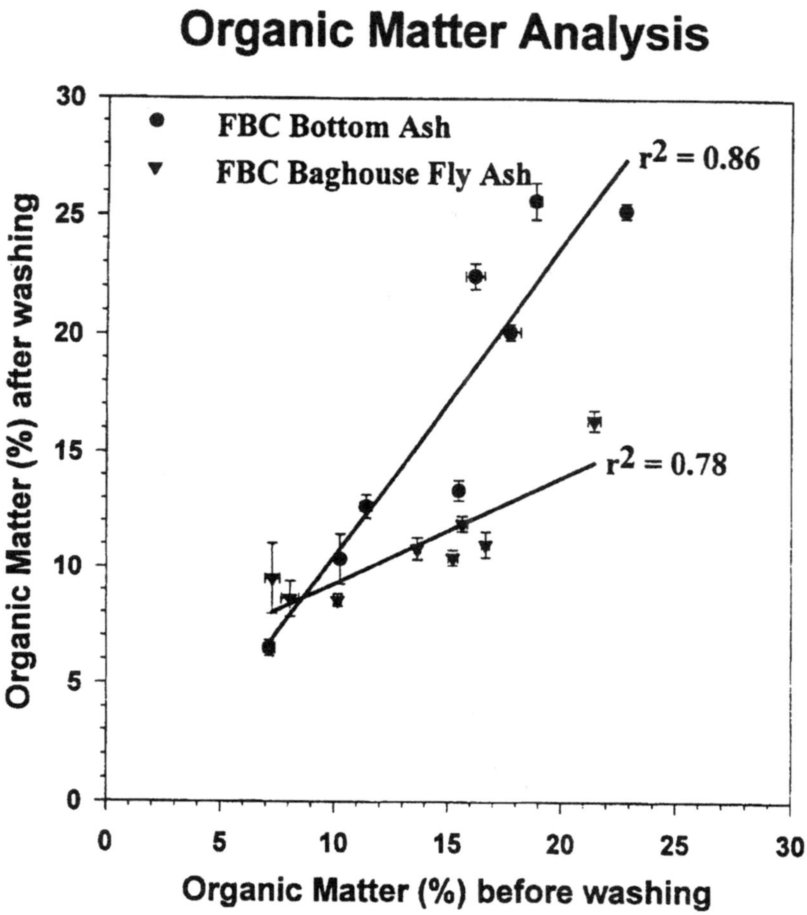

Figure 7. Relation between area from fTIR spectra andorganic matter content of ORIS and Ash mistures.

other states and therefore be regulated for land application. By mixing FBCBA with ORIS the pH was reduced below this level so the mixtures have the potential to be land applied The results were not as good for the mixtures of BFA+ORIS. However, these mixtures are less attractive for land application because BFA does contain heavy metals (data not shown), does not release electrolytes and has less stable organic material than the FBCBA+ORIS mixtures.

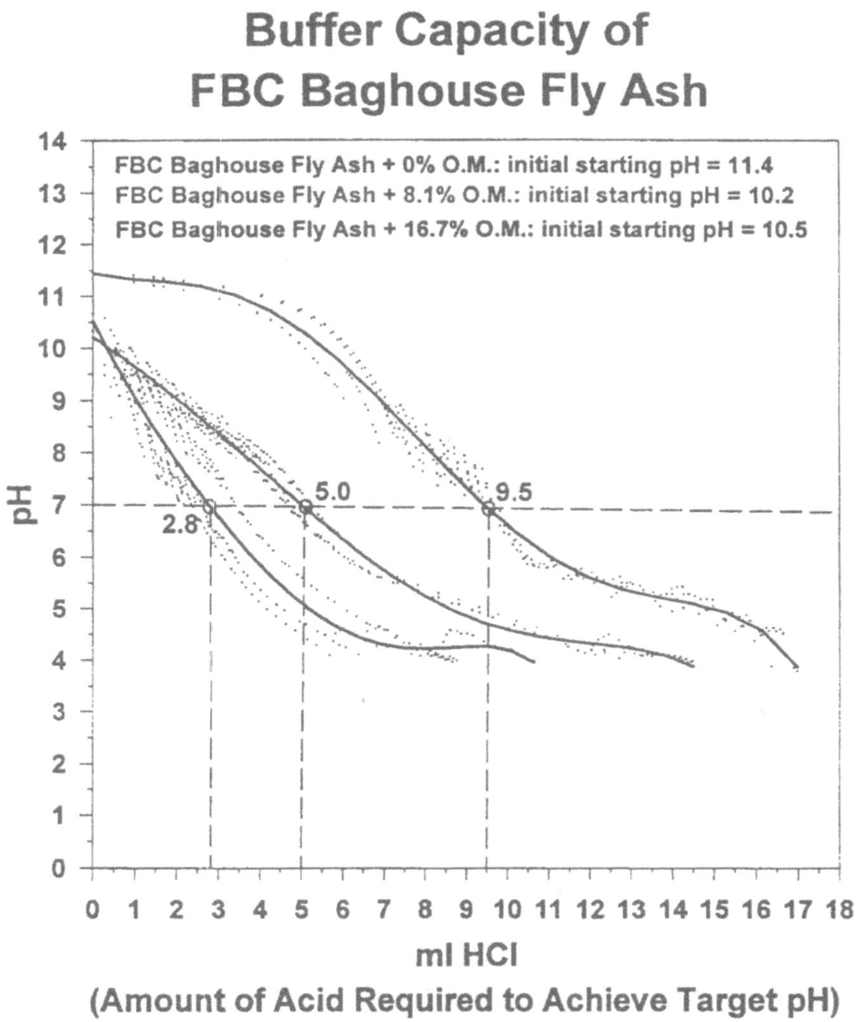

Figure 8. Titration curves for themixtures with ORIS and Baghouse Fly Ash.

CONCLUSIONS

Mixtures of ORIS with the ashes were made and the mixed materials were similar in appearance, structure and organic matter content to natural topsoil. FTIR proved to be a fast and accurate method to characterize the organic matter content of the mixtures. The mixtures were more environmentally as a result of reduced pH and alkalinity of the mixtures relative to the ashes alone and creating a stable coating of organic material on the ash. The ORIS served as a proton donor to both reduce the intial pH and the buffering

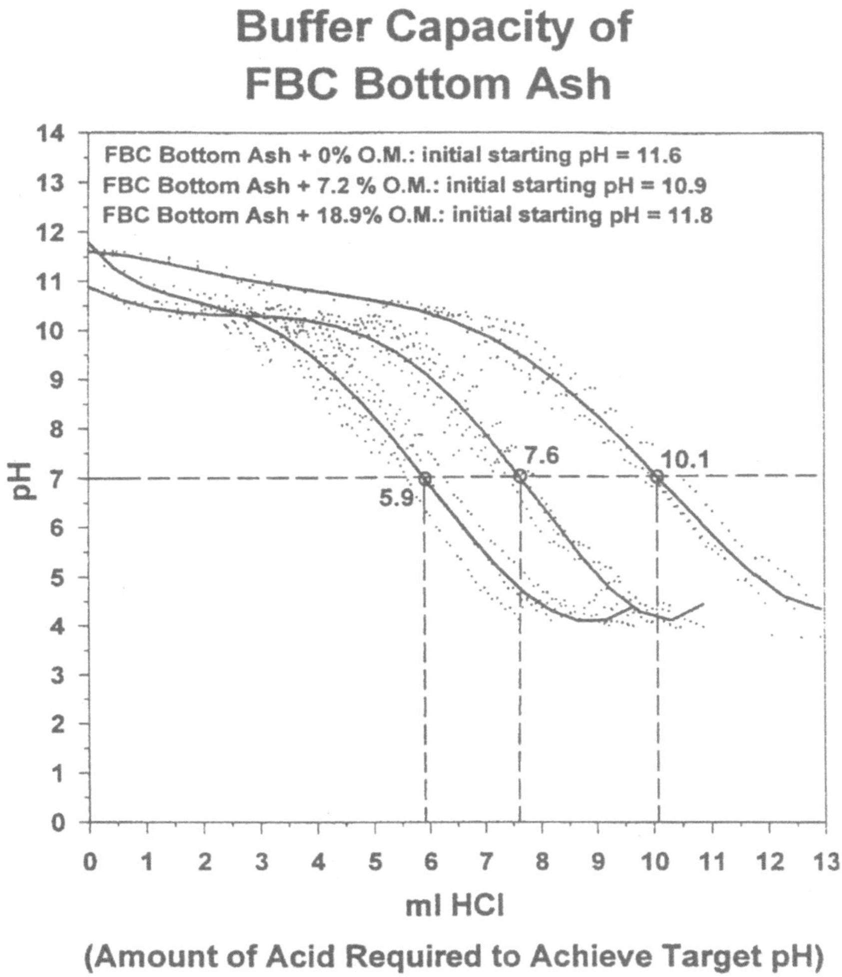

Figure 9. Titration curves for the mixtures with ORIS and FBCBA.

capacity of the mixtures particularly for the FBCBA+ORIS mixtures so they could be used as an unregulated soil amendment. The addition of the ash to the ORIS provided a structural backbone and gave the ORIS structural integrity. This gave the mixtures good handling and soil-like properties which would make it easy to spread and use as a soil amendment for reclamation, while providing for re-vegetation and potential erosion control through the release of electrolytes. The mixtures of FBCBA+ORIS were of lesser in pH and of greater stability during washing compared to those of the BFA+ORIS. The mixtures of ORIS and ashes had the appearance and handling properties of synthetic soil and had less adverse properties than each waste stream alone.

REFERENCES

Adriano DC, Page AL, Elseewi AA, Chang AC & Straughan I (1980) Utilization and disposal of fly ash and other coal residues in terrestrial ecosystems; A review. J. Environ. Qual. 9:333-344.

Aitken RL & Bell LC (1985) Plant uptake and phytotoxicity of boron in Australian fly ashes. Plant and Soil 84:245-257.

Andren AW, Klein DH & Talmi Y (1975) Selenium in coalfired steam plant emissions. Environ. Sci. Technol. 9:856-858.

Baham J & Sposito G (1994) Adsorption of dissolved organic carbon extracted from sewage sludge on montmorillonite and kaolinite in the presence of metal ions. J. Environ. Qual. 23:147-151.

Bennet OL, Hern JL, Perry HD, Reid RL, Stout WL, Edwards JH & Smedley KO (1985) Agricultural uses of atmospheric fluidized bed combustion residue (AFBCR) - A seven year study. Proc. 2nd Annual Pittsburgh Coal Conference (p.558-576).

Campbell DJ, Fox WE, Aitken RL & Bell LC (1983) Physical characteristics of sands amended with fly ash. Aust. J. Soil Res. 21:147-154.

Campbell JA, Smiith RD & Davis LE (1978) Application of X-ray Photoelectron Spectroscopic to the study of fly ash. Appl. Spectrosc. 32(3):316-319.

Chang AC, Lund LJ, Page AL & Warneke JE (1977) Physical properties of fly ash-amended soils. J. Environ. Qual. 6:267-270.

Eary LE, Rai D, Mattigod SV & Ainsworth CC (1990) Geochemical factors controlling the mobilization of inorganic constituents from fossil fuel combustion residues: II. Review of the minor elements. J. Environ. Qual. 19:202-214.

Elseewi AA & Page AL (1984) Molybdenum enrichment of plants grown on fly ash-treated soils. J. Environ. Qual. 13:394-398.

Garau MA, Dalmau JL & Felipo MT (1991) Nitrogen mineralization in soil amended with sewage sludge and fly ash. Biol. Fertil. Soils 12:199-201.

Gu B, Schmitt J, Chen Z, Liang L & McCarthy JF (1994) Adsorption and desorption of natural organic matter on iron oxide: mechanisms and models. Environ. Sci. Technol. 28:38-46.

Hogg DS, McLaren RG, Swift RS (1993) Desorption of copper from some New Zealand soils. Soil Sci. Soc. Am. J. 57:361-366.

Jardine PM, Weber NL & McCarthy JF (1989) Mechanism of dissolved organic carbon adsorption on soil. Soil Sci. Soc. Am. J. 53:1378-1385.

Mattigod SV, Rai D, Eary LE & Ainsworth CC (1990) Geochemical factors controlling the mobilization of inorganic constituents from fossil fuel combustion residues: I. Review of major elements. J. Environ. Qual. 19:188-201.

Norton LD, Shainberg I & King KW (1993) Utilization of gypsiferous ammendments to reduce surface sealing in some humid soils in the eastern USA. In: Poesen JWA & Nearing MA (Eds) Catena Supplement 24 (pp.77-92).

Page AL, Elseewi A & Straughan I (1979) Physical and chemical properties of fly ash from coal-fired power plants with reference to environmental impacts. Res. Rev. 71:83-120.

Page AL, Logan TG & Ryan JA (1987) Land application of sludge. Lewis Publ., Chelsea. MI.

Roy WR, Griffin RA, Dickerson DR & Schuller RM (1984) Illinois basin Coal fly ashes. 1. Chemical characterization and solubility. Environ. Sci. Technol. 18(10):734-742.

Schaumberg GD, LeVesque-Madore CS, Sposito G & Lund LJ (1980) Infrared spectroscopic study of the water-soluble fraction of sewage sludge-soil mixtures during incubation. J. Environ. Qual. 9:297-303.

Stout WL, Hern JL, Korcak RF & Carlson CW (1988) Manual for applying fluidized bed combustion residue to agricultural lands. U. S. Dept. of Agric., Agric. Res. Serv., ARS-74, 15pp.

Wischmeier WH, Johnson CB & Cross BV (1971) A soil erodibility nomograph for farmland and construction sites. J. Soil Water Cons. 26:189-193.

Zhon JL, Rowland S, Mantoura RFC & Braven J (1994) The formation of the humic coatings on mineral particles under simulated estuarine conditions. A mechanism study. Wat. Res. 28:571-579.

REMEDIATION OF SOILS CONTAMINATED WITH TOXIC ORGANIC COMPOUNDS.

MICHAEL A. COLE
Department of Natural Resources and Environmental Sciences, University of Illinois, Urbana, Illinois 61801, USA
E-mail mcole@ax1.sco.uiuc.edu

ABSTRACT

Decontamination or removal of soil that is contaminated with toxic organic compounds can be very costly, as illustrated by the values in Figure 1. In numerous situations, clean-up activities do not take place because the cost of clean-up may exceed the value of the property or dramatically decrease the willingness of the property owner to initiate remediation. Therefore, inexpensive, but effective, remedial methods are needed because the number of U.S. sites that are contaminated by toxic organic chemicals, toxic inorganic chemicals, or a mixture of the two is quite large. In addition to the nearly 1300 locations on the National Priorities List ("Superfund") and subject to RCRA (Resource Conservation and Recovery Act) there are a much larger number of sites of smaller size that may pose a threat to adjacent populations. For example, Brown, et al. estimated that 75,000 to 100,000 or more of below-ground petroleum storage tanks were leaking. Glass, et al. (1995) estimated that, in the US alone, about 37,000 candidate sites for Superfund, 80,000 sites for RCRA, 1.5 million leaking underground storage tanks, and about 25,000 Department of Defense sites in need of remediation exist. Similar situations exist in developed countries of Europe and in less developed nations. In the U.S., sale of contaminated property is difficult or impossible, a situation which may lead to abandonment by the owner and lost opportunities to redevelop contaminated, but potentially very valuable, urban sites into productive uses. Long-term use of property for military operations has also resulted in contamination with organic solvents, petroleum hydrocarbons, and explosives. Wilson and Clarke (1994) estimated that remediation costs for NPL and RCRA sites alone could be as much as $750 billion, or an amount equal to the U.S. military budget for about 15 years. The cost estimate for European sites (including the European Union) is between $300 and $400 billion. At these costs, it is unlikely that more than a small fraction of the most critical sites will ever be remediated.

One solution to the high cost of clean-up it to switch from expensive remedial methods to alternative methods that are significantly less expensive. On average, bioremediation is among the lowest cost methods for detoxification of soils contaminated with organic compounds (Figure 1) and composting is intermediate in cost among bioremediation technologies (Figure 2). When comparing the total budget for cleanup of a large site, the

S. Brown, J.S. Angle and L. Jacobs (eds.), Beneficial Co-Utilization of Agricultural, Municipal and Industrial By-products, 175–194.
© 1998 *Kluwer Academic Publishers.*

savings associated with use of bioremediation instead of chemical or physical based technologies give bioremediation an overwhelming monetary advantage (Table 1).

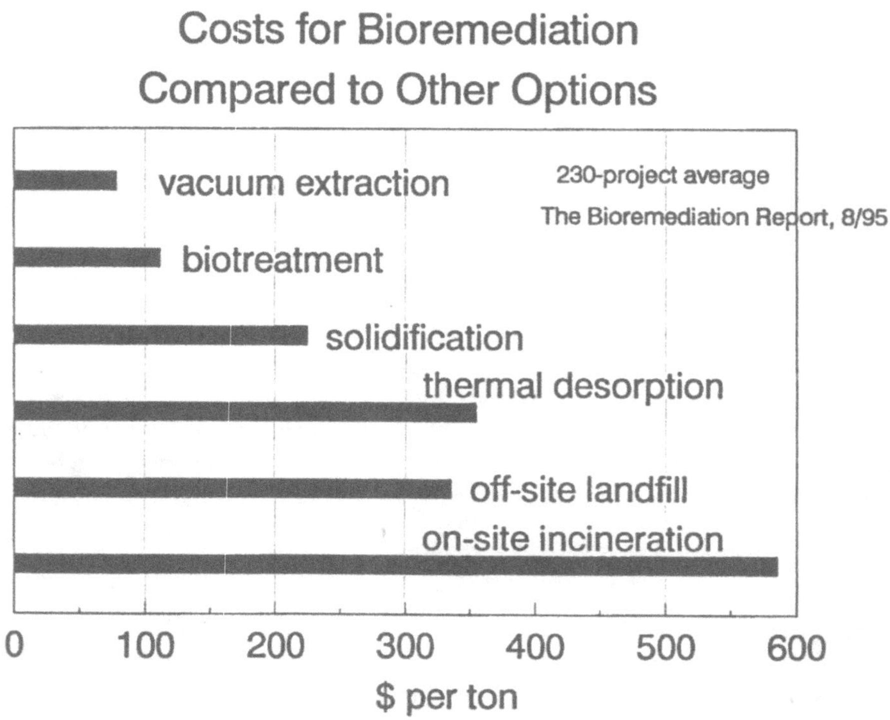

Figure 1. Comparative costs of remedial options for soils or hazardous wastes. (Data obtained from the Bioremediation Report, August, 1995)

Table 1. Total project costs for various remedial options. Costs are based on a site of 1 acre area, 20 feet deep (about 32,000 cubic yards).

Remedial Technology	Total Project Costs
Vacuum extraction	$2.5 million
Compost-based	$3.6 million
Solidification	$7.3 million
Thermal desorption	$11.4 million
Off-site landfill	$10.8 million
On-site incineration	$18.9 million

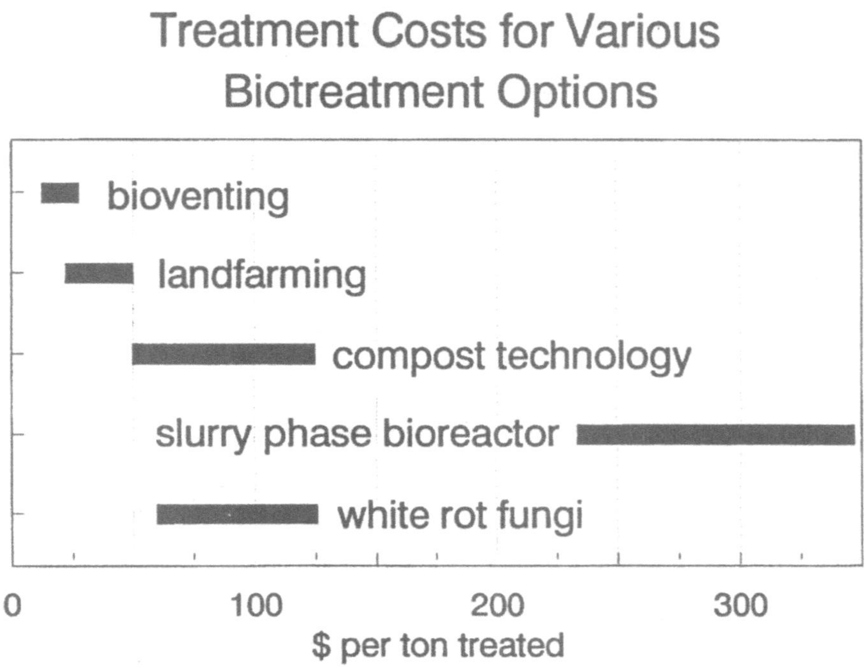

Figure 2. Comparative costs of bioremediation options for soils or hazardous wastes.

In this chapter, I will discuss three compost-based remedial technologies: (1) Use of the composting process for destruction of organic contaminants, (2) Use of mature compost as a soil amendment to accelerate contaminant degradation, and (3) Use of a hybrid system that employs the combined activities of plants and microorganisms; in this case, the compost allows plant establishment and growth in contaminated matrices where plant growth is poor or non-existent in the absence of compost.

APPLICATIONS OF THE COMPOSTING PROCESS TO SOIL REMEDIATION

A wide range of common environmental contaminants have been demonstrated to degrade rapidly in compost, as summarized in Table 2 and Figure 3. In some cases, the composting process was used, and in other cases, mature compost was added to accelerate contaminant degradation. Of the three methods, adding mature compost with or without planting will nearly always be less expensive than composting because less handling and processing of materials is required.

Table 2. Contaminants that degrade in compost or during the composting process

General class of contaminant	Examples
Petroleum hydrocarbons (TPH)	Gasoline, diesel fuel, jet fuel, oil and grease
Polynuclear aromatic hydrocarbons (PAH)	Wood preservatives, coal gasification wastes, refinery wastes
Pesticides	Insecticides and herbicides
Explosives	TNT, RDX, nitrocellulose

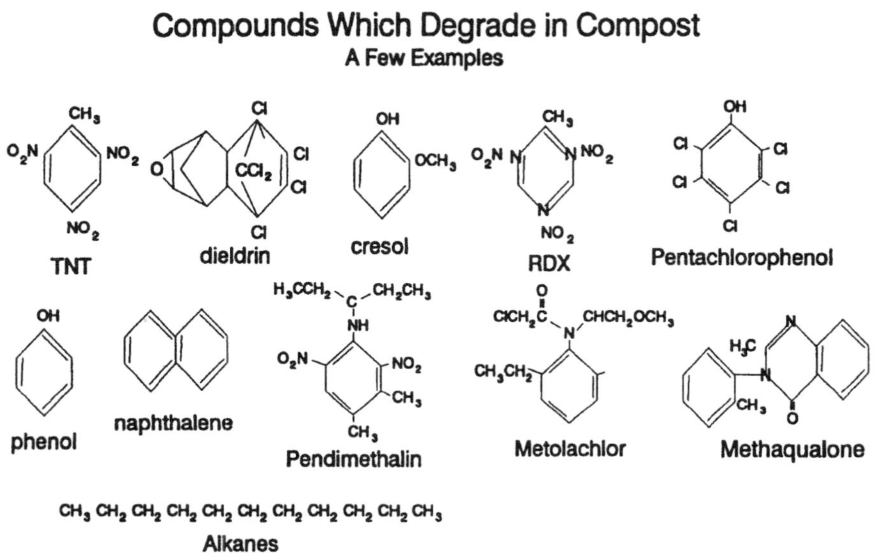

Figure 3. Structures of organic compounds that have been shown to degrade during composting or in soil amended with mature compots.

Among the compounds shown in Figure 3, the explosives TNT and RDX have been more widely studied than other compounds in experiments ranging from bench (laboratory) scale to large pilot studies. Most of the work has focused on the composting process, with typical results shown in Figure 4. The results are noteworthy for the rapid degradation; these explosives are stable for decades in contaminated surface materials because they are toxic to both plants and microorganisms. Consequently, "natural" or intrinsic remediation does not occur. Destruction of parent explosives is accompanied by highly significant declines in mutagenicity and toxicity of the contaminated matrix (Figure 5).

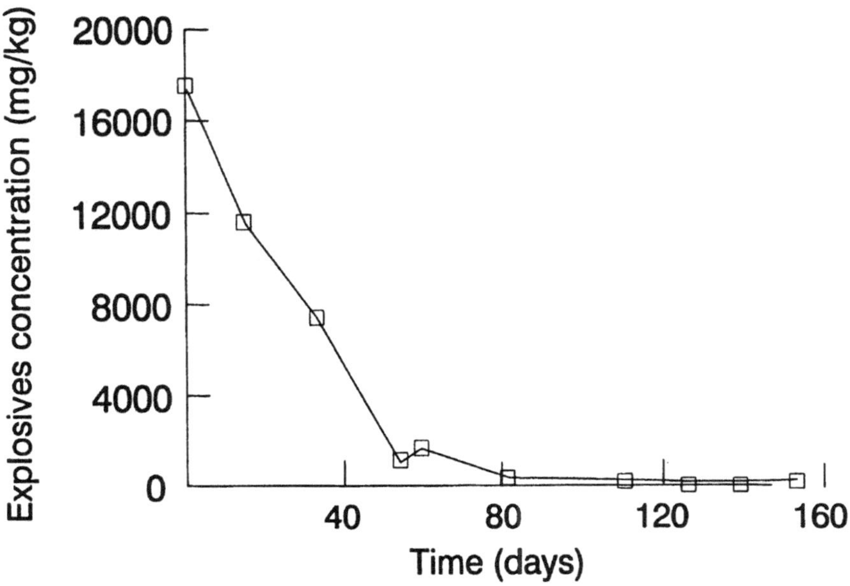

Figure 4. Degradation of the explosive, TNT, during composting (Williams and Keehan, 1993).

Composting is typically a thermophilic process, with temperatures reaching 55°C or higher during the early stages of degradation. When soil is mixed with raw organic materials, substantial quantities of heat are required to raise the soil temperature into the thermophilic range; if the proportion of soil to raw organic materials is too high, thermophilic conditions cannot be achieved. Brinton, et al. (1994) examined the effects of percentage of soil on temperature rise during composting of contaminated soil and found that up to 30% soil by volume could be mixed with compostable materials and still achieve thermophilic conditions. In a similar study, Williams and Marks (1991) found that inclusion of 40% soil in a composting mix resulted in sub-thermophilic temperatures and reduced degradation of explosives. Both of these studies indicate that a mixing volume of 30% soil with 70% initial compost feedstock provided the best results. Volume loss of feedstock is typically about 50% of initial, so the final, decontaminated mix would have about twice the volume of contaminated soil. If contaminants degrade completely, disposal of the extra volume should not be a problem. However, if contaminant degradation is incomplete, one has a substantially larger volume of contaminated material to further treat or dispose of. This problem should be avoidable by following a gradualistic approach of bench scale to pilot scale to full scale projects to ensure that reliable degradation of contaminants can be achieved (Saber, 1995; USEPA, 1989). One difficulty with this approach when using the composting process is that laboratory-scale composting units may not provide results that are similar in either extent or time scale to results obtained in large-scale composting. For example, Kaplan and Kaplan (1982) found relatively poor degradation of the explosive, 2,4,6-trinitrotoluene (TNT) in laboratory reactors (< 1 liter of material), whereas other researchers found good degradation of TNT

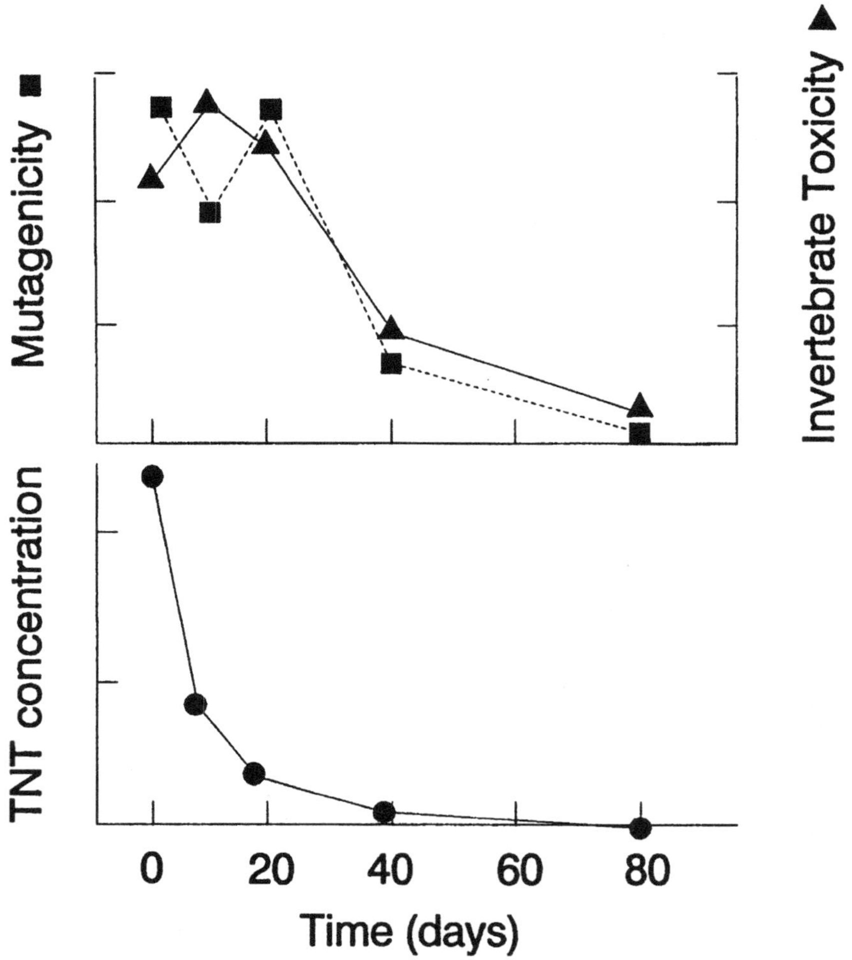

Figure 5. Reduction in total TNT content, leachate TNT, toxicity, and mutagenicity of explosives-contaminated soil during in-vessel composting (Griest, et al., 1993).

in pilot-scale studies when cubic meter quantities are composted. Based on these comparisons, even partial degradation under laboratory test conditions may be justification for conducting larger scale pilot studies.

A common complaint about solid-phase bioremediation methods is that they are too slow. For example, commonly-used procedures for bioremediation of petroleum-contaminated soils require several months to a year to achieve cleanup, a time scale that may be in excess of deadlines or owner's patience. Yare compared the time required to degrade a mixture of volatile organic solvents, polynuclear aromatic hydrocarbons (PAH), and phenanthrene in a solid-phase system (biopile) compared to a slurry-phase reactor. Biopile treatment time was 94 days, at which time >99% of initial volatiles, 91% of PAH, and 87% of phenanthrene had been degraded. In contrast, a 10-day treatment in a slurry-phase reactor gave degradation of >99% of volatiles, 63% of PAH, and 58% of phenanthrene. In this case, the biopile took substantially longer, but resulted in greater contaminant degradation and was achieved at much less cost than the slurry-phase reactor. Extended time periods increase cost, since the site must be monitored and operated for an extended period. However, use of the composting process or adding mature compost to biopile-type operations may dramatically decrease time to cleanup, as shown in the following examples.

Dooley et al. (1995) studied degradation of the herbicide, dicamba, during the composting process. Their results (Figure 6) indicate that successful remediation was achieved in only 52 days. Without the compost, typical degradation rates of dicamba in soil are 1-2 mg/kg/month (Goring, et al., 1975). Hence, treatment time for the high concentration material without composting would have been a year or more.

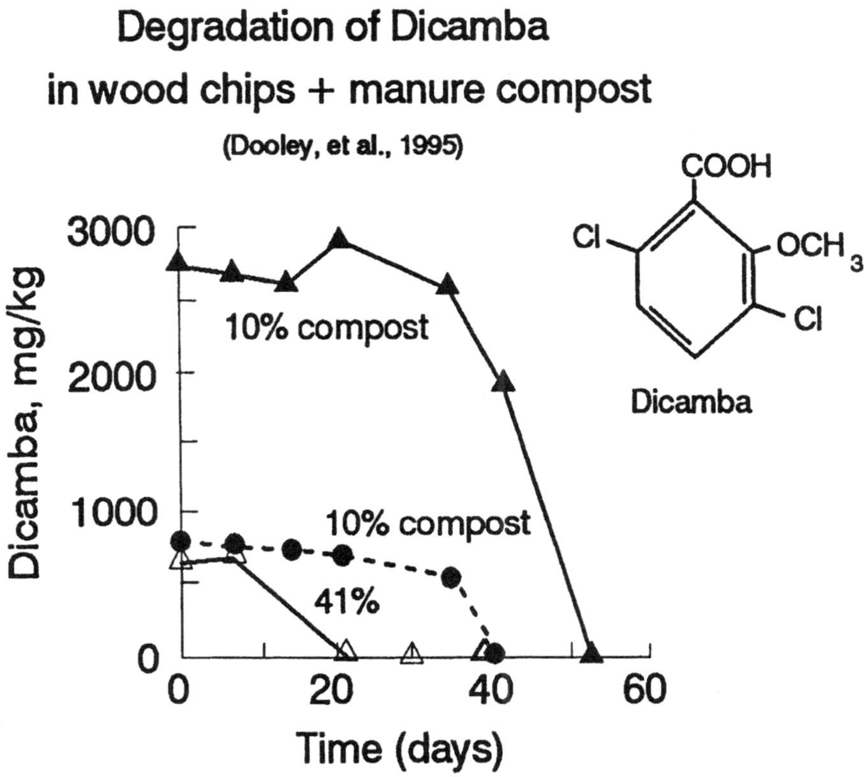

Figure 6. Degradation of the herbicide, dicamba, during composting (Dooley et al., 1995).

Beaudin, et al. (1996) composted a mixture of soil containing mineral oil and grease (35% v/v), maple leaves(20%, v/v), and alfalfa (35%, v/v) and other ingredients. The hydrocarbon mixture was a highly weathered material, which is often resistant to biodegradation. After a rapid initial period, degradation rate of residual material ceased

(Figure 7). The authors also described degradation during land-farming and reported only 30% degradation of contaminants in 180 days. In contrast, 50% degradation was achieved by composting in 105 days and 73% degradation in 287 days. Persson, et al. (1995) composted oily sludges containing hydrocarbon mixes in the lubricating oil and diesel oil molecular weight range and obtained 85% reduction from an initial concentration of 45 g/kg to 7 g/kg in 60 days. Mesophilic (25 - 35° C) conditions using decomposed horse manure were used.

The focus of this discussion has been on biological degradation, but the high temperatures achieved during composting will also accelerate chemical reactions that are relatively slow in soils, whose temperatures are only 15° to 30° C in most temperate climates. By comparison, typical temperatures during composting are 50°C or higher. Humic materials can catalyze degradation of atrazine (Li and Felbeck, 1972) and other compounds (Stevenson, 1994). Since the humic content of mature compost can be as high as 30% by weight, whereas typical soils contain less than 5%, compost also provides a much higher concentration of reactive material than is found in soil. From a thermodynamic standpoint, the substantially higher chemical reactivity of materials at higher temperatures combined with a large increase in reactive sites would be expected to accelerate abiotic processes quite substantially. This subject has not been explored to a significant extent at this time.

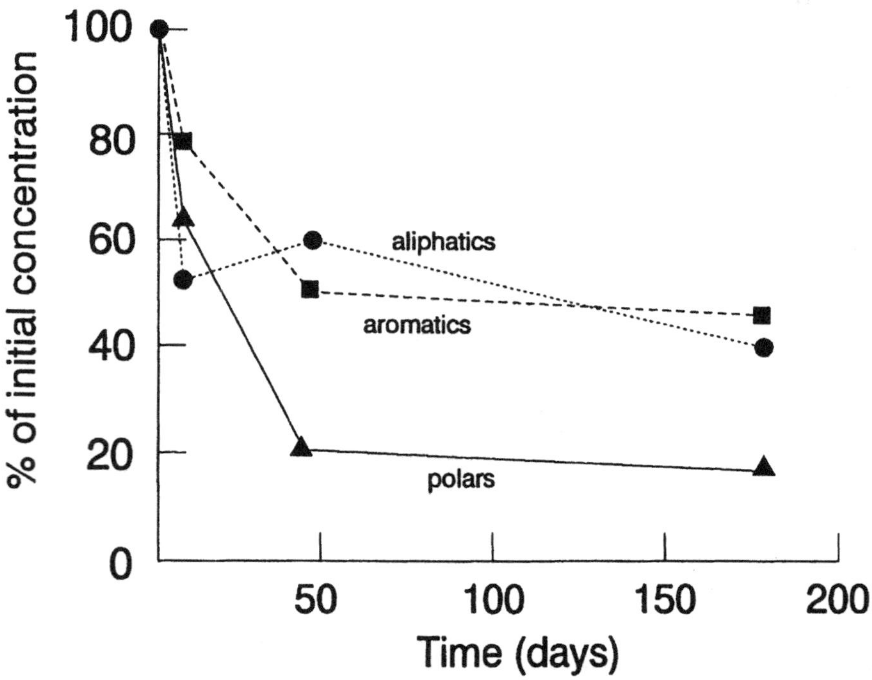

Figure 7. Degradation of mineral oil + grease during composting (Deaudin, et al., 1996)

Degradation of various aromatic compounds has been studied in composting systems, including chlorophenols, pesticides, and polynuclear aromatic hydrocarbons. In many cases, degradation occurred quite rapidly. For example, Civilini, et al. (1996) found > 90% degradation in 5 days of easily-degraded naphthylene, as well as 1-, and 2-methyl naphthalene during composting of wood-preservative contaminated soil and > 80% degradation of slowly-degraded PAH such as chrysene and pyrene in 15 days. Benoit and Barriuso (1995) studied degradation of 2-chloro- and 2,-dichlorophenol during composting and found rapid loss of parent compounds (Figure 8). Mineralization (formation of carbon dioxide) of substrate-derived carbon was relatively limited, with most of the carbon going into a bound residue fraction (Figure 9). The authors suggested that the bound residues were the result of oxidative coupling of the chlorophenols or their metabolites to humic materials in the compost. Similar behavior of chlorophenols has been reported in soil (Stott, et al., 1983).

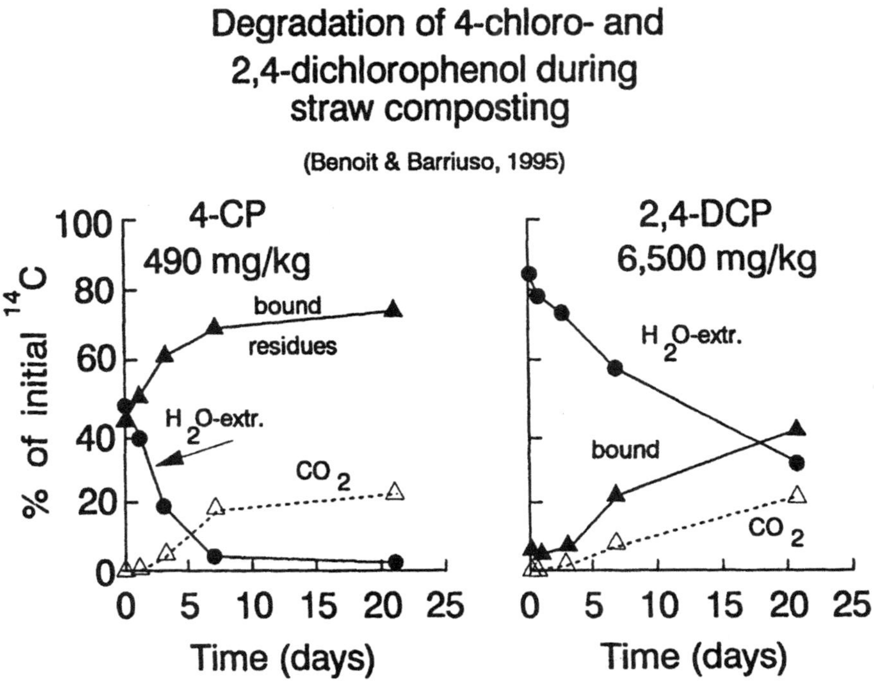

Figure 8. Degradation of 2-chlorophenol (2-CP) and 2,4-dichlorophenol (2,4-DCP) during composting (Benoit and Barriuso, 1995).

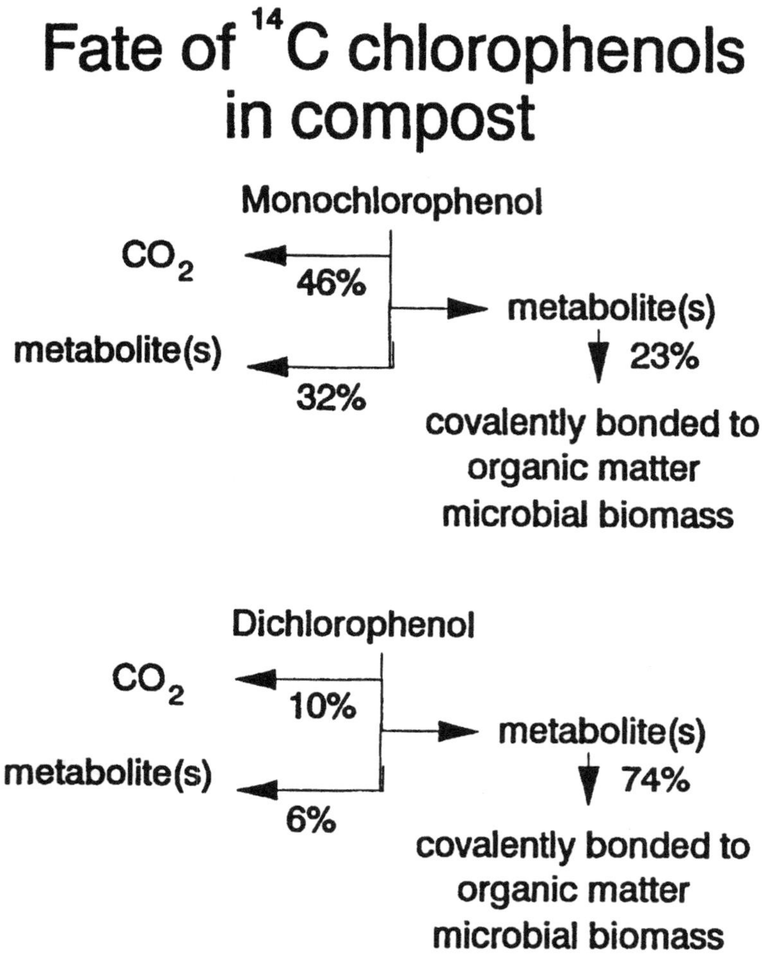

Figure 9. Distribution of 14C derived from 2-CP and 2,4-DCP after composting (data from Benoit and Barriuso, 1995).

USE OF MATURE COMPOST TO ACCELERATE CONTAMINANT DEGRADATION

Stegmann, et al. (1991) and Hupe, et al. (1996) studied the effects of mature compost on hydrocarbon degradation in soil-compost mixes in laboratory reactors. In this work, the authors had the best results by mixing mature, 6-month-old compost with petroleum (TPH)-contaminated soil and found degradation rates of about 375 mg TPH/kg/day. These values are much higher than those reported for *in situ* biodegradation of about 40 mg/kg/day (Atlas, 1991). TPH-contaminated soils frequently contain 5,000 to 20,000 mg TPH/kg. Based on rates shown in Figure 10, these materials could be remediated in only

2 weeks to 2 months, in contrast to 6 months or more for typical land-farming operations. Mass balance studies (Table 3) indicated that during a 21-day treatment period, substantial mineralization and bound residue formation occurred. The chemical nature of the bound residue was not determined; this material could be either strongly sorbed hydrocarbon or partially degraded hydrocarbon which was coupled to humic materials in the compost. In a field scale study, Bartusiak (1984) recorded oil degradation rates of about 110 mg/kg/day with a steel mill sludge containing primarily relatively high molecular weight (and therefore, relatively slowly degraded) hydrocarbons (Westlake et al., 1984).

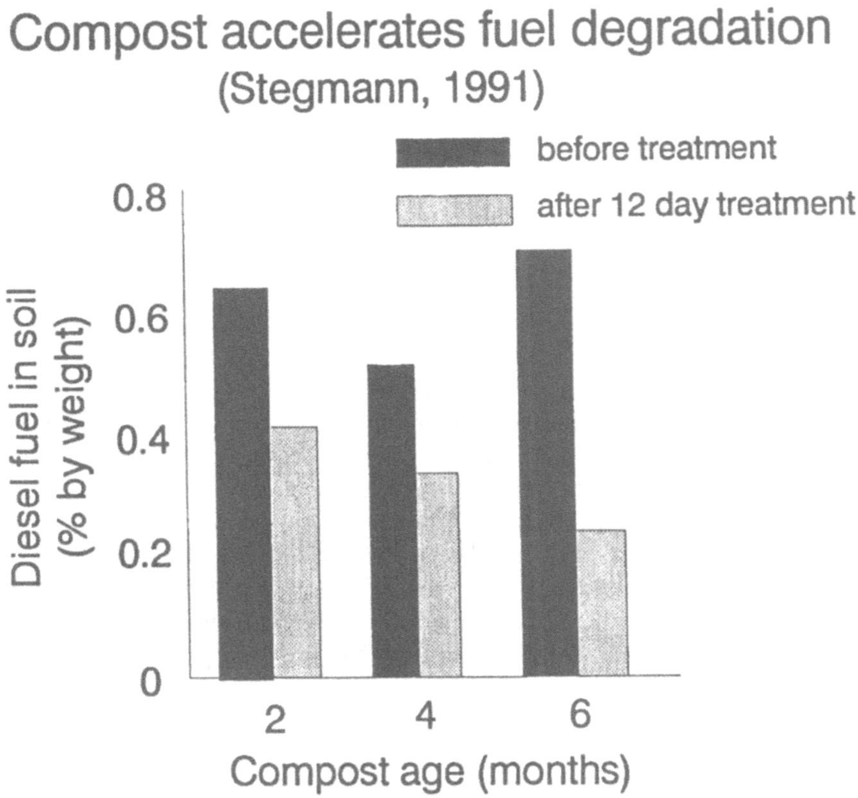

Figure 10. Degradation of petroleum hydrocarbons in compost-amended soil (Stegmann, et al., 1991).

In addition to direct use of composting or compost to accelerate contaminant degradation, various researchers (Civilini, et al., 1996; Civilini and Sebastianutto, 1996; Castaldi, et al., 1995) have used microorganisms isolated from compost for both basic biochemical studies and as inoculants in remediation projects.

Table 3. Mass balance for carbon from petroleum hydrocarbons during incubation of a soil-compost mixture. (Source: Hupe, et al., 1996.)

Fraction	% of initial-C in fraction
Extractable TPH	8
Volatilized	4
Converted to CO_2	59
Not accounted for (bound residue)	24
Microbial biomass	4

HYBRID SYSTEMS WHICH USE A COMBINATION OF PLANT AND MICROBIAL ACTIVITIES

Phytoremediation is a developing technology in which plants and root-associated microorganisms are the agents for degradation of xenobiotics. In its most widely used form, plants are established in contaminated soil and, as a result of water flow through the soil to the plant, a very dynamic non-equilibrium situation is created. Under these conditions, desorption of adsorbed organics is favored because contaminated water is constantly removed by the plants. As the organic contaminants enter the root zone (the rhizosphere, a zone of high microbial activity and populations in comparison to root-free soil), degradation may occur. Shann and Boyle (1994) demonstrated that degradation of the herbicides, 2,4-D and 2,4,5-T, was more rapid in rhizosphere soil than in root-free soil and that the extent of mineralization was greater with rhizosphere soil (Figure 11). Phytotoxicity of contaminated soil is a severe limitation to widespread application of phytoremediation; many common contaminants are more toxic to plants than they are to microorganisms. We addressed this problem in prior work (Cole, et al., 1994, 1995; Liu and Cole, 1996) by comparing growth of a mixture of grassy and broad-leaf weed species in herbicide-contaminated soil obtained from an agrichemical retail facility. When the contaminated soil was diluted with uncontaminated soil (as would be done for land-farming of the material), weed growth was very poor, even when the mixtures contained only 1.5% contaminated soil (Figure 12). In contrast, there was good weed growth in mixtures of mature compost + contaminated soil (Figure12). The goal in these studies was also to determine the relative contributions of microbial degradation and plant degradation of the herbicides. The data in Table 4 indicate that the presence of plants contributed significantly ($P < 0.05$) to degradation of pendimethalin and trifluralin in some cases, and the plants accelerated degradation of metolachlor in mixtures containing uncontaminated soil, but not in mixes of contaminated soil + compost.

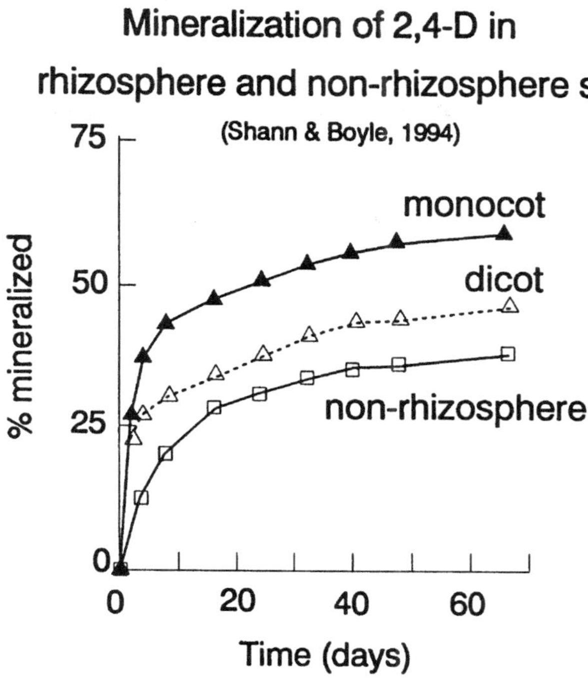

Figure 11. Relative degradatin rates and extent of mineralization of 2,4-D and 2,4,5-T in rhizosphere and non-rhizosphere soils. (Redrawn from Shann and Boyle, 1994, Figure 1).

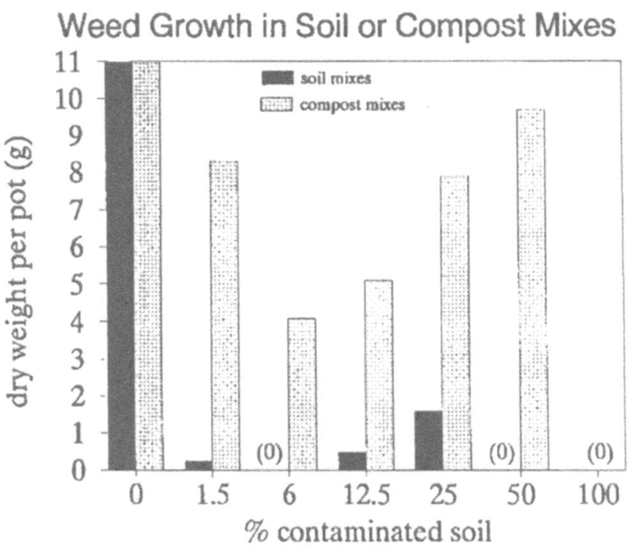

Figure 12. Weed growth in mixtures compost or uncontaminated soil with herbicide-contaminated soils. (From Cole, et al., 1995, Figure 2).

Table 4. Effects of mix composition and planting on pesticide degradation in contaminated soil mixed with compost or uncontaminated soil, following 40 d of plant growth. Values are means ± standard deviations of duplicate extractions of four replications per treatment. (Data from Cole, et al., 1995, Table 1).

Mixture	Treatment	Trifluralin	Metolachlor $mg\ kg^{-1}$ soil	Pendimethalin
Initial concentation	None	2.2 ± 0.9	3.0 ± 0.2	11.8 ± 5.1
100% contaminated	Planted	0.80 ± 0.82	3.4 ± 5.0	1.6 ± 0.4
100% contaminated	Not Planted	0.48 ± 0.77	0.99 ± 1.4	1.8 ± 0.4
50:50 soil	Planted	nd[b]	nd	0.5 ± 0.6 (0.01)
50:50 soil	Not Planted	0.52 ± 0.53 (0.07)	0.18 ± 0.16 (<0.001)	1.0 ± 0.2 (0.02)
50:50 compost	Planted	0.36 ± 0.33 (0.02)	nd	1.5 ± 0.6 (0.02)
50:50 compost	Not Planted	0.44 ± 0.69 (0.08)	2.8 ± 3.4 (0.29)	2.6 ± 3.4 (0.12)

[a] Values in parentheses indicate the probability that the values are less than expected from dilution alone (based on a one-tailed T-test for means of unequal variance).
[b] nd, not detected.

SYSTEM DESIGN AND APPROPRIATE SITUATIONS FOR COMPOST-BASED TECHNOLOGIES

Composting of contaminated materials or incubation of mixtures of mature compost and contaminated soil (with or without plants) could be done on a field-scale using simple designs such as the one shown in Figure 13. The design is mechanically simple and the unit can be constructed from readily available and relatively inexpensive materials. It provides full containment of materials while preventing washing away of material by rain. If volatile compounds are being processed, air flow would be set to draw air into the pile and pass it through a biofilter to remove the volatiles; the original Beltsville process for composting of sewage sludge (biosolids) that was described by Willson, et al. (1980) is an excellent example of a well-tested system design. In this case, the complexity is in the biological component, not the physical components and the only moving parts are the microbes and the ventilation system. The result is likely to be an effective, fast-acting, and inexpensive remedial system.

No remedial technology is appropriate for all contaminants and situations. Guidelines for best use of composting or addition of compost for remediation include:

1. Contaminants are less than 20 feet deep.
2. Contaminants are biodegradable and/or strongly adsorbed to the compost.
3. Soil is toxic to plants and microbes and the compost is able to reduce toxicity.

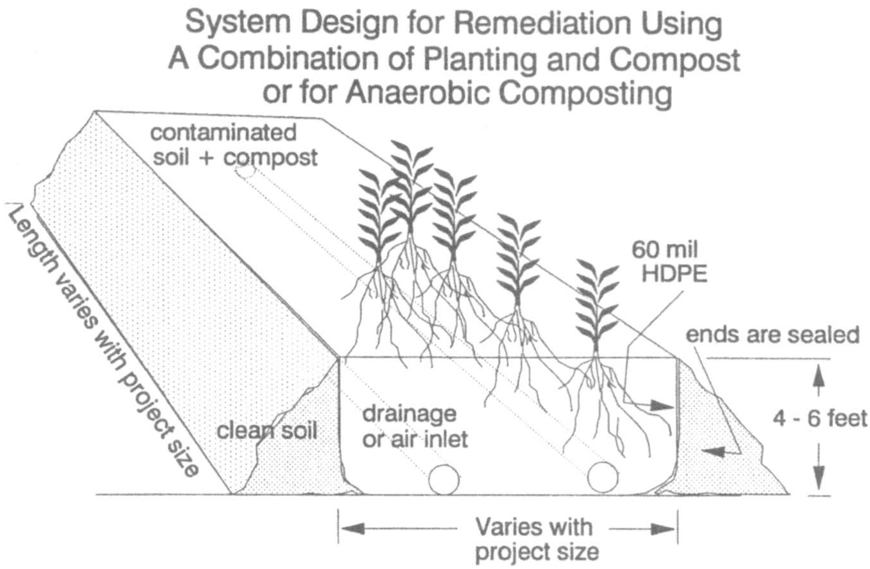

Figure 13. A simple contained system for composting of hazardous waste or treatment of mixtures of compost and contaminated soil (Cole, unpublished).

INFORMATIONAL GAPS THAT LIMIT POTENTIAL USE OF COMPOST-BASED TECHNOLOGIES

Before composting or use of mature compost can be widely accepted as remedial technologies, several issues need to be resolved. First, there is substantial anecdotal evidence that the degradation rate of specific contaminants may be affected by the materials that are being composted. For example, Semple and Fermor (1995) reported 16% mineralization of ^{14}C-labeled pentachlorophenol during 60 days of incubation with laboratory-produced compost or spent mushroom medium (quite similar in composition to compost), but found about 30% mineralization in mushroom medium of a lesser degree of stabilization.

Second, nearly all researchers have found a relatively low extent of mineralization of aromatic compounds in compost, and in some cases, formation of water-extractable metabolites.. Kaplan and Kaplan (1982) found that potentially toxic intermediates were formed during laboratory composting of explosives and raised the question of whether or not composting could be safely used for remediation of explosives. Griest, et al. (1993) reported >98% transformation of TNT during composting, but the material retained about 12% of its original mutagenicity and the aqueous leachate still had about 10% of its toxicity to an aquatic invertebrate (Figure 5). The other critical issue is whether or not the lack of full degradation and formation of non-extractable metabolites is a satisfactory endpoint of remediation. The behavior of aromatic compounds in compost is similar to behavior of hydroxylated or amino aromatic compounds in soils (eg., Figure 9 and Michel, et al., 1995), where partial degradation occurs, followed by covalent coupling of the metabolite to humic substances (Bertin, et al., 1991; Calderbank, 1989; Richnow, et al., 1994; Haider, et al., 1994; Sjoblad and Bollag, 1981), as shown in Figure 14. Hydroxylated metabolites are formed during degradation of nearly all aromatic compounds (Kelley, et al., 1993). In some cases, coupling of chlorinated phenols to humic materials is accompanied by dehalogenation (Dec and Bollag, 1994). This process, referred to as formation of "bound residues" results in long-term immobilization of the metabolites, but not complete destruction. The bound residues typically are very slowly degraded (Wolf and Martin, 1976; Völkel, et al., 1994).

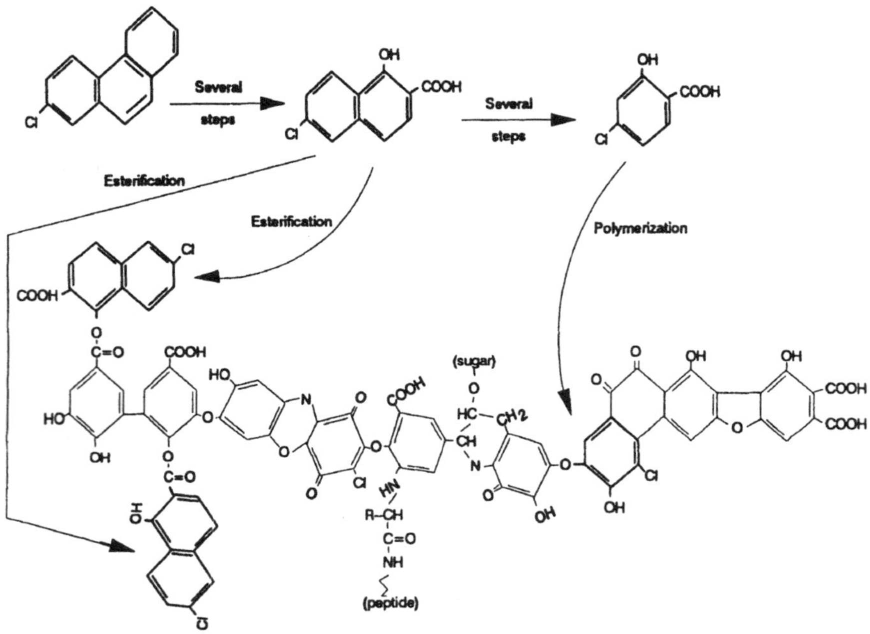

Figure 14. Possible mechanism for formationof bound residues during composting of soil containing aromatic contaminants (Humic acid structure from Stevenson, 1994; reactions adapted from Richnow, 1994).

Bound residues are defined by the International Union of Pure and Applied Chemistry (IUPAC) as "chemical species originating from pesticides, used according to good agricultural practice, that are unextracted by methods which do not significantly change the chemical nature of these residues" (Völkel, et al., 1994). In practice, loss of extractability by organic solvents is suggestive of bound residue formation (Haider, et al., 1994). The process is not simply adsorption (Piccolo, 1994), since sorbed low-molecular weight metabolites often remain solvent-extractable. Bound residue formation can be the result of synthesis of relatively labile bonds, such as ester groups, in which case, long-term stability would be relatively low. On the other hand, formation of ether linkages between humic materials and metabolites would result in relatively long-term stabilization of the metabolite in a form of low bioavailability. If the metabolite is actually incorporated into the core structure of the humic acid (Stevenson, 1994), the residence of the metabolite-derived carbon will be decades to centuries. Michel et al. (1995) found that substantial amounts of ^{14}C derived from ^{14}C-labelled 2,4-dichlorophenoxyacetic acid (2,4-D) were incorporated into humic and fulvic acids during composting of yard trimmings containing 2,4-D, but the nature of the bonds joining the bound residues to humic materials was not investigated. At the present time, the idea that partial degradation and immobilization of the metabolites is a satisfactory outcome is not widely accepted because complete destruction of the target compound has not occurred, but this may become acceptable if appropriate toxicological studies are performed to ensure that the metabolites are not hazardous and to establish that degradation products do not become bioavailable over time.

A number of studies on xenobiotic degradation in compost were conducted by measuring only loss of parent compound, but did not adequately measure volatilization or adsorption of compounds to vessel components such as plastics. At thermophilic temperatures, volatilization losses can be significant. Petruska, et al. (1982) found that nearly 50% of added chlordane was volatilized, but only about 5% was converted to bound residues; the balance was recovered as parent compound. Silviera and Ganjo (1995) reported 17% volatilization, 45% adsorption to vessel materials and 25% biodegradation of ^{14}C-naphthalene in laboratory reactors. If the authors had measured only loss of naphthalene, 87% of the naphthalene would have been apparently degraded. The work by Snell (1982) is the single most comprehensive examination of degradation of synthetic organics during composting, but the value of the work is reduced because of poor capture of volatiles. On a practical scale, it is evident that composting of soils contaminated with volatile organics should be done by passing the air that exists windrows through a filter to escape of contaminated air.

A third issue that requires resolution is the fact that the outcome of remediation experiments may vary depending on the scale of the experiment. For example, bench-scale results may not transfer well in terms of rate of degradation to pilot-scale or field-scale experiments. In several cases, better results were obtained in larger scale experiments when compared to very small scale laboratory experiments. Part of the difficulty in this case is probably the result of the inability to generate typical and authentic composting conditions in small laboratory containers. For pilot-scale composting studies, a volume of at least 10 to 20 cubic meters of material is required to achieve the typical thermal profiles that are seen in large windrows. Hence, the results from a pilot study of only a cubic meter may not transfer to a larger system.

REFERENCES

Atlas RM (1991) Bioremediation of fossil fuel contaminated soils. In: Hinchee RE & Olfenbuttel RF (Eds) *In Situ* Bioreclamation (p. 15). Battelle Press, Columbus, OH.

Bartusiak RD (1984) Microbial destruction of industrial solid waste. In: Majumdar SK & Miller EW (Eds) Solid and Liquid Wastes: Management Methods and Socioeconomic Considerations (pp. 93-103). The Pennsylvania Academy of Science, Easton, PA.

Beaudin NR, Caron F, Legros R, Ramsey J, Lawlor L & Ramsay B (1996) Cocomposting of weathered hydrocarbon-contaminated soil. Compost Science and Utilization 4:37-45.

Benoit P & Barriuso E (1995) Effect of straw composting on the degradation and stabilization of chlorophenols in soil. Compost Science and Utilization 3(3):31-37.

Bertin G, Schiavon M, Andreux F & Portal JM (1991) Contribution to the study of nonextractable pesticide residues in soils: incorporation of atrazine in model humic acids prepared from catechol. In: Berthelin (Ed) Diversity of Environmental Biogeochemistry (pp. 105-110).

Brown RA, Norris RD & Bruaker GR (1985) Aquifer restoration with enhanced bioreclamation. Pollution Engineering 17:25-28.

Calderbank A (1989) The occurrence and significance of bound pesticide residues in soil. Reviews of Environmental Contamination and Toxicology 103:71-103.

Brinton WF, Evans E & Droffner ML (1994) Amendment Selection and Process Monitoring for Contaminated Soil Composting. U.S. Army Environmental Center. Final Report ENAEC-TS-CR-8109.

Castaldi FJ, Bombaugh KJ & McFarland B (1995) Thermophilic slurry-phase treatment of petroleum hydrocarbon waste sludges. In: Hinchee RE, Brockman FJ & Vogel CM (Eds) Microbial Processes for Bioremediation (pp. 231-250). Battelle Press, Columbus, OH.

Civilini M, Domenis C, de Bertoldi M & Sebastianutto N (1996) Composting and selected microorganisms for bioremediation of contaminated materials. In: de Bertoldi M, Sequi P, Lemmes B & Papi T (Eds) The Science of Composting (pp. 913-923). Blackie Academic and Professional, London.

Civilini M & Sebastianutto N (1996) Degradation of naphthalene by microorganisms isolated from compost. In: de Bertoldi M, Sequi P, Lemmes B & Papi T (Eds) The Science of Composting (pp. 870-883). Blackie Academic and Professional, London.

Cole MA, Liu X & Zhang L (1994) Plant and microbial establishment in pesticide-contaminated soils amended with compost. In: Anderson TA & Coates JR (Eds) Bioremediation through Rhizosphere Technology (pp. 210-222). American Chemical Society, Washington, DC.

Cole MA, Liu X & Zhang, L (1995) Effect of compost addition on pesticide degradation in planted soils. In: Hinchee RE, Anderson DB & Hoeppel RE (Eds) Bioremediation of Recalcitrant Organics (pp. 183-190). Battelle Press, Columbus, OH.

Cole MA, Zhang L & Liu X (1995) Remediation of pesticide contaminated soil by planting and compost addition. Compost Science and Utilization 3:20-30.

Dec J & Bollag JM (1994) Dehalogenation of chlorinated phenols during binding to humus. In: Anderson TA & Coats JR (Eds) Bioremediation through Rhizosphere Technology (pp. 102-111). American Chemical Society, Washington, DC.

Dooley MA, Taylor K & Allen B (1995) Composting of herbicide-contaminated soil. In: Hinchee RE, Anderson DB & Hoeppel RE (Eds) Bioremediation of Recalcitrant Organics. Battelle Press (pp. 190-207), Columbus, OH.

Glass DJ, Raphael T, Valo R & van Eyk J (1995) International activities in bioremediation: Growing markets and opportunities. In: Hinchee RE, Kittel JA & Reisinger HJ (Eds) Applied Bioremediation of Petroleum Hydrocarbons (pp. 11-33). Battelle Press, Columbus, OH.

Goring CAI, Laskowski DA, Hamaker JW & Meikle RW (1975) Principles of pesticide degradation in soil. In: Haque R & Feeed VH (Eds) Environmental Dynamics of Pesticides (pp. 135-172). Plenum Press, New York.

Griest WH, Stewart AJ, Tyndall RL, Caton JE, Ho C-H, Ironside KS, Caldwell WM & Tan E (1993) Chemical and toxicological testing of composted explosives-contaminated soil. Environmental Toxicology and Chemistry 12:1105-1116.

Haider K, Spiteller M, Wais A & Fild M (1994) Interaction of humic substances with xenobiotics--an analytical approach to "bound" organic residues. In: Senesi N & Miano TM (Eds) Humic Substances in the Global Environment and Implications on Human Health (pp. 1071-1073). Elsevier, Amsterdam.

Hupe K, Luth JC, Heerenklage J & Stegmann R (1996) Enhancement of the biological degradation of contaminated soils by compost addition. In: de Bertoldi M, Sequi P, Lemmes B & Papi T (Eds) The Science of Composting (pp. 913-923). Blackie Academic and Professional, London.

Kaplan DL & Kaplan AM (1982) Thermophilic biotransformations of 2,4,6-trinitrotoluene under simulated composting conditions. Applied and Environmental Microbiology 44:757-760.

Kelley I, Freeman JP, Evans FE & Cerniglia CE (1993) Identification of metabolites from the degradation of fluoranthene by mycobacterium sp. Strain PYR-1. Applied and Environmental Microbiology 59:800-806.

Li G-C & Felbeck GT Jr. (1972) Atrazine hydrolysis as catalyzed by humic acids. Soil Science 114:201-209.

Liu X & Cole MA (1996) Minimum effective compost addition for remediation of pesticide-contaminated soil.

In: de Bertoldi M, Sequi P, Lemmes B & Papi T (Eds) The Science of Composting (pp. 903-912). Blackie Academic and Professional, London.

Michel FC Jr., Reddy CA & Forney LJ (1995) Microbial degradation and humification of the lawn care pesticide 2,4-dichlorophenoxyacetic acid during the composting of yard trimmings. Applied and Environmental Microbiology 61:2566-2571.

Persson A, Quednau M & Ahrne S (1995) Composting oily sludges: Characterizing microflora using randomly amplified polymorphic DNA. In: Hinchee RE, Douglas GS & Ong SK (Eds) Monitoring and Verification of Bioremediation (pp. 147-155). Battelle Press, Columbus, OH.

Petruska JA, Mullins DE, Young RW & Collins ER Jr. (1985) A benchtop system for evaluation of pesticide disposal by composting. Nuclear and Chemical Waste Management 5:177-182.

Piccolo A (1994) Interactions between organic pollutants and humic substances in the environment. In: Senesi N & Miano TM (Eds) Humic Substances in the Global Environment and Implications on Human Health (pp. 961-979). Elsevier, Amsterdam.

Richnow HH, Reidt C, Seifert R & Michaelis W (1994) Chemical cross-linking of xenobiotics and mineral oil constituents to humic substances derived from polluted environments. In: Senesi N & Miano TM (Eds) Humic Substances in the Global Environment and Implications on Human Health (pp. 1043-1048). Elsevier, Amsterdam.

Saber DL (1995) Hierarchy of treatability studies for assured bioremediation performance. In: Hinchee RE, Douglas GS & Ong SK (Eds) Monitoring and Verification of Bioremediation. Battelle Press (pp. 157-163), Columbus, OH.

Semple KT & Fermor TR (1995) Composting systems for the bioremediation of chlorophenol-contaminated land. In: Hinchee RE, Anderson DB & Hoeppel RE (Eds) Bioremediation of Recalcitrant Organics (pp. 93-100). Battelle Press, Columbus, OH.

Shann JR & Boyle JJ (1994) Influence of plant species on *in situ* rhizosphere degradation. In: Anderson TA & Coates JR (Eds) Bioremediation through Rhizosphere Technology (pp. 70-81). American Chemical Society, Washington, DC.

Silviera AE & Ganho RB (1995) Composting wastes contaminated with naphthalene. Compost Science and Utilization 3(4):78-81.

Sjoblad RD & Bollag J-M (1981) Oxidative coupling of aromatic compounds by enzymes from soil microorganisms. In: Paul EA & Ladd JN (Eds) Soil Biochemistry (pp. 113-452), Volume 5. Marcel Dekker, Inc., New York.

Snell JR (1982) Rate of Biodegradation of Toxic Organic Compounds While in Contact with Organics Which are Actively Composting. PB84-193150. National Technical Information Service, Washington, DC. 90 pp.

Stegmann R, Lotter S & Heerenklage J (1991) Biological treatment of oil-contaminated soils in bioreactors. In: Hinchee RE & Olfenbuttel RF (Eds) On-Site Bioreclamation (pp. 188-208). Battelle Press, Columbus, OH.

Stevenson FJ (1994) Humus Chemistry. John Wiley & Sons, New York. 496 pp.

Stott DE, Martin JP, Focht DD & Haider K (1983) Biodegradation, stabilization in humus, and incorporation into soil biomass of 2,4-D and chlorocatechol carbons. Soil Science Society of America Journal 47:66-70.

USEPA (1989) Guide for Conducting Treatability Studies under CERCLA. EAP/540/2-89/058. 118 pp.

Völkel W, Choné T, Mansour M, Scheunert I & Andreux F (1994) MnO_2-induced addition of 3,4-dichloroaniline to catechol rings: a model for the formation of pesticide bound residues in humus. In: Senesi N & Miano TM (Eds) Humic Substances in the Global Environment and Implications on Human Health (pp. 1201-1206). Elsevier, Amsterdam.

Westlake DWS, Jobson A, Phillippe R & Cook FD (1974) Biodegradability and crude oil composition. Canadian Journal of Microbiology 20:915-928.

Williams RT & Keehan KR (1993) Hazardous and industrial waste composting. In: Hoitink HAJ & Keener HM (Eds) Science and Engineering of Composting (pp. 363-382). Renaissance Publications, Worthington, OH.

Williams RT & Marks PJ (1991) Optimization of Composting of Explosives Contaminated Soil. U.S. Army Corps of Engineers. Final Report. CETHA-TS-CR-91053. Approx. 250 pp.

Wilson DJ & Clarke AN (1994) Hazardous Waste Site Soil Remediation. Marcel Dekker, Inc., New York. p. v.

Willson GB, Parr JF, Epstein E, Marsh PB, Chaney RL, Colacicco D, Burge WD, Sikora LJ, Tester CF & Hornick S (1980) Manual for Composting Sewage Sludge by the Beltsville Aerated-Pile Method. EPA-600/8-80-022. 65 pp.

Wolf DC & Martin JP (1976) Decomposition of fungal mycelia and humic-type polymers containing carbon-14 from ring and side-chain labeled 2,4-D and chlorpropham. Soil Sci. Soc. Am. J. 40:700-704.

Yare BS (1991) A comparison of soil-phase and slurry-phase bioremediation of PNA-containing soils. In: Hinchee RE & Olfenbuttel RF (Eds) On-Site Bioreclamation (pp. 173-187). Battelle Press, Columbus, OH.

COMPARISON OF COMMERCIAL FERTILIZER AND ORGANIC BY-PRODUCTS ON SOIL CHEMICAL AND BIOLOGICAL PROPERTIES AND VEGETABLE YIELDS

M.R. BROSIUS, G.K. EVANYLO, L.R. BULLUCK and J.B. RISTAINO
Department of Crop and Soil Environmental Sciences, Virginia Tech University, Blacksburg, Virginia 24061, USA
E-mail gevanylo@vt.edu

ABSTRACT

Organic wastes may substitute for commercial fertilizers and enhance soil chemical and biological properties. This study compared the effects of commercial fertilizer and organic wastes on soil properties and crop yields on three organic and three conventional vegetable farms in Virginia and Maryland. The organic farms produced sweet corn and the conventional farms produced muskmelon or watermelon in 1996. Commercial N, P and K fertilizers were applied according to Virginia Cooperative Extension (VCE) recommendations. The organic fertility sources, composted cotton gin trash (3 farms), composted yardwaste and livestock manure (2 farms), or beef manure (1 farm), were applied to meet the N needs of each crop according to estimated N mineralization rates of the waste after subtracting estimated cover crop N contributions. The concentrations of most nutrients and soil pH, organic matter, and total C were higher after amending with organic wastes than fertilizer. Populations of *Trichoderma* and thermophilic microorganisms, known antagonists of pathogenic soilborne fungi, were initially greater in the organically-managed soils, and *Trichoderma* was higher in organic waste-amended than commercially-fertilized soils at harvest. The beneficial species *Trichoderma* and *Pseudomonas* were higher and plant pathogenic oomycete fungi were lower in the organic waste-amended than the commercially-fertilized soils regardless of cultural history. The plant pathogenic fungi increased with time and were greater in plots treated with inorganic fertilizers than organic amendments at harvest. Beneficial soil microorganisms appear to be favored by long term organic amendments, but some groups can be increased with organic amendments within a single season. Yields were similar between treatments, which demonstrates that organic amendments can provide adequate amounts of required plant nutrients in addition to enhancing soil chemical and biological attributes.

INTRODUCTION

The success of long term crop production and maintenance of environmental quality is

S. Brown, J.S. Angle and L. Jacobs (eds.), Beneficial Co-Utilization of Agricultural, Municipal and Industrial By-products, 195–202.
© 1998 *Kluwer Academic Publishers*,

dependent on soil quality. Indicators of soil quality include cation-exchange capacity (CEC), available nutrients, total and available N, organic matter, total C, pH, and the number and community structure of certain soil organisms (Arshad and Coen, 1992; Visser and Parkinson, 1992).

Soil quality can be improved or degraded by management. Since the 1950's, mainstream agriculture has attempted to optimize soil fertility through the application of commercial fertilizers. The access of farmers to inexpensive fertilizers permitted short-term amelioration of nutrient-deficient soils; however, increasing soil nutrient supplying capacity may better be accomplished by improving the soil's biological activity, not adding more nutrients (King, 1990). Long term use of commercial fertilizers may reduce soil organic matter and biological activity (Fauci and Dick, 1994).

Plant and animal-based wastes may substitute for commercial fertilizers and enhance chemical and biological attributes of soil quality in agricultural production systems. Organic matter increases the soil's abilities to hold and make available essential plant nutrients and to resist the natural tendency of soil to become acid (Cole et al., 1987). Build-up of organic matter through additions of crop and animal residues have been shown to increase the population and species diversity of microorganisms and their associated enzymatic activity and respiration rates (Kirchner et al., 1993; Weil et al., 1993).

Compost, specifically, provides certain advantages over uncomposted wastes. In comparison to manures, N release and availability from compost is slower; however, crop yields are frequently higher because of enhanced soil properties and increased supply and availability of other nutrients (Maynard, 1993; McConnell et al., 1993). Compost also increases beneficial soil fungi populations and decreases soilborne pathogens (Hoitink and Grebus, 1994). The objective of this research was to assess the effects of organic and inorganic soil amendments on vegetable yields and selected soil biological and chemical properties indicative of soil quality on three organic and three conventional farms.

MATERIALS AND METHODS

Field plots were established in fall 1995 at three organic and three conventional production farms in Virginia and Maryland. Fertility treatments were: 1) N, P & K fertilizer applied at rates recommended by VCE, and 2) compost, animal manure, and cover crops designed to provide required plant available N (Table 1). The mineralizable portion of TKN was conservatively estimated as 10% from compost and 35% from beef manure. Plant available N from hairy vetch and spring oats were estimated at 112 and 56 kg/ha, respectively. Treatment plots were 7.5 m^2 and replicated 3X in a randomized complete block.

Amendments were incorporated into the soil immediately after application. Sweet corn seed was planted on the organic farms and muskmelon or watermelon were transplanted on the conventional farms within one week of soil amending. Producers carried out production and harvesting operations.

Organic amendments were analyzed for Mehlich 1 (M1)-extractable P, K, Ca, Mg, Mn, Zn, Cu, and B; total Kjeldahl N (TKN); NH_4- and NO_3-N; and volatile solids/total C to estimate nutrient availability (Table 2). Ten soil cores (3 cm in diameter) per plot were sampled to a 0.3 m depth when melon vines began to run or at corn silking for M 1-nutrients, pH, TKN, NH_4- and NO_3-N, total C, organic matter and cation exchange capacity (CEC).

Table 1. Rates of nutrients from organic amendments and commercial fertilizer applied to alternative and conventional treatments.

Farm	Organic amendment Type	Rate	Est'd N[a]	Commercial fertilizer N	P	K
		Mg/ha	kg/ha	----------kg/ha----------		
		Organic Farm				
Polishuk	Composted yardwaste & beef manure + spring oats	42	114	157	90	112
Heller	Beef manure + hairy vetch	9	137	157	45	112
Smith	Composted yardwaste & poultry manure	30	145	100	27	27
		Conventional Farm				
Copeland	Composted cotton trash	26	95	157	112	90
Faulk	Composted cotton trash	26	95	112	56	196
Whitener	Composted cotton trash	35	128	112	28	84

[a] Estimated N from mineralization of organic wastes and cover crops.

Table 2. Total nutrient analysis of organic amendments applied to farms in 1996.

Amendment & farm	Volatile solids	TKN	P	K	Ca	Mg	S
	------------------------------------g/kg------------------------------------						
Yard waste compost							
Polishuk	173	6.9	2.5	6.7	17.4	6.4	1.3
Smith	561	24.1	8.8	12.5	25.8	5.6	3.4
Beef manure	----	11.0	2.6	9.9	1.6	1.8	1.3
Cotton trash compost	437	18.3	4.9	10.3	19.1	3.4	2.4

Plots were sampled for biological indicators by collecting 25 soil cores (1.9 cm in diameter) to a 0.2 m depth from each plot when melon vines began to run or at corn silking and after harvest (September). Soil dilutions using 0.25% sterile water agar solution were conducted, and ten-fold serial dilutions were made. Ten g soil in 90 ml sterile water agar solutions were utilized to determine populations of total fungi (Potato Dextrose Agar), total bacteria (Difco Tryptic Soy broth 30 g/l, Difco Bacto-agar 15 g/l), plant pathogenic *Fusarium* species (Glucose 100 g/l, Yeast Extract 2 g/l, Rose Bengal 0.5 g/l, Difco Bacto-agar 20 g/l), *Trichoderma* and *Gliocladium* species (Trichoderma Medium E; Papavizas and Lewis, 1989), oomycetes *Pythium* and *Phytophthora* species (Masago's Agar without hymexazol; Masago, et al., 1977), fluorescent *Pseudomonas* species (King's Medium B amended with Novobiocin, Pencillin G, and Cyclohexamide; Sands and Rovira, 1970), thermophilic microorganisms (Yeast Extract Glucose Agar), and enteric bacteria (Endo Agar). Dilution factors varied depending on the assay performed. Three replicate plates were used for each assay per dilution factor. Colonies on the plates were enumerated, and means of raw data were calculated according to the following formula:

$CFU/g = (CN_1 + CN_2 + CN_3)/3 \times 10^{dl} \times dw$, where
CFU = Total number per g dry soil
CN_i = Number of colony forming units per plate
dl = Dilution factor
dw = Soil dry weight (determined gravimetrically)

Fresh yield was determined by weighing corn ears and melons from 3.6 m sections of each of two center rows from each plot. Corn yields for Polishuk and Heller were determined by a single time harvesting. Corn yields for Smith were determined as total stalk dry weight due to extensive corn earworm damage. Melon yields were determined during a two week sampling period. Data were statistically analyzed using ANOVA and LSD's at the 0.05 probability level (SAS Institute).

RESULTS

Chemical Properties

The organic soil amendments either increased or did not affect the concentrations of available soil nutrients compared to commercial fertilizer (Table 3). Phosphorus and K were increased in several soils because the organic wastes were applied based on crop N, not P or K, requirements. Micronutrients were not supplied by the commercial N-P-K fertilizer, but they are plentiful in cotton trash, yard waste, and livestock manure; thus, many of the organically-fertilized soils exhibited increased concentrations of micronutrients, especially Mn and B. The addition of essential micronutrients and Ca, Mg and S are benefits of organic wastes because crop requirements, which are often difficult to predict, may be satisfied without knowing absolutely which nutrients are insufficient.

Other soil chemical properties were not affected by treatments to the same extent as were soil nutrient concentrations (Table 4). Soil pH was increased by the organic soil amendment in several instances, probably due to addition of base cations and the buffering effect of organic matter on soil acidity. Soil organic matter or total C and CEC were increased by the organic amendments at two sites, but there was a trend toward increased soil organic matter, total C, CEC and soil pH with organic soil amendment at nearly every site.

Soil TKN was higher (2 sites) or the same with the organic amendments than with commercial fertilizer at every farm, but early season NO_3-N (plant-available form) was the same or lower with the organic wastes (especially cotton trash compost) than with commercial fertilizer. Thus, plant available N may have been limiting in the organic treatments early in the growing season. Conversely, later mineralization may have provided adequate N for plant growth without the potential for early season nitrate leaching.

Table 3. Effects of fertility treatment on Mehlich I soil extractable nutrient concentration.

Treatment	P	K	Ca	Mg	Mn	Zn	Cu	B
	Mehlich I extractable nutrient							
	mg/kg soil							
				Heller				
Alternative	80	274	601	140	11.1	1.0	0.5	0.19
Conventional	73	135	597	125	5.7	0.7	0.5	0.11
LSD (.05)	ns	24	ns	ns	3.5	0.2	ns	0.02
				Polishuk				
Alternative	49	220	1401	131	28.3	1.9	0.5	0.56
Conventional	56	134	1247	93	19.8	1.3	0.5	0.37
LSD (.05)	ns	ns	123	27	5.7	ns	ns	0.18
				Smith				
Alternative	160	394	1126	191	41.5	8.5	1.3	0.60
Conventional	90	244	878	140	25.8	4.9	1.3	0.34
LSD (.05)	28	7	ns	ns	12.1	ns	ns	0.14
				Whitener				
Alternative	92	175	499	78	8.7	2.5	0.6	0.28
Conventional	72	93	440	73	6.8	2.8	0.6	0.16
LSD (.05)	ns	39	ns	ns	ns	ns	ns	0.05
				Faulk				
Alternative	64	166	512	82	7.6	1.6	0.4	0.27
Conventional	43	100	314	55	3.7	0.7	0.4	0.08
LSD (.05)	11	ns	79	ns	1.9	ns	ns	0.10
				Copeland				
Alternative	51	146	275	63	5.5	3.2	0.7	0.20
Conventional	55	86	208	50	4.0	3.1	0.8	0.08
LSD (.05)	ns	41	ns	ns	0.4	ns	ns	0.10

Biological Properties

The beneficial *Trichoderma* spp. and thermophilic microorganisms (mostly *Actinomyces* and *Streptomyces* spp.) and the plant pathogenic *Fusarium* spp. exhibited significant "history by time of sampling" interactions (Table 5). Populations of *Trichoderma* spp. and thermophilic microorganisms were initially greater in the historically organic than the conventionally-managed soils; however, the populations of *Trichoderma* spp. and thermophiles remained constant in the historically organic soils while the numbers of organisms in the historically conventional soils increased during the growing season. *Fusarium* spp. populations were initially equal in both historically organic and conventional soils; however, they remained the same where commercial fertilizer was applied, but increased in the soils amended with organic wastes by the end of the growing season.

Table 4. Effects of fertility treatment on early season soil chemical parameters.

Treatment	pH, H_2O	CEC	Organic matter	Total C	TKN	NO_3-N
		cmol+/kg	g/kg	g/kg	mg/kg	mg/kg
			Heller			
Alternative	6.00	7.7	17.2	11.7	1.6	14.0
Conventional	5.53	7.4	13.6	10.0	1.5	16.8
LSD (.05)	ns	ns	ns	ns	ns	ns
		Polishuk				
Alternative	6.90	11.6	24.9	16.1	2.7	16.3
Conventional	6.50	10.0	19.1	12.5	2.0	27.3
LSD (.05)	ns	ns	ns	2.2	0.5	ns
		Smith				
Alternative	5.97	9.6	38.0	22.9	3.1	54.2
Conventional	5.57	7.5	29.5	18.3	2.4	62.8
LSD (.05)	ns	ns	ns	ns	ns	ns
		Whitener				
Alternative	6.23	5.1	21.3	15.2	1.7	3.2
Conventional	5.67	4.7	21.7	14.7	1.6	23.1
LSD (.05)	ns	ns	ns	ns	ns	17.1
		Faulk				
Alternative	6.27	3.5	18.9	14.3	1.4	14.4
Conventional	5.23	2.4	12.7	9.3	1.0	43.6
LSD (.05)	0.29	0.4	4.4	1.0	0.1	12.2
		Copeland				
Alternative	6.37	2.7	14.1	10.7	1.4	0.0
Conventional	5.63	2.1	11.4	9.7	1.0	27.5
LSD (.05)	0.14	0.2	ns	ns	ns	27.2

Table 5. Effect of cultural history by time of sampling interaction on mean population densities of biological communities.

	Early season		Post harvest		History x time
	Conventional	Organic	Conventional	Organic	(P value)
	-------------Colony forming units x 10^3/g soil-------------				
Trichoderma spp.	0.40	32.4	16.5	24.4	0.0205
Thermophilic species	3.60	14.0	32.8	21.6	0.0126
Fusarium spp.	31.3	34.6	28.1	62.8	0.0302

The beneficial species *Trichoderma* and *Pseudomonas* were higher and the plant pathogenic oomycete fungi (*Pythium* and *Phytophthora* spp.) were lower in the organic than the commercially-fertilized soils regardless of cultural history (Table 6). Enteric bacteria populations were higher in the organic treatments, probably due to the addition of animal manure or manure-based compost (Table 6). None of the fields had significant levels of recognizable soilborne diseases.

Table 6. Effect of fertility treatments on mean population densities of biological communities.

	Alternative	Conventional	Significance
	No. of colony forming units x 10^3/g soil		(P value)
Trichoderma spp.	24	13	0.004
Pseudomonas spp.	370	280	0.084
Phytophthora and *Pythium* spp.	0.07	0.10	0.051
Enteric bacteria	18,500	10,800	0.041

Table 7. Effects of fertility treatments on yield.

		Fertility Treatment		
Farm	Crop	Alternative source	Commercial fertilizer	LSD (.05)
	Mg/ha	Mg/ha	Mg/ha	
Heller	sweet corn	6.89	8.81	ns
Polishuk	sweet corn	6.66	10.95	2.53
Smith	sweet corn[a]	2.58	2.78	ns
Whitener	muskmelon	14.88	23.07	ns
Faulk	muskmelon	16.25	15.41	ns
Copeland	watermelon	3.96	4.82	ns

[a] Total stalk dry matter

Despite the beneficial effects of the organics amendments on soil chemical properties, no yield increases above those achieved with the commercial fertilizer occurred (Table 7). Corn yield was lower with yard waste compost than commercial fertilizer at one location, possibly due to inaccurately estimating the plant available N contributions from the spring oats and compost. The early season soil NO_3-N analyses indicate that plant available N from the organic sources was limiting.

DISCUSSION

The nutrient values of waste-amended soils can be raised above those of commercially-fertilized soils, and yields with wastes can be comparable to those of commercially-fertilized soils. Organically-managed soils had higher populations of beneficial organisms and lower populations of pathogenic organisms. However, some soil microorganisms known to play a role in control of soilborne pathogens (*Trichoderma* and *Pseudomonas* species) increased within a single season with organic amendments. The use of organic soil amendments (especially compost) may result in a soil that has greater capacity to resist the spread of plant pathogenic organisms. The improvement in overall soil quality may produce more vigorous-growing and high-yielding crops.

ACKNOWLEDGMENTS

Thanks to the United States Department of Agriculture Sustainable Agriculture and Education Program for funding this research, to the farmer participants Ellen Polishuk, John Smith and Emily Chewning, Joel Copeland, Arthur Whitener, and Nell Faulk; to the Chesapeake Bay Foundation Claggett Farm Manager Michael Heller; and technical specialists Michael Faulk (Natural Resource Conservation Service) and Cliff Slade (Virginia Cooperative Extension).

REFERENCES

Arshad MA & Coen GM (1992) Characterization of soil quality: Physical and chemical criteria. American Journal of Alternative Agriculture 7:25-32.

Cole CV, Williams J, Shaffer M & Hanson J (1987) Nutrient and organic matter dynamics as components of agricultural production systems models. In: Follett RF (Ed) Soil Fertility and Organic Matter as Critical Components of Production Systems (pp 147-166). SSSA Spec. Publ. 19. ASA, CSSA, and SSSA, Madison, WI.

Fauci MF & Dick RP (1994) Soil microbial dynamics: Short- and long-term effects of inorganic and organic nitrogen. Soil Science Society of America Journal 58:801-806.

Hoitink HAJ & Grebus ME (1994) Status of biological control of plant diseases with composts. Compost Science and Utilization 2(2):6-12.

King LD (1990) Sustainable soil fertility practices. In: Francis CA, Butler CB & King LD (Eds) Sustainable Agriculture in Temperature Zones (pp 144-177). John Wiley & Sons, Inc., New York.

Kirchner MJ, Wollum II AG & King LD (1993) Soil microbial populations and activities in reduced chemical input agroecosystems. Soil Science Society of America Journal 57:1289-1295.

Masago H, Yoshikawa, Fukada M & Nakanishi N (1977) Selective inhibition of Pythium spp. on a medium for direct isolation of Phytophthora spp. from soils and plants. Phytopathology 67:425-428.

Maynard AA (1993) Evaluating the suitability of MSW compost as a soil amendment in field-grown tomatoes. Compost Science and Utilization 1(2):34-36.

McConnell DB, Shiralipour A & Smith WH (1993) Compost application improves soil properties. BioCycle 34(4):61-63.

Papavizas GC & Lewis JA (1989) Improved medium for isolation of Trichoderma from soil. Plant Disease 66:1019-1020.

Sands DC & Rovira AD (1970) Isolation of fluorescent pseudomonads with a selective medium. Applied Microbiology 20:513-514.

Visser S & Parkinson D (1992) Soil biological criteria as indicators of soil quality: Soil microorganisms. American Journal of Alternative Agriculture 7:33-37.

Weil RR, Lowell KA & Shade HM (1993) Effects of intensity of agronomic practices on a soil ecosystem. American Journal of Alternative Agriculture 8:5-14.

FATE AND POTENTIAL OF XENOBIOTICS

GEORGE A. O'CONNOR
University of Florida, Gainesville, Florida 32611, USA
E-mail gao@gnv.ifas.ufl.edu

ABSTRACT

The potentially toxic organics (TOs) that can be present in agricultural, municipal, and industrial wastes number in the thousands. Extensive data bases show that TO concentration in biosolids are low; much less data is available for other wastes, but the consensus is that TO concentrations are similar. TO reactions in wastes/soils also are expected to be similar to soil/biosolid - TO reactions. Thus, the sizable data, research, and risk assessment base for biosolids seems applicable to many other wastes. If TO variety and concentrations in a waste are similar to those in biosolids, risk to humans, animals, and the environment is minimal. Various fate processes of TO in soil/waste mixtures, as well as risk assessment for TOs in biosolids, are reviewed to support the conclusion. Research needs are identified to confirm the conclusion.

INTRODUCTION

Prevalence/Toxicity

The potentially hazardous organics that may be present in agricultural, municipal, and industrial by-products number in the thousands. Of particular concern, however, are the so-called priority pollutants, herein referred to as toxic organics (TOs). Fortunately, data from numerous surveys suggest that the concentrations of TOs in municipal biosolids are almost always low, and the most recent survey of US municipal sludges, the National Sewage Sludge Survey (USEPA, 1990) confirmed this generalization (Table 1). The most common TO (phthalate esters) reported in biosolids (and many other wastes) is, fortunately, relatively non-toxic (Naylor and Loehr, 1982).

Data on the priority pollutants for other by-products are not nearly as complete as for biosolids, but some generalizations are possible (Hansen and Schaeffer, 1995):

1. TO concentrations (particularly PAHs) in ash from coal and municipal waste incineration are generally greater than median biosolids, but still ≤ 10 mg/kg.
2. TO concentrations in manures and composts are usually less than in biosolids and ashes, but some unique problems may occur:

S. Brown, J.S. Angle and L. Jacobs (eds.), Beneficial Co-Utilization of Agricultural, Municipal and Industrial By-products, 203–217.
© 1998 *Kluwer Academic Publishers*,

a) Although transient in nature, the presence of certain biocides and other medications in animal manures may render them temporarily unsuitable for land disposal unless well-diluted with manures from untreated animals.

b) Several pesticides are retained in composts (made from yard trimmings) at relatively low, but significant concentrations. Most pesticides were < 1 mg/kg in mature composts, but 2, 4-D was = 56 mg/kg. PAH concentrations can reach 2-4 mg/kg (see below).

c) Significant (2-5 fold) differences in TO concentrations are reported for source-separated versus mixed MSW composts (Cook and Beyea, 1995). Thus, source-separated composts had 2 mg PAH/kg, 20 ng TEQ dioxins/furans/kg, 0.5 mg PCBs/kg versus 4.4 mg PAH/kg, 45-100 ng TEQ dioxins/furans/kg, and 1-2 mg PCBs/kg of mixed MSW compost.

3. As with sludge, ashes, manures, and composts, TO occurrence and concentrations in other final products depend partly on TO amounts in the raw materials. The following is a partial list of TOs associated various wastes. Detail is available in individual chapters in Parr et al. (1983), or as cited.

a) Nitroaromatic compounds in munition and explosive wastes.

b) Chlorophenols (and a host of other organics) in pesticide and organic chemical manufacturing wastes.

c) n-alkanes, alkenes, branched and cyclic alkanes, and various aromatics in petroleum wastes.

d) Acids, bases, solvents, cyanides, bacteriostats, disinfectants, and numerous other organics from pharmaceutical manufacturing wastes.

e) Resin acids, unsaturated fatty acids and their derivatives, chlorinated phenolics, halomethane, aldehydes, etc. from the pulp and paper industry.

f) Surfactants, and various "fortifiers" from soap and detergent industries.

g) Dyes from textile wastes.

h) Creosote and PCP from wood preservative wastes, and more recently,

i) Olestra, (Allgood, et al., 1997) and polydimethylsilicones (PDMS) polymers associated with various consumer products wastes.

Clearly the variety of TOs possible in wastes is enormous, but if we restrict our attention to by-products of non-hazardous wastes the list is narrowed significantly. Regulatory restrictions on land application of hazardous wastes are rigorous, and discourage widespread use of land application of by-products made from such materials. Further, if we take advantage of what we know about soil-plant reactions of TOs in typical by-products applied at reasonable rates and using good "agronomic" practices, we can make significant progress in assessing the fate and potential effects of xenobiotics. Consider the following.

Table 1. Examples of toxic organics in sludge and representative concentrations (National Sewage Sludge Survey, 1990).

Compound	Typical Conc
	mg kg^{-1} (dw)
Phthalate Esters	
Bis(2-ethylhexyl)phthalate	≥100
Polynuclear Aromatic Hydrocarbons (PAHS)	
Benzo[a]pyrene	1-10
Fluroanthene	1-10
Polyhalogenated Biphenyls	
Polychlorinated biphenyl (Aroclor 1248,1254)	0.5-1
Polybrominated biphenyls	<<1
Chlorinated Pesticides and Hydrocarbons	
Aldrin/dieldrin	<1
DDT/DDE/DDD	<1
Chlordane	<1
Acid-Extractables	
Pentachlorophenol	1-5
Dinitrophenol	<1
Volatile Aromatics	
Benzene	1-10
Trichloroethylene (TCE)	1-10
Alkyl Amines	
Nitrosamines	<<1

Quantities

The occurrence and concentration of most TOs in most by-products are not expected to be very different from those shown in Table 1 for biosolids. Biosolids typically contain TOs in the <1-100 mg/kg (dry wt.) range, with ≥ 90% present at concentrations ≤ 10 mg/kg. When such materials are applied at agronomic rates (5-10 Mg/ha), the resulting soil-waste mixtures (0-15 cm depth) have TO concentrations at least 100 fold less than in the waste alone. Thus, TO concentrations in well-managed (normal) waste-amended soils are similar to (or lower than) routine pesticide-soil concentrations of 0.01-1 mg/kg. Even in heavily waste-amended soils, amended at 100-1000 Mg waste/ha, most TOs will be present in the soil-waste mixture at concentrations ≤ 1-10 mg/kg.

Recognizing the expected low concentration is important for several reasons:

1. Low initial soil concentrations means low concentrations are available for various transfer pathways (to plants, animals, to the air, to leaching, etc.).
2. The similarity to routine pesticide-soil concentrations allows one to utilize the vast body of literature characterizing pesticide-soil-plant (-animal) interactions, and

3. It forewarns of analytical difficulties that should be expected in analyzing soils, plants, water, animal and human tissues, and the need for careful experimental design to demonstrate compound transfer (or the lack thereof).

FATE IN SOIL

Once added to the soil, waste-borne TOs are subject to a variety of processes (Weber and Miller, 1989) that affect their lability.

Adsorption/Desorption

The adsorption of an organic chemical by soil is routinely described by a linear relationship, $C_s = K_d C_L$, or by the empirical Freundlich relationship, $C_s = K C_L^{1/n}$ - where C_s is the sorbed concentration, C_L is the equilibrium solution concentration, K_d is the distribution coefficient, and K and 1/n are empirical constants. At low initial TO concentrations, adsorption is usually linear, whereas at greater initial TO concentration, adsorption can be non-linear, and the Freundlich expression is frequently needed to adequately describe TO retention. (Note that retention determined at high initial concentrations may over predict the expected behavior at low concentration).

Values of K_d (or K) are often normalized to a soil's fractional organic carbon (OC) content (f_{oc}): a practice shown to reduce variation in TO retention characterization across soils. This yields K_{oc} values, calculated as K_d (or K) divided by f_{oc}. This manipulation tacitly assumes that all OC has the same binding tendency for a given TO, i.e., that all SOM-OC or waste-OC sorbs TOs similarly. It also assumes that the organic fraction of soil (or waste-amended soil) is the dominant sorbing solid phase of a TO in soil. Neither assumption is correct for all TOs or all soils, but is an acceptable first approximation in most cases.

K_{oc} values have been repeatedly shown to be generally related to the K_{ow} of various compounds, most recently by Baker, et al. (1997). The exact mathematical relationship (regression) often varies with chemical group, soil (or soil-waste characteristic), but a gross 1:1 relationship ($\log K_{oc} \simeq \log K_{ow}$) is often reported. Thus, based on a TO's chemical properties (K_{ow}, widely available), a first approximation of TO retention in soil (K_{oc}) can be estimated, and K_d (or K) can be calculated for a particular soil from easily measured OC contents. This allows estimation of a TO's sorption without having to measure each chemical's retention in each soil or soil-waste combination.

Sorption is perhaps the single, most important factor affecting the behavior of TOs in the soil environment. Adsorption affects rates of volatilization, diffusion, and leaching, as well as the availability of TOs to biotic or abiotic degradation, and TO uptake by plants or other organisms (e.g., Kaufman, 1983). Consider the effect on plant uptake.

To have the greatest potential for plant uptake, a TO must remain in the vicinity of the plant root (not be quickly leached away) in a labile form. For example, most residual soil-acting herbicides have K_d values in the range of 1-20, with values up to 40 being satisfactory for most soil applications (Graham-Bryce, 1984). Compounds with K_d's >1000 tend to become inactivated (non-labile) by soil-sorption. In a soil with $f_{oc} = 0.0125$

(OM=2-2.5%) K_d's of 1, 20, 40 and 1000 would represent K_{oc}'s (and K_{ow}'s) of ~ 2.3, 3.6, 4, and 5, respectively. Thus, soil applied chemicals with log K_{ow} values ≥ 5 are expected to be so strongly retained by soil as to be unavailable to plants (or to leaching). An extension of this argument is that even soluble TOs can be retained in non-retentive (sandy, low OM) soils by the addition of sorbing OC to the soil- a phenomenon repeatedly observed in practice (e.g. Weber and Mrozek, 1979). Toxic organics contained in organic wastes are expected to be much less labile than TOs added without additional OC.

The partitioning expressions (K_d or K) assume equilibrium and complete reversibility between the solid (sorbed) and liquid phases. Numerous authors (e.g., Di Torro and Horzempa, 1982; Scheunert et al., 1985), have demonstrated that retention reactions of numerous TOs are at least partially irreversible over time scales lasting weeks to months. The irreversibility is especially prominent with TOs that have log $K_{ow} > 5$, but chemicals with log K_{ow} values of 2-3 can also remain unleached from sandy soils months to years after application (Pignetallo, 1989; Frink and Bugbee, 1990). Thus, TOs prominent in many by-product wastes (especially biosolids, composts, manures) tend to either be hydrophobic (high log K_{ow}) already, or are strongly retained in/on the high OC by-product, and minimally desorbed. Such behavior severely limits the bioavailability and lability of organic waste-associated TOs in real world situations. Even wastes not normally considered to be organic, e.g., fly ash can have significant retention capacity for organic chemicals of low $K_{ow} \leq 3$, apparently via physical adsorption processes (Banesee et al., 1995).

Degradation

Plant uptake, and lability in general, is concentration dependent; hence, a compound's persistence has an obvious impact on potential plant uptake, leaching volatilization, etc. Gillett (1983) distinguished priority pollutants on the basis of their soil half-lives: < 10d (Class A); 10-50d (Class B); and > 50d (Class C). Class A compounds would likely be lost from the soil-plant system quickly (ie. before they could be taken up). Indeed, significant degradation of such Class A TOs can occur in sewage treatment plants, compost piles, lagoons, etc. so these TOs are often present in wastes at low concentrations. Co-composting wastes or adding readily decomposable organic wastes to soils can enhance biodegradation of TOs in soils or other wastes. Gillett considered compounds with half lives > 14d of sufficient stability to be of concern in the environment.

Soil sorption is now widely recognized to affect microbial degradation of many compounds (Scow et al., 1986; Alexander, 1994). Strongly adsorbed chemicals are apparently less available to microbes because only low concentrations are desorbed into solution and are available for microbial uptake and intracellcular metabolism.

Close association between TOs and wastes could, thus, be expected to slow the degradation of some TOs, but is it likely that Class A compounds would still be lost too quickly to impact plants. Class C TOs, however, might be expected to be even more recalcitrant when waste-borne or - sorbed. Indeed, recalcitrant TOs, e.g. PAHs, PCBs, etc. have large K_{ow} values and are strongly (and irreversibly) bound in/on biosolids. Such chemicals can be expected to persist for years, and will likely become part of soil humus, with effective half-lives of perhaps decades to centuries.

Volatilization

Vapor phase partitioning (volatilization) of a TO in the soil influences the spread of the compound through the soil and loss to the atmosphere. Even chemicals with relatively low vapor pressures can be transported significant distances by volatilization. Chemicals with high vapor pressures usually partition readily from the soil solution to the soil air, where they can move throughout the soil and across the soil surface. Chemicals in the vapor phase may be taken up by plant roots, or by above-ground portions of the plant.

The distribution of chemicals between the liquid and vapor phases is frequently described by Henry's law:

$$HC = \frac{16.04\ PM}{TS}$$

where P is the vapor pressure of pure solute, M is the molecular weight of solute, T is absolute temperature, S is solubility in water, and HC is the dimensionless Henry's constant.

Note that compounds (e.g. PCBs, dioxins) with relatively low vapor pressures (P) can nevertheless be significantly volatilized because they have low aqueous solubilities (S). Ryan et al. (1988) reasoned that compounds with HC values $> 10^{-4}$ should be readily lost from soil. However, soil (or waste) sorption can significantly reduce chemical volatilization (Fairbanks et al., 1987). Thus, the arbitrary value of 10^{-4} for HC may overestimate the importance of volatilization in high OC soils or volatilization of waste-associated TOs.

The expected easy volatilization of highly volatile chemicals (e.g. toulene) can be apparently impeded in some biosolids at high moisture contents (e.g. Jacobs et al., 1987). However, once applied to soils and incubated at reasonable moisture contents, volatile TOs (e.g. toulene, DCE, TCE, ethylbenzene) are rapidly lost and become undetectable in a few weeks (Jin and O'Connor, 1990; Webber and Goodin, 1989).

Shoot and lower leaf contamination with PCBs has been demonstrated (Fries and Marrow, 1981) and attributed to vapor transport of volatilized PCBs from surface deposits. Fairbanks et al. (1987), however, demonstrated significantly reduced PCB volatilization following sludge additions to PCB-spiked soils. Further, there was no PCB vapor contamination of crops grown in soil amended with sludge-borne PCBs. Surface incorporation of PCB-contaminated sludges reduces PCB volatilization (Strek et al., 1981) and thus, possible vapor contamination of crops. Similar minor volatilization (and subsequent crop contamination) is expected from sludge-borne (or waste associated) PAHs, chlorinated pesticides and hydrocarbons and dioxins (Hsu et al., 1993; O'Connor et al., 1991). Care must be taken to distinguish between plant contamination with TOs vaporized from growth media (or vaporized from other sources) and contamination brought to the plant in the air (Witte et al., 1988; Harms and Sauerback, 1983).

Plant Uptake

Plant uptake of waste-associated TOs can occur by:

1. Uptake from the soil solution (liquid phase transfer) with translocation from roots to shoots (e.g., Briggs et al., 1982; Topp et al., 1986),
2. Absorption by roots of volatilized TOs (vapor phase transfer) from the soil (e.g., Topp, et al., 1986; Travis and Hattener - Frey, 1988), and
3. Partitioning of bound TO directly into plant tissue from soil particles or aerosols deposited on leaves (e.g., Patterson, et al., 1990).

Factors that reduce the concentration or bioavailability of TOs in the growth media obviously reduce the potential for plant uptake. Thus, low concentration of TOs in sludge - (or waste-) amended soils, strong adsorption of TOs to soil or waste, rapid loss of TOs from soil (normal or enhanced degradation, (Bellin and O'Connor, 1990), incorporation of waste into soils - reduced volatilization, etc. all act to reduce the initial potential for plant contamination. These, and other factors (O'Connor, 1996) lead us to conclude that the potential for plant uptake of biosolids-borne and compost-borne TOs is small (Chaney et al., 1996; Chaney and Ryan, 1993; O'Connor et al., 1991). Even if some TOs are taken up by plants, the TOs may be metabolized within the plants to innocuous compounds, or accumulated on plant parts (e.g. carrot peels) that are usually removed in normal culinary practices such as washing and peeling (O'Connor, 1996).

Leaching

Leaching tendency of a TO can be expressed as a first-order leaching coefficient (USEPA, 1992) as:

$$K_{lec} = \frac{NR}{(BD * K_d + \theta_w + H\theta_a)\ d_i}$$

where K_{lec} = loss rate coefficient for leaching (y^{-1})
NR = annual recharge to ground water (m/y)
BD = bulk density of soil (kg/m^3)
K_d = equilibrium partition coefficient for TO in soil (m^3/kg)
θ_w = water-filled porosity in soil (unitless)
θ_a = air-filled porosity in soil (unitless)
H = dimensionless Henry's constant
d_i = depth of waste incorporation in soil (m)

Note the obvious effect of soil-waste adsorption of the TO on the leaching coefficient. [Also note that if the chemical is highly volatile, the amount of chemical available for other losses is reduced so leaching coefficients are small].

For a particular chemical, its K_{ow} can lead to an estimated K_{oc} from which K_d can be calculated $K_d = K_{oc} \cdot f_{oc}$. Increasing the amount of organic carbon via waste addition, thus, increases K_d for a specific TO -soil combination. The greater K_d, the less TO leaching. When K_d >1000 (common for TOs like PCBs, PAHs, dioxins, DDT, chlordane etc.), leaching losses are minuscule, and ground water quality is not predicted to be impacted. Soluble TOs (e.g. 2, 4-D, nitrosodimethyamine, some phthalate esters, etc.) with K_ds < 20 are predicted to leach and to impact groundwater quality if they are not volatilized or degraded. Fortunately, the other dissipation processes are often prominent, and few chemicals in the wastes we are discussing significantly impact ground water.

POTENTIAL EFFECTS

Risk Assessment

Over the last 5 years, or so, the U.S. EPA has undertaken a rigorous risk assessment of sludge-borne toxicant behavior (EPA, 1989 and 1992) involving numerous exposure pathways for so-called high exposed individuals (Table 2). Other nations and scientists have also judged the (pathway) considerations appropriate for evaluation (Wild and Jones, 1992). The U.S. effort was the basis for the final rules on the use or disposal of sewage sludge (US EPA, 1993), known as the "503" regulations.

The various pathways (Table 2) considered were intended to allow risk assessment for animals, humans, and the environment. All of the processes discussed previously were considered and more. The analysis is data-intensive (sometimes exceeding available data bases), and challenges our understanding of toxicant behavior in real-world systems. Fortunately, sewage sludge had been studied in the US and the world for > 20 years, and significant progress could be made in the risk assessment associated with each pathway for sludge. Such data bases for other soil amendments, and the numerous by-products considered in this conference are not nearly as complete. Nevertheless, Chaney and Ryan (1993) and Logan (1996) were able to apply the risk assessment approach to composts using the limited data available and by making reasonable assumptions of compost-borne toxicant behavior (fate) like those described above. Even with limited data, judicious risk assessment identifies the most critical pathways for various toxicants and the research needed to make the assessment more complete. Such an approach will be necessary for a thorough assessment of TO risk associated with individual by-products, but much can be learned from the assessment for sewage sludge. I believe the assessment approach is applicable to other soil amendments, and that will be my approach today.

I will not detail the various assessments, improvements to the original EPA effort, nor the intracies of each pathway. You are instead, referred to the numerous thorough papers we have written on the subject (Chaney et al., 1996 and 1991; Chaney and Ryan, 1993; O'Connor, 1996 and 1991). The reviews are also valuable sources of citations to a wealth of the kind of excellent data collected by dedicated scientists from around the world over several decades.

Table 2. Environmental pathways of concern identified for application of sewage sludge to agricultural land.

	Pathway	Description of HEI
1.	Sewage Sludge→ Soil→ Plant→ Human	Human ingesting plants in sewage sludge-amended soil.
2.	Sewage Sludge→ Soil→Plant→Human	Residential home gardener.
3.	Sewage Sludge→ Human	Children ingesting sewage sludge.
4.	Sewage Sludge→ Soil→Plant→ Animal→Human	Farm households producing a major portion of the animal products they consume. It is assumed that the animals eat plants grown in soil amended with sewage sludge.
8.	Sludge→Soil→ Plant→Human	Plants grown in sewage sludge-amended soil.
10.	Sewage Sludge→ Soil→Soil Organism→ Soil Organism Predator	Animals eating soil organisms living in sewage sludge - amended soil.
12.	Sewage Sludge→Soil→ Surface Water→ Human	Water Quality Criteria for receiving water for a person who consumes 0.04 kg/day of fish and 2 liters/day of water.
13.	Sewage Sludge→Soil→ Air→Human	Human breathing volatile pollutants from sewage sludge.

We were able to show experimentally, from careful literature reviews, and from basic understanding of TO fates in soil that the bioavailability of sludge-borne TOs is low. Thus, for example, pathways 1, 2, and 8 (plant uptake based) were assessed as presenting vanishingly small risk. Similarly, most other pathways were shown to represent small risks (absolutely), or to be small in comparison to three important pathways.

The important pathways (in order) of TO impact on the environment are,

1. Direct ingestion of sludge by children (pathway 3),
2. Human consumption of meat from animal ingesting biosolids (pathway 4), and
3. Predators consuming biota living in sludge amended soil (pathway 10).

Note the importance of direct contact of the risk target (Highly Exposed Individual, HEI) with sludge-borne TO. Pathways involving exposure to TOs that had to pass through several transfers (e.g. sludge →soil → plants → humans, or sludge → soil → surface (or ground) water) were much less risky. Chaney et al., (1996 and 1991) conducted a thorough assessment of each pathway with the most risk-sensitive TOs (PCBs and PAHs). They concluded that because sludges contain low levels of these TOs, the estimated cancer risk of the HEI was << 10^{-4}. Furthermore, actual low sludge PCB contents and the (low) probability of simultaneously meeting all the constraining criteria used to define an HEI suggest a lifetime risk of $<10^{-7}$.

Such assessments resulted in organics being largely excluded from the final USEPA study rule (40 CFR, part 503); the exception is incineration. Officially, the agency used three criteria to deselect organic from the final sludge regulation (USEPA, 1993). To be deleted from regulation, one of the following criteria had to be met:

1. Pollutant banned for use, has restricted use, or is not manufactured for the US (low concentration expected in sludge now, and even lower concentration expected over time),
2. Pollutant has a low (<5%) frequency of detection in sludge (low concentration actually occurs in sludge), and
3. The limit for the pollutant, as derived from exposure assessment is less than that expected in a "99th percentile" sludge (minimal risk).

The only chemical to consistently fail criteria 2 and 3 was PCBs. It is scheduled for re-evaluation in Round II review of the regulation. In fact, however, if a more reasonable percentile concentration (e.g. 90th vs. 99th) is used, even PCBs meet criteria 3 for land application. The recent review by Chaney et al. (1996), further strengthens the minimal risk assumption. Thus, for various reasons in addition to the EPA criteria tests, we conclude that it is reasonable (and safe) to delete the original list of (and many other) TOs from regulation. Similar conclusions of minimal risk are likely with other wastes if TO variety and concentration are similar to those in sludge (e.g. Chaney and Ryan, 1993; Logan, 1996).

There are other TOs, however, that deserve greater research attention. Notable among an almost infinite list are 1) chlorinated dibenzo-p-dioxins (CDDs) and dibenzo furans (CDFs), 2) aromatic surfactants, and 3) other, unknown antropogenic or natural organic compounds.

Of the various CDDs and CDFs, 2, 3, 7, 8-tetrachlorodibenzo-p-dioxin (TCDD) is the most notorious. It is a persistent xenobiotic that is widely dispersed in the environment, and is acutely and chronically toxic to laboratory animals. Epidemiological studies have failed to unequivocally document significant TCDD-related health effects in humans, but concern persists about its transfer to humans. The concern potency for factor (q_1^*) that EPA uses for TCDD is 4 orders of magnitude larger than any other TO of original concern. Thus, any risk assessment involving TCDD will likely allow extremely low loading rates. Such consideration will also dictate very sophisticated and expensive analytical techniques to confirm risks.

Few reports of CDDs and CDFs in municipal sludges or composts are available (e.g. O'Connor et al., 1991; Cook and Beyea, 1995). If detected (often in specialty sludges like wood processing waste), the compounds occur at low ug kg^{-1} or even ng kg^{-1} concentrations. Remember that these very low initial (waste) concentrations would be diluted at least 100 fold when the waste is land applied at agronomic rates and mixed with soil. Further, the OM added by waste would increase soil sorption capacity for the compounds and likely reduce volatilization and leaching losses. The bioavailability to plants and other organisms of the waste-borne compounds is expected to be vanishingly small. Given the wide-spread presence of CDDs and CDFs in the environment, it is likely that no significant difference in plant (or animal) contamination between waste-amended and control soil grown crops will be detected.

Given these arguments, why do I believe compounds deserve research attention? Animal carcinogens, even if they're not only suspected human carcinogens, frighten people and are easy prey for well-intentioned, but scientifically naive, regulators. We must generate solid scientific data to guide sound regulatory decisions. We learned this lesson repeatedly in our consideration of the original "503" regulations, even for "toxics of concern" with much less notoriety (PRC, 1989).

At the other end of the concentration scale of TOs in sludge from CDDs and CDFs are the aromatic surfactants. The substances are mainly detergent-derived and occur in sludges at concentrations orders of magnitude higher than the originally targeted TOs. Chemicals such as linear alkylbenzesulphonates, nonyl phenol, nonyl phenol monoethoxylate and diethyloxylate are often present in sludges in the g kg^{-1} range (rather than $\mu g\ kg^{-1}$ or mg kg^{-1}). Some of the compounds are apparently refractory, and have been reported in sludge-amended soils at the mg kg^{-1} level.

There is conflicting data in the literature (at least partially associated with different chemical forms) regarding surfactant degradation rates (O'Connor et al., 1991). Some laboratory studies suggest rapid degradation (half-life as low as 5 days), whereas some field studies suggest that significant residues persist well beyond the time expected for complete degradation. Further, some plant studies suggest uncharacteristically high plant accumulation from soil.

Modern homemakers seem to like the superconcentrated detergents recently introduced, and this has increased demand for surfactants as ingredients. Thus, it is not unreasonable to expect even greater concentrations of these (or similar) compounds in municipal sewage sludges. Manufacturers are reportedly switching to more biodegradable surfactants, but little is known of the surfactants, environmental behavior - especially in sludge systems. Work is underway to characterize sludge-borne surfactant behavior (M. Overcash, personal communication, 1994).

Other substances that can occur in sludges or composts at high concentrations include olestra, PDMS, various hydrocarbons, etc. Environmental assessments of some of these compounds have been elucidated (e.g. Allgood et al., 1997 for olestra), but compound/waste - specific reactions have not been thoroughly studied for others. These compounds are not likely anymore environmentally hazardous in and of themselves, but their concentrations (g/kg) in some wastes may magnify small risks. The elevated concentrations mean that the protection offered by dilution with soil and strong adsorption of TOs at low concentrations are possibly voided. We must examine soil/waste reactions of any TO found at high concentrations in the waste to confirm risk, no matter how small we may expect the risk to be.

Research Needs

The confidence espoused herein that TOs represent minimal, or manageable, environmental and human risk is firmly based in research and risk assessment primarily on sludge-borne TOs. Such confidence must be tempered, however, by several considerations, especially when other wastes are involved. Such considerations form the basis for the following research needs.

1. Verify the paradigm used in the risk assessment for TOs in biosolids.
   The pathway model used in the sludge risk assessment was extensive, and represented our best understanding of TO fate and tranpsort at the time. As our understanding of soil-waste reactions mature, however, the model will no doubt have to change. Environmental risk assessment is, in particular, a developing art/science. Those wishing to apply the paradigm to other wastes/chemicals/scenarios must verify that the model is appropriate and/or develop new paradigm.

2. Identify new pathways of import.
   Not all possible routes of TO impact on human, animals, and the environment could be modeled; partly because of data gaps, and partly because we didn't (and still don't) thoroughly understand all possible impacts. Thus, we must continually consider other pathways of possible impact, (e.g. endocrine disruption) and more sophisticated analysis of impact (e.g. new dioxin risk).

3. Fill data gaps in some existing pathway.
   One set of pathways that we acknowledged limited confidence in was wildlife sensitivity to TOs. Few data are available in this area, especially as compared to data on plant, human or domesticated animals. Successful environmental risk assessment will require considerably broader wildlife databases.

4. Uniqueness checks for particular waste/soil combinations.
   Accompanying constituents (e.g. solvents, acids, etc.) in wastes can significantly affect various transfer/transformation processes of TOs. Thus, solvents can affect adsorption/desorption reactions, solute transport through soil, TO volatilization and/or susceptibility to biotic and abiotic degradation. Experience with pesticides has already taught us to consider chemical carrier or adjuvant effects, and we should continually check for similar effects in the host of waste/soil combinations possible.

5. Continued analysis of wastes.
   TOs are no longer regulated in biosolids because the National Sewage Sludge Survey (NSSS) showed minimal contamination with the "toxicants of concern" and because low soil/waste concentration promote TO assimilation to low risk levels. The makeup of biosolids, however, can logically be expected to change as our diets, household chemical use, etc. change. The huge variety of wastes considered at this conference should be expected to differ in TO composition and concentration from biosolids. We must develop a solid database of individual waste TO contents rather than blindly accept generalities from "experts" like me. Incidentally, I still favor annual composite biosolids analyses for TO to continually confirm the low concentrations we assumed.

6. Supply data <u>before</u> regulators act.
   Risk assessment offers many opportunities for conservative assumptions,

default values for important parameters, and policy decisions. Often these "opportunities" are buried deep in the technical support documents, unavailable to casual readers. These "opportunities" can result in multiple conservation (worst-case) assumptions that result in unrealistic appraisal of risk and bad policy. The best way to counter those potentialities is to generate needed data before the analysis is done. Because we had 20+ years of field data to work with, we were able to correct several simplifying assumptions or data selection "opportunities" in EPA's initial sludge rule offerings, and produce a scientifically sound rule. Industries and federal/state agencies must recognize the need for fundamental research/analysis before trying to regulate wastes if similarly sound rules are to be developed for other wastes. If they don't, we scientists need to help them realize the need.

ACKNOWLEDGMENTS:

I acknowledge the cooperation, encouragement and sharing of data and insight of numerous colleagues, especially, R.L. Chaney, J.A. Ryan, M.B. Overcash, M.C. Jones, M.B. Webber, and T.J. Logan.

REFERENCES

Alexander M (1994) Biodegradation and Bioremediation. Academic Press, San Diego, CA.

Allgood GS, McAvoy DC & Woltering DM (1997) Environmental assessment of a new food ingredient, the fat replacer olestra. Environ. Toxicol. Chem. 16:586-600.

Baker JR, Mihelcic JR, Luehrs DC & Hickey JP (1997) Evaluation of estimation methods for organic carbon normalized sorption coefficients. Water Environ. Res. 69:136-145.

Banerjee K, Cheremisinoff PN & Cheng SL (1995) Sorption of organic contaminants by fly ash in a single solute system. Environ. Sci. Technol. 29:2243-2251.

Bellin CA & O'Connor GA (1990) Plant uptake of pentachlorophenol. J. Environ. Qual. 19:598-602.

Briggs GG, Bromilow RH & Evans AA (1982) Relationship between lipophilicity and root uptake and translocation of non-ionized chemicals by barley. Pestic. Sci. 13:495-504.

Chaney RL, Ryan JA & O'Connor GA (1996) Organic contaminants in municipal biosolids: risk assessment, quantitative pathways, analysis, and current research priorities. The Sci. Total Environ. 185:187-216.

Chaney RL & Ryan JA (1993) Heavy metals and toxic organic pollutants in MSW-composts: research results on phytoavailability, bioavailability, fate, etc. In: Hiotink HAJ & Keener HM (Eds) Science and Engineering of Composting: Design, Environmental, Microbiological and Utilization Aspects. OSU, Columbus, OH.

Chaney RL, Ryan JA & O'Connor GA (1991) Risk assessment for organic micropollutants: U.S. point of view. In: Hermite PL (Ed) Proceedings of an EEC Symposium on Treatment and Use of Sewage Sludge and Liquid Agricultural Wastes (Athens, Sept. 1990).

Cook J & Beyea J (1995) Potential toxic and carcinogenic chemical contaminants in source-separated municipal solid waste composts: Review of available data and priorities for action. Compost Council, Alexandria, VA.

DiTorro DM & Horzempa LM (1982) Reversible and resistant components of PCB adsorption-desorption isotherms. Environ. Sci. Technol. 16:594-602.

Fairbanks BC, O'Connor GA & Smith SE (1987) Mineralization and volatilization of polychlorinated biphenyls in sludge-amended soils. J. Environ. Qual. 16:18-25.

Fries GF & Marrow GD (1981) Chlorobiphenyl movement from soil to soybean plants. J. Agric. Food Chem. 29:757-759.

Frink CR & Bugbee GJ (1990) Ethylene dibromide: persistence in soil and uptake by plants. Soil Sci. 148:303-307.

Graham-Bryce IJ (1984) Optimization of physicochemical and biophysical properties of pesticides. In: Mayee PS, Kohn GK & Henn JJ (Eds) Pesticide Synthesis Through Rational Approaches. American Chemical Society, Washington, D.C.

Gillett JW (1983) A comprehensive prebiological screen for ecotoxicological effects. Environ. Toxicol. Chem. 2:463-476.

Hansen LG & Schaeffer DJ (1995) Livestock and domestic animals. In: Rechcigl JE (Ed) Soil Amendments Impacts on Biotic Systems (pp. 41-79). Lewis Publishers, Boca Raton, FL.

Jin Y & O'Connor GA (1990) Behavior of toluene added to sludge-amended soils. J. Environ. Qual. 19:573-580.

Kaufman DD (1983) Fate of toxic organic compounds in land-applied wastes. In: Parr JF, Marsh PB & Kla JM (Eds) Land Treatment of Hazardous Wastes (pp. 77-151). Noyes Data Corporation, Park Ridge, NJ.

Kissel JC, Henry CL & Harrison RB (1992) Potential emissions of volatile and odorous organic compounds from municipal solid waste composting facilities. Biomass and Energy 3:181-191.

Logan TJ (1996) An assessment of health and environmental risks of trace elements and toxic organics in land-applied municipal solid waste compost. Compost Council, Alexandria, VA.

Logan TJ, Harrison BJ, McAvoy DC & Greff JA (1996) Effects of olestra in sewage sludge on soil physical properties. J. Environ. Qual. 25:153-161.

Naylor LM & Loehr RC (1982) Priority pollutants in municipal sewage sludge. Part II. Biocycle 23(6):37-42.

O'Connor GA (1996) Organic compounds in sludge-amended soils and their potential for uptake by crop plants. The Sci. Total Environ. 185:71-81.

O'Connor GA, Chaney RL & Ryan JA (1991) Bioavailability to plants of sludge-borne toxic organics, Rev. Environ. Toxicol. 121:129-165.

Patterson S, Mackay D, Tam D & Shou WY (1990) Uptake of organic chemicals by plants: a review of processes, correlations and models. Chemosphere 21:297-331.

Parr JF, Marsh PB & Kla JM (1983) Land Treatment of Hazardous Wastes. Noyes Data Corporation, Park Ridge, NJ.

Pignatello JJ (1989) Sorption dynamics of organic compounds in soils. In: Sawhney BL & Brown K (Eds) Reactions and Movement of Organic Chemicals in Soil. Soil Science Society of America Spec. Publ. 22, Madison, WI.

PRC (1989) Peer Review Committee - Peer review of standards for the disposal of sewage sludge (U.S. EPA Proposed Rule 40 CFR Parts 257 and 503). USDA-CSRS, W-170 Regional Research Committee, Washington, D.C.

Ryan JA, Bell RM, Davidson JM & O'Connor GA (1988) Plant uptake of non-ionic organic chemical from soil. Chemosphere 17:2299-2323.

Scow KM, Simkins S & Alexander M (1986) Kinetics of mineralization of organic compounds at low concentrations in soil. Appl. Environ. Microbiol. 51:1028-1035.

Scheunert I, Topp E, Schmitzer J, Klein W & Korte F (1985) Formation and fate of bound residues of ^{14}C benzene and ^{14}C chlorobenzenes in soils and plants. Ecotoxicol. Environ. Saf. 9:159-170.

Strek HJ, Weber JB, Shea PJ, Mrozek E Jr & Overcash MR (1981) Reduction of polychlorinated biphenyl toxicity and uptake of carbon-14 activity by plants through the use of activated carbon. J. Agric. Food Chem. 29:288-293.

Topp E, Scheuner I, Attar A & Korte F (1986) Factors affecting the uptake of ^{14}C labeled organic chemicals by plants from soil. Ecotoxicol. Environ. Saf. 11:219-228.

Travis C & Hattener-Frey HM (1988) Uptake of organics by aerial plant parts: A call for research. Chemosphere 17:277-284.

U.S. Environmental Protection Agency (1990) National sewage sludge survey: Availability of information and data, and anticipated impacts on proposed regulations; proposed rule. Federal Register. 55:47210-47283.

U.S. Environmental Protection Agency (1992) Technical Support Document for Land Application of Sewage Sludge. Vol. I. PB 93-110575. Vol. II. PB 93-110583. NTIS, Springfield, VA.

U.S. Environmental Protection Agency (1989) Development of Risk Assessment Methodology for Land Application and Distribution and Marketing of Municipal Sludge. EPA/600/6-89/001.

U.S. Environmental Protection Agency (1993) 40 CFR Part 257 et al., Standards for the Use or Disposal of Sewage Sludge; Final Rules. Fed. Reg. 58:9248-9415.

Weber JB & Miller CT (1989) Organic chemical movement over and through soil. In: Sawhney BL & Brown K (Eds) Reactions and Movement of Organic Chemicals in Soils (p 305). Soil Science Society of America, Spec. Publ. 22, Madison, WI.

Weber JB & Mrozek E (1979) Polychlorinated biphenyls: phytotoxicity, absorption and translocation by plants and inactivation by activated carbon. Bull. Environ. Contamin. Toxicol. 23:412-417.

Webber MD & Goodin JD (1989) Persistence of volatile organic contaminants (VOC) in sludge treated soils. In: Alternative Uses for Sewage Sludge, an International Workshop Organized by the European Commission, New York.

Wild SR & Jones KC (1992) Organic chemicals entering soils in sewage sludge: screening for their potential to transfer to crop plants and livestock. Sci. Total Environ. 119:86-119.

Witte H, Langenohl T & Offenbacher G (1988) Investigation of the entry of organic pollutants into soils and plants through the use of sewage sludge in agriculture. Part A. Organic pollutant load in sewage sludge, Part B. Impact of the application of sewage sludge on organic matter contents in soils and plants Korrespondenz Abwasser 13:118-136.

FATE AND POTENTIAL EFFECTS OF TRACE ELEMENTS: ISSUES IN CO-UTILIZATION OF BY-PRODUCTS

J.A. RYAN and L.T. BRYNDZIA
National Risk Management Research Laboratory, US Environmental Protection Agency, Cincinnati, Ohio 45224, USA
E-mail ryan.jim@epamail.epa.gov

INTRODUCTION

Starting with the earliest records, we find that soils have been altered by man's activities and have been the receptors (directly or indirectly) for most, if not all, of his discards. As affirmed by the areas of land which have become ecological wastelands from man's smelting and mining activity, this has not been without cost in terms of changes in soil and its utility. Attempts at remediation of these soils has proven to be a long term and expensive proposition that often results in destruction of the soil rather than its remediation. These observations and the realization that soil resources are not infinite has caused society to question the use of soil resources and their capacity to continue to act as receptors of waste streams. However, even the most unobservant must be aware that the ultimate decision on the uses of soil will be a societal decision and may be more emotional (e.g. political, economic, etc.) than technical. Thus, those who want the technical issues utilized in the decision making process must realize that the better informed society becomes, the more likely the decision will include the technical concerns. With this understanding it becomes necessary not only to solve the technical problem, but to make certain that the general population understands its solution.

An example of this balancing act of risk and benefits is the standards for the disposal of sewage sludge (biosolids) promulgated under the Clean Water Act (CWA) Section 503 by the U.S. Environmental Protection Agency (U.S. EPA, 1993a). This example illustrates that successful co-utilization of by-products in agricultural systems is dependent upon understanding the economic and societal benefits gained by the practice, as well as the environmental effects and risk of the addition of materials to the soil. Thus, the information about fate and effects of metals obtained in the development of the CWA 503 serve as an example of what we need to know if co-utilization of by-products is to be acceptable. Additionally it illustrates what is needed to resolve the concerns over the long term fate of metals in waste amended soil systems.

The organic and nutrient content of biosolids makes it a valuable resource. Its beneficial uses are not limited to the production of agricultural commodities. Biosolids are used in silviculture to increase forest productivity and to revegetate and stabilize harvested forest land and forest land devastated by fires, land slides, or other natural disasters.

S. Brown, J.S. Angle and L. Jacobs (eds.), Beneficial Co-Utilization of Agricultural, Municipal and Industrial By-products, 219–233.
© 1998 *Kluwer Academic Publishers.*

Biosolids are productively used to stabilize and revegetate areas destroyed by mining, dredging, and construction activities. Air-dried biosolids are frequently used to fertilize highway median strips, clover leaf exchanges, and for covering expired landfills. Currently one-third of the biosolids generated in the United States are effectively reused by applying it to the land, or sold or given away for use in home gardens. In comparison, Japan uses 42 percent of its biosolids for coastal reclamation and home garden or farming uses. The United Kingdom applies 51 percent of its biosolids to the land. Thus, we conclude that biosolids provide a benefit, and that successful co-utilization of by-products will depend upon demonstrating that they provide a benefit.

The benefits of using biosolids to improve land productivity are substantial but not without risks. Biosolids can contain high levels of pathogenic organisms (e.g., viruses, bacteria) or high concentrations of pollutants which, if improperly handled, could contaminate soil, water, crops, livestock, fish, and shellfish. The U.S. Environmental Protection Agency adopted regulations designed to protect human health and the environment from reasonably anticipated adverse effects from pathogens, organics and metals contained in the material. (U.S. EPA, 1993a). These regulations are based on an assessment of the available data from soils amended with biosolids. The basic understanding of contaminants developed during evaluation of this data resulted in a paradigm shift where the material being applied becomes as important as the soil or contaminant in determining the environmental fate of the contaminant. This understanding has been the subject of much debate (Ryan and Chaney, 1993; Chaney and Ryan, 1993 and 1994; and McBride, 1995). For discussion and rationalization of the 503 data analysis see the EPA Technical Support Document for Land Application of Sewage Sludge (1993b).

TRACE ELEMENTS

Our concern here is limited to understanding the environmental effects of trace elements during co-utilization of by-products. In defining trace elements we recognize that ten elements (Al, Ca, Fe, K, Mg, Na, O, P, Si, and Ti), make up 99% of the composition of soil. Thus, most of the elements of the periodic table are considered trace elements. The individual trace elements occur as trace constituents of primary minerals, where they have been incorporated by isomorphic substitution in the crystal lattice, and their individual concentrations do not normally exceed 1000 mg/kg and in most cases are less than 100 mg/kg. The environmental fate and effect of only a few (As, B, Cd, Cr, Cu, Mo, Ni, Pb, Zn) have been extensively studied. Thus, these elements serve as our focus, as we must conclude that other elements are not of concern.

The basis for concern about metals in soil is their potential for adverse effects on humans or the environment. The pathways by which soil metals cause harm are related to:

- Plant uptake - harm to plant and/or movement of metals into agricultural or ecosystem food chain.
- Ingestion of soil by animal - harm to animal and/or movement of metals into agricultural or ecosystem food chain.

- Ingestion of soil by humans - direct ingestion of soil or household dust by children.
- Loss to water systems - leaching into groundwater and/or into surface water as well as erosion of soil material into surface waters where metals can be released to cause ecosystem harm.
- Inhalation of metals by humans - volatilization and/or dust movement.

For each metal, one pathway provides the highest probability of adverse effects to some receptor and therefore becomes the limiting pathway. For example, direct ingestion of soil Pb by children; plant phytotoxicty for Cu, Zn, Ni; and human health from food chain transfer of Cd are the pathways of concern. These understandings have come about through a comprehensive or holistic view of contaminants in the environment.

The Development of the "Soil-Plant Barrier" concept provided structure to the knowledge which had been generated, and focused attention on the major data gaps. In this concept it must be recognized that plants not only serve as a mechanism for transfer of soil metals to higher tropic levels, but also act as a filter to limit transfer. It is common knowledge that some elements are readily absorbed by plant roots and translocated into shoots where the tissue concentration can poison livestock (e.g. Se and Mo) and/or become a concern for human health (e.g. Cd). In contrast, other elements readily absorbed by plant roots and translocated into shoots, may adversely affect the plant, which suffers severe yield reductions before the concentration in the tissue represent a threat to humans, livestock or wildlife from chronic lifetime consumption (e.g. Zn, Ni,and Cu). Other elements are so insoluble in the soil or fibrous plant root, that they are not translocated to the shoot (e.g. Pb and Cr). A summary of these concepts and understandings are present in Table 1.

MASS BALANCE

It must be recognized that none of these elements has an infinite supply and that if all of the world reserves of the elements were mixed in soils, a finite concentration would be obtained. Mixing the world reserves of the elements into the plow layer of the pasture, range and arable lands allows for an average increase in soil metal composition to be estimated (Table 2). A comparison of the average increase with the average soil background concentration indicates that the increase is less than is present in the average soil. The exception is Cu, where the increase would be double the present background. Thus, it would seem that Cu would be the element of most concern. As illustrated by the concentration of Pb in urban soils, utilization of average increase is not indicative of specific situations, but it does provide a realization that not all soils can be excessively increased in elemental composition. Thus, aberrant elemental composition of a soil becomes an element and situation specific consideration.

It is important to recognize that the metals added to the soil as by-products are not concentrated soluble metal salts, but are by-products which contain the metal as a trace impurity. Thus, in addition to the metal of concern, a number of other materials are added to the soil when by-products are applied. The material is applied to the soil surface where it is either left unincorporated or is mixed into a finite soil depth, typically 15 cm. As the

Table 1. Foliar analysis (mg/kg dry weight) for metals in plant productivity or food chain safety.

Element	Plant Conc. mg/kg dry weight.	Toxic Plant Conc.[a] mg/kg dry weight	Animal[b] mg/kg dry diet
$As_{(inorganic)}$	0.01 - 1	> 10	50
B	20 - 70	> 75	150
Cd	0.2 - 0.8	5 - 700	0.5[c]
Cr	0.1 - 1	> 20	3000
Cu	3 - 20	25 - 40	25
Mo	0.1 - 3	> 100	10
Ni	0.1 - 5	50 - 100	50
Pb	2 - 5	-	30[c]
Zn	15 - 150	500 - 1500	300

[a] Summarized by Chaney (1983)
[b] Based on chronic animal tolerance of elements summarized by NRC (1980)
[c] Maximun levels tolerated are based on Cd and Pb in liver, kidney, and bone in foods for humans rather than simple animal tolerance by animals; ignored role of Zn in inhibition of Cd uptake by plants or animals.

Table 2. Average potential increase in soil metal composition.

Element	World Reserves (kg)[a]	Average Soil Increase (mg/kg)[b]	Background Soil Conc. (mg/kg)[c]
As	4.4×10^{9}	0.5	5.2
Cd	5.4×10^{8}	0.05	0.18
Cu	3.1×10^{11}	32	18
Ni	4.7×10^{10}	5	16.5
Pb	6.9×11^{10}	7	10.6
Zn	1.4×10^{11}	15	42.9

[a] USGS Minerals Commodity Summary (1996).
[b] 4.8×10^{9} ha of pasture, range and arable land.
[c] Shacklette and Boerngen (1984) and Holmgren et al. (1992).

same volume of soil is mixed with each application, repeat applications of by-products should dilute the original soil volume in the mixing zone. Conservation of mass dictate that metal composition of the soil by-product mixture not exceed the concentration of the most concentrated material. The ultimate composition however, can exceed the concentration contained in the most concentrated material if it has a large biodegradable fraction. Even then, however, its final concentration will not exceed the concentration of metal based on the ash content of the material. Thus, assuming that there is continuous application of the by-product to the soil system, the soil will eventually reach the concentration of the by-product or its decomposition product when the zone of mixing is all by-product. Obviously, at this point the chemical characteristic of the soil and its ability to react with the added metal would have no affect on the environmental fate of the metals added by the by-products. Rather, the environmental fate of the metal would be entirely dependent upon its reaction with the by-product matrix. From a mass perspective this can be visualized by considering common horticultural operations (e.g. raised bed gardens, flower gardens, and golf greens) where a material is either mixed into the soil or added to the surface of the soil and the upper horizon of the system ultimately assumes the chemical and physical characteristics of the amendment. This same situation occurs if by-products are continuously mixed into the upper horizon of the soil. The time required for the upper horizon (zone of incorporation) to become essentially 100% by-products is a function of rate of application and removal by erosion but is estimated to be hundreds of years (Ryan and Chaney, 1994). It is under this condition that we want to define the fate of the contaminants in the system.

If the issue is metal activity rather than total metals, the consideration of mass or concentration would be equivalent to consideration of metal activity. If metal activity supported by the by-product is greater than that of the soil, the soil mechanisms responsible for binding will reduce metal activity. With increased by-product addition, the soil binding sites will be saturated and metal activity will approach that supported by the by-product. Kinetic consideration of the reactions and changes in soil and by-products composition with time, can alter this simple approach, but do not change the basic concept. Changes in composition (e.g. humus formation from the fresh organic fraction, ageing of added amorphous surfaces) change their adsorption characteristics and alter thermodynamic considerations. It is clear that the addition of a readily decomposable substrate causes a change in the biomass and organic fraction of the system. These decomposition products can increase the soluble organic concentration in the solution phase of the system, which could alter the total soluble metal content of the solution. Additionally, the release of cations and anions which were part of the organic matrix also impact the solution composition and the fate of the metals. Thus, the chemistry of the mixed system is more complex than just a simple mass balance, but the principles are the same. The concerns with the system is how it influences metal activity as it weathers, and what it behaves like at steady state.

METAL REACTIONS IN SOIL

Numerous reviews of metal reactions in soils are available (Hodgson, 1963; Ellis and

Knezek, 1972; Lindsay, 1972; Alloway, 1990; and McBride 1994). From these reviews it is apparent that many soil components (clays, organic matter, hydrous oxides, carbonates, etc.) have the ability to bind metals in soils and reduce the amount of soluble metal present. Several different mechanisms are involved in the adsorption of metal ions: cation exchange (or non-specific adsorption), specific adsorption, organic complexation, precipitation and coprecipitation. The chemical matrix of soil consists of numerous primary and secondary minerals in various stages of weathering, as well as organic material in various stages of decomposition. However, the lack of knowledge of the solid phases and their frequent amorphous nature result in a tendency to discount their importance in specific adsorption, precipitation and dissolution reactions and consider exchange reactions (cation exchange) as the major adsorption mechanism in soil systems. The extent of adsorption is measured and adsorption isotherms calculated. However, little precision about which particular process is responsible for adsorption is provided, and ways for direct measure of the mechanism or small change in solid mineral species are not apparent. Yet, it is knowing the specific adsorption mechanism, its long term stability, and the impact of the adsorption mechanism on the soluble metal, or, more specifically, the environmental significant ionic form that is needed if we are to understand these complex systems.

This lack of knowledge on specific adsorption mechanism and precipitation products leads to a simplistic view of environmental risk from metal contamination where basic soil properties (e.g. CEC and pH) and total metal content are all that are considered. Further, utilization of conservative assumptions in assessment of metal addition results in the conclusion that allowable metal concentrations in soil are not greater than values found in normal soil. However, the implication that change in soil metal concentration leads to unacceptable risks has insidious implications. For example, it causes limited addition of materials (fertilizers, municipal compost, biosolids, etc.) containing metals to soil, resulting in added cost for their disposal. Additionally, costs associated with heroic attempts to remove metals from soils which are deemed to be a risk by these conclusions must be borne by society. Thus, it is incumbent that we understand whether changes in soil composition caused by addition of by-products has an effect on the chemistry of the system and what this means to the risk of metals in the system.

It is recognized that total metal composition is not the dominant issue, rather it is the free metal cation in solution that is of environmental concern. Metal ions chelated with organic ligands or complexed with inorganic ligands in the soil solution are not adsorbed directly by plants, but must be separated from the ligand. Only the free ion (e.g. Cd^{2+}) crosses the plasmalemma and enters the cytoplasm Thus, chemical processes controlling the concentrations of metal ions and complexes in the soil solution (those concerned with the adsorption of metals from the liquid phase onto the solid phase) are of importance and exert a major influence on metal uptake by plant roots and movement to ground water, as well as bioavailability.

Recent advances in our understanding of environmental soil chemistry have demonstrated that consideration of specific mineralogical species, their reaction kinetics and thermodynamics are important if the risk from increasing metal concentration are to be correctly addressed. Additionally, it must be recognized that the source of the metal addition can add new solid phases to the system which affects the stability of metal complexes, solid phase precipitation and metal chelates, adding further complexities to soil chemistry. Thus, it is apparent that not only is the issue of adsorption by soil components important, but adsorption by the matrix added with the metal, and the long- term persistence of these new phases, may be of equal or greater concern. As different sources of metals are added to the soil system through co-utilization of by-products, the complexity of the system increases. It is in this "Co utilization of by products in soil", that we wish to know the fate of metals.

IMPLICATION OF ADDED SOLIDS

Research with pure metal salts addition at field scale had illustrated yield reductions by Cu (Walsh et al., 1972) and As (Steevens et al., 1972), whereas comparable additions of biosolids (Hinesley and Hanson, 1984) failed to show yield reductions. The new metal bearing phase of biosolids was altering bioavailability, but was generally considered to be organic in nature, and thus of only short term benefit. In the mid 70's available information illustrated that the relationship between plant concentration of metals and soil concentration of metals in biosolids amended soil was not simple. Cunningham et al. (1975a, 1975b, 1975c) showed that the relationship was a function of source of metal added to the soil. The availability of metals added to soil by metal salt addition was greater than the availability of metal salts mixed with biosolids added to soil. The availability of which, in turn, was greater than that of metals contained in biosolids added to soil. This observation has been confirmed by subsequent studies which evaluated the extractability and phytoavailability of biosolids metals when the metals were added to the biosolids before anaerobic digestion, or after digestion (Bloomfield and McGrath, 1982; Davis and Carlton-Smith, 1981, 1984). In each case, adding the metals after biosolids digestion (immediately before application to soil) caused the metals to be much more phytoavailable than metals added during sewage treatment or before biosolids stabilization. However, metals added to biosolids were less phytoavailable than metal salts added to the soil without the biosolids. This difference between salt metal addition and biosolids metal addition was confirmed in field experiments as early as the mid 70's (Figure 1, Giordano and Mays, 1977).

Similarly, when amounts of metals required to reduce yields of barley or vegetables were determined with salts in greenhouse pots, with mixtures of high metal biosolids in pots in the greenhouse, or with average quality biosolids in the field, the salts and high metals biosolids caused phytotoxicity (Davis and Carlton-Smith, 1984). The average quality biosolids caused only yield increase (Johnson et al., 1983). It should have been apparent that the biosolids were providing a surface for metal adsorption or precipitation, causing a reduction in availability of metals to plants. We should have strived to determine what this phase was, rather than assume Beckett's postulated time-bomb hypothesis was correct (Beckett et al., 1979).

Beckett assumes that the responsible binding phase was organic, and, that as the organic material decomposes, its complexing nature would be lost, with a subsequent release of metal to the inorganic system where it behaves as a metal salt added to the soil. Thus, metal adsorption would be controlled by the soil. This concept should have

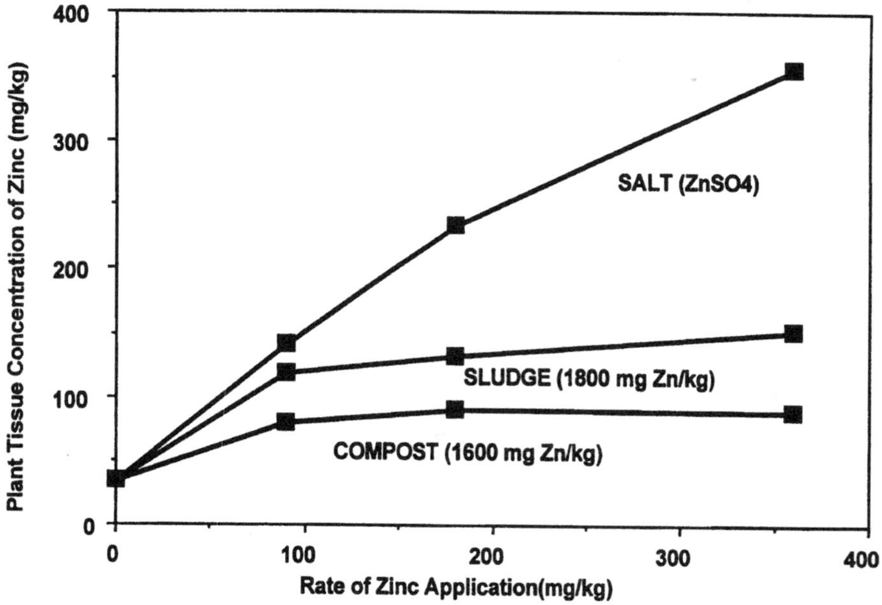

Figure 1. Effect of source of metal addition on plant tissue concentration from a field study (average of 4 years). From Giordano and Mays, 1977.

disappeared when it was illustrated that plant response was greatest the first year after application and decreased in subsequent years (Hinesly and Hansen, 1984; Bidwell and Dowdy, 1987; and Chang et al., 1987). Thus, early recommendations for limits on biosolids metal application were directly related to basic soil properties (e.g., CEC and pH).

In a series of experiments utilizing 11 paired sets of soils (control soil and soil from long-term biosolids application site) Mahler et al. (1987, 1988a, 1988b) illustrated that the biosolids added adsorptive capacity to the soil-biosolids system. In these studies, plants grown in soils which had received biosolids exhibited less response to added Cd than control soils which had not received biosolids. This was even true for the samples of similar organic matter content, illustrating that the additional adsorptive surface added by biosolids was not organic, or at least, if it was, it resided in a very recalcitrant organic fraction. These sites had received biosolids applications from 5 to 30 years and yield of sensitive crops was unaffected by the previous biosolids application. Cadmium salt addition, at rates of 0, 5, and 10 mg/kg, resulted in a linear response to plant Cd concentration regardless of the initial soil Cd level, without changes in yield. The linearity of the individual response curves, coupled with the yield data, indicated that the soil/biosolids systems, rather than plant systems, controlled the plant concentration of Cd for these series of experiments. Within each of the paired soils the slope of the response curve was generally lower for the soil which had a history of biosolids application. The exception to this general result was an unlimed Chard experiment in a Salinas soil, where a greater response was observed for the biosolids treatment. This aberration could be attributed to the lower pH of the soil which had a history of biosolids application (pH 5.8

vs 6.8). In the limed experiment where the pH values were comparable (pH 7.3 vs 7.2) the soil with a history of biosolids application exhibited 70 % of the response of the control soil. In a limed Chard experiment, all but two of the soils (Burkhardt and Pembroke) exhibited the lower response curve effect of biosolids application on plant response to added soil Cd. In these two soils the response was the same for the control and biosolids treated soils. The elemental composition of these two soils would cause one to question if these soils had received biosolids. It is also apparent that the plant response to soil Cd added via biosolids (response associated with control and biosolids amended soil without inorganic Cd addition) was less than that added via the Cd salt addition.

The effect of addition of biosolids on the chemistry of the system and differences between different sources of biosolids was further illustrated with samples collected from a long term field experiment (Figure 2). The samples were collected from a 15 year field experiment where anaerobically digested sewage biosolids was annually applied for 14 years (1968-1981) by furrow irrigation on plots of Blount silt loam (Aeric ochraqualf, fine, illitic, mesic) continuously planted to corn (*Zea mays* L.) (Hinesly and Hanson, 1984). As often as weather permitted, liquid biosolids (Chicago liquid digested sewage biosolids) were applied at depths of 25.4 mm (maximum), 12.7 mm (one-half maximum), and 6.4 mm (one-fourth maximum) to each of the appropriate plots on the same day. Total cumulative loading on the maximum plot was 765.1 mt/ha. Load to load variability in percent solids and variation in dewatering caused by weather conditions caused variability in the year to year loading rates, ranging from 25.61 to 128.37 mt/ha. Samples collected from these replicated plots were amended with another aerobically produced biosolids to reproduce the Cd loading equivalent to the previous field experiment. Several crops including a single cross corn, lettuce (*Lactuca sativa*), and swiss chard (*Beta vulgaris*) were produced under growth chamber conditions. Within a crop, yield was not affected by treatment, however plant tissue concentration was. The results clearly illustrate that the plant response was dependent upon material applied (response curve for aerobic biosolids addition to control soil (Figure 2) vs. response curve for the anaerobic biosolids addition from the field experiment (Figure 3)). Further, comparisons of response curves for the aerobic biosolids addition to the control, 1/4 max, 1/2 max and max treated samples illustrate that the rate of addition of the anaerobic biosolids from the field experiment affected the response from addition of the aerobic biosolids (Figure 2). As plant composition was not limited by the ability of the plant to accumulate Cd (i.e. linear Cd plant tissue response in the control soil with addition of the aerobically stabilized biosolid allowed greater plant Cd concentration than any of the other treatments) it is concluded that the differences are associated with soil chemical changes altering the phytoavailability of soil Cd in biosolids amended soils. Further it is apparent that the alteration in Cd phytoavailability was directly related to the addition of anaerobic stabilized biosolids. The reason for this difference in phytoavailability could not be discerned from this experiment, but it was clearly related to the amount of biosolids added to the soil.

In the case of two soils (Wea and Morley) from the Mahler et al. 1987 study the control soil had more organic C than the biosolids amended soil, implying that the added biosolids organic C had degraded. As illustrated by the paired soils; response to inorganic Cd addition, the apparent loss of organic matter from biosolids addition did not cause the plant response to approach that of the control, or these soils to behave differently than the

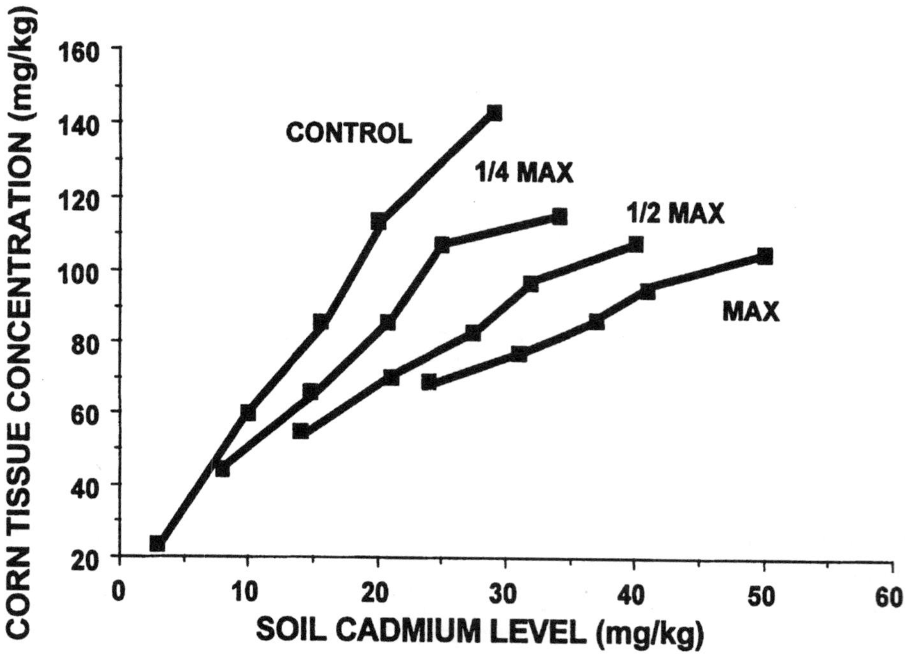

Figure 2. Effect of addition of aerobically digested biosolids to Illinois Long Term Field Experiment on tissue Cd concentration.

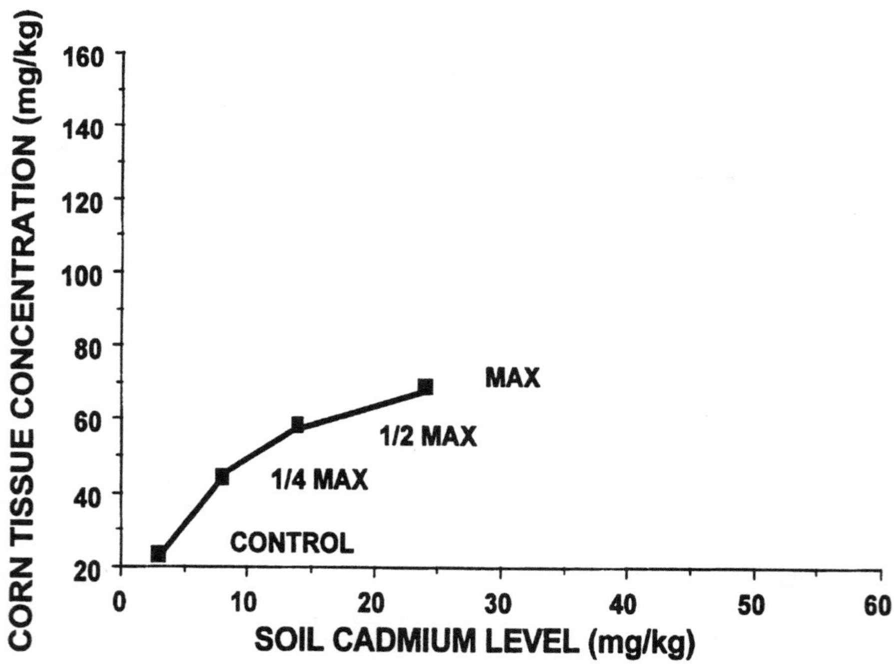

Figure 3. Plant tissue concentration from Illinois Long Term Field Experiment.

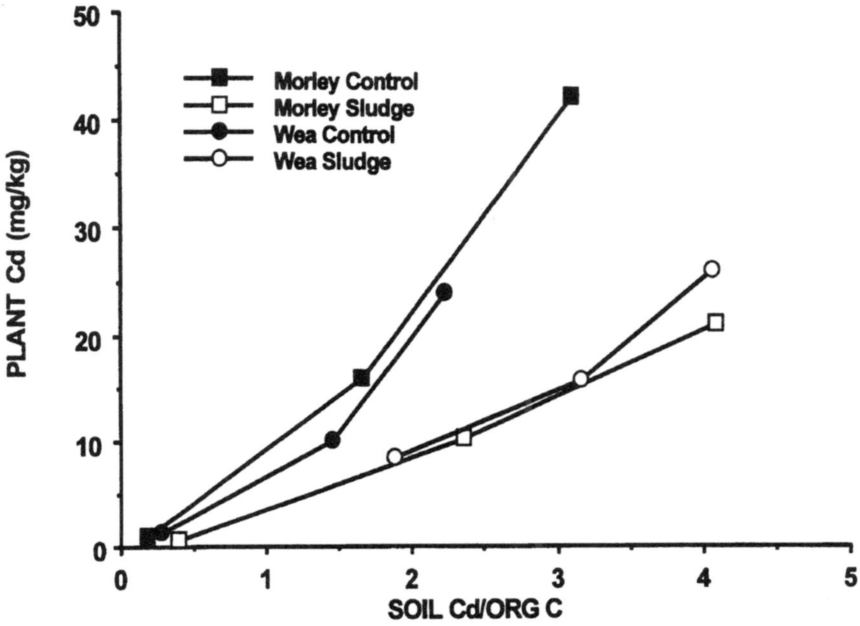

Figure 4. Effect of previous sludge addition on plant tissue response to added cadmium (Mahler et al. 1987).

others (Figure 4). Therefore, it is apparent that the inorganic phase of the biosolids and/or the organic fraction resistant to decomposition was primarily responsible for the alteration in chemical behaviors of these soils. Further, the Wea soil (Dayton OH) had not had biosolids added to it for 15-20 yrs. prior to this sampling and contained a total Cd content of 4.5 mg/kg soil. A field study on this site illustrated that the plant uptake of the metals from the previous biosolids applications were not as high as anticiated. In fact the Swiss Chard plant Cd composition based on the addition of inorganic Cd to the control soil in the growth chamber experiment should be 12.3 mg/kg whereas the field experiment found it to be 1.51 mg/kg.

It is apparent from this body of work, that the addition of biosolids to soil is, in fact, altering the chemical phases in the soil system, beyond the simple addition of greater quantities of trace elements. It is also apparent that this alteration does not require large additions of biosolids and that the changes are of a permanent nature. A review and interpretation of this information was presented by the Corey et al. (1987b) workgroup. In short, the specific metal adsorption capacity of biosolids persistently increases the ability of the soil-biosolids mixture to adsorb metals, thereby reducing the phytoavailability/ bioavailability of biosolids-borne metals. Corey et al. predicted that biosolids adsorption chemistry should control the activity of free metal ions in the soil solution of biosolids-amended soils after reaching the biosolids application rate which saturated the soil metal binding sites of the soil system. This increase in sorption capacity is augmented by the long "equilibration" times between the introduction of trace metals into the waste stream and time of land application of processed biosolids. These long equilibration times facilitate maximal sorption of the trace metals by reactive surface functional groups or into low solubility solids. Based on these understandings, researchers have attempted to characterize the chemical aspects of biosolids which made metals so much less available to plants (phytoavailable) or biological systems (bioavailability) than were metal-salts.

Addition of by-products to soils have not always illustrated this binding of metals. For example, unpublished information from swine waste application to sandy soils showed incidences of yield reductions associated with the Cu and Zn. The *Seattle Times* has reported that the use of fertilizers containing waste products (foundry sands and slags) as bulking agents has resulted in yield reductions. Thus, it is most probable that the ability of each by-product to provide metal binding is different and will have to be evaluated. This can be done on a case-by-case evaluation or if the basic chemistry of the binding is known, its probability of occurrence within the by-product can be predicted.

In the case of one metal, Pb, research indicates that the mineral form of the metal can be changed with a subsequent change in the bioavailability of the metal. Results of experiments on the thermodynamics and kinetics of the dissolution of Pb minerals, forms of soil-Pb and P minerals have been published, [see Ma et al., 1993, 1994; Laperche et.al., 1996; and Zhang et al., 1997]. These results have shown that Pb is rapidly and effectively precipitated from solution by orthophosphate, aqueous P, hydroxyapatite, or phosphate rock to form a series of Pb phosphates, that attenuate Pb in aqueous solution, exchangeable form and contaminated soil material, to below the U.S. EPA action level of 15 μg/L dissolved Pb. Results suggest that, in the presence of P, or high phosphate materials such as biosolids, the formation of a highly insoluble mineral is responsible for the observed decrease in Pb bioavailability. It has also been shown that, hydroxyapatite can effectively immobilize aqueous Pb in the presence of common soil solution anions: NO_3, Cl, F, SO_4, and CO_3 as well as cations: Zn, Cd, Ni, Cu, Fe, and Al; Further, that hydroxyapatite also removed Pb from Pb-EDTA solution in the presence of excess EDTA, indicating that basic Ca-phosphates can sequester Pb even in the presence of strongly complexing organic ligands; that goethite adsorbed Pb is readily converted to pyromorphite in the presence of apatite; confirmed formation of pyromorphite from Pb-contaminated soil (4,000-50,000 mgPb/kg), paint chips (30% Pb, w/w), cerrusite, anglesite, galena,and Pb-humic complexes. Addition of appatite to Pb contaminated soil reduces the plant uptake of Pb and SEM and XRD analyses indicate that apatite reacted with the Pb to form pyromorphite on root surfaces. Animal feeding experiments have illustrated that Pb bioavailability followed the order: Pb-acetate >> contaminated soil> pyromorphite = control and that the addition of apatite or rock phosphate to the contaminated soil reduced the bioavailability of the contaminated soil Pb. This illustrates that the formation of phyromorphite in soils not only reduce the solubility of the soil Pb, but also reduces its bioavailability.

These results demonstrate that both hydroxyapatite and phosphate rocks were effective in reducing Pb solubility and bioavailability through dissolution of hydroxyapatite or phosphate rocks and precipitation of pyromorphite. The effective and rapid immobilization of Pb from solution and contaminated soils by hydroxyapatite or phosphate rock, the limited effects from other minerals, anions, and cations, the apparent environmental stability of the reaction products, along with the ready availability and low-cost of hydroxyapatite or phosphate rock suggest that this approach might have great merit for cost-effective *in situ* immobilization of Pb contaminated water, soils, and wastes. Research efforts to confirm these conclusions and evaluate the efficacy of *in situ* treatments are moving forward.

CONCLUSIONS

It is apparent from the above discussions of metals that their effects are dependent upon metal, and source of metal. Thus, in the complex system of co-utilization of by-products there are two solutions: 1) field data illustrating the effect of the material being discussed and/or 2) basic information on mechanism of metals in these complex systems. If the amount of data required to provide convincing evidence of the safe use of biosolids is an indication of what will be required for each by-product, then we have a work program which will keep a number of scientist busy for an extended time. If, however, an attempt is made to understand the mechanism, the knowledge gained will provide the ability to predict how by-products behave and how to make by-products safe. As illustrated by the identification of a possible mechanism for Pb, the payoff is not only advantageous from the perspective of the use of the material, but offers the possibility to solve problems with contaminated soil. The concept of the use of waste streams to do this is certainly appealing. Its success will depend upon the basic knowledge of the reactions of the contaminant and the waste product.

REFERENCES

Alloway BJ (1990) Heavy Metals in Soils. Johy Wiley and Sons, NY, USA.

Beckett PHT, Davis RD & Brindley P (1979) The disposal of sewage sludge onto farmland: The scope of the problem of toxic elements. Water Pollution Control 78:419-445.

Bidwell AM & Dowdy RH (1987) Cadmium and zinc availability to corn following termination of sewage sludge applications. J. Environ. Qual. 16:438-442.

Bloomfield C & McGrath SP (1982) A comparison of the extractabilities of Zn, Cu, Ni, and Cr from sewage sludges prepared by treating raw sewage with the metal salts before or after anaerobic digestion. Environ. Pollut. B3:193-198.

Chaney RL (1983) Potential effects of waste constituents on the food chain. In: Parr JF, Marsh PB & Kla JM (Eds) Land Treatment of Hazardous Wastes (pp. 152-240). Noyes Data Corp., Park Ridge, NJ.

Chaney RL & Ryan JA (1994) Risk Based Standards for Arsenic, Lead and Cadmium in Urban Soils. DECHEMA- Fachgespräche Umweltschutz. ISBN 3-926959-63-0.

Chaney RL & Ryan JA (1993) Heavy metals and toxic organic pollutants in MSW-compost: Research results on phytoavailability, bioavailability, fate, etc. In: Hoitink HAJ et al. (Eds) Science and Engineering of Composting: Design, Environmental, Microbiological and Utilization Aspects. The Ohio State University, Columbus, OH.

Chang et al. (1987) In: Page AL, Logan TJ & Ryan JA (Eds.) Land Application of Sludge : Food Chain Implications. Lewis Publishers Inc., Chelsea, MI.

Corey RB, King LD, Lue-Hing C, Fanning DS, Street JJ & Walker JM (1987) Effects of sludge properties on accumulation of trace elements by crops. In: Page AL, Logan TJ & Ryan JA (Eds) Land Application of Sludge - Food Chain Implications (pp. 25-51). Lewis Publishers Inc., Chelsea, MI.

Cunningham JD, Keeney DR & Ryan JA (1975a) Phytotoxicity in and metal uptake from soil treated with metal-amended sewage sludge. J. Environ. Qual. 4:455-460.

Cunningham JD, Keeney DR & Ryan JA (1975b) Yield and metal composition of corn and rye grown on sewage sludge-amended soil. J. Environ. Qual. 4(4):448-454.

Cunningham JD, Keeney DR & Ryan JA (1975c) Phytotoxicity and uptake of metals added to soils as inorganic salts or in sewage sludges. J. Environ. Qual. 4:460-461.

Davis RD & Carlton-Smith CH (1981) The preparation of sewage sludges with controlled metal content for experimental purposes. Environ. Pollut. B2:167-177.

Davis RD & Carlton-Smith CH (1984) An investigation into the phytotoxicity of zinc, copper and nickel using sewage sludge of controlled metal content. Environ. Pollut. B8:163-185.

Ellis BG & Knezek BD (1972) Adsorption reactions of micronutrients in soils. pp 59-78. In: Mortvedt JJ, Giordano PM & Lindsay WL (Eds) Micronutrients in Agriculture. Soil Sci. Soc. Amer., Madison, WI.

Giordano PM & Mays DA (1977) Effect of land disposal applications of municipal wastes on crop yields and heavy metal uptake. EPA-600/2-77-014.

Hinesly TD & Hanson (1984) Effects of using sewage sludge on agricultural and disturbed lands. EPA-600/S2-83-113

Hodgson JF (1963) Chemistry of micronutrient elements in soil. Adv. Agron. 15:119-159.

Holmgren GGS, Meyer MW, Chaney RL & Daniels RB (1993) Cadmium, lead, zinc, copper, and nickel in agricultural soils of the United States of America. J. Environ. Qual. 22:335-348.

Jing J & Logan T (1992) Effects of sewage sludge cadmium concentration on chemical extractability and plant uptake. J. Environ. Qual. 21:73-81.

Johnson NB, Beckett PHT & Waters CJ (1983) Limits of zinc and copper toxicity from digested sludge applied to agricultural land. In: Davis RD, Hucker G & L'Hermite P (Eds) Environmental Effects of Organic and Inorganic Contaminants in Sewage Sludge (pp. 75-81). D. Reidel Publ., Dordrecht.

Laperche V, Logan TJ, Gaddam P & Traina SJ (1996) Effect of apatite amendments on plant uptake of Pb in contaminated soil.Environ. Sci. Technol. 30:3321-3326.

Lindsay WL (1972) Inorganic phase equilibrium of micronutrients in soils. In: Mortvedt JJ, Giordano PM & Lindsay WL (Eds) Micronutrients in Agricultural (pp. 41-57). Soil Sci. Soc. Amer., Madison, WI.

Ma QY, Traina SJ, Logan TJ & Ryan JA (1993) In situ lead immobilization by apatite. Environ. Sci. Technol. 27:1803-1810.

Ma QY, Logan TJ, Traina SJ & Ryan JA (1994) Effects of NO^{3-}, Cl^-, F^-, SO_4^{2-}, and CO_3^{2-} on Pb^{2+} immobilization by hydroxyapatite. Environ. Sci. Technol. 28:408-418.

Ma QY, Traina SJ, Logan TJ & Ryan JA (1994) Effect of aqueous Al. Cd, Cu, Fe(II), Ni, and Zn on Pb immobilization by hydroxyapatite. Submitted to Environ. Sci. Technol.28:1219-1228

Mahler RJ, Ryan JA & Reed T (1987) Cadmium sulfate application to sludge-amended soils. I. Effect on yield and cadmium availability to plants. Sci. Total Environ. 67:117-131.

Mahler RJ & Ryan JA (1988a) Cadmium sulfateapplication to sludge-amended soils: II. Extraction of Cd, Zn, and Mn from solid phases. Commun. Soil Sci. Plant Anal. 19:1747-1770.

Mahler RJ & Ryan JA (1988b) Cadmium sulfate application to sludge-amended soils: III. Relationship between treatment and plant available cadmium, zinc, and manganese. Commun. Soil Sci. Plant Anal. 19:1771-1794.

McBride MB (1994) Environmental Chemistry of Soils. Oxford University Press. New York.

McBride MB (1995) Toxic metal accumulation from agricultural use of sludge: Are USEPA regulations protective. J. Environ. Qual. 24, 5-18.

NCR (National Research Council) (1980) Mineral Tolerance of Domestic Animals. National Academy of Sciences, Washington D.C. 577pp.

Ryan JA & Chaney RL (1993) Regulation of municipal sludge under clean water act section 503: A model for exposure and risk assessment for MSW-compost. In: Hoitink HAJ et al. (Eds) Science and Engineering of Composting: Design, Environmental, Microbiological and Utilization Aspects. The Ohio State University, Columbus, OH.

Shacklette HT & Boerngen JG (1984) Element concentrations in soils and other surficial materials of the conterminous United States. U.S. Geol. Surv. Prof. Papers 1270:1-105.

Steevens DR, Walsh LM & Keeney DR (1972) Arsenic phytotoxicity in a plainfield sand as affected by feric sulfate or aluminum sulfate. J. Environ. Qual. 1:301-303.

U.S. Environmental Protection Agency (1993a) 40-CFR Part 257. Standards for the Use or Disposal of Sewage Sludge:Final Rules. Fed. Reg. 58,9248-9415.

U.S. Environmental Protection Agency (1993b)Technical support document for land application of sewage sludge. Vol.I. PB93-110575, Vol.II PB#-110583. NTIS.

Walsh LM, Erhardt WH & Seibel HD (1972) Copper toxicity in snapbeans (Phaseolus vulgaris L). J. Environ. Qual. 1:197-200.

Zhang P, Ryan JA & Bryndzia TS (1997) Pyromorphite formation from goethite adsorbed lead. Environ. Sci. Technol. 31:2673-2678.

IMPACT OF BIOSOLIDS AND CO-UTILIZATION WASTES ON RHIZOBIA, NITROGEN FIXATION AND GROWTH OF LEGUMES

J. SCOTT ANGLE
Department of Natural Resources Sciences and Landscape Architecture, University of Maryland, College Park, Maryland 20742, USA
E-mail ja35@umail.umd.edu

ABSTRACT

The application of co-utilization products to soil, especially biosolids contaminated with high concentrations of heavy metals, can potentially affect soil microbial populations. Organic matter and micro and macronutrients can all enhance the number and activity of microbes in soil. Alternatively, toxic organics, heavy metals and soluble salts can have deleterious effects on number and activity. Some reports from Europe indicate that biosolids-borne heavy metals can reduce the number, diversity and metabolic activity of rhizobia present in metal-amended soil. Metal toxicity is believed to select for ineffective mutants of rhizobia that fix little or no nitrogen. Studies at the University of Maryland, College Park, MD, and the Environmental Chemistry Lab of the USDA, ARS, Beltsville, MD, have shown that many of the parameters noted above are enhanced with application of biosolids to soil. Numbers of rhizobia capable of nodulating alfalfa, white clover and soybeans were found to increase following application of biosolids to soil. This was observed despite the fact that the metal content of the soil was high enough to produce visible signs of toxicity in the respective macrosymbiont. Rhizobia isolated from plants experiencing severe phytotoxicity were capable of fixing nitrogen at rates similar to rhizobia isolated from unamended plots. Diversity of rhizobia was also increased in metal contaminated soil compared to rhizobia isolated from control soils. Enhanced diversity was observed using both serological analysis of rhizobial isolates and REP-PCR genetic fingerprints from soil isolates. When we examined survival of rhizobia and growth of white clover in chelate-buffered nutrient solution, zinc and cadmium were found to be approximately ten times more toxic to the plant compared to the rhizobial microsymbiont. At zinc and cadmium activities where plants experienced severe phytotoxicity, most plants were well nodulated and nodules were capable of fixing adequate amounts of nitrogen. The only observed metal related effect was a delay in nodulation. Nodulation was typically delayed for 24 to 48 hours compared to plants from control plots. Observation of young roots revealed that delayed development of root hairs was responsible for the observed difference in nodulation. Once root hairs were produced, nodulation was found to proceed at a normal rate even in the presence of high concentrations of heavy metals. It is our belief that where adverse effects of biosolids on rhizobia are reported, effects

S. Brown, J.S. Angle and L. Jacobs (eds.), Beneficial Co-Utilization of Agricultural, Municipal and Industrial By-products, 235–245
© 1998 *Kluwer Academic Publishers,*

were related to past reductions in soil pH. Numerous reports indicate that when soil pH declines below 5.5, rhizobia that survive lack capacity to fix atmospheric nitrogen. It is therefore our opinion that unless metal loading rates to soil greatly exceed permissible levels, no adverse effects on clover will result from metals alone.

INTRODUCTION

Assessing the impact of biosolids or co-utilization wastes on soil microbial populations is difficult since these products contain many compounds that affect soil microbial populations. It is extremely difficult to identify and separate confounding effects on the soil microbial population when observed. Biosolids, co-utilization products, composts and many other organic waste products contain heavy metals (which may be either toxic elements or essential metal cofactors), organic materials (which also can be toxic or essential for growth), and soluble salts. Application of organic wastes to soil can further affect soil water holding capacity, aeration, pH, competition from other organisms as well as a host of other biotic and abiotic factors.

It is rare that a single factor alone affects the soil microbial population. Typically, an organic byproduct may contain materials that both enhance and reduce microbial populations. A further complicating potential effect is the fact that waste products and metals can affect both microbial numbers in soil and metabolic activity of existing populations. While numbers and activity are often related, this may not always be the case when waste toxicants affect just one aspect of the life cycle of an organism.

The focus of this paper is on the effects of heavy metal contaminants in biosolids and co-utilization products on a highly specific, yet economically important component of the microbial ecosystem - rhizobia. *Rhizobium, Bradyrhizobium* and *Sinorhizobium* are gram negative bacteria that form symbiotic relationships with legumes. This symbiotic relationship is capable of fixing atmospheric nitrogen, thus eliminating the plants requirement for large, exogenous inputs of fertilizer nitrogen. Loss of rhizobia from the soil ecosystem is a significant concern.

In the early 1980's, McGrath, Giller et al. reported that *Rhizobium leguminosarum* bv *trifolii* isolated from long term-sludge plots at Woburn, UK, were unable to fix atmospheric nitrogen. The loss of ability to fix atmospheric nitrogen was attributed to exposure to metals contained within the biosolids applied to the field.

Based upon purported effects on soil microbes from the UK, limits have been set for maximum metal loadings to soil. Rates vary for the UK and other parts of Europe. The US has also established maximum permissible concentrations of metals in soil. Once defined concentrations are achieved, no further addition of metals to soil is permitted. (Table 1)

Table 1. Maximum permissible soil concentrations for Zn and Cd in US and Europe.

Country	pH	Zn	Cd
UK	6.0-7.0	300	3.0
UK	5.5-6.0	250	3.0
UK	5.0-5.5	200	3.0
Germany	>6.0	200	1.5
Germany	5.0-6.0	150	1.0
US (EPA)	---	1400	19.5

METAL AND BIOSOLID EFFECTS ON:

Survival of Free Living Rhizobia

There is little doubt that metals can affect rhizobia in soil. McGrath et al. first observed that nodules of white clover grown in the long-term plots of Woburn were small and white (McGrath et al., 1988). Upon further testing, it was reported that the nodules were not capable of fixing atmospheric nitrogen. Nodules from control plots that received only manure were fully effective. *Rhizobium leguminosarum* bv. *trifolii* isolated from ineffective nodules were genetically similar (Hirsch et al., 1993) suggesting that heavy metals selected for metal-resistant isolates incapable of fixing nitrogen.

While it is clear that metals in soil can adversely affect soil populations of rhizobia, concentrations where an adverse effects occurs is difficult to define. Only free metal ions adversely affect soil microbes. Since most metals in soil are either chelated or complexed, the free metal ion concentration in soil may or may not be related to the total metal concentration (Chaney, 1988). Setting specific limits for maximum metal concentrations is therefore not generally useful to protect microbial populations. This was the rationale for attempting to define metal loading by soil pH.

Chaundri et al. (1993) indicated that metals in solution and soil rapidly reduced viable numbers of rhizobia. With a soil Zn concentration of about 250 mg kg^{-1}, the number of *Rhizobium leguminosarum* bv. *trifolii* declined to undetectable levels relatively quickly. This adverse effect on survival was observed below the UK limit for maximum concentrations. Similar results were reported by Giller et al. (1993) and Martensson and Witter (1990). However, in each of these studies, the pH of the soil in which the rhizobia population rapidly declined was lower than in the unamended control plots. Thus, it is difficult to separate effects of pH and metals on free living populations of rhizobia. These authors also demonstrated that rhizobia from metal contaminated soil were capable of tolerating higher levels of metals than rhizobia from noncontamianted soil (Chaudri et al., 1992).

Comparable studies by Angle and Chaney as well as others, however, found no adverse effects on soil populations of rhizobia. In fact, early studies by Kinkle et al. (1987) showed that the number of *Bradyrhizobium japonicum* was enhanced following application of high rates of biosolids to soil. Enhanced numbers were attributed to the positive chemical and physical factors associated with the application of biosolids to soil. In the single study where Angle et al. found an adverse effect of biosolids on rhizobia, the effect was attributed to high concentrations of soluble salts in soil, and not to heavy metals (Madariaga and Angle, 1992). Angle and Chaney (1991) examined the effects of zinc and cadmium from smelter emissions on *Rhizobium meliloti*. Although soils examined contained from low to extremely high (1540 mg Zn kg^{-1} and 32 mg Cd kg^{-1}) concentrations of metals, there was no relationship between metal concentration and rhizobial numbers or nitrogenase activity. We hypothesize that the intrinsic level of metal tolerance of these organisms precluded their need to respond the presence of high concentrations of metals. El-Aziz et al. (1991) showed that the extractable concentration of Zn from a smelter-contaminated soil averaged 0.75 and 0.40 ug ml^{-1} when extracted with 10 mM $Ca(NO_3)_2$ and water, respectively. Isolates of *Rhizobium meliloti* from control soil lacking metal enrichment exhibited Zn tolerance of approximately 15 ug ml^{-1}. Although the study of El-Aziz et al. could not precisely estimate and compare metal activities in soil and media, it is clear that the difference between intrinsic metal tolerance and available metal concentrations in soil was more than an order of magnitude different from one another.

More recent studies by (Ibekwe et al., 1997a) reported that even when relatively high concentrations of metals were present in soil, there was no negative effect of the metals on the population. Adverse effects on the population of *R. leguminosarum* bv *trifolii* were found when soil pH was below 5.8 - regardless of whether metal concentrations were elevated. In a control soil containing normal concentrations of metals, the population of *Rhizobium leguminosarum* bv. *trifolii*, averaged 2,044 and 0 colony forming units per g soil for soil with a pH of 6.1 and 4.4, respectively. When the soil concentration (as determined by aqua regia digest) of Zn and Cd averaged 140 and 4.4 mg kg^{-1}, respectively, the populations were nearly identical compared to the control soil. No metal effects were noted - all adverse effects were directly related to low soil pH. Obbard and Jones (1993) have also reported similar observations, noting that effective populations of rhizobia were found in soil exceeding UK permissible limits. Where a reduction in numbers of *Rhizobium leguminosarum* bv. *trifolii* was observed, the decline was attributed to absence of a host plant, not to high concentrations of heavy metals.

Diversity

Loss of rhizobial diversity in metal contaminated soil is another concern related to metal additions. Hirsch et al. (1993) showed that metals in long-term biosolids disposal plots selected for a non-diverse population of *R. leguminosarum* bv. *trifolii*. She attributed this observation to metal pressure on the metal sensitive population and subsequent selection for a metal resistant, ineffective population. However, Ibekwe et al. (1997c), using a more powerful genetic approach (REP-PCR), observed that rhizobial diversity in soil was actually increased in metal contaminated soil. He attributed this observation to the inherent metal tolerance of rhizobia and the uneven distribution of metals in contaminated soil. Where loss of diversity was observed, the loss was directly attributed low soil pH. In an earlier study, Kinkle et al. (1987), using a serological technique for typing of *Bradyrhizobium japonicum*, reported that diversity was unaffected following long-term exposure of this organism to heavy metals. The observations of Kinkle et al. were attributed to the intrinsic high metal tolerance of *Bradyrhizobium*. Nearly all slow-growing rhizobia have been reported to be capable of surviving in soils with high metal concentrations - possibly due to their ability to increase niche pH and thus reduce metal availability (Angle et al., 1992). The thick polysaccharide capsule surrounding *Bradyrhizobium* may also help to exclude metals from entering the cell.

The observations of Kinkle et al. and Ibekwe et al. are somewhat contrary to the common belief that metal pressure decreases soil microbial diversity. In addition to the work of Hirsch et al., many other studies conducted in water and sediment have shown that metals significantly reduce microbial diversity. We hypothesize that the observations of Kinkle et al. and Ibekwe et al. are related to two factors. First, nearly all soil bacteria are intrinsically resistant to concentrations of available metals at a level typically much higher than found in even highly contaminated soil (Angle et al., 1993; El-Aziz et al., 1991). Secondly, neither metal concentration nor availability are well distributed within the fabric of the soil. Some microsites in soil contain very high concentrations of metals. Within this "zone" of high metal concentration, a shift in the population may occur. Other areas of the soil may have "normal" concentrations of metals, thus precluding the need for a shift in the population. Metal availability may also vary within the fabric of the soil. When bulk soil samples are collected, areas of both high and low metal concentrations are mixed, along with the corresponding populations in each. Thus, the "observed" population includes both rhizobia from very high metal concentration and availability and rhizobia from areas of low soil metal concentration and availability. The resulting "bulk diversity" thus appears to increase.

Nodulation

If effective rhizobia are present in soil, formation of nodules with the roots of legumes typically occurs. A number of studies have shown that high metal concentrations in soil delay nodulation. El-Kenawy et al. (1997) reported that the very earliest stages of nodule initiation were affected by high soil concentrations of zinc and cadmium. Bacterial attachment to root hairs, root hair curling, infection thread formation, and first signs of visible nodules were all delayed by one to three days (Table 2). No nodules were observed on plants grown at very high metal concentrations and at a low soil pH. Low soil pH, in the absence of metal contamination, also slowed all of the parameters noted above. However, by the conclusion of the study, there were no differences between any of the treatments. All plants (except at high metals and low pH) were well nodulated. Rates of nitrogen fixation were also similar. Martensson and Witter (1990) have also reported that metals from sewage sludge were capable of delaying nodulation. From a review of available information we attribute this delay in nodulation to direct root hair damage and not to any effect on the rhizobial population or on the actual process of nodulation. Martensson (1992) also suggested that reduced nodulation associated with high concentrations of Zn in soil were potentially related to effects on the plant, although the exact plant parameter was not discussed.

Ibewke et al. (1997a) showed that the addition of metal rich biosolids to soil with a pH near neutral had no adverse effect on nodulation (Table 3). In fact, the number of effective nodules per plant was increased by biosolids. However, at low soil pH, few effective

Table 2. Effect of different concentrations of heavy metals and pHs on the early stages of *Rhizobium*-White Clover symbiosis.

Stage of Symbiosis	Soil pH	Soil Metal Concentration	Days after seedling appeared above the soil					
			1	2	3	5	7	10
Bacterial attachment	5.5	High						
	6.8	High		+				
	5.8	Low	+					
	7.2	Low	+					
Root hair curl	5.5	High						
	6.8	High				+		
	5.8	Low			+			
	7.2	Low		+				
Infection thread formation	5.5	High						
	6.8	High					+	
	5.8	Low				+		
	7.2	Low			+			
Nodule formation	5.5	High						
	6.8	High						+
	5.8	Low					+	
	7.2	Low				+		

Table 3. Response of white clover grown in Leonard jars when inoculated with soils from the control and biosolids-amended plots.

Biosolids Treatment and pH	Shoot Weight	Number of Effective Nodules	Effective Nodule Weight	Number of Ineffective Nodules	Ineffective Nodule Weight
	g dry weight jar^{-1}	number jar^{-1}	mg dry weight jar^{-1}	number jar^{-1}	mg dry weight jar^{-1}
Control - high pH	0.49	34	20	28	10
Control - low pH	0.10	0	0	6	3
Nu-Earth - low pH	0.08	3	2	0	0
Nu-Earth - high pH	1.20	100	70	12	1
$LSD_{0.05}$	0.32	21	20	9	1

nodules were observed; whether or not biosolids were present in the soil. These observations were reflected in changes in shoot weight. Shoot weight was increased at high soil pH and with the addition of biosolids. The metals in the biosolids had little adverse effect on nodulation and nitrogen fixation while the improved soil physical and chemical properties associated with biosolids enhanced shoot weight above that observed without biosolids.

Nitrogen Fixation

Heckman et al. (1987) studied rates of nitrogen fixation in soil amended with high rates of metal contaminated biosolids. Although the content of organic matter in biosolids amended soil had declined to background levels 15 years after application, concentrations of nitrogen in soil remained elevated. Past reports have shown that elevated concentrations of soil nitrogen repress rates of nitrogen fixation. Mechanisms for repression are well known as it has been repeatedly shown that soil nitrogen inhibits the process of nodulation. However, in the study of Heckman et al., the rate of nitrogen fixation in the biosolids-amended plots was higher than in the unamended control. It was clear that the relatively high concentrations of metals in soil had no adverse effect on nitrogen fixation. In fact, rates of nitrogen fixation in some biosolids-amended plots were higher than with plants from control plots. The improved physical and chemical properties of the biosolids-amended soil enhanced plant growth; these same factors also supported a higher population of rhizobia in the soil (Kinkle et al., 1987) thus allowing for enhanced nitrogen fixation.

McGrath et al. (1988), on the other hand, reported that nitrogen fixation was nil from white clover collected from biosolids-amended plots at Woburn in the UK. Loss of ability to fix nitrogen was attributed to exposure to metals. This loss could either be due either to selection for an ineffective subpopulation or to mutation of the effective population resulting in a loss of effectiveness. Another explanation, although undocumented, could be that the pH of the soil at Woburn many years prior to the assay of effectiveness, declined (as a result of nitrification or some other process) to a level that could have caused loss of effectiveness. At a later date when a decision was made to grow legumes at Woburn, soil pH was increased. This "masked" the effect of pH on the rhizobial population, resulting in the inability to relate soil pH and effectiveness.

Ibekwe et al. (1995) showed that plants grown in an unamended control soil with low pH often exhibited low rates of nitrogen fixation. These authors attributed this observation to the loss of rhizobial effectiveness associated with low soil pH. When high concentrations of metals were present, but soil pH was near neutral, high rates of nitrogen fixation were observed. Ibekwe et al. (1995) found that the presence of Zn (303 mg kg^{-1}) and Cd (1.24 mg kg^{-1}) had little effect on the rate of nitrogen fixation. Since the incorporation of biosolids into soil enhanced the number of effective nodules present on white clover, shoot weight was significantly increased.

DISCUSSION

It is clear from the foregoing that it has yet to be resolved whether high concentrations of heavy metals in soil or other soil factors have the greatest capacity for adversely affecting survival of free living rhizobia, nodulation, nitrogen fixation, and plant growth. It is certainly possible that different factors have different effects on each component of the process of nitrogen fixation. McGrath et al. have argued that excessive metal soil concentrations are responsible for the occasional observation that white clover rhizobia have lost their ability to fix atmospheric nitrogen. Metal loadings to soils in the UK and Europe were partially based upon this argument. However, there are equal numbers of studies at similar to higher metal concentrations where plants were found to adequately fix nitrogen. The cause of this descrepency is important to understand.

Our results suggest that soil pH may be the overriding factor in determining rhizobial effectiveness. However, when studying effects of soil pH on the ecosystem and plant growth, any such studies are confounded by the fact that soil pH affects metal solubility and availability. Metals in soil can precipitate as free ions, with OH-, carbonates, phosphates, molybdates, humates, fulvates and with a number of other common ligands (Schmidt, 1997). Each of the reactions noted above are affected by soil pH.

When pH is low in metal enriched soil it is difficult to determine whether adverse effects on rhizobia are due to metals or low pH. There is ample evidence in the literature, that indicates however, that low soil pH may adversely affect rhizobia, including a loss of the ability to fix nitrogen (Munns, 1977). Mulder et al. (1977) showed that low soil pH reduced rhizobial activity and ability to multiply. Holdings and Lowe (1971) further demonstrated that low soil pH increased the number of ineffective rhizobia in soil.

Since these studies of the 1970's, many other studies have attempted to isolate rhizobia that could survive in low pH soil and retain the ability to actively fix nitrogen (Glenn and Dilworth, 1991). The purpose of these studies was to identify rhizobia that can survive and form effective symbiosis with legumes in acidic soils, including reclaimed mine soils. In future studies, it would be interesting to determine whether rhizobial isolates tolerant to low pH are also tolerant to high concentrations of soil metals.

In addition to our work, numerous other studies have shown that effective rhizobia can be found in soils containing concentrations of metals in excess of the EC limits. Smith and Giller (1992) isolated effective *R. leguminosarum* bv. *trifolii* from dedicated biosolids disposal sites that contained metal concentrations of 310 to 5400 mg kg^{-1} Cr, 41 to 145 mg kg^{-1} Ni, 83 to 1050 mg kg^{-1} Zn, 2.6 to 19.2 mg kg^{-1} Cd, and 230 to 3600 mg kg^{-1} Pb. Roots of *Trifolium repens* were well nodulated and actively fixing nitrogen when grown in soil containing 216 mg kg^{-1} Cd and 20,500 mg kg^{-1} Zn (Rother et al., 1983). Obbard (1993) also showed that effective rhizobia were present in soil containing very high concentrations of metals.

Additionally, ineffective rhizobia can be found in soil containing "normal" concentrations of metals. We propose that where ineffective rhizobia are found, the loss of ability to fix atmospheric nitrogen was usually due to exposure to low pH and not to the presence of metals. In fact, rhizobia in soil are the protected by greater sensitivity of plants to metals compared to bacteria. Legumes will experience phytotoxicity and typically die long before rhizobia populations are affected. The work of Ibekwe et al.

(1997b), in the first direct comparison of macro and microsymbiont metal sensitivity, clearly showed that white clover is much more sensitive to the toxic effects of metals than is it's corresponding microsymbiont (Figure 1). At a cadmium activity where plants failed to survive, the number of rhizobia in solution was not affected. Further, of those rhizobia at the high cadmium activity, there was no adverse effect on their ability to fix nitrogen. This work was preformed in nutrient solution where metal activities were controlled with the use of a chelate buffer and were estimated with the use of GEOCHEM (Parker, et al., 1995). Plants died at metal activities much lower than activities where rhizobia were affected. Similar results were reported for alfalfa and it's microsymbiont where the plant was found to be much more sensitive to Zn and Cd than *Rhizobium meliloti* (Ibekewe et al., 1996). Hence, it will be impossible to economically cultivate legumes on soils that could potentially have a deleterious effect on rhizobia, thus to a large extent making the issue of metal effects on rhizobia a moot concern.

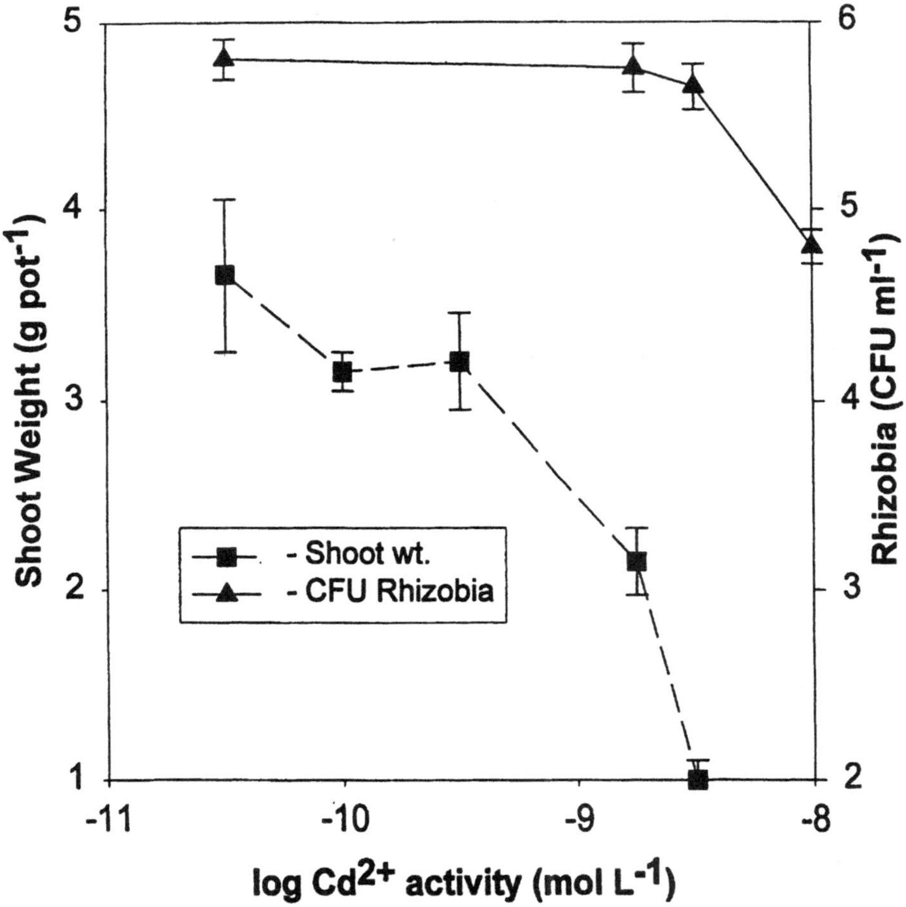

Figure 1. Effect of increasing Cd activities on shoot weight of white clover and rhizobia. Bars at each point represent the standard error of the mean.

In conclusion, where ineffective rhizobia have been found in metal contaminated soil, we propose that the loss of ability to fix nitrogen was due to present or past exposure to low soil pH. Where ineffective isolates are found in soil enriched in metals, but with near neutral pHs, it is possible that past (now corrected) exposure to low soil pH selected for an ineffective population of rhizobia. At a later date when legumes were cultivated, it was found that the plants failed to thrive unless soil pH was increased. Most legumes can not grow at low pH (<5.2) and thus liming of soil is a routine and necessary practice (Glenn and Dilworth, 1991). However, at this point in time, only ineffective rhizobia are present in soil, leading to the appearance of ineffective rhizobia on the roots of the newly planted legume.

REFERENCES

Angle JS & Chaney RL (1991) Heavy metal effects on soil populations and heavy metaltolerance of *Rhizobium meliloti*, nodulation and growth of alfalfa. Water, Air, and Soil Pollution 57-58:597-604.

Angle JS, Chaney RL & Rhee R (1993) Bacterial resistance to heavy metals is related to extractable concentrations in soil and media. Soil Biology and Biochemistry 25:1443-1446.

Angle JS, McGrath SP & Chaney RL (1992) Effects of media components on toxicity of Cd to rhizobia. Water, Air and Soil Pollution 64:267-633.

Chaney RL (1988) Metal speciation and interactions among elements affect trace element transfer in agricultural and environment food chains In: Kramer JR & Allen HE (Eds) Metal Speciation: Theory, Analysis and Application (pp. 219-260). Lewis Publ., Chelsea, MI.

Chaudri AM, McGrath SP & Baker KE (1992) Metal tolerance of isolates of *Rhizobium leguminosarum* bv. *trifolii* from soil contaminated by past applications of sewage sludge. Soil Biology and Biochemistry 24:83-88.

Chaudri AM, McGrath SP, Giller KE, Rietz E & Saurbeck DR (1993) Enumeration of indigenous *Rhizobium leguminosarum* biovar *trifolii* in soil previously treated with metal contaminated sewage sludge. Soil Biology and Biochemistry 25:301-309.

El-Aziz M, Angle JS & Chaney RL (1991) Metal tolerance of *Rhizobium meliloti* isolated from heavy metal contaminated soil. Soil Biology and Biochemistry 23:795-798.

El-Kenawy Z, Angle JS, Gewaily EM, El-Wafai NA, Van Berkum P, Chaney RL & Ibekwe MA (1997) Zinc and cadmium effects on the early stages of nodulation of white clover. Agronomy Journal (In press).

Giller KE, Nussbaum R, Chaudri AM & McGrath SP (1993) *Rhizobium meliloti* is less sensitive to heavy-metal contamination in soil than *R. leguminosarum* bv. *trifolii* or *R. loti*. Soil Biology and Biochemistry 25:273-278.

Glenn A R & Dilworth M J (1991) Soil acidity and the microbial population: Survival and growth of bacteria at low pH. In: Wright RJ et al. (Eds) Plant-Soil Interactions at Low pH (pp. 576-579). Kluwer Academic Press, Dordrecht.

Heckman JR, Angle JS & Chaney RL (1987) Residual effects of sewage sludge on soybean: I. Accumulation of soil and symbiotically fixed nitrogen. Journal of Environmental Quality 16:118-124.

Hirsch PR, Jones MJ, McGrath SP & Giller KE (1993) Heavy metals from past applications of sewage sludge decrease the genetic diversity of *Rhizobium leguminosarum* biovar *trifolii* populations. Soil Biology and Biochemistry. 11:1485-1490.

Holdings AJ & JF Lowe (1971) Some effects of acidity and heavy metals on the *Rhizobium*-leguminous plant association. Plant and Soil Special Volume:153-166.

Ibekwe MA, Angle JS, Chaney RL & Van Berkum P (1995) Sewage sludge and heavy metal effects on nodulation and nitrogen fixation of legumes. Environmental Science and Technology 24:1199-1204.

Ibekwe MA, Angle JS, Chaney RL & van Berkum P (1996) Zinc and cadmium toxicity to alfalfa and its microsymbiont. Journal of Environmental Quality 25:1032-1040.

Ibekwe AM, Angle JS, Chaney RL & VanBerkum P (1997a) Enumeration and nitrogen fixation potential of *Rhizobium leguminosarum* biovar *trifolii* grown in soil with varying pH values and heavy metal concentrations. Agriculture, Ecosystems and the Environment 61:103-111.

Ibekwe AM, Angle JS, Chaney RL & vanBerkum P (1997b) Assessment of zinc and cadmium effects on *Rhizobium leguminosarum* bv. *trifolii* and while clover (Trifolium repens cv. Regal) using chelator-buffered nutrient solution. Soil Science Society of America Journal (In press).

Ibekwe AM, Angle JS, Chaney RL & vanBerkum P (1997c) Differentiation of clover *Rhizobium* from biosolids amended and control soils with varying pH. Soil Science Society of America Journal (In press).

Kinkle BK, Angle JS & Keyser HH (1987) Long-term effects of metal rich sewage sludge applications on soil populations of *Bradyrhizobium japonicum*. Applied and Environmental Microbiology 52:81-86.

Madariaga GM & Angle JS (1992) Sludge borne salt effects on survival of *Bradyrhizobium japonicum*. Journal of Environmental Quality 21:276-280.
Soil Biology and Biochemistry 24:435-445.

Martensson AM & Witter E (1990) Influence of various soil amendments on nitrogen fixing soil microorganisms in a long-term field experiment, with special reference to sewage sludge. Soil Biology and Biochemistry 22:977-982.

McGrath SP, Brooks PC & Giller KE (1988) Effects of potentially toxic metals in soil derived from past applications of sewage sludge on nitrogen fixation by *Trifolium repens*. Soil Biology and Biochemistry 20:415-424.

Mulder EG, Lie TA & Houwers A (1977) The importance of legumes under temperate conditions. In: Hardy RWF & Gibson HA (Eds) Treatise on Dinitrogen Fixation. IV Agronomy and Ecology (pp. 221-242). John Wiley and Sons, New York.

Munns DN (1977) Soil acidity and related factors. In: Vincent JM et al. (Eds) Exploring the Legume-*Rhizobium* Symbiosis in Tropical Agriculture. Vol. 45 (pp. 211-236). Misc. Publication, College of Agriculture, Univ. of Hawaii.

Obbard JP & Jones KC (1993) The effects of heavy metals on Dinitrogen fixation by *Rhizobium*-while clover in a range of long term sewage sludge amended and metal contaminated soils. Environmental Pollution 79:105-112.

Parker DR, Norvell WA & Chaney RL (1995) GEOCHEM-PC. A chemical speciation program for IBM and compatible personal computers. In: Loeppert R et al. (Eds) Chemical equilibrium and reaction models (pp. 253-269). Soil Science Society of America Special Publication 42. Soil Science Society of America and American Society of Agronomy, Madison, WI.

Rother JA, Millbant JW & Thornton T (1983) Nitrogen fixation by white clover (*Trifolium repens*) in grasslands on soil contaminated with cadmium, lead and zinc. Journal of Soil Science 34:127-136.

Schmidt JP (1997) Understanding phytotoxicity thresholds for trace elements inland-applied sewage sludge. Journal of Environmental Quality 26:4-10.

Smith SR & Giller KE (1992) Effective *Rhizobium leguminosarum* bv. *trifolii* present in five soils contaminated with heavy metals from long-term applications of sewage sludge or metal mine soil. Soil Biology and Biochemistry 24:781-788.

USEPA (1993) Clean water act. Section 503. Vol. 58. No. 32, United States Environmental Protection Agency, Washington DC.

MICROBIAL EFFECTS ON ENVIRONMENTAL HEALTH AND PRODUCT QUALITY ASPECTS OF RECOVERY AND CO-UTILIZATION OF BIO-MINERAL PRODUCTS

PATRICIA D. MILLNER
Soil Microbial Systems Laboratory, Agricultural Research Service, US Department of Agricultural, Beltsville, Maryland 20705, USA
E-mail pmillner@asrr.arsusda.gov

ABSTRACT

The environmental and economic incentives to recycle and recover useful resources from various organic materials, such as animal manures, biosolids, landscape trimmings and food processing and marketing residuals, continue to increase. The same incentives are stimulating interest among scientists, horticulturists, entrepreneurs, and the minerals industries in the potential for utilization of various inorganic by-products, from cement, rock, aggregate, power plant and steel manufacturing operations, as beneficial additives to these organics. The resulting 'bio-mineral' products will provide stabilized organic materials augmented with mineral components useful in agriculture, horticulture, landscaping and reclamation. Some form of treatment will be required to remove pathogenic organisms that are present in primary organic constituents of such bio-mineral products. Existing federal standards ("40CFR503 for Code of Federal Regulations") for biosolids include criteria for assessing pathogen destruction achieved by several treatment processes. The applicability and appropriateness of these criteria to organic and bio-mineral products, not containing biosolids, is discussed along with the need for test refinements appropriate to new technologies. Potential disease transmission pathways and case reports involving emerging pathogens, e.g., *E. coli* O157:H7, salmonellae, *Cryptosporidium parvum*, suggest that environmental health implications of farm and food production practices involving animal manure be reviewed and evaluated in a manner equivalent to that of biosolids. Specific information on pathogen survival and destruction relative to current and developing organics and bio-minerals recycling processes is needed to support development of guidelines for appropriate practices. Depending on the treatment process and the targeted mineral composition of the final product, inorganic constituents may be added prior to, during, or after organics processing. The sequence of mineral augmentation in organics processing has the potential to influence the quality of the product, but more research is urgently needed to optimize these benefits and to support recommendations that promote environmental and public health.

S. Brown, J.S. Angle and L. Jacobs (eds.), Beneficial Co-Utilization of Agricultural, Municipal and Industrial By-products, 247–258.
© 1998 *Kluwer Academic Publishers,*

INTRODUCTION

For centuries agricultural and urban communities have land applied (recycled) organic residues that contain human pathogens. Such residues include night soil, sewage, biosolids, household and processing by-products, liquids and solids from anaerobic digestors, and animal manures. In this century, with the explosive growth in the human population and the intensification of agricultural and industrial production, recycling of some materials, notably biosolids and certain industrial by-product sludges, have been regulated to protect public health and the environment from problems that can result from inappropriate uses and management. As a consequence of this intensification, the complexity and scale of recycling has increased and new technologies have been developed to deal with the ever-changing situation.

Strong environmental and economic incentives have also stimulated interest in recycling and recovery of resources from various inorganic by-products, from cement, rock, aggregate, power plant and steel manufacturing operations. Scientists, horticulturists, entrepreneurs, and the minerals industries have explored the possibility of mixing organic materials with mineral components to produce 'bio-mineral' blends. The potential to add value to recycled organics to create new products for agriculture and horticulture can realistically be achieved as long as issues relative to public health and safety, economics and public acceptability (such as annoyances from odors, noise, and traffic) are addressed effectively. This report focuses on issues relative to public health and food safety.

PUBLIC HEALTH ISSUES

Protection of public health from food borne diseases has always been a primary concern of agriculture and the food industry. Now with recycling and other industries producing materials for use in the production of animal feed and/or food for human consumption, these enterprises will also need to concern themselves with the role their products could have in disease transmission. Food produced on soils amended with untreated or inadequately treated human and animal by-products, or irrigated with water that may have come into contact with these untreated by-products, is subject to potential contamination by a variety of public health pathogens. Adequate pathogen reduction measures would be needed prior to distribution and use of by-products in crop production and/or prior to and during food processing. Recently, several trends relative to food production, distribution, processing, and consumption have emerged simultaneously with the rapid expansion in organics recycling. These trends have actually, not just *potentially*, accentuated the impact of pre-harvest practices have on food safety, as evidenced by the incidence of food and waterborne illness outbreaks. These major trends are:

- Increased emphasis on the health and nutritional benefits of consuming more fresh fruits and vegetables;
- Increasing demand by both consumers and the food service industry for fruits and vegetables that are packaged and sold in a minimally-processed, pre-cut form

(called "fresh-cut");

- Increased demand for and production of organic foods, which rely on extensive use of recycled organic matter in the crop production practice;
- Use of animal manures and other organic materials in crop production systems;
- Food and waterborne disease outbreaks involving pathogens not previously or extensively associated with food contamination;
- Extensive interstate and international (import and export) distribution of fresh fruits and vegetables; and
- Increased numbers of susceptible individuals: immunosuppressed; elderly; infants and children; malnourished, unhealthy, homeless.

FOOD AND WATERBORNE DISEASE INCIDENTS

Several recent food and waterborne illness outbreaks have been traced to contamination of either fresh fruits and vegetables or water by animal manures containing pathogenic or parasitic agents. The microorganisms of concern include: *Escherichia. coli* O157:H7 and verotoxigenic, enterohemorrhagic strains, *Salmonella typhi* and *S. typhimurium*, *Yersinia enterocolitica* (Morris and Feeley, 1979), *Campylobacter jejuni* (Franco, 1988), *Listeria monocytogenes* (van Renterghem et al., 1991), viruses, *Giardia lamblia, Cryptosporidium parvum*, and *Cyclospora*.

Notable Cases

Escherichia coli O157:H7 has caused approximately 60 extremely severe, food borne illness outbreaks in the U.S. Most of these have involved undercooked ground-beef products, but several involved raw milk, or person-to-person transmission of infection (Feng, 1995). However, several recent outbreaks of *E. coli* O157:H7 have implicated novel vehicles of transmission: acidic foods, fruits, vegetables, yogurt, and water.

- An outbreak of enterohemorrhagic diarrhea (1 fatality) in rural Maine was traced back (Cieslak et al., 1993) to consumption of vegetables fertilized with manure from animals seropositive for *E. coli* O157:H7. The soil also contained *E. coli* O157:H7.

- A multistate outbreak of *E. coli* O157:H7 infections was traced back to a mesclun lettuce mix (Hilborn et al., 1997). Cattle found near the lettuce processing facility were confirmed as carriers of *E. coli* O157:H7; lettuce wash water and processed lettuce had non-O157:H7 strains of *E. coli*. Subsequent studies show that serotype O157:H7 can survive under various controlled atmospheres as might be used in packaging of lightly processed vegetables such as the lettuce mix (Abdul-Raouf et al., 1993; Hao and Brackett, 1993).

- Outbreaks of *E. coli* O157:H7 infection and cryptosporidiosis were traced back to drinking unpasteurized apple cider (Besser et al., 1993) in several locations in North America, i.e., Connecticut and New York (CDC, 1997), and British Columbia, California, Colorado, and Washington (CDC, 1996). The two routes of disease transmission suspected both involve cattle manure. The first case (Besser et al., 1993) involved cattle grazing in the field adjacent to the cider mill; whereas other cases have involved the use of "drop" apples that had fallen on the manured soil in the orchards. Some years ago, an outbreak of *Salmonella typhimurium* was traced back to consumption of commercial, but unpasteurized apple cider (CDC, 1975), which included drop apples. In 1993, an outbreak of cryptosporiodsis (Millard et al., 1994) was also suspected of resulting from manure contamination from the adjacent field in which cattle grazed. *Cryptosporidium* was detected in the feces of the implicated herd.

- In a series of restaurant cases involving about 48 people in 1993, contaminated mayonnaise with a pH of 3.6-3.9 (Weagant et al., 1994) was suspected as the vehicle of transmission . Subsequent studies showed that *E. coli* O157:H7 can survive, but not multiply, in such acidic mayonnaise, even when stored at 5°C for 55 days (Zhao and Doyle, 1994).

- In 1991, 16 people became infected with *E. coli* O157:H7 after eating yogurt made from pasteurized milk (Morgan et al., 1993) that is suspected of becoming contaminated after pasteurization (Upton and Coia,1994). During follow-up study of this outbreak Willshaw et al. (1993) showed that verotoxigenic, non-O157:H7 serotypes of *E. coli* are also common in asymptomatic dairy cattle and these may be a source of infection for humans.

- A massive outbreak of *E. coli* O157:H7 in Missouri involving more than 240 people and four fatalities is suspected to have resulted from backflow contamination of the water supply during a water main break (Swerdlow et al., 1992).

- Over the past several years large outbreaks of cryptosporidiosis caused by *Cryptosporidium parvum* have occurred with both small and large water supplies (Fox and Lytle, 1996; Hayes et al., 1989; LeChevallier, et al., 1991a, b; MacKenzie et al., 1994; Solo-Gabriele and Neumeister, 1996).

UNUSUAL AND COMPLEX ROUTES OF DISEASE TRANSMISSION

The fact that acid foods were associated with infections shows that the traditional view that low pH foods are safe can no longer be accepted without reservation (Conner and Kotrola, 1995; Miller and Kaspar, 1994; Zhao and Doyle, 1994; Zhao et al., 1993). Furthermore, the involvement of nonbovine materials, such as water, fruits, and vegetables, indicates that other pathways of disease transmission are also likely. Since animal manures are a well known source of enteropathogenic *E. coli* (Armstrong et al., 1996; Bettleheim, 1996; Hancock et al., 1994; Mawdsley, 1995; Mechie et al., 1997; Wang et al., 1996; Zhao et al., 1995), *C. parvum* (Ghiorse, 1996; Pell, 1996) and a variety of other enteric bacteria (Snowdon et al., 1989; Strauch, 1991), protozoans, helminths (Feachem et al., 1983; Shuval et al., 1984) and viruses (Pesaro et al., 1995), possible disease transmission routes might involve overland flow of water carrying pathogens originating in animal manures to production fields on adjacent farms producing fruits or vegetables. Alternatively, overland flow could transport pathogens from animal production farms to water sources used to irrigate crops sold as fresh market produce. Likewise, pathogens from animal production units might be vectored to nearby fruit or vegetable production farms by wild birds and small or large mammals that glean and browse at farm sites. In addition, a false sense of process adequacy could result when heat shocked or otherwise environmentally injured bacteria escape detection (Ahmed and Conner, 1993; Droffner and Brinton, 1996; Plym-Shell and Ekesbo, 1993; Yanko et al., 1995) unless techniques that allow for recovery are carefully selected and used to periodically confirm that pathogen reduction processes are performing adequately.

The advent of fresh-cut fruits and vegetables for convenience has raised a new food safety issue. Cutting fresh produce before marketing removes normal barriers to contamination by pathogens and exposes cut surfaces to an environment that is rich in nutrients that can support growth of bacteria pathogenic to humans. Obviously, the use of untreated manures, or other recycled organics that have not undergone pathogen reduction processes, in cropping systems opens the way for potential pre-harvest contamination of fresh produce. Table 1 lists some of the types of pathogens that are of concern and the natural sources and reservoirs that cause human diseases. The infectious dose of a pathogenic organism varies with the organism and, particularly for bacteria and viruses, can vary widely. Table 2 reports infective dose data, showing values as low as 100 for *Salmonella* (various species), < 1 Plaque Forming Unit for Poliovirus, and 1 egg for helminths. This information coupled with what little is generally known about the survival times of pathogens on soil and plants (Kowal, 1985) provided a framework for 40 CFR 503, *Standards for the Use or Disposal of Sewage Sludge* (U.S. EPA, 1993).

Applicators (typically farmers or farm managers) will need to be made aware of this potential source of contamination and the appropriate interventions that should be used if pathogen laden material is land applied or used to prepare foliar sprays for use in biocontrol of plant pathogens. In terms of bio-mineral blends for use in production of crops, especially those marketed as fresh or lightly processed, some form of treatment will be required to destroy pathogenic organisms that enter the product with the primary organic constituents. Agricultural extension agents and the composting/recycling industries with their corresponding training programs can be essential partners in educating applicators in appropriate uses for materials treated to varying degrees to reduce pathogens.

INTERVENTIONS TO REDUCE PATHOGENS IN RECYCLED ORGANICS

Fortunately, in the U.S., guidelines for reduction of pathogens in biosolids have not only been established, all major regions in the U.S. have successfully used the EPA standards in field situations, such that no disease outbreaks have ever been reported in association with land application of biosolids that met the specific regulations. These pathogen standards for biosolids could serve as a guide for development of pathogen reduction processes and land application recommendations for other types of by-products, particularly animal manures, that contain many of the same types of pathogens. In fact, the federal rules (U.S. EPA, 1993) for biosolids provide several options for pathogen reduction processes, all of which could be adapted to bio-mineral blends.

Table 1. Selected Pathogens in Municipal Wastewater, Solid Waste, Sewage Sludge, and Manure, and the corresponding natural sources, reservoirs, and human diseases/symptoms.

Pathogen	Sources & Reservoirs	Human Disease/Symptoms
Bacteria		
Salmonella spp.	Poultry, swine, cattle - feces	Salmonellosis (food poisoning), typhoid, carriers
Shigella spp.	Infected humans or primates	Bacillary dysentery
Yersinia spp	Poultry, swine, cattle - feces Humans, primates, wild animals, dogs	Acute gastroenteritis (diarrhea, abdominal pain)
Vibrio cholerae	Wastewater and sewage	Cholera
Campylobacter jejuni	Poultry, swine, cattle - feces; Domestic animals & birds-feces	Gastroenteritis
Escherichia coli (patho- or toxigenic strains)	Swine, cattle, rarely poultry - feces Humans, domestic and wild animals, birds, and reptiles	Gastroenteritis, diarrhea, toxemia, septicemia, meningitis
Listeria monocytogenes	Cattle, poultry, rarely swine - feces, Aborted fetuses and fluids	Gastroenteritis
Viruses		
Poliovirus	Humans	Poliomyelitis
Coxsackievirus, Echovirus	Humans	Meningitis, pneumonia, hepatitis, fever, etc.
Hepatitis A virus	Humans, possibly other primates	Infectious hepatitis
Rotavirus	Humans (children)	Acute gastroenteritis with severe diarrhea
Norwalk Agents	Humans	Epidemic gastroenteritis with severe diarrhea
Reovirus	Humans	Respiratory infections, gastroenteritis
Protozoa		
Cryptosporidium parvum	Livestock, wildlife, humans	Gastroenteritis, diarrhea
Entamoeba histolytica	Humans, dogs, cats	Acute enteritis, diarrhea
Giardia lamblia	Humans, deer, beaver	Giardiasis (diarrhea & abdominal cramps)
Balantidium coli	Mammals especially swine and rats	Diarrhea and dysentery
Toxoplasma gondii	Cats and humans	Toxoplasmosis
Helminth Worms		
Ascaris lumbricoides	Humans	Digestive disturbances, abdominal pain.
Trichuris trichiura	Humans, pigs, monkeys, lemurs	Whipworm disease - Abdominal pain, diarrhea, anemia, weight loss
Taenia saginata; T. solium	Humans, cattle, buffalo, giraffe, llama, reindeer; swine, dogs, cats	Taeniasis - nervousness, insomnia, anorexia
Necator americanus	Humans	Hookworm disease- Anemia

Information Source: Feachem et al., 1983.

Table 2. Reported Infective Dose Data.

Organism	Infective Dose	Range	Reference
Bacteria			
Clostridium perfringens	10^6	10^6 to 10^{10}	Kowal, 1985
Escherichia coli	10^2 to 10^4	10^4 to 10^{10}	Kowal, 1985; Keswick, 1984
Salmonella species	10^2	10^2 to 10^{10}	Kowal, 1985
Shigella dysenteriae	10 to 10^2	10 to 10^9	Kowal, 1985; Keswick, 1984; Levine, 1973
Shigella flexneri	10^2	10^2 to 10^9	Kowal, 1985
Streptococcus faecalis	10^9	10^9 to 10^{10}	Kowal, 1985
Vibrio cholerae	10^3	10^3 to 10^{11}	Kowal, 1985; Keswick, 1984
Viruses			
Echovirus 12	HID_{50} 919 PFU HID_1 17 PFU est=d	17-919 PFU	Kowal, 1985
Poliovirus	1 $TCID_{50}$, <1 PFU	4 x 10^7 $TCID_{50}$ for infants 0.2 to 5.5 x 10^6 PFU for infants	Kowal, 1985
Rotavirus	HID_{50} 10 ffu HID_{25} 1 ffu est=d	0.9 to 9 x 10^4 ffu	Ward et al., 1986
Parasites			
Entamoeba coli	1-10 cysts	1-10 cysts	Kowal, 1985
Cryptosporidium	10 cysts	10-100 cysts	Casemore, 1991
Giardia lamblia	1 cyst estimated	NR	Kowal, 1985
Helminths	1 egg	NR	Kowal, 1985

HID = human infective dose $TCID_{50}$ = tissue culture infectious dose for 50 % response
PFU = plaque forming units ffu = focus forming units NR = not reported

The major pathogen reduction mechanisms applicable to organics recycling rely on a combination of high temperature (>55°C), time, alkaline pH (pH> 11.5), and reduced moisture content (<30%). Beta or gamma irradiation also is effective, but is typically too expensive for practical purposes. A number of practices, e.g., composting, alkaline stabilization, thermophilic anaerobic digestion, and heat drying, utilize various combinations of these mechanisms to meet the pathogen reduction requirements for production of Exceptional Quality biosolids (EPA, 1994). This means that the biosolids have been treated sufficiently to permit their distribution and use to the public without restriction.

The approach taken by the U.S. EPA in developing pathogen and vector attraction reduction standards for biosolids provides an appropriate basis for development of similar recommendations and guidelines for other recyclable organics and in the production of bio-mineral products that will be used in feed and food production.

The blending of mineral by-products with organic residues has been approved for use with biosolids in the N-Viro process (N-Viro International) and the EnVessel Process (RDP Technologies, Inc.), both of which can utilize various alkaline by-products such as fly ash and cement kiln dust. Both increase pH to $\geq$ 11.5 and generate ammonia concentrations that assist in destroying pathogens injured by heat, and thereby effectively meet pathogen reduction criteria described in the Part 503 rule.

There are basically several critical points in the production and distribution of mineral-organic blends when the benefits of addition of mineral components may improve the process or enhance the product. If the mineral by-product size can increase the porosity or reactive surface areas in a mixture with organics, then composting may be facilitated. This needs to be evaluated. If the mineral by-product chemically reacts throughout the organic mass, as is the case with alkaline by-products, then blending could assist in destroying pathogens present in organic residues. Depending on the ultimate targeted use and ability of the producers to store product at various stages of production, blending may be more beneficial early in the process rather than later. To some extent, when to blend will also depend on whether composting will be part of the process used to stabilize the organic materials. For certain uses and for cases in which the mineral by-products are minimally reactive with the organics in the short-term, blending just prior to bulk or bagged distribution may suffice. Of course, a prudent producer would have implemented a suitable pathogen reduction process, such a composting, with regard to the organic constituents, prior to blending and unrestricted distribution. In the use of any of the various approaches to pathogen reduction, a prudent producer would also put in place an appropriate quality assurance program.

SUMMARY

As society continues its efforts to recycle nutrients and preserve and protect the natural resource base, the opportunity and need to blend various by-product streams to meet specialized nutrient and market needs will expand. Along with this expansion, there will be a corresponding need to evaluate and account for the source, fate, and distribution of public health pathogens that can adversely impact the quality and safety of foods, especially those intended for the fresh or lightly-processed marketplace. There is an urgent need to develop science-based recommendations for safe re-use of manure and organics in cropping systems producing fruits and vegetables for the fresh and lightly processed markets. Improper land application of untreated or inadequately treated manures, biosolids, septage, sewage, or other pathogen laden residues can lead to human disease outbreaks and a means of maintaining disease reservoirs of the pathogenic agents involved. Suitable practices, many of which were developed to meet U.S. EPA pathogen standards for biosolids use, could be adapted to eliminate pathogen survival and movement from other organic by-products into soil, water, crops, especially produce likely to be consumed fresh. Practices and guidelines for eliminating pathogens at pre-harvest as well as post-harvest stages is needed to account for the effects of environmental parameters within the framework of present-day on-farm and marketing practices and novel characteristics of emergent pathogens. Such recommendations are needed to satisfy the pre-harvest aspects of farm-to-table food safety as well as a holistic approach to organics and minerals recycling. Implementation of sound organics re-use practices on-farm will sustain the public confidence in the safety of food and water supplies. Science, technology, and management will need to work collaboratively to advance the development and use of recycling in ways that appropriately interface with the interstate and international food production and distribution systems that have evolved.

REFERENCES

Abdul-Raouf UM, Beuchat LR & Ammar MS (1993) Survival and growth of *Escherichia coli* O157:H7 on salad vegetables. Appl. Environ. Microbiol. 59:1999-2006.

Ahmed N and Conner DE (1993) Evaluation of different media for recovery of thermally injured *Escherichia coli* O157:H7. J. Food Protection 56:913.

Armstrong GL, Hollingsworth J & Morris JG Jr. (1996) Emerging foodborne pathogens: *Escherichia coli* O157:H7 as a model of entry of a new pathogen into the food supply of the developed world. Epidemiol. Rev. 18:29 - 51.

Besser RE, Lett SM, Weber JT, Doyle MP, Barrett TJ, Wells JG & Griffin PM (1993) An outbreak of diarrhea and hemolytic uremic syndrome from *Escherichia coli* O157:H7 in fresh-pressed apple cider. JAMA 269:2217-2220.

Bettelheim KA (1996) Enterohaemorrhagic *Escherichia coli*: A new problem, an old group of organisms. Austral. Vet. J. 73:20-26.

Casemore DP (1991) The epidemiology of human cryptosporidiosis and the water route of infection. Wat. Sci. Tech. 24(2):157-164.

CDC (1997) Outbreaks of *Escherichia coli* O157:H7 infection and cryptosporidiosis associated with drinking unpasteurized apple cider - Connecticut and New York, October 1996 Morb. Mort. Wkly Rep 46:4-8.

CDC (1996) Outbreak of *Escherichia coli* O157:H7 infections associated with drinking unpasteurized commercial apple juice - British Columbia, California, Colorado, and Washington, October 1996 Morb. Mort. Wkly Rep. 45:975.

CDC (1975) Epidemiologic notes and reports: *Salmonella typhimurium* outbreak traced to a commercial apple cider -New Jersey. Morb. Mort. Wkly Rep. 24:87-88.

Cieslak PR, Barrett TJ, Griffin PM, Gensheimer KF, Beckett G, Buffington J & Smith MG (1993) *Escherichia coli* O157:H7 infection from a manured garden. Lancet 342:367.

Conner DE & Kotrola JS (1995) Growth and survival of *Escherichia coli* O157:H7 under acidic conditions. Appl. Environ. Microbiol. 61:382-385.

Droffner ML & Brinton WF (1996) Occurrence and detection of viable *Listeria* in food scrap compost. Abl. Hyg. 199:51-59.

Feachem RG, Bradley DJ, Garelick H & Mara DD (1983) Sanitation and disease: health aspects of excreta and wastewater management. World Bank Studies in Water Supply and Sanitation 3. John Wiley & Sons, New York.

Feng P (1995) *Escherichia coli* Serotype O157:H7: Novel vehicles of infection and emergence of phenotypic variants. Emerg. Infect. Dis. 147-52.

Fox KR & Lytle DA (1996) Milwaukee's crypto outbreak: investigation and recommendations. J. Amer. Water Works Assoc. 88:87-94.

Ghiorse WC (1996) *Cryptosporidium* and *Giardia*: Update, pp. 56-59. Proc. Animal Agriculture and the Environment, No. Amer. Conf., Rochester, NY, Dec. 11-13, 1996. Northeast Regional Agricultural Engineering Service, Cooperative Extension (NRAES-96), 152 Riley-Robb Hall, Cornell Univ., Ithaca, NY.

Hancock DD, Besser TE, Kinsel ML, Tarr PI, Rice DH & Paros MG (1994) The prevalence of *Escherichia coli* O157:H7 in dairy and beef cattle in Washington State. Epidemiol. Infect. 113:119-207.

Hao YY & Brackett RE (1993) Growth of *Escherichia coli* O157:H7 in modified atmosphere. J. Food Protection 56:330-332.

Hayes DB, Matte TD, O'Brien TR, et al. (1989) Large community outbreak of cryptosporidiosis due to contamination of a filtered public water supply. N. Engl. J Med. 320:1372-1376.

Hilborn ED, Mermin J, Mshar P, Slutsker L, Mshar R, Farrar J, Vance M, Lambert-Fair M, Furgalack T, Vugia D & J. Hadler J (1997) A multistate outbreak of *Escherichia coli* O157:H7 infections associated with mesclun mix lettuce. Epidemic Intelligence Service, 46th Annual conference, April 14-18, 1997.

Keswick BH (1984) Sources of groundwater pollution In: Bitton G & Gerba G (Eds) Groundwater Pollution Microbiology (pp. 39-64). John Wiley and Sons, New York.

Kowal NE (1985) Health Effects of Land Application of Municipal Sludge. U.S. EPA, Research Triangle Park, NC. EPA-600/1-85-015.

LeChevallier MW, Norton WD & Lee RG (1991a) Occurrence of *Giardia* and *Crytosporidium* spp. in surface water supplies. Appl. Environ. Microbiol. 57:2610-2616.

LeChevallier MW, Norton WD & Lee RG (1991b) Occurrence of *Giardia* and *Crytosporidium* spp. in filtered drinking water supplies. Appl. Environ. Microbiol. 57:2617-2621.

Levine MM, DuPont HL & B. Formal SB (1973) Pathogenesis of Shigella dysenteriae (Shiga) dysentery. J. Infect. Dis. 127(3):261-270.

MacKenzie W, Neil M, Hoxie N, et al. (1994) A massive outbreak in Milwaukee of *Cryptosporidium* infection transmitted through the public water supply N. Engl. J. Med. 331:161

Mawdsley JL, Bardgett RD, Merry RJ, Pain BF & Theodorou MK (1995) Pathogens in livestock waste, their potential for movement through soil and environmental pollution. Review Appl. Soil Ecol. 2:1-15.

Mechie SC, Chapman PA & Siddons CA (1997) A fifteen month study of *Escherichia coli* O157:H7 in a dairy herd. Epidemiol. Infect. 118:17-25.

Millard PS, Gensheimer KF & Addiss DG (1994) An outbreak of cryptosporidiosis from fresh-pressed apple cider. JAMA 272:1592-1596.

Miller LG & Kaspar CW (1994) *Escherichia coli* O157:H7 acid tolerance and survival in apple cider. J. Food Prot. 57:460-464.

Morgan D, Newman CP, Hutchinson DN, Walker AM, B. Rowe B & Majid F (1993) Verotoxin producing *Eshcerichia coli* O157 infections associated with the consumption of yoghurt. Epidemiol. Infect. 111:181-187.

Morris GK & Feeley JC (1979) *Yersinia enterocolitica*: A review of its role in food hygiene. Bull. World Health Organization 54:57-85.

Pell AN, Bryant JE & Anguish L (1994) *Giardia* and *Cryptosporidium parvum*: contagion and containment. Proc. Liquid Manure Application Systems Conf. (pp. 85-94). Rochester, NY, Dec. 1-2, 1994. Northeast Regional Agricultural Engineering Service, Cooperative Extension, (NRAES-79) 152 Riley-Robb Hall, Cornell University, Ithaca, NY.

Pesaro F, Sorg I & Metzler A (1995) In situ inactivation of animal viruses and a coliphage in nonaerated liquid and semiliquid animal wastes. Appl. Environ. Microbiol. 61:92-97.

Plym-Forshell L & Ekesbo I (1993) Survival of Salmonellas in composted and not composted solids animal manures. J. Vet. Med. B 40:654-658.

Shuval HI, Yekutiel P & Fattal B (1984) Epidemiological evidence for helminth and cholera transmission by vegetables irrigated with wastewater: Jerusalem - A case study. Water Sci. Technol. 17:433-442.

Snowdon J, Cliver D & Converse JC (1989) Land disposal of mixed human and animal waste; A review. Waste Management Res. 7:121-134.

Solo-Gabriele H & Neumeister S (1996) The Las Vegas cryptosporidiosis outbreak. J. Amer. Water Works Assoc. 88:76-86.

Strauch D & Ballarini G (1994) Hygienic aspects of the production and agricultural use of animal wastes. J. Vet. Med. B 41:176-228.

Strauch D (1991) Survival of pathogenic microorganisms and parasites in excreta, manure and sewage sludge. Rev. Sci. Tech. Off. Int. Epizoot. 10(3):813-846.

Swerdlow D, Woodruff BA, Brady RC, et al. (1992) A waterborne outbreak in Missouri of *Escherichia coli* O157:H7 associated with bloody diarrhea and death. Ann. Intern. Med 117: 812-819.

Upton P & Coia J (1994) Outbreak of *Escherichia coli* O157 infection associated with pasteurized milk supply. Lancet 34:1015.

U.S. EPA (1993) The Standards for the Use or Disposal of Sewage Sludge (Title 40 of the Code of Federal Regulations [CFR], Part 503). Federal Register 58:9248-9404.

U.S. EPA (1994) A Plain English Guide to the EPA Part 503 Biosolids Rule. U.S. EPA, Office of Wastewater Management, EPA/832/R-93-003, September.

Van Renterghem B, Huysman F, Rygole R & Verstraete W (1991) Detection and prevalence of *Listeria monocytogenes* in the agricultural system. J. Appl. Bacteriol. 71:211-217.

Wang G, Zhao T & Doyle M (1996) Fate of enterohemorragic *Escherichia coli* O157:H7 in bovine feces. Appl. Environ. Microbiol. 62:2567-2570.

Ward RL, Berstein DI, Young EC, Sherwood JR, Knowlton DR & Schiff GM (1986) Human rotavirus studies in volunteers: Determination of infectious dose and serological response to infection. J. Infect. Dis. 154(5):871-880.

Weagant SD, Bryant JL & Bark DH (1994) Survival of *Escherichia coli* O157:H7 in mayonnaise and mayonnaise-based sauces at room and refrigerated temperatures. J. Food Prot 57:629-631.

Willshaw GA, Cheasty T, Jiggie B & Rowe B (1993) Vero cytotoxin-producing Escherichia coli in a herd. Vet. Rec. 132:46-51.

Yanko WA, Walker AS, Jackson JL, Libao LL & Gracia AL (1995) Enumerating Salmonella in biosolids for compliance with pathogen regulations. Water Environ. Res. 67(3): 364-370.

Zhao T & Doyle MP (1994) Fate of enterohemorrhagic *Eshcerichia coli* O157:H7 in commercial mayonnaise. J. Food Prot. 57:780-783.

Zhao T, Doyle MP & Besser RE (1993) Fate of enterohemorrhagic *Escherichia coli* O157:H7 in apple cider with and without preservatives. Appl. Environ. Microbiol. 59:2526-2530.

Zhao R, Doyle MP, Shere J & Garber L (1995) Prevalence of enterohemorrhagic *Escherichia coli* O157:H7 in a survey of dairy herds. Appl. Environ. Microbiol. 61:1290-1293.

AN ENVIRONMENTAL MANAGEMENT SYSTEM FOR BIOSOLIDS

JOHN M. WALKER and JAMES G. HORNE
Office of Wastewater Management, US Environmental Protection Agency, Washington, DC, 20460
E-mail walker.john@epamail.epa.gov

INTRODUCTION

The U.S. Environmental Protection Agency's (EPA's) 40 CFR Part 503 rule (US EPA, 1994) provides comprehensive requirements for the management of biosolids generated during the process of treating municipal wastewater. This risk-based regulation is designed to protect public health and the environment from reasonably anticipated adverse effects of pollutants that may be present in biosolids. The Part 503 rule is designed to manage organic and inorganic pollutants in biosolids, pathogens and vectors. The Part 503 rule is self-implementing, meaning that persons who generate and land apply biosolids must follow its provisions whether or not a permit has been issued. Most states have additional rules that govern biosolids recycling.

This science-based Part 503 rule has been subjected to many different reviews including an extensive review by the National Research Council of the National Academy of Science (1996). Reviewers have confirmed the appropriateness or the development process of the rule and have been found it to be sound and adequately protective.

The proper implementation of the Part 503 rule is dependent upon adherence to its requirements by biosolids generators and land appliers. The regulated entities must keep track of the quality of the biosolids, and, depending on the level of pollutants in the biosolids (i.e., if greater than the pollutant concentration limits in Table 3 of the Part 503 rule), of cumulative quantities of pollutants added to the land. In addition the biosolids generators and land appliers must certify what form of pollutant and vector control they are achieving and that they are applying the biosolids at an agronomic rate that will not pollute ground and surface water.

No matter how sound the requirements of the Part 503 rule, they are not effective unless complied with, and, there are not an infinite number of inspectors and enforcement personnel to continually track the compliance of each land application project. In addition, the Federal and state rules do not generally regulate nuisance factors such as odor, noise, traffic, and dust, nor the management of nutrients other than nitrogen, which must be applied at the agronomic rate, all of which fuel public concern.

Since the Part 503 rule was promulgated in 1993, the experience has been that most generators and land appliers of biosolids are usually in compliance with state and the Federal Part 503 rule pollutant requirements. The area where some biosolids generators

S. Brown, J.S. Angle and L. Jacobs (eds.), Beneficial Co-Utilization of Agricultural, Municipal and Industrial By-products, 259–267.
© 1998 *Kluwer Academic Publishers*,

and land appliers have gotten into trouble is with monitoring and record- keeping requirements. And, virtually all projects have had difficulties at one time or another because of nuisances which draw the attention of the public. The most common nuisance problem has been with odor. Once such a nuisance problem has drawn attention to a project where biosolids from another jurisdiction are being land applied, the public believes there are many other problems, e.g., disease from pathogens in the biosolids and food poisoning and/or ground and surface water contamination from the city's sewage toxins.

Recognizing that there will always be a limited number of inspectors, and that there is a need to improve the user- and environmental-friendliness of biosolids recycling operations, biosolids stakeholders have become interested in the potential benefits of developing and adopting an Environmental Management System (EMS) for biosolids.

PURPOSE

This paper discusses the potential structure and useability of environmental and quality management systems for biosolids beginning with a description of the internationally developed ISO 14,001 EMS and the ISO 9,001 quality management system (QMS). An evaluation of the ISO 14,001 EMS as a potential model for a biosolids EMS by a group of National Biosolids Stakeholders follows. A discussion of the recommendations of the Stakeholders for the development of a National Code of Good Practice for biosolids follows. A model streamlined EMS for biosolids to be authenticated by various forms of third-party verification is outlined.

COMPONENTS OF AN ISO 14,001 EMS

The ISO 14,001 EMS is an international standard that has been established for environmental management systems. The ISO 14,001 sets a framework by which participants can more effectively fufill their environmental obligations and reduce impacts on the environment. The framework allows for meeting applicable statutes and regulations and, in addition, moving beyond compliance and improving overall performance through pollution prevention. Those entities adhering to the principles of ISO 14,001 do so voluntarily and set their own objectives and measures of performance. These must, however, include compliance with applicable rules and provisions for prevention of pollution and for continual improvement. In short, the framework allows for maximizing the environmental benefits and mitigating negative environmental impacts of their system.

The ISO 14,001 EMS includes a number of procedural steps that each participant goes through. When establishing the EMS, persons who are involved with all aspects of the program should be members of the EMS planning and implementation teams. Once these steps have been taken, the environmental benefits can be realized. Components of an ISO 14,001 EMS include:

1. *A Policy Statement* -- what the participant voluntarily commits to achieve,

including:

- Commitments at a minimum:
  - -- Prevention of pollution
  - -- Continual improvement of the EMS leading to improved performance and positive environmental results
  - -- Compliance with applicable statutory and regulatory requirements
- Other commitments relevant to biosolids that are not subject to regulatory requirements could include:
  - -- Minimization of nuisances like odors, traffic, and dust
  - -- Management of nutrients

2. *Analysis of Environmental Aspects and Impacts*

   - Includes all of the organizations' activities and products and services (regulated and nonregulated)
   - Identifies positive and negative impacts of operations
   - Identifies negative aspects that have a significant impact [It is up to each individual participant to decide what is significant and what "aspects" they will address.]

3. *Setting Performance Objectives and Targets for Individual EMSs* which link back to the three commitments established in the organizations' policy by determining what factors are important in those three areas.
   For example:

   - Maintaining full compliance;
     - -- While being in full compliance with Part 503 is required for the use or disposal of biosolids, being in full compliance is not a prerequisite for having an EMS under ISO 14001.
   - Continual improvement of the EMS
   - Prevention of pollution

4. *Implementing the EMS to Meet the Objectives,* including:

   - Training of employees
   - Establishing work instructions and practices
   - Developing mechanisms for measuring the attainment of objectives and targets

5. *A Program for Periodic Audit of the EMS Operation*

   - Establish procedures for an internal audit to determine how you meet your performance objectives. This auditing system includes:
     - -- Evaluating compliance with applicable regulatory requirements
     - -- Meeting previously agreed upon nonregulatory performance measures
     - -- Establishing procedures for documentation and reporting noncompliance
     - -- Establishing procedures to look for causes of noncompliance and other problems

6. *Checking and Taking Corrective and Preventive Actions* when deviations from the EMS occur
7. *Establishing a Process for Responding to Inquires from External Stakeholders*
8. *Undertaking Periodic Reviews of the EMS by Top Management* to ensure its continued performance
   - In essence a feedback loop such as total quality management (TQM) for getting feedback from top management and modifying the EMS as necessary
   - The management review is comprehensive, i.e., it goes over each component of the EMS

Note: For the ISO 14,001 and 9,001 EMS and QMS accreditation, each element of an organization's system is audited every three years. The accreditation can be withdrawn at any time based on the results of these audits.

COMPONENTS OF AN ISO 9,001 QMS

Some entities may also wish to follow the provisions of the international standard for Quality Management Systems (QMS) -- ISO 9,001. An ISO 9,001 standard could be helpful to a firm that wanted to market a heat-dried, composted, or an alkaline-stabilized product across state lines and particularly internationally. An ISO 9,001 QMS does not guarantee that an organization's product will meet certain quality standards. Rather, like the ISO 14.001, it establishes practices and procedures that may help achieve and maintain the desired level of quality. Many of the components of an ISO 9,001 QMS are similar to those under ISO 14,001 and will not be discussed further in this paper.

DISCUSSION OF APPLICABILITY OF THE ISO 14,001 EMS TO BIOSOLIDS

The ISO 14,001 EMS were discussed during a National Biosolids Stakeholders Meeting in Potomac, MD, on May 15 and 16, 1997. These stakeholders had assembled to build on the recommendations of a National Stakeholder's vision for a comprehensive national biosolids program for managing biosolids that had been developed two years earlier. One of the high priorities of the participants in the 1997 National Biosolids Stakeholders meeting was to consider and make recommendations for a stakeholder-based program that would help ensure the user-friendly and environmentally sound management of biosolids.

A straw persons view of a biosolids EMS was prepared for consideration by the stakeholders. The following two sub-sections list the potential components of a biosolids EMS and potential incentives for adopting a biosolids EMS.

Potential Components of a Biosolids EMS

1. Preparing for your biosolids EMS

- Preparing a policy statement
- Analyzing environmental aspects and impacts (This process should include other key representative stakeholders.)
- Setting measurable performance objectives and targets (This process should also include other key representative stakeholders.)

2. Implementing the EMS to meet the objectives through training, work practices, EMS and compliance auditing, external communications, and management review.

3. Mandatory and potential elements of a code of good practice

   - Compliance with Part 503 and State regulations -- mandatory
   - Field storage and land application guidances
   - Odor Control
   - Traffic and spill management plans
   - Dust and noise control plans
   - Nutrient management plans
   - A communications program for working with the press, public, and farmers
   - Train to meet requirements of the code
   - Plan for continual improvement
   - Plan for self audit and management review

4. A Program for Periodic EMS and Compliance Audits

   - A plan for periodic review of the EMS by top management including stakeholders
   - A program to check on the performance of previously established objectives and targets
   - A program for taking preventive and corrective actions as needed based on a system's performance

5. Possible Approaches for Third-Party Verification of the EMS
   - Accredited third-party verification of adherence to ISO 14001 or ISO 9001
   - State verification of adherence to code
   - Verification by other biosolids groups or associations
   - Self verification or declaration

Potential Incentives

- Greater acceptance of biosolids program by the public and community leaders
- Possible yield financial benefits
- Competitive advantage
- Could potentially assist municipalities in gaining qualified application contracts more economically
- Better compliance
- Participating entities have a more organized view of the impacts of their system and can better control and articulate those impacts
- Enhancing acceptability of biosolids products for export if a biosolids EMS meets

the requirements of an internationally recognized EMS and QMS
- Less need for regulatory oversight

Additional Critical Questions

Stakeholders were asked to first understand and then consider a number of questions about the applicability of the ISO EMS to biosolids. Questions that were considered included:

- Will it be necessary for the EMS/QMS set up for biosolids to be based on the components of ISO 14,001/9,001?
- Will there be a system in place for certifying those entities desiring ISO 14,001 and 9,001 certification?
- Will there be an opportunity for those entities that do not wish formal ISO 14,001 and/or 9,001 accreditation to still benefit by being recognized as following a code of good practices with less formal third party oversight?
- How do we verify adequate performance in the new EMS system -- alternatives?
- What entities could be recognized as acceptable for third party accreditation of the EMS and QMS?
- What is the need for and roles of enforcement as it relates to entities with or without an EMS?
- How can we avoid inappropriate competition between municipalities and practioners who subscribe to the EMS?
- What incentives are there for participants, e.g., less regulatory oversight and/or reduced frequency of monitoring and record keeping? How do we enhance or tailor incentives for participants?
- What disincentives are there for participants and how can these be addressed?
- What hurdles are there to overcome and how do we overcome them?
- Where do you set the standards -- compliance or compliance plus?
- What activities do we want to include beyond compliance?
- What kinds of training would be needed?
- Will an EMS reduce the importance of state authorization?
- Is it desirable to obtain several volunteers that would pilot the following: (i) all the steps necessary to receive EMS certification, and, (ii) a code of good practices without obtaining the more formal EMS certification? If so, how do we encourage these volunteers?
- Are there volunteers willing to test the different forms of program verification for self-declaration to complete EMS third-party certification? If so, how do we encourage these volunteers?

Stakeholder Involvement

The National Biosolids Stakeholders agreed on the importance of allowing outside stakeholders to have input as appropriate in the development and implementation of key components of the EMS, especially analysis of aspects and impacts, setting objectives and targets, and management review. Imbedded in this desire is the challenge of finding stakeholders to represent the public point of view, who are willing to participate and work for the establishment of an implementable EMS for biosolids.

A primary concern of the stakeholders, who met in Potomac, MD, was an ISO-based biosolids EMS in which there would not be an even level of performance for all biosolids managers. Under ISO 14,001 each generator and land applier can set their own separate performance objectives and targets in accordance with what their organization determines to be important. This could result in considerable confusion among the public who are likely to be concerned that some biosolids managers, even though meeting their individual commitments, were not doing as well as others in managing the environment. A second important concern of the stakeholders was about how third-party verification might work.

While the manner in which an organization or group chooses to establish and adhere to an ISO 14,001 EMS is voluntary, the stakeholders believed that it would be helpful if there was an EMS tailored especially for biosolids that contained certain uniform user-friendly and environmentally-sound management goals that each participating entity would pledge to meet. Then, each facility could tailor its operations as they desired to meet the uniform goals of the EMS, and the public would view the attained level of performance of evaluated entities against a standard set of performance goals. Such a generic EMS might be especially helpful for smaller facilities and groups which might want to establish and follow sound practices.

The consensus of the participants in the Potomac, MD, National Biosolids Stakeholders meeting was to recommend the development of a national code of sound practice for biosolids that, at a minimum, requires compliance with applicable state and Federal biosolids rules and that sets additional goals for user and environmentally friendly management of biosolids. The stakeholders also decided that should be an opportunity for pilot-testing by volunteering municipalities and land appliers of various forms of third-party verification to determine how well different entities are meeting their performance objectives and targets.

A CURRENT STREAMLINED BIOSOLIDS EMS

Subsequent to that stakeholder meeting in Potomac, MD, there has been considerable efforts to further develop an appropriate EMS for biosolids. The California Water Environment Association has been working to develop a Manual of Practices that establishes compliant user-friendly and environmentally-sound practices for land application of biosolids in California. In addition, the U.S. Department of Agriculture (USDA), under the provisions of an Interagency Agreement with EPA, has led the development of a Field Storage Guidance for Biosolids. Most recent discussions have pictured a streamlined EMS for biosolids consisting of four or five components. These components are:

1. A ***one-page Code of Sound Practices*** for biosolids which would list agreed upon goals of the entity pledging to follow the code. The goals would include:

    - Identifying those points that are critical to user-friendly and environmentally sound management and, instituting a system to manage those critical points

- Complying with Part 503 and with State regulations
- Minimizing odor
- Managing traffic and spills
- Controlling dust and noise
- Managing nutrients
- Having effective communications with the press, public, and farmers
- Training to meet the requirements of the code and to achieve continual improvement
- Planning for self audit, management, and potentially some form of third-party review
- Continuing to improve

2. A ***national guidance*** containing sound practices (based on the California, USDA/EPA and other relevant guidances.) These guidances will

   - Identify areas that are critical to providing user-friendly, environmentally-sound practices for biosolids management within each part of the code. For example, one of the most critical areas for achieving user-friendly management of biosolids is the management of odor. One very critical point for odor control is at the treatment plant, where effective partnership is required between the generator and the land applier. Another critical area for odor management would be at a processing facility, such as a composting facility, which would involve a knowledge of meteorological conditions, sound processing, and scrubbing and/or dispersion of processing air and fugitive sources of odor as necessary to meet a satisfactory odor control standard. Other actual critical points for controlling odors would be at the field storage and land application sites. There would also need to be one or more satisfactory back up methods for management of malodorous biosolids.

   - Provide alternate methods for attaining the goals of the Code

3. An ***action plan*** for achieving the goals of the Code that is prepared by each entity that pledges to abide by the code. This plan for achieving the goals also includes a corrective action plan for addressing areas that need improvement as identified in audits. Many of the principles of the ISO 14,001 will provide a useful framework for the entity developing the action plan.

4. A ***training program***

5. ***Self and third-party verification systems*** for assuring that the action plan is being followed by participating entities

CONCLUSION

The outlook for the development of an effective national biosolids environmental management system is good. Both national and regional biosolids management organizations are now being formed which should greatly facilitate this process. This management system offers the potential for greater public acceptance of the beneficial use of biosolids. The current EPA 503 regulations mandate that all biosolids applied to land conform to standards. These science based standards attempt to assure that there will be no negative health impacts associated with land application. A national EMS would go the necessary next step to reduce any perceptions of negative effects.

REFERENCES

US EPA (1994) A plain English guide to the EPA part 503 biosolids rule. Office of Wastewater Management. Washington, DC. EPA/832/R-93/003.

National Research Council/National Academy of Science (1996) Use of Reclaimed Water and Sludge in Food Crop Production. National Academy of Science. Washington, DC.

INTRODUCTION OF TEST METHODS FOR THE EXAMINATION OF COMPOSTING AND COMPOST

PHILIP B. LEEGE
The U. S. Composting Council, Bethesda, Maryland 20814 and The Procter & Gamble Company, Cincinnati, Ohio 45224, USA
E-mail leege.pb@pg.com

ABSTRACT

The composting industry depends on sampling and analytical test methods borrowed from allied industry sources. These sources include assorted solid waste management, water and wastewater management, and the soil, peat, food, feed and fertilizer industries. Under these circumstances, users in the composting industry have modified and adapted test methods to suit a variety of conditions. It is not unusual to have differing interpretations of test results, to disagree on test application suitability, or to substitute one test for another even without demonstrated correlation among tests. A common set of standards for consistent and reliable testing by compost producers, marketers, users, researchers and regulators, and for their clients, does not exist.

The Composting Council's Standards and Practices Committee undertook a long-term project to develop a standard set of sampling and test methods for composting feedstocks, composting and compost. The first step of the project was undertaken in cooperation with the Minnesota Office of Environmental Assistance, and the University of Minnesota Agricultural Extension Center, Research-Extension Analytic Laboratory. The United States Department of Agriculture (USDA), Agriculture Research Service, Beltsville, MD, has joined in this effort. Numerous other Agricultural Extension Centers and independent laboratories will become engaged during the introduction, peer review and acceptance process. The initial set of about 40 laboratory tests undertaken in Minnesota is expanding, along with sampling procedures, with the help of more laboratories.

The sampling and test methods being documented are suitable for several uses. Some methods are "quick tests" for composting process monitoring and control, and for approximation of product attributes. Other more rigorous and laborious tests are for regulatory compliance, product safety and data for product marketing claims.

The First Edition of *Test Methods for the Examination of Composting and Compost* was released in December 1997 by the Standards and Practices Committee for introduction, familiarization and comments, and for solicitation of additional methods. A

S. Brown, J.S. Angle and L. Jacobs (eds.), Beneficial Co-Utilization of Agricultural, Municipal and Industrial By-products, 269–282.
© 1998 *Kluwer Academic Publishers,*

subsequent step will involve team evaluations of methods, and agreement. The USDA plans to follow this step with greenhouse and field testing to help evaluate different tests for the same compost attribute, and correlate compost attribute measurements to growth response. The concluding step is a proposed joint long-term effort with a recognized national standards-setting organization to standardize the protocols, and insure for their vitality.

INTRODUCTION

The composting industry in the United States and around the globe is growing, along with use of a diversity of composting feedstocks for compost, improved understanding of composting process management requirements and options, and increasing commercial acceptance for compost products. The composting industry, like all users of processing feedstocks and producers of a product, needs to have sampling and testing protocols. The composting industry has depended on the use of sampling and test methods developed for similar uses in allied industries, including solid waste management, water and wastewater management, and the soil, peat, food, feed and fertilizer industries. This dependence has not been uniformly disciplined, and has served to illustrate a lack of compatibility for composting technologies and for materials with high organic matter content. It is not unusual to have differing interpretations of test results, to disagree on test application suitability, or to substitute one test for another even without demonstrated correlation among tests. Testing to support process monitoring is often neglected, and test results are often not considered reliable. Consensus for sampling and test methods has not been undertaken, and a common set of standards for consistent and reliable testing by compost producers, marketers, users, researchers and regulators, and for their clients, does not exist.

A number of needs have been identified for a standardized approach which are driving development of standard sampling and test methods for the composting industry. A standardized approach provides the analytical basis for the following:

- composting feedstock definition and composting process management,
- composting process and equipment design and development,
- compost testing in lieu of those developed for soils, water, wastes, peat, food, feed, or fertilizers,
- product regulation development,
- safety regulation limits and compliance demonstration for processes and products,
- compost product development,
- compost marketing data credibility,
- support for development of degradable/compostable products,
- data required to develop soil nutrient management planning, and
- relating compost use to its attributes.

To support these identified needs, three types of testing are required which include the following:

1. Quick tests for feedstock definition and process management, and for approximating consistency of compost quality;
2. Routine tests for marketing data; and
3. Legal tests for regulatory safety and product marketing claims.

The objectives of a standardized approach for compost sampling and testing include:

- Enable consistency and reliability of test methodology, for
  - basing regulatory limits;
  - demonstrating regulatory compliance; and
  - quantifying market data.
- Enable evaluation of the variability of compost characteristics.

DESCRIPTION OF "TEST METHODS FOR THE EXAMINATION OF COMPOSTING AND COMPOST"

Test Methods for the Examination of Composting and Compost (TMECC) is both a catalog and a laboratory manual. In those cases where test methods, already being used in allied industries and maintained by a standards setting organization, have been shown to be suitable in precisely the same form for the composting industry, they are referenced, their application is explained, and the *TMECC* is a catalog. In those cases where it has

The ASTM format and style have been used throughout the TMECC, as follows:

Test Method

1. Scope
2. Referenced Documents
3. Terminology
4. Summary of Test Methods
5. Significance and Use
6. Interference and Limitations
7. Sample Handling
8. Apparatus
9. Reagents and Materials
10. Procedure
11. Calculations
12. Flow Diagrams

Methods Summary

13. Report
14. Precision and Bias
15. Keywords

been necessary to modify the method or develop a new method, the *TMECC* is a laboratory manual that follows the American Society for Testing and Materials (ASTM) format and style.

The Table of Contents lists the following chapters:

Chapter 1. General Introduction, presents the basic Composting Unit Operations Model, shown by Figure 1, which is the basis of sampling and testing for process monitoring. The basic composting unit operations model is related to the composting key process variables in Table 1. The check marks indicate which key process variables must be managed

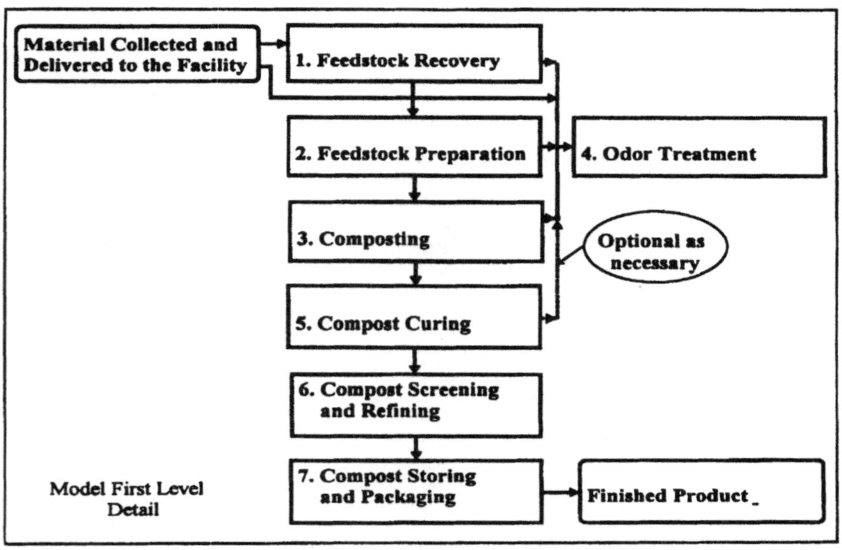

Figure 1. Composting Unit Operations Model for each step to have successful composting.

Table 1. Management of key process variables.

KEY PROCESS VARIABLES	FEED-STOCK	PROCESSING STEPS						
	Feedstock Collected and Delivered	**Step 1:** Feedstock Recovery	**Step 2:** Feedstock Preparation	**Step 3:** Composting	**Step 4:** Odor Treatment	**Step 5:** Compost Curing	**Step 6:** compost Screening and Refining	**Step 7:** Compost Storing and Packaging
Pile structure	✓	✓	✓	✓	✓	✓	✓	✓
Nutrient balance	✓	✓	✓					
Oxygen and pH			✓	✓	✓	✓		✓
Moisture	✓	✓	✓	✓	✓	✓	✓	✓
Temperature				✓	✓	✓		
Time			✓	✓	✓	✓		✓

The Composting Products Model (Figure 2) shows sources of potential products from the composting process and is the basis for sampling and testing of these products. Compost Product Attribute Determination (Table 2) shows feedstock choices and processing steps that determine certain safety and market attributes of compost products. The check marks indicate which process unit operation will impact each product attribute when properly managed.

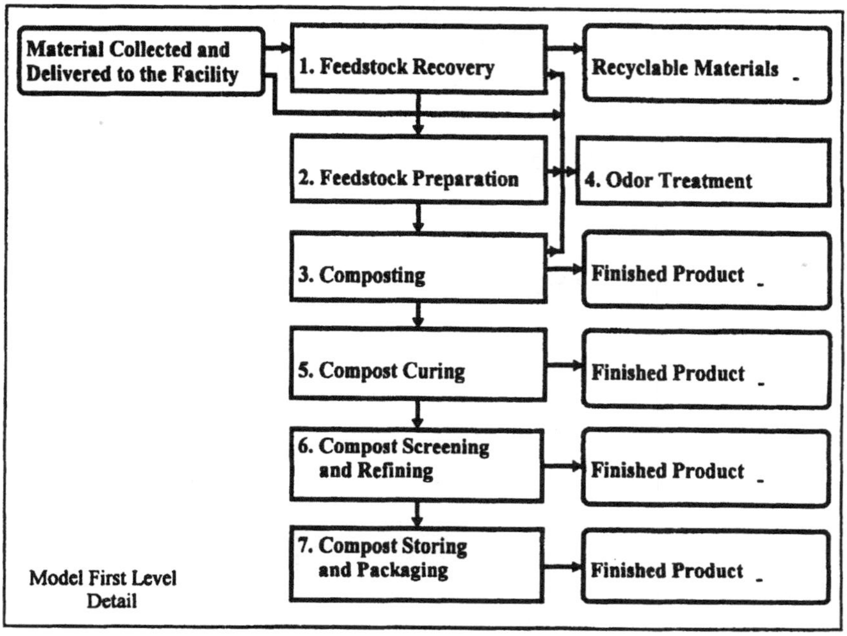

Figure 2. Composting Products Model check marks indicate which process unit operation will impact each product attribute when properly managed.

The preceding instructional material is an aid for selection of the correct sampling and test method. Additional guidance is found in the Test Method Banner (Table 3) for each method which shows unit operations, along with the test method name, measurement units, and test method designation for selection of the proper method. The banner for the "Air Capacity" test method is reproduced below, as an example of banner layout and content.

Table 3 shows application of Chapter 7 methods 07.01-A, 07.01-B, and 07.01-C which are all Air Capacity tests but have different application requirements. At the present time, the *TMECC* contains the following methods:

Table 2. Compost product attribute determination.

PRODUCT ATTRIBUTES	Feed-stock: Feedstock Collected and Delivered	PROCESSING STEPS: Step 1: Feedstock Recovery	Step 2: Feedstock Preparation	Step 3: Composting	Step 4: Odor Treatment	Step 5: Compost Curing	Step 6: Compost Screening and Refining	Step 7: Compost Storing and Packaging
SAFETY ATTRIBUTES								
Regulated chemicals	✓	✓	✓	✓			✓	
Pathogens			✓	✓		✓		✓
MARKET ATTRIBUTES								
Man-made inerts	✓	✓					✓	
Growth screening	✓	✓	✓	✓		✓		✓
Stability	✓	✓	✓	✓		✓		✓
Organic matter content	✓	✓	✓	✓		✓		
pH	✓		✓	✓		✓		✓
Soluble salt content	✓							
Water-holding capacity	✓		✓	✓		✓	✓	
Bulk density			✓	✓		✓	✓	✓
Particle size and texture			✓	✓		✓	✓	
Moisture percent				✓		✓		✓
Plant food content	✓		✓	✓		✓		✓

Table 3. Test method banner.

Test Method:	Air Capacity						Units: % v v^{-1}	
Test Method Applications								
Process Management							Product Attributes	
Step 1: Feedstock Recovery	Step 2: Feedstock Preparation	Step 3: Composting	Step 4: Odor Treatment	Step 5: Compost Curing	Step 6: compost Screening and Refining	Step 7: Compost Storing and Packaging	Safety Standards	Market Attributes
	07.01-A	07.01-A	07.01-A	07.01-A				
		07.01-B		07.01-B		07.01-B		
								07.01-C

<u>Chapter 10 Pathogens</u>
10.01 Pathogens
10.02 Escherica Coli
10.03 Salmonella
10.04 Fecal Coliform Bacteria
10.05 Fecal Streptococcus
10.06 Total Coliform
10.07 Total Heterotrophic Plate Count
10.08 Yeast and Mold
10.09 *Ascaris*
10.10 Aspergillus Fumigatis
10.11 Enterococcus

<u>Chapter 11 Organic Contaminants</u>
11.01 Chlorinated Herbicides
11.02 Dioxin/Furans
11.03 Organochlorine Pesticides
11.04 Organophosphorus Pesticides
11.05 Polychlorinated Biphenyls
11.06 Volatile Organic Compounds

Development Process for "Test Methods for the Examination of Composting and Compost"

Three of seven steps for developing consensus acceptance for the *TMECC* by the composting industry have been completed to-date. The seven steps are described in more detail below.

Step 1. Project Concept Development

The Standards and Practices Committee and Market Development Committee of The U.S. Composting Council developed definitions for composting key process variables, composting feedstock, composting process, product regulatory compliance, and safety and market attributes for which verification protocols are needed by the composting industry.

Step 2. Minnesota Compost Utilization Project (MN CUP) pilot and initial application of field sampling, and laboratory preparation and analyses.

Objective: The Standards and Practices Committee of The U.S. Composting Council intended to adapt existing sampling and test methods and to document methods used by working cooperatively with the Minnesota Office of Environmental Assistance, the University of Minnesota, and Malcolm Pirnie, Inc. Over a two year period, compost was sampled and analyzed from eight compost production plants in Minnesota.

Discussion:

Activities during this two-year cooperation included three phases.

1. Phase I, initial project start-up -- For the first two months, three compost samples were collected at each of three Minnesota composting facility locations to document variability at a site, within one batch. One of the three samples from each location was subdivided into ten subsamples at the laboratory to evaluate within-sample variability.
2. Phase II -- Sampling continued on a monthly basis through one year for the three previously mentioned sites, and five additional Minnesota composting facilities were included for the duration of the monthly sampling phase.
3. Phase III -- Sampling frequency was decreased to a quarterly basis during the second year of sampling for seven facilities and one facility was lost to fire.

Test Parameters from MN-CUP

During the MN-CUP, approximately 40 parameters were considered. Methods developed for other materials were modified and adapted, including existing methods by ASTM, American Society of Agronomy (ASA), U.S. Environmental Protection Agency (EPA) Solid Waste-846, and Association of Official Analytic Chemists (AOAC). Parameters considered included:

1. Chemical analyses using US EPA 3051 digest (slightly modified for compost, high organic matter conditions) and inductively-coupled plasma atomic emission spectrometric (ICP-AES) determinations for metals and salts, wet combustion determination for total Kjeldahl N (TKN, micro-digest technique), colorimetric determination for NO_3-N and NH_4-N, and cation exchange capacity (CEC) using a modified ammonium displacement technique.
2. Physical analyses for total solids and moisture (wet basis), ash, man-made inerts (plastics, metal, glass), bulk density, water-holding capacity, and air-capacity.
3. Biological analyses for stability (O_2-Uptake, Hoitink's Respirometry), growth and germination - a direct seeding technique was developed, organic carbon using dry combustion (LECO analyzer).
4. Organics included volatile fatty acids and PCB's, which were determined at private laboratories outside the University of Minnesota system.

Step 3. Draft a Sampling and Analysis Manual

Objective: Provide a scientifically based catalog and laboratory manual of methods that may be used for feedstocks and compost analysis for regulatory and market requirements, and management of the composting process.

Discussion: Development of the First Edition of *TMECC* included reformatting, enhancing and critiquing those methods devised and modified at the University of Minnesota and from outside laboratories for MN CUP. Other methods that complement MN CUP, separately funded by The U.S. Composting Council and

The Procter & Gamble Company were incorporated.

Step 4. Introduction, Familiarization, Expansion and Feedback (To begin Winter 1997/'98)

Objectives: Introduce the book, and establish credibility and familiarization of its content for industry acceptance and use.

Discussion: The U.S. Composting Council in collaboration with the USDA is undertaking a process to develop consensus acceptance for the sampling and test methods initially through an introductory round of sharing the documentation. The introductory round is targeted at private and public laboratories, along with compost producers willing on a voluntary basis to carefully read, evaluate and comment on the methods. The Council expects to involve about 65-70 laboratories who will form into sub-groups of four to five labs each, but not less than three, who will select a given set of methods in which they have expertise. The sub-groups will work cooperatively among themselves to read the method, try the method using commercially available bagged compost, and comment and critique on the method. In addition participants are asked to volunteer test methods with which they are familiar for addition to the *TMECC*. The sub-groups will be formed during the winter of 1997/=98 and report back by mid-summer 1998. Feedback from the sub-groups will be used in editing the book in preparation for Step 5, Collaborative Evaluation.

Laboratory Selection: Laboratories who agree to participate will select tests based upon a review of the draft methods and on lab expertise. Laboratories will need to have the apparatus required by the test method, and labs are urged to suggest identical substitute apparatus.

Step 5. Collaborative Evaluation (Winter 1998/'99)

Objectives: Develop statistical basis for validation based on AOAC (or equivalent) criteria, and define variability associated with each test method and relative accuracy.

Discussion: The feedback from Step 4 will be used to edit and make additions and deletions to the *TMECC* for a Second Edition. The Second Edition *TMECC* will be provided, along with standard composts, for sub-group use in developing data on the compost provided. A minimum of three labs will again form a sub-group and be asked to develop consensus on the test method definition based on reproducibility of test results.
Three standard samples will be obtained from operating facilities where process management is well established and routine. The composting feedstocks will be:

- urban yard debris
- urban yard debris cocomposted with biosolids
- source separated household/institutional/ commercial discarded organic material

The collaborative evaluation step will be undertaken during the winter of 1998/'99 and feedback solicited in spring of 1999. The feedback will be used to again edit, revise and make additions to the text. When the collaborative evaluation step 5 is complete, the revised text will be the Third Edition and become The U.S. Composting Council's recommended test methods.

Step 6. USDA Greenhouse and Field Testing (Enhancement)

An enhancement step is planned after recommended test methods are in place. The USDA will conduct an extensive series of greenhouse and field growth tests to understand the correlation between similar tests for the same attribute and the correlation between test data and growth results.

A "*Best Practices in Field Research for Soil Organic Amendments*" is under development for use in the USDA studies. Data from "*Best Practices*" testing will become part of a body of data to be considered in understanding the true potential commercial value of compost and those feedstocks, processing practices and compost attributes that give value to compost.

Step 7. Standardization

This is a proposed joint long-term effort with a recognized, national standards-setting organization to standardize the protocols and insure for their vitality.

CONCLUSION

For many in the composting industry release of *Test Methods for the Examination of Composting and Compost* has been long awaited. More and more states have been developing regulatory limits on various compost attributes that have validity in-so-far-as uniformly accepted and credible verification protocols and science-based correlations to risk and to benefits exist. The use of compost is finding increased acceptance in a broad variety of applications and its proper use, based on accurate measurement of economic benefits, is becoming increasingly important. Composting process management strategies to minimize nuisance odor generation, manage odors, and deal with heavy metal, pathogen, product stability/maturity and other issues, such as throughput rates, capital investment, and predictable product definitions, must rely on simple and dependable testing.

In releasing *Test Methods for the Examination of Composting and Compost,* the U. S. Composting Council is making a very important contribution at this juncture to establish one of the foundations that help assure continuing development of a sound and sustainable composting industry.

ACKNOWLEDGMENTS

The author wishes to thank Maurice Watson and Wayne Thompson for assistance in preparing this paper.

Members of the Standards and Practices Committee collaborated with the Minnesota Compost Utilization Project team headed by Sigurd Scheurle with Dawn Gorres and David Benke assisting to develop the initial project plan, and worked with George Johnson, Robert Munter and Wayne Thompson to develop planning for sampling and testing.

In addition to the people already named, contributors in the development of sampling and test methods to-date include Paul Bloom, John Bouwkamp, Gary Breitenbeck, William Brinton, Bruce Cook, Gerald Croteau, William Dancer, Joann Ernst, Brenda Farrell, George Fitzpatrick, William Fonteno, Frank Gouin, Marcia Grebus, Tom Halbach, Robert Harrison, Martha Mamo, Mindy Miller, Robert Miller, Patricia Millner, Michael Overcash, Aziz Shiralipour, Frank Shields, and Irene Watson-Craik.

NUTRIENT MANAGEMENT PLANNING FOR CO-UTILIZATION OF ORGANIC BY-PRODUCTS

LEE W. JACOBS
Department of Crop and Soil Sciences, Michigan State University, East Lansing, Michigan 48824, USA
E-mail jacobsl@pilot.msu.edu

INTRODUCTION

A large quantity of organic by-products are generated in the U.S. annually. Categories of organic materials include animal manure; crop residues and green manures; sewage sludge (biosolids) and septage; municipal solid waste (MSW), yard waste; food processing; industrial organics; and logging and wood manufacturing residues (USDA, 1978). A USDA report (1978) listed annual production of these organic materials in the United States. It showed that crop residues, animal manure, and MSW were produced in the largest quantities. If one excludes crop residues (that largely remain on crop land where they are produced), the two largest quantities of organic materials produced that must be "managed" in our society are animal manure and MSW.

While agriculture has recently been challenged to manage animal manure nutrients better than in the past, management of the two major municipal waste streams (biosolids and MSW) have reached a critical stage (Parr and Hornick, 1993). Due to the escalating costs of landfilling and incineration, interest in recycling more of these two municipal waste streams has been increasing. For biosolids, the percent that is applied to land for beneficial reuse has increased from 23% in 1978 to 36% in 1990 (USEPA, 1993). While only about 1% of MSW was estimated to have been applied to land in 1978, some increase can be expected as many states have banned yard wastes from landfills.

For MSW, and particularly yard wastes, composting is currently being viewed as a means of treating the organic portions of this waste stream prior to beneficial reuse. Due to this interest, the Solid Waste Composting Council conducted a study to estimate the potential supply of and demand for compost in the U.S. (Slivka et al., 1992). This compost study provides some useful insights into where organic composts, or organic by-products in general, would be utilized. Compost from MSW as well as compost from biosolids, horticultural/silvicultural residues, and animal manure were considered.

The total potential supply of these four types of compost that could be produced annually was estimated to be 46 million metric tons/yr, which included 27 million metric tons for MSW, 14 million metric tons for horticultural and 2.7 million metric tons each for biosolids (sewage sludge) and animal manure. In contrast, the potential utilization of this compost was estimated to be about 470 million metric tons/yr with this utilization being

S. Brown, J.S. Angle and L. Jacobs (eds.), Beneficial Co-Utilization of Agricultural, Municipal and Industrial By-products, 283–287.
© 1998 *Kluwer Academic Publishers,*

predominantly on agricultural land -- 86% on crop land, 10% on forestland, and 2% for sod production.

These data illustrate that, if the U.S. is going to beneficially reuse a significant portion of the organic by-products generated in our society, a portion of these materials will need to be applied to agricultural land. Other potential applications (which include landscaping, topsoil, bagged/retail products, landfill final cover, container and field nurseries, and surface mine reclamation) will contribute less than 2% to the potential demand/utilization of compost. Therefore, agriculture can play a significant role in helping our society utilize the soil resource to manage its organic waste streams.

NUTRIENT CONTRIBUTIONS FROM ORGANIC BY-PRODUCTS

Most organic by-products contain low contents of nutrients, are of biological origin, and can be safely returned to the land resource as a waste management alternative. If their distribution is not economically limited by the costs of transporting them to available land, organic by-products can often be applied to crop land as a low nutrient fertilizer. However, because they are organic and therefore decomposable, organic by-products can be a source of odor and and attract flies if not managed properly.

While organic by-products can make a contribution to the nutrients required for crop production, the nutrient value may or may not be significant compared to fertilizer nutrients. Data for Michigan is presented in Table 1 as an example to illustrate this point. While animal manure can potentially contribute a significant proportion of the total nutrients needed for crop production, contributions from biosolids and MSW are minimal. While MSW does contain significant quantities of nutrients, the relative nutrient value of the material is low. Most MSW is still going to landfills and not yet utilized on land.

Crops need sixteen essential plant nutrients for growth and reproduction, thirteen of which are generally provided by the soil in sufficient quantities. These nutrients include

Table 1. Estimated nutrients present in organic waste materials compared to fertilizer nutrients sold in Michigan.

Material	Amount	N	P_2O_5	K_2O
	Tons/year			
Animal Manure	18,500,000 (wet)	96,800	56,000	96,400
Biosolids				
Total produced	299,000 (dry)	8,970	13,100	1,670
Land applied	62,000 (dry)	1,860	2,710	350
Mun. Solid Waste[a]	7,310,000 (dry)	51,200	33,600	26,300
Fertilizers[b]	1,279,000	250,000	122,000	228,000

[a] Quantity of MSW estimated by using 1990 MI population (9,312,000) and USEPA per capita MSW generation rate (4.3 lb/day or 1,570 lb/yr). Average nutrient content taken from Parr and Hornick (1993) paper.
[B] TVA, 1992.

three major (N, P, K), three secondary (Ca, Mg, S), and seven micronutrients (B, Cl, Cu, Fe, Mn, Mo, Zn). Quantities of N-P-K are usually applied in the greatest amounts to supplement the nutrients available from the soil to meet the needs of crops. Therefore, the greatest value of organic by-products for application to crop land will usually be for their N-P-K content.

Unlike commercial fertilizers that can be blended to obtain varying ratios of N-P_2O_5-K_2O to match fertilizer recommendations and the needs of crops, the ratio of these nutrients in organic by-products will generally be fixed. Crops do not remove nutrients in the same ratio as the nutrients in organic by-products, particularly for N:P. Table 2 shows the relative N-P-K content and N:P ratio of crops and organic by-products. These data show that the average N:P ratio for crops is about 6.8:1, while the average N:P ratio for organic by-products is about 3.8:1.

Table 2. Approximate contents of some nutrients in selected crops (CAST, 1995)[a] and selected organic by-products (Parr and Hornick, 1993).

Crop or By-Product	Yield	N	P	K	N:P
Grains					
Corn, grain	9,400	150	26	36	5.8
Corn, stover	10,000	110	18	130	6.1
Wheat, grain	2,700	55	12	14	4.6
Wheat, stover	3,400	20	2	30	10
Rice, rough	4,000	55	10	9	5.5
Rice, straw	5,600	33	5	60	6.6
Sorghum, grain	3,800	55	12	17	4.6
Sorghum, stover	6,700	73	10	86	7.3
Hay					
Alfalfa	9,000	200	20	162	10
Coastal Bermuda	17,900	340	34	243	10
Red clover	5,600	110	12	90	9.2
Timothy	5,600	67	12	86	5.6
Fruits and vegetables					
Dry bean	2,000	84	12	5	7.0
Cabbage	45,000	146	34	117	4.3
Potato (tuber)	27,000	90	15	135	6.0
Tomato (fruit)	45,000	134	20	144	6.7
Others					
Cotton (seed)	1,680	45	10	14	4.5
Cotton (stalks, leaves, burs)	2,240	39	5	32	7.8
Soybean	2,700	169	17	50	9.9
Sugar cane	67,000	108	26	243	4.2
		- - - - - per cent - - - - -			
Organic By-Product					
Cattle manure		4.4	1.1	2.4	4.0
Crop residues		1.1	0.2	2.0	5.5
Sewage sludge		4.0	2.0	0.4	2.0
Municipal solid waste		0.7	0.2	0.3	3.5

[a] Original source of crop nutrient contents from Eaken, 1972.

Therefore, when manure and biosolids (the most common organic by-products applied to crop land) are spread at rates based on meeting the N needs of crops, excess P will be added. Since the N fertilizer recommendation will usually be greater than the P_2O_5 fertilizer recommendation, applying organic by-products on the basis of N has the advantage of allowing higher rates per hectare that require smaller land bases for a given quantity of by-product. The disadvantage of using N application rates is the potential accumulation of P in soils and an associated increase in bioavailable P that may also increase the risk of nonpoint source losses of P.

MANAGEMENT OF ORGANIC BY-PRODUCT NUTRIENTS

The application of organic by-products to crop land as a source of nutrients requires good nutrient management to avoid excess additions that can cause nutrient imbalances, or increase the risk of polluting water resources. Excess NO_3-N can potentially contaminate groundwater. Therefore, the plant available N (PAN) present in organic by-products must be estimated, so as to avoid excess N loadings to the soil-plant system. This will normally involve the use of worksheets or the help of computer programs to estimate the amount of organic N that will be mineralized and the amount of ammonia loss during and following application. The resulting PAN/metric ton of organic by-product can then be used to determine the appropriate rate that will satisfy the fertilizer N recommendation and minimize the potential for NO_3-N leaching.

Over the long-term, rates of application based on N will need to be reduced to avoid additions of excess P that will accumulate in surface soils. As soil test levels of P become high, the risk of nonpoint source losses of P to surface waters by runoff and/or erosion increases (Sharpley et al., 1993). Over-application of manures in many states has led to the restriction of further P additions, when soil fertility test levels for P reach critical values (Sharpley et al., 1998). Therefore, to accomplish good nutrient management, organic by-products should be applied at agronomic rates that will provide adequate nutrients for crop growth, without causing pollution to surface and groundwater resources.

To determine the agronomic rate for an application site, the fertilizer recommendations are needed for the crop that is to be grown. These recommendations will usually be based on the soil fertility test results, taking into account the expected crop yield, and will identify what additional quantities of nutrients are needed beyond what the soil can provide. The concentration of nutrients in an organic by-product must next be determined by sampling and analysis. Most often the content of the primary nutrients (i.e., N, P_2O_5 and K_2O) will be satisfactory. When any secondary and micronutrients are being recommended in addition to N-P-K, amounts of these in the compost can also be measured and credited against the fertilizer recommendations.

Once the nutrient content of the organic by-product is known, the rate of application can be determined, based on the quantity of recommended nutrients. Since the by-product will not contain the exact combination of N-P-K needed for each crop-soil combination, the by-product nutrients will need to be supplemented by commercial fertilizers or another by-product. A number of management tools are available to assist with these calculations, including worksheets and computer programs. The amount of organic by-product PAN applied should not exceed the recommended N fertilizer rate, and P inputs from by-product and fertilizers should be managed to avoid high soil test levels and the potential for nonpoint source losses of P.

To complete the management of organic by-product nutrient additions, suitable applicators must then be used to apply the calculated rate in the field in a uniform manner. Recordkeeping is another recommended management practice to help accomplish the nutrient management plan.

In summary, agriculture can assist with waste management in our society and help U.S. citizens manage their by-product residuals in an environmentally responsible manner. To accomplish this, from a technical point-of-view, involves primarily good nutrient and odor management. However, the non-technical aspects of applying by-products to land, like education and public acceptability, must not be forgotten, so generators, potential users and the general public can gain a better understanding of land application as a waste management alternative.

REFERENCES

CAST (1995) Waste management and utilization in food production and processing. Task Force Report No. 124 (Boersma LL & Murarka IP -- Cochairs), Council for Agricultural Science and Technology, Ames, IA. 125 p.

Eaken JH (1972) Food and fertilizer. Chap.1 In: The Fertilizer Handbook, The Fertilizer Institute, Washington, DC.

Parr JF & Hornick SB (1993) Utilization of municipal wastes. In: Metting FB Jr. (Ed) Soil Microbial Ecology (pp. 545-559). Marcel Dekker, Inc., New York.

Sharpley AN, Daniel TC & Edwards DR (1993) Phosphorus movement in the landscape. J. Production Agriculture 6:492-500.

Sharpley A, Meisinger JJ, Breeuwsma A, Sims JT, Daniel TC & Schepers JS (1998) Impacts of animal manure management on ground and surface water quality. In: JL Hatfield JL & Stewart BA (Eds) Animal Waste Utilization: Effective Use of Manure as a Soil Resource (pp. 173-242). Ann Arbor Press, Chelsea, MI.

Slivka DC, McClure TA, Buhr AR & Albrecht R (1992) Compost: United States supply and demand potential. Biomass Bioenergy 3:281-299.

USDA (1978) Improving soils with organic wastes. Report to the Congress in response to Section 1461 of the Food and Agriculture Act of 1977 (PL 95-113). U.S. Government Printing Office, Washington, DC.

USEPA (1993) Standards for the use and disposal of sewage sludge. Federal Register 58:9248-9415.

MANGANESE DEFICIENCY INDUCED BY LIME RICH CO-UTILIZATION PRODUCTS

SALLY BROWN and RUFUS CHANEY
Environmental Chemistry Laboratory, Agricultural Research Service, US Department of Agriculture, Beltsville, Maryland 20705, USA
E-mail sbrown@asrr.arsusda.gov

ABSTRACT

Lime addition to biosolids during the wastewater treatment process controls odors, aids in dewatering and reduces pathogens. Pathogen reduction as a result of lime addition can be sufficient to meet PSRP (Process to Significantly Reduce Pathogens) standards set by EPA 503 regulations (USEPA, 1993). When used properly, lime stabilized biosolids will increase soil pH as well as add nitrogen and other nutrients to soil for agronomic crops. However, under certain conditions (light-textured Coastal Plain soils which are prone to Mn deficiency) application of lime-treated biosolids has resulted in Mn deficiency in susceptible crops. Manganese deficiency is the result of several soil, plant and environmental factors and will not necessarily reoccur in the same field from year to year. However, deficiencies following biosolids applications have occurred on long-term research plots as well as on farmer's fields. As a result of these deficiencies, the Maryland Department of the Environment (MDE) considered limiting the use of limed biosolids on susceptible soils to meet the lime requirement of the crop.

Research in our lab has identified many ways to alleviate these deficiencies. Field experiments have shown that the addition of 200 kg Mn ha^{-1} as $MnSO_4$ was sufficient to alleviate deficiency in wheat (*Triticum aestivum*) on long-term field plots. Foliar Mn spray was also able to correct deficiency. Addition of Mn to biosolids prior to land application was able to increase Mn concentration in diagnostic leaves of wheat on short-term field plots. Application of sulfur to farmer's fields was able to lower soil pH and eliminate deficiency.

Variation in the Mn content of biosolids, in the lime content of the biosolids, as well as the difficulties in predicting when Mn deficiencies will occur, make addition of Mn to biosolids prior to land application an impractical alternative. All other identified solutions appear to be feasible. However, biosolids applicators are meeting with resistance to these solutions from both extension and regulatory personnel. This illustrates the potential for problems when all of the potential agronomic implications associated with the use of a co-utilization product are not taken into consideration before a beneficial reuse program is instituted.

S. Brown, J.S. Angle and L. Jacobs (eds.), Beneficial Co-Utilization of Agricultural, Municipal and Industrial By-products, 289–298.
© 1998 *Kluwer Academic Publishers,*

INTRODUCTION

Land application of municipal biosolids (sewage sludge) can be a beneficial agronomic practice that provides nutrients, organic matter, and limestone to soils and crops. In addition, it reduces the costs of crop production and reduces the costs of biosolids management. Before being land-applied, biosolids must be stabilized to kill pathogenic organisms, reduce malodors, and reduce vector attractiveness (40 CFR 503; COMAR). Stabilization processes are used in both Class B (PSRP: Process to Significantly Reduce Pathogens) and Class A (PFRP: Process to Further Reduce Pathogens) biosolids.

Hydrated lime [$Ca(OH)_2$] and/or CaO are often added to biosolids at many Public Owned Treatment Works (POTW). Low levels may be added to reduce odor generation during storage, hauling, and application, while high levels may be added to achieve Class B Treatment so biosolids can be directly land-applied (Lime-Treated Biosolids). When combined with heating or with drying, lime treatment can also produce Class A biosolids. Class A treated biosolids are pasteurized and comprise no significant pathogenic risk to highly exposed individuals or the environment. Public acceptance of PFRP-treated biosolids is greater because of the inherently greater safety of these materials. Lime-treatment or lime stabilization of biosolids also offers an additional incentive for the farmer considering land application. Limed biosolids can replace limestone normally added to soils to maintain agronomically desirable soil pH levels.

Application of high rates of lime-stabilized biosolids has also been found to alleviate sub-soil acidity on long-term research plots (Brown et al., 1997). Soil pH was raised to over 7.00 to a depth of 1 m with the application of 224 Mg ha^{-1} of lime stabilized biosolids to a Galestown sand. Similar results have been observed in other studies (Tan et al., 1985; Tester, 1989). Although there is no quantitative evidence, fulvic acid in the biosolids may have formed water soluble complexes with Ca that travel through the soil profile. As these complexes break down, the decomposition of the organic acid in conjunction with the released Ca ion are able to alleviate the acidity in the lower soil horizons. This work suggests that the use of lime stabilized biosolids may have benefits that have only been partially understood.

Only one limitation of limed biosolids has been identified in field research: land-application of limed biosolids may induce Mn deficiency in sensitive crops when applied to soils with inherently low Mn levels. When lime-stabilized biosolids are applied to meet the N requirements of a crop, the added lime may be in excess of the lime requirements of the soil. Overliming can result in Mn deficiency in light-textured, acidic, leached soils when limestone is applied such that soil pH reaches 6.5 or higher. Mn deficiencies have often been reported on over-limed Atlantic Coastal Plain soils that are naturally low in available Mn (Anderson and Mortvedt, 1982; Martens, personal communication). Most of the previously reported biosolids induced Mn deficiency has been associated with very high biosolids application rates (see review in Chaney and Ryan, 1993). Although occasional reports of biosolids-induced Mn deficiency were noted in the 1980s with biosolid application at agronomic rates, no detailed studies were made to scientifically characterize these reports. In general, Mn deficiencies do not result immediately after application of lime stabilized biosolids to low Mn soils. The breakdown of biosolids-added organic matter solubilizes sufficient Mn to avert deficiencies. A particular combination of a number of factors needs to occur for deficiencies to be observed.

Plant and Soil Factors

A number of plant and soil factors contribute to the occurrence of Mn deficiency. Some information on these variables is necessary to understand why deficiencies are not always possible to predict. The most important function of Mn in all plants is its role as a cofactor in the splitting of H_2O molecules and generation of O_2 in the light stage of photosynthesis. Manganese deficiency has been associated with reduced photosynthesis and chlorophyll in wheat (Ohki, 1985). Manganese deficiency in wheat has also been associated with dry weight reduction, reduced tillering, and delayed grain development and ripening (Gallagher and Walsh, 1943; Ohki, 1976).

Certain plants are more susceptible to Mn deficiency than others (Moraghan and Mascagni, 1991). A plant's efficiency at taking up Mn from the rhizosphere may be a factor in its ability to tolerate low Mn soils. Rapid removal of Mn from the rhizosphere may increase the rate of reduction and diffusion of soil Mn to the Mn depleted zone (Graham, 1988). Another consideration in the Mn efficiency of a genotype may be the extent of root/soil contact. Barley, which is significantly more tolerant of low Mn status soils than wheat, has the same rate of Mn uptake per unit root length as wheat (Marcar and Graham, 1987). However, barley shows much greater lateral root development and thus greater root surface area.

Throughout the growing season, a plant may experience fluctuations in the fraction of soil Mn that is phytoavailable. A plant may translocate excess Mn during periods of high availability. For example, concentrations in xylem sap vary from 1-3500 mmol (Kochian, 1991; Welch, 1995). This variation can be explained by a number of factors including Mn status of the soil and plant, plant species, transpiration rate, and the time of day. Translocation of excess Mn does not provide an effective reserve since excess Mn in older leaves is not mobile and there is limited ability to translocate Mn from roots and stems in cases of deficiency (Welch, 1995; Kochian, 1991; Lonegran, 1988b).

Soybeans and wheat are two crops that are commercially important in Maryland and are also prone to Mn deficiency. Critical tissue concentrations of Mn have been identified for both wheat and soybean. Mascagni and Cox (1985) determined that Mn in the youngest fully developed leaf of soybean at the R2 growth stage provided the most consistent indicator of deficiency. A critical level of 22 mg kg^{-1} Mn was consistent for the different cultivars tested. They noted that this level did not vary substantially with plant age as long as the same diagnostic leaf (youngest open trifoliate) was sampled. The critical concentration of Mn in the YEB or YML (youngest mature leaf) in wheat has been identified as being between 11 and 13 mg kg^{-1} (Ohki, 1984, Hannam and Ohki,1988). This concentration seems to be consistent between cultivars. Testing plant tissue is currently the only way to confirm Mn deficiency.

Manganese in the Soil

Within the soil system, Mn can exist in several oxidation states (+2, +3, and +4). These states govern Mn solubility and phytoavailability. In general, Mn^{2+} is the most stable species in the solution phase with Mn^{3+} and Mn^{4+} existing primarily as Mn oxide precipitates. As soil pH increases, the relative abundance of Mn^{2+} decreases and soil Mn exists primarily in a solid phase.

Reduction and oxidation reactions have also frequently been used to explain the transitions of Mn from one oxidation state to another. In well aerated, coarse textured soils such as those prevalent in the Atlantic Coastal Plain region, the relevance of extended periods of reducing conditions in determining the phytoavailability of soil Mn is questionable except for the spring and winter. Two studies, to develop models of O_2 diffusion, noted the importance of water in the intercrumb pore space, anaerobic regions within soil aggregates, and the areas of rapid organic matter decomposition in defining limits to gas diffusion and describing anaerobic microsites (Renault and Stengal, 1994, Renault and Sierra, 1994). These studies also indicate that the time necessary to achieve anaerobic equilibrium for the general soil profile is greater than the typical time required for water to drain through the soil profile in many Coastal plain soils (Renault and Sierra, 1994). Manganese deficiencies will only occur when a susceptible species is experiencing a normal demand for the micronutrient and sufficient microsites do not exist in the soil to supply this demand. Particular combinations of soil moisture, organic matter status, and plant growth rate are necessary for deficiencies to actually occur. This is the case even when plants prone to Mn deficiency are grown on low Mn soils at soil pH above 6.5.

Manganese deficiencies generally do not occur immediately following land application of biosolids, even though soil pH rises rapidly. The decomposition of the biosolids added organic matter potentially provides sufficient microsites of high Mn availability to limit the occurrences of Mn deficiency. Generally, only during the second or third year following biosolids application does the rate of OM decomposition slow sufficiently, with soil pH remaining sufficiently elevated to permit Mn deficiency to occur. In many cases of biosolids applied at agronomic rates, soil pH will fall as the organic matter decomposes and Mn deficiencies will not occur. Deficiencies have also been observed in long-term field plots at the University of Maryland Beltsville facility, that received very high rates of biosolids application. These were initially noted 15 years after initial application.

Correction of Biosolids Induced Mn Deficiency

Foliar spray of a water soluble Mn compound, applied with or without a surfactant has generally been the most successful means to alleviate deficiency (Gettier et al., 1985; Heckman et. al, 1993; Lewis and McFarlane, 1986; Reuter et al., 1988; Mascagni and Cox, 1985; Mascagni and Cox, 1984). Although foliar spray is effective and relatively inexpensive (0.5 kg Mn ha^{-1}), it may not be an appropriate solution for deficiencies that result from biosolids application. Repeated foliar application may be necessary over a single growing season. Deficiencies may occur for several seasons following biosolids application. A biosolids applicator may require a 'one shot' solution to maintain good relations with both regulators and farmers.

An ideal remedy for biosolids induced deficiency would involve application of a single amendment that would be effective over the course of the elevated soil pH levels resulting from biosolids application. An alternative would be to amend biosolids with excess Mn to avert deficiencies altogether. Manganese enriched biosolids could be seen as a potential alternative fertilizer source for this micronutrient. In fact the value of biosolids as a micronutrient source has been calculated (Forste, 1993).

Work in our lab has been done to address both of these possibilities. Soil applied Mn is generally not as consistently effective as foliar applied Mn. However, it provides the option for a one shot remedy for biosolids induced deficiencies. Soil applied Mn becomes phytoavailable as it dissolves into soil solution. The rate at which soil applied Mn dissolves and reprecipitates affects its phytoavailability: availability decreases as it gradually reprecipitates. Manganese sulfate is the most common form of soil applied Mn. Other forms of Mn (MnO and MnO_2) have also been tested, however, they are generally less reactive than $MnSO_4$. Surface area, particle size, and reactivity of the oxide will strongly influence the solubility of any Mn fertilizer (Jones and Leeper, 1951; Mascagni and Cox, 1984). A wide range of soil applied Mn (8-112 kg ha^{-1}) has been reported to effectively eliminate deficiency (Reuter et al., 1988; Gettier et al., 1984). Effectiveness and persistence will partially depend on the variety of soil and plant factors that have been discussed.

Research

An initial study was conducted on a long-term biosolids research experiment in Beltsville, MD, on plots established using lime stabilized biosolids in 1976 on a Galestown sand (sandy, siliceous, mesic Psammentic Hapludult) (Brown et al., 1997). The plots began to exhibit Mn deficiencies in 1990. The year before the current study was set up, deficiencies were severe enough to preclude any growth or yield. An randomized complete block (RCB) experiment was set up in the original plot areas to test the effectiveness of high rates of soil applied Mn to alleviate this deficiency. Soil Mn amendments tested included MnO at 50, 100, and 200 kg ha^{-1}, and $MnSO_4$ at 200 kg ha^{-1}. Foliar Mn spray ($MnSO_4$ at 0.5 kg ha^{-1} Mn) was included as a subplot treatment. MnO was used in addition to $MnSO_4$ as it was considered to be a potentially cost effective alternative to $MnSO_4$.

Two susceptible crops, soybeans and wheat were double cropped for two seasons (1993-1995). Manganese deficiencies were not observed for soybeans through the course of the study. Wheat exhibited severe deficiency for both seasons. Deficiency was alleviated by both the foliar Mn spray and the soil applied $MnSO_4$. Response to amendments was clear from all of the indices used in the study (Table 1). Visual observations in the years following the study indicate that the soil applied $MnSO_4$ remains phytoavailable. Application of $MnSO_4$ at this rate would cost approximately $400/A. This is a case of severe deficiency, caused by application of 224 Mg ha^{-1} of biosolids. Standard agronomic loading rates are a factor of 20 less than the rate used in the study. It is difficult to predict if deficiency would have been observed if application had been made over a 20 year interval. It is possible that lower rates of $MnSO_4$ would be able to alleviate deficiencies induced by biosolids applied at agronomic rates. Although the study effectively eliminated deficiency, further research is necessary to determine the rate of

Table 1. Response of wheat to soil applied Mn fertilizer on long-term biosolid amended field plots at the University of Maryland Research Facility in Beltsville, MD. Results indicate that a variety of plant parameters are acceptable tools for measuring effectiveness of Mn fertilizers.

Soil Treatment	Plant Chlorophyll	Leaf Manganese	Yield
	----mg kg^{-1}---	---kg ha^{-1}---	
Control	25.9 a[a]	5.97 a	738 a
MnO			
50 kg ha^{-1}	31.9 a	9.48 a	1435 b
100 kg ha^{-1}	26.6 a	9.33 a	1117 ab
200 kg ha^{-1}	31.3 a	9.94 a	1560 b
$MnSO_4$			
200 kg ha^{-1}	39.0 b	19.37 b	2520 c

[a] Means followed by the same letter are not significantly different using the Waller Duncan procedure with $P< 0.05$.

$MnSO_4$ that is necessary to eliminate deficiency caused by agronomic application of biosolids. If a lower application rate is effective, this may be a cost effective alternative to foliar Mn Spray.

In addition to high rates of $MnSO_4$ application to soil, sulphur addition has been used to reduce soil pH and thereby alleviate Mn deficiency. On a farm that had severe yield reduction following application of biosolids, pH titration curves were made using field soils (Brown et al., 1997b). Sulphur applications were made to the fields on the basis of these curves and soil pH was sufficiently reduced to eliminate deficiency. However, the soil testing Lab at the University of Maryland does not consider sulphur addition to be a standard practice and will not make recommendations on sulphur requirements.

An additional study was conducted to determine if adding Mn to biosolids prior to land application could increase plant available Mn and alleviate Mn deficiency in wheat (*Triticum aestivum* L.) and soybean (*Glycine max* L.). Two experiments were established on a Galestown sand. The first study tested three sources of Mn ($MnSO_4$, MnO 'Granusol', and MnO_2 'Amanore') added to biosolids at 10,000 mg Mn kg^{-1} with biosolids applied to soil at 10 Mg ha^{-1}. The addition of the Mn to the biosolids prior to land application could potentially increase the solubility of the materials. Wheat and soybeans were double cropped for two growing seasons. Although visible Mn deficiency was not observed during the course of the study, the $MnSO_4$ treatment supplied significantly higher Mn to wheat and soybean than the other Mn products (Table 2). In the second study, $MnSO_4$ was added to biosolid at 0, 1,000, 2,000, 5,000 and 10,000 mg Mn kg^{-1}. Biosolids were applied at 10 and 50 Mg ha^{-1}. Although Mn deficiency was not observed, the addition of 5,000 mg Mn kg^{-1} appeared to be effective at increasing plant Mn concentrations (Table 3). Confirmation of this observation in a case of actual Mn deficiency would be necessary before the potential for this technology can be confirmed.

The absence of a Mn deficiency during the course of these studies may be the result of variation in the biosolids used. Variation in biosolid Mn concentrations may have been responsible for the observed discrepancy between desired levels of Mn addition and observed Mn concentration in biosolids (Table 4). This variation may also have some

Table 2. Wheat and soybean Mn concentrations (mg kg^{-1}) for leaf and grain samples for plants grown on limed-biosolid amended soils that had been enriched with different Mn salts. Biosolids had been applied at 10 Mg ha^{-1}, Mn had been added to biosolids at 10,000 mg kg^{-1}.

Plant	Control	Biosolid Control	$MnSO_4$	MnO	MnO_2
Wheat (1994)					
YEB[a]	53.3 a†	62.5 a	68.5 a	53.8 a	53.9 a
Grain	49.3 a	44.6 bc	47.7 ab	37.3 d	41.5 bc
Soybean (1995)					
YEL	80.5 a	63.3 bc	71.6 ab	60.3 c	59.1 c
Grain	30.9 a	28.4 ab	29.5 ab	27.4 b	27.7 ab
Wheat (1995)					
YEB	50.2 a	36.9 b	46.9 a	36.9 b	31.1 b
Grain	34.3 a	31.1 a	33.3 a	32.0 a	31.2 a
Soybean (1995)					
YEL	155.7 a	126.5 a	131.7 a	118.7 a	133.1 a

[a] Means followed by the same letter are not significantly different using the Duncan Waller procedure with P< 0.05.

Table 3. Wheat and soybean leaf and grain Mn (mg kg^{-1}) from plants grown on limed-biosolid amended plots on a Galestown Sand (sandy, siliceous, mesic psammentic Hapleudult) that had received different levels of $MnSO_4$ addition. Samples were collected during the 1994-5 season.

Soil Treatment	Wheat		Soybean	
	Yeb	Grain	Yel	Grain
Control	17.26 c[a]	29.39 ab	71.76 a	29.07 a
Limed Control	17.30 c	26.03 bcd	44.93 cd	23.55 b
Biosolids @ 10 Mg ha				
Control	16.84 c	22.83 d	47.53 bcd	24.77 b
1,000 mg kg^{-1} Mn	16.76 c	24.82 cd	50.40 bc	25.75 b
2,000 mg kg^{-1} Mn	19.96 c	26.01 bcd	47.85 bcd	24.55 b
5,000 mg kg^{-1} Mn	23.77 bc	26.94 abcd	54.03 b	25.67 b
10,000 mg kg^{-1} Mn	30.20 b	31.08 a	70.28 a	26.42 ab
Biosolids @ 50 Mg ha^{-1}				
Control	21.60 bc	23.53 d	35.20 e	24.70 b
1,000 mg kg^{-1} Mn	29.18 b	26.36 bcd	34.19 e	23.41 b
2,000 mg kg^{-1} Mn	28.79 b	29.17 abc	42.15 d	23.69 b
5,000 mg kg^{-1} Mn	42.71 a	31.21 a	47.86 bcd	26.48 ab

[a] Means followed by the same letter are not significantly different using the Duncan Waller procedure with P< 0.05.

Table 4. Desired and actual Mn concentrations (mg kg^{-1}) for Mn enriched limed-biosolids that were applied to a Galestown sand to test the effectiveness of adding Mn to biosolid prior to application to eliminate the potential for limed-biosolid induced Mn deficiency. Values for titratible alkalinity are also included (g kg^{-1} $CaCO_3$) to illustrate the range of lime addition during biosolid treatment.

Mn Source	Desired Treatment Concentration	Actual Mn Concentration	Titratible Alkalinity
Control		1146 a[a]	53.3
MnSO4	1000	1451 a	72.3
	2000	2955 b	54.6
	5000	4232 c	29.3
	10000	5274 c	80.4
Amanore	10000	6784 d	80.4
Granusol	10000	5780 cd	90.3

[a] Means followed by the same letter are not significantly different using the Duncan Waller procedure ($P < 0.05$).

relationship to the occurrence of Mn deficiency in the field. Variation in titratible alkalinity (Table 4) may also influence the occurrence of Mn deficiency. If the Mn concentration in biosolids and the rate of lime addition varies sufficiently, Mn addition may be an unnecessary expense. The failure of the 10 Mg ha^{-1} biosolid application rate to bring the soils to calcareous levels may have resulted from the use of a plot area that had not recently been limed (pH of the control plot was 5.6). It may also have resulted from varying levels of lime addition during biosolid treatment. Soil pH in the treatments that received this rate of biosolid application ranged from 5.97 to 6.63.

The addition of 5,000 mg kg^{-1} Mn to biosolids as $MnSO_4$ however, would increase production costs by approximately $10/Mg biosolids. The observed variation in both the biosolids Mn concentrations and the titratible alkalinity, coupled with the range of soil and plant factors involved in Mn deficiency, indicate that this may not be the most cost effective method to eliminate Mn deficiency. As a large portion of the biosolids generated may have either relatively high total Mn, low CCE, or a combination of the two, it is unlikely that Mn deficiency will result following biosolids application. Although a negative impact on acceptance of biosolids will result from every case of biosolids induced deficiency, from an economic standpoint, it may be more practical to correct deficiencies as they occur rather than trying to prevent them.

Regulatory Reactions

The reaction of regulatory agencies to biosolids induced Mn deficiencies has primarily been to limit biosolids applications in susceptible areas. For example, the State of Virginia established regulations on application of limed biosolids. Virginia prohibits application of limed biosolids to lower Coastal Plain soils which have a pH of 6.0 or higher. Although soil formation processes that affect the probability of lime-induced Mn deficiency were taken into account in the regulation, this regulation was not verified by field research (Little, 1986; Martens, personal communication).

In Maryland, the initial regulatory reaction was to recommend that application of limed biosolids be restricted by soil series. The Maryland Department of the Environment worked with the Natural Resource Conservation Service (NRCS) to determine what soil series would be most susceptible to Mn deficiency. As an interim measure, the MDE required that farmers receiving biosolids sign an agreement alerting them to the potential for Mn deficiency as a result of biosolids application. This agreement requires biosolid appliers to correct the Mn deficiency if it occurs for up to 5 years following biosolid application. The final recommendation of the NRCS indicate that Mn deficiencies may occur across most of Maryland's eastern shore, one of the primary areas for biosolids application in the state. The MDE, in cooperation with the USDA and the University of Maryland Nutrient Management program, determined that the Mn agreement was preferable to restricting application based on soil series. All parties involved are generally comfortable with the advisory.

Although Mn deficiencies have been reported on the Eastern shore, without biosolids application, their occurrence as a result of biosolids application has caused a decrease in the number of new farmers requesting application. The delay between the first reports of biosolids-induced deficiency and the issuance of the advisory are partially responsible for this growing mistrust. In addition, the reluctance of the University Nutrient Management Program and Soil Test Lab to work with applicators on methods to remedy deficiencies indicates that a proper level of cooperation between the two groups does not exist.

Although most parties involved now appear to be aware of the potential for the problem to occur and liabilities have been clearly defined, the failure to anticipate Mn deficiency and the slow response to occurrences of deficiency have had a detrimental effect on the public perception of biosolids within Maryland. Public perception is often the primary limitation to the establishment of a successful program. The damage that can be done by not preemptively addressing these problems is far greater than the actual cost of proper research into materials.

REFERENCES

Anderson OE & Mortvedt JJ (1982) Soybeans: Diagnosis and correction of manganese and molybdenum problems. Southern Coop. Serv. Bull. 281. Univ. Georgia.

Brown SL, Angle JS & Chaney RL (1997a) Subsurface liming and metal movement in soils amended with lime-stabilized biosolids. J. Environ. Qual. 26:724-732.

Brown SL, Angle JS & Chaney RL (1997b) Correction of limed-biosolid induced Mn deficiency on long-term field plots. J. Environ Qual. (In press).

Chaney RL & Ryan JA (1993) Heavy metals and toxic organic pollutants in MSW-composts: Research results on phytoavailability, bioavailability, etc. In: Hoitink HAJ et. al. (Ed) Proc. International Composting Research Symposium. (In press).

Forste JB (1994) Gaining public acceptance for biosolids. In: Clapp CE, Larson WE & Dowdy RH (1994) Sewage Sludge: Land Utilization and the Environment (pp. 81-85). SSSA, Inc. Madison, WI.

Gallagher PH & Walsh T (1943) The susceptibility of cereal varieties to manganese deficiency. J. Agr. Sci. 33:197-203.

Gettier SW, Martens DC, Hallock DL & Stewart MJ (1984) Residual Mn and associated soybean yield response from $MnSO_4$ application on a sandy loam soil. Plant and Soil 81:101-110.

Gettier SW, Martens DC & Brumback TB Jr. (1985) Timing of foliar manganese application for correction of manganese deficiency in soybean. Agron. J. 77:627-630.

Graham RD (1988) Genotypic differences in tolerance to manganese deficiency. In: Graham RD et al. (Eds)

Manganese in Soils and Plants (pp. 261-276). Kluwer Academic Publishers.
Hannam RJ & Ohki K (1988) Detection of manganese deficiency and toxicity in plants. In: Graham RD et al. (Eds) Manganese in Soils and Plants (pp. 243-259). Kluwer Academic Publishers.
Heckman JR, Ingerson-Mahar JM, Lee DL & Prostko EP (1993) Alfalfa responses to foliar and soil application of manganese. Commun. Soil Sci. Plant Anal. 24 (7&8):559-572.
Jones LHP & Leeper GW (1951) The availability of various manganese oxides to plants. Plant and Soil 3:141-153.
Kochian LV (1991) Mechanisms of micronutrient uptake and translocation in plants. In: Mortvedt JJ, Cox FR, Shuman LM, & Welch RM (Eds) Micronutrients in Agriculture (pp. 229-285). Soil Science Society of America, Madison, WI.
Lewis DC & McFarlane JD (1986) Effect of foliar applies manganese on the growth of safflower and the diagnosis of manganese deficiency by plant tissue and seed analysis. Aust. J. Agric. Res. 37:567-572.
Little DA Jr. (1986) The potential of lime-stabilized and chemically-fixed municipal sewage sludges for land application as lime, and nitrogen sources. M.S. Thesis. Virginia Polytechnic Institute and State University, Blacksburg, VA.
Loneragan JF (1988) Distribution and movement of manganese in plants. In: Graham RD (Eds) Manganese in Soils and Plants (pp. 113-124). Kluwer Academic Publishers.
Marcar NE & Graham RD (1987) Tolerance of wheat, barley, triticale, and rye to manganese deficiency during seedling growth. Aust. J. Agric. Res. 38:501-511.
Mascagni HJ & Cox FR (1985) Critical levels of manganese in soybean leaves at various growth stages. Agron J. 77, 373-375.
Mascagni HJ & Cox FR (1985) Effective rates of fertilization for correcting manganese deficiency in soybeans. Agron. J. 77:363-366.
Mascagni HJ & Cox FR (1985) Evaluation of inorganic and organic manganese fertilizer sources. Soil Sci. Soc. Am. J. 49:458-461.
Ohki K (1984) Manganese deficiency and toxicity effects on growth, development, and nutrient composition in wheat. Agron J. 76, 213-218.
Ohki K (1985) Manganese deficiency and toxicity effects on photosynthesis, chlorophyll, and transpiration in wheat. Crop Science. 25:187-191.
Renault P & Stengel P (1994) Modeling oxygen diffusion in aggregated soils: I. Anaerobiosis inside the aggregates. Soil Sci. Soc. Am. J. 58:1017-1023.
Renault P & Sierra J (1994) Modeling oxygen diffusion in aggregated soils: II. anaerobiosis in topsoil layers. Soil Sci. Soc. Am. J. 58:1023-1030.
Reuter DJ, Alston AM & McFarlane JD (1988) Occurence and correction of manganese deficiency in plants. In: Graham RD et al. (Eds) Manganese in Soils and Plants (pp. 205-244). Kluwer Academic Publishers.
Tan KH, Edwards JH & Bennett OL (1985) Effect of sewage sludge on mobilization of surface applied calcium in a greenville soil. Soil Sci. 139:262-268.
Tester CF (1990) Organic amendments effects on physical and chemical properties of a sandy soil. Soil Sci. Soc. Am. J. 54:827-831.
Welch RM (1995) Micronutrient nutrition of plants. Critical Reviews in Plant Sciences 14(1):49-82.
USEPA (1993) Standards for the use of disposal of sewage sludge. Fed. Regist. 58:9248-9415.

SUSTAINABLE SOIL, WATER AND AIR QUALITY: MANKIND'S ULTIMATE CHALLENGE AND OPPORTUNITY IN THE 21ST CENTURY

J. PATRICK NICHOLSON
N-Viro International Corporation, Toledo, Ohio 43603, USA

In the historic novel "A Tale of Two Cities", the author Charles Dickens tells us "It was the best of times and it was the worst of times." As time runs out on the 20th Century, no words better describe the 20th Century's historic impact on civilization.

In the 20th Century, we witnessed unparalleled advances in science and technology, in the quality of life, in education, in communications, in medicine, and indeed in the very seeds of democracy.

Yet, in the 20th Century we also witnessed more bloodshed of man by man, more terror and the development of the tools of terror, more destruction of family life and human discipline, more crime, more drugs of all kinds, and finally, in the end, more greed and avarice than ever before witnessed in this country, if not in the world. Most importantly, we witnessed the unparalleled destruction by man of man's very home and environment. We witnessed man's greed and power allowing man to pollute and harm the air we breathe and the soil and water so essential to our survival. We began in this century to recognize the insanity of our actions, but these calls to action have been blunted and delayed and deliberately confused by the power of special interests to maintain the status quo which is so profitable to so few and so destructive to so many.

In today's society can any world leader build that bridge to the 21st Century? Can this President or any Chief of State provide the independent political leadership and courage to do what is critically necessary to sustain this planet earth for our children, our children's children and their children? This indeed is the ultimate challenge and opportunity in the 21st Century.

" ... Much of the land we continue to farm is losing its inherent productivity because of unsound agricultural practices and overuse. ...More than 550 million hectares (one third of all farmland) are losing topsoil or undergoing other forms of degradation as a direct result of poor agricultural methods."

" ... the huge amounts of fertilizers and pesticides now routinely used in agriculture frequently drain off into the groundwater beneath the fields, contaminating them for many centuries to come." (Earth in the Balance: Ecology and the Human Spirit, Vice President Al Gore, 1992).

In 1993, the U.S. Department of Agriculture developed an excellent report titled "Agricultural Utilization of Municipal, Industrial and Animal Waste". In that report,

S. Brown, J.S. Angle and L. Jacobs (eds.), Beneficial Co-Utilization of Agricultural, Municipal and Industrial By-products, 299–304.
© 1998 *Kluwer Academic Publishers*,

USDA stated that "annual animal manure production exceeds 2.2 billion tons." This is 40-50 times more manure than human sludge or biosolids waste. Moreover, the report showed that BOD levels from such wastes were 10-100 times higher than from treated biosolids. In other words, manures are 500-5,000 times a bigger problem or opportunity than biosolids. However, in all reality, manure management is non-existent because non-point source water pollution regulation is non-existent. We have spent billions on point source pollution prevention. And yet we have done practically nothing on non-point source water pollution. Why not? What special interests are preventing sound and scientific environmental and agricultural policies and practices? Why are these issues being ignored? All we seek is the truth!

The U.S. Department of Agriculture report states:

> "Waste utilization problems present a challenge and an opportunity for U.S. agriculture. We are currently confronted with the long-term goal of developing crop production practices that promote sustainability. Animal wastes and many municipal and industrial wastes have substantial potential value for agricultural utilization. The development of methods to optimally integrate waste utilization into sustainable agricultural practices could provide a major part of the solution to urban and industrial waste disposal problems."

Two billion tons of manure are being generated annually. Three major concerns are obvious. They are public health, air emissions and water pollution. How can USDA, USEPA, U.S. Dept. of Health and the White House ignore these concerns?

1. How dangerous are farm-generated pathogens? The March 21, 1997 issue of Science tells us that indeed, as long suspected, the 1918 killer virus, that resulted in 675,000 deaths in the USA alone, "was a classical swine flu." Robert Webster, virologist at St. Jude Children's Research Hospital in Memphis, Tennessee states: "What this says is we had better watch what is happening in the pig population of the world."

On Sunday, March 23, 1997, Dateline and in Tuesday, March 25, 1997's Science Section of the New York Times we read and saw the wonderful work of Dr. JoAnn M. Burkholder of North Carolina State University. Dr. Burkholder has long been fighting a lonely battle against the "microscope killer", *Pfiesteria piscida*, and against a North Carolina power structure that has fought her heroic efforts in every possible way including threats to her life. The Time's story states "*Pfiesteria* is nourished by runoff from urban development and industries like hog farming." A CBS 60 Minutes December story was equally critical of pork power practices in North Carolina.

The excellent discussion draft for the president's conference on food safety discussed the well known problems of *Salmonella* and *E. coli*. Last year in Florida, the orange juice industry was severely impacted by these pathogens. Many of those orange groves were fertilized by untreated chicken manure. A significant percent of oranges are harvested from the ground.

Perhaps the greatest immediate fear is the parasite protozoan, *Cryptosporidium parvum*. A recent USDA survey showed that 100% of all states surveyed have a problem, 59% of all farms have a problem, and 22% of all calves have a problem. In 1993, *Cryptosporidium* infection impacted over 50% of the human population of Milwaukee.

How can animal manure disposal or utilization be ignored? Is the next case of "mad cow disease" about to take place in our USA? The Wall Street Journal recently reported a possible case in Indiana. With 2 billion tons of untreated manure disposed annually near our water supplies, how are we, our children and our children's children protected without national reliance on demonstrated science and technology?

2. Manure disposal sites are a huge source of unrestricted emissions of hydrocarbons, methane, nitrous oxides and CO_2. Why are these point source emissions ignored?

3. Unenforced CAFO regulations only provide standards for NPDES water discharge permits. No federal laws or regulations now exist to provide manure management standards. Two billion tons of manure (50 times the volume of municipal sludge) containing organic and nutrient levels 10-100 times per ton greater than municipal sludge, are generated annually. Today there are no federal laws, no regulations, no standards, and precious little guidance or technology transfer relative to two billion tons of manure! Why?

The challenge and the opportunity have been well defined by other international authorities besides the U.S. Department of Agriculture.

"Manure supplies nitrogen, phosphorus, and other nutrients for crop growth; adds organic matter and improves soil structure and tilth; and increases the soil's ability to hold water and nutrients and to resist compaction and crusting. Disposal of manure as a waste often leads to both surface water and groundwater degradation. Improved manure management can effectively capture the benefits of manure as an input to crop production and can reduce the environmental problems associated with manure disposal." Soil and Water Quality: An Agenda for Agriculture, National Research Council, 1993.

To summarize societies most pressing concerns:

1. Our soils worldwide are losing their sustainability due to many factors, including an over dependence on chemical fertilizers and pesticides, soil erosion, mismanagement, and diminishing organic and mineral content.

2. A great opportunity exists through proven established technology to utilize the huge quantities of organic and mineral wastes generated annually to compliment, not supplement, chemical fertilizers and pesticides, and to ensure worldwide sustainable soil fertility. However, the current uncontrolled use of such waste materials, creates immense water quality, sociological, and public health concerns and problems. Land application regulations of biosolids and manures must require safe and inaccessible storage, pathogen reduction until time of use, responsible odor controls, and management practices and technologies that control leaching to ground water and runoff to surface waters. Without enforcement there is no compliance. Without compliance, public health, social responsibility, and environmental protection are all seriously endangered with current land application practices.

In 1988 USEPA published a report titled "Solid Waste Disposal in the USA." This report identified over 3 billion dry tons of organic wastes generated annually in the USA. Pulp and paper products, food and kindred products and animal manures constituted over 90% of this total. A vast majority of these organic wastes are disposed in landfills, lagoons, and surface impoundments. Regulations are either minimal or non-existent! Tragically, these organic wastes, as currently disposed, create immense quantities of CO_2, methane, and nitrous oxide emissions which are the primary causes of global warming and ozone depletion, as well as other social problems as so well documented recently by CBS (60 Minutes - "Pork Power"). Moreover, these disposal facilities generate large amounts of soluble solutions with high BOD levels that overflow and leach into America's groundwater, waterways and wetlands, greatly increasing non-point source water pollution.

Today, over 200,000,000 tons of boiler ash, fluidized bed ash, resource to energy ash, scrubber ash, lime kiln dust, cement kiln dust, and wood ash are generated and wasted annually. This is tragic! These materials have tremendous soil fertility value. They offer the very best way to pasteurize, stabilize, and immobilize organics so that organics can be effectively utilized to reduce chemical dependency, increase soil fertility and greatly improve soil and water quality. The importance of minerals in soils is well understood. The N-Viro process utilizes these by-products for pH elevation, exothermic heat, and drying. The N-Viro process is patented and N-Viro will defend its intellectual property.

The challenge is clear. Potential opportunities include the following.

1. A classic example of what can be accomplished involves the sugar industry in the Pacific and Taiwan Sugar company. Last year while visiting Hawaii, I learned that Taiwan Sugar was dominating sugar sales in the Pacific to such extent that most sugar plantations in Hawaii were shutting down. While some blamed low labor rates in Taiwan, I questioned this position as Taiwan has a reasonably good standard of living. I visited Taiwan to seek the truth. Taiwan Sugar maintains a herd of over 500,000 hogs. They compost the hog manure and use the compost almost exclusively as their source of both fertilizer and pesticide, thus greatly reducing their chemical input costs. Moreover, the stabilization of composting allowed Taiwan Sugar to use the manure without damaging either their precious surface or ground water resources. This company is on the ball. Believe me, our challenge for economic leadership in the 21st Century is coming from the Pacific. Science and technology are our best ways of staying competitive, but we too had better be on the ball. Science and technology, not political clout, must determine America's future agricultural, environmental, and economic policies and programs, or we can simply hand over world leadership to the Pacific Rim.

2. In order for organic and mineral wastes to be utilized so as to not pollute either the water, the land, or the air, two requirements are absolutely necessary.

   a. Sound soil nutrient management practices, including seasonal application, no till, zoning, crop rotation, etc., must be developed and implemented.

b. Technologies, such as compost and N-Viro Soil^{tm}, that immobilize and stabilize organics and nutrients so that they provide "slow release" soil fertility through controlled mineralization, must be recognized and implemented. In seeking solutions it is important to remember the words of Pope John Paul II: "We are involved in a quest along with our fellow men ... let us avoid moralizing or suggesting that we have a monopoly on the truth." Indeed, we welcome and encourage the development of alternative concepts or technologies that ensure results comparable to compost or N-Viro Soil^{tm} and their ability to increase soil fertility while concurrently improving soil and water quality.

3. Technology transfer must be an essential component of that magnificent visionary bridge into the 21st Century. That bridge must be wide enough to allow the environmental, agricultural and public communities to work together to do what is right. Political courage and leadership, together with scientific truth, are the absolutely vital components of that bridge structure.

The House Agricultural Appropriations Sub-Committee, under the leadership of Chairman Joe Skeen and Minority Leader Marcy Kaptur, has appropriated $900,000.00 to the U.S. Department of Agriculture to provide demonstration funds to the Rodale Institute, the Compost Council, and N-Viro International Corporation. The demonstration project is showing the ability of compost and N-Viro Soil^{tm} technologies to utilize manure and biosolids in such a manner as to provide both sustainable soil fertility and reduced non-point source water pollution. The two processes, i.e. compost and N-Viro Soil^{tm}, stabilize and immobilize nutrients and organics so that they are only available through the "slow release" mechanism of mineralization. Moreover, they help reduce the leaching of chemical fertilizers, thus increasing their efficiency and they reduce the need for chemical pesticides. Compost is an established biological technology. N-Viro Soil^{tm} is a new process with an established reference base and significant public recognition, that combines biological, chemical, and physical processes to pasteurize organic wastes, and convert to a stable, storable product that is capable of providing "slow release" soil fertility through immobilization and mineralization.

N-Viro International Corporation's patented N-Viro Soil^{tm} and L-B Soil processes combine organic by-products with mineral by-products. The resultant products, which result from chemical, physical, and biological processes, are disinfected (N-Viro Soil achieves true pasteurization including total destruction of *Cryptosporidium* parasites), stabilized (immediate and long term odor suppression), and immobilized (nutrients, organics and metals remain insoluble and air emissions are greatly reduced).

Both N-Viro Soil and L-B Soil provide significant mineral benefits to the soil. Calcium and other critically need soil minerals such as salt, copper, selenium, magnesium and boron are readily available in N-Viro products. It is well recognized that agricultural soil mineral deficiencies are a major factor in human dietary problems and sustainable soil fertility concerns. N-Viro provides an aggressive answer to these concerns.

A sustainable national program of technology transfer is absolutely vital if such technologies are going to be understood, accepted, and utilized.

The issue of air pollution caused by animal manures is clearly an EPA responsibility. However, beyond air pollution, we urge the Administration to encourage USDA to accept responsibility and authority for improved manure management and safety. Incentives, guidelines and standards, not regulations, are needed. No one is more concerned about farm-generated diseases and water pollution than the American farmer. Moreover, we have been most impressed by the concern and leadership at the American Farm Bureau and the Pork Producer's Council on this issue. USDA has the people, the technology, the credibility, and the trust of the American farmer to solve this problem and to convert these organic resources into safe sustainable soil fertility products. All they need now is direction.

The problems raised in this talk of public health, huge unregulated air emissions and non-point source water pollution are no more critical than the need for long term sustainable soil fertility. The use of compost and bio-mineral technologies can create an unlimited worldwide opportunity to safely use animal and other organic by-products to provide fertility. The technology now exists to do the job and do it right. What is desperately needed is technology transfer and leadership from Washington.

PRODUCTION AND MARKETING OF POTTING AND LANDSCAPE SOILS CONTAINING COAL COMBUSTION BY-PRODUCT

FRANK FRANCIOSI[1], TED BILDERBACK[2], BILL LORD[2] and LEO BENNETT[3]
[1]*RT Soil Sciences, a Division of ReUse Technology, Inc., Rocky Mount, North Carolina 27802, USA and* [2]*NC Cooperative Extension Service, and* [3]*Cedar Creek Nursery*
E-mail FF22@aol.com

ABSTRACT

Gardening is the number one leisure time activity in the United States. During 1994, Americans spent over $25 billion on gardening and landscaping. In North Carolina alone, potting and landscaping soils are consumed at a rate of 2 million cubic yards each year according to a recent survey by North Carolina State University.

Coal combustion by-products (CCB's) can be used in the production of superb quality, compost based, synthetic soils. The process involves the composting of source separated organic materials that are by-products generated by agricultural and industrial operations. These include livestock operations, cotton gins, wood manufacturing plants, and land clearing debris. This natural biological process is optimized by using proven scientific techniques and equipment. Coal combustion by-products can be used in many ways to improve the characteristics of these compost based soils. Fly ash may be used as a replacement for sand, providing water holding capacity as well as improved drainage characteristics. Flue gas desulfurization residues can be used as a source of elemental calcium and sulfur which are required by certain plant types. These and other potential benefits from CCB usage will be discussed. This paper will present scientific data demonstrating the advantages of compost/CCB blend soils in comparison to conventional potting and landscape soils. This paper will also describe some of the challenges in gaining regulatory approvals and market acceptance and the strategies employed to over come these barriers.

INTRODUCTION

For centuries ash has been used as a soil amendment in agriculture and the home garden. Wood fly ash is used to increase soil K, Ca, and soil pH. More recently CCB's have been utilized in improving soil structure and water holding capacity. RT Soil Sciences has spent the last three years researching and developing value added uses for these by-products in

S. Brown, J.S. Angle and L. Jacobs (eds.), Beneficial Co-Utilization of Agricultural, Municipal and Industrial By-products, 305–314.
© 1998 *Kluwer Academic Publishers*.

the agricultural and horticultural marketplace. RT Soil Sciences uses select source separated organic by-products to produce ***ENCORE!*** ® brand horticultural products.

Composting is a natural biological process of decomposition. RT Soil Sciences optimizes this process by monitoring compost conditions and maximizing optimum conditions for active composting. Optimum conditions exist at 50-60 ^{0} C, 45-50% moisture, oxygen at 5% or greater, and carbon to nitrogen ratio of 2:1.

Composting is performed by windrow method. Windrows are formed by using a front-end loader and recipes are mixed by volume. Feedstocks are tested prior to determine base lines for each component of the recipe. RT Soil Sciences uses agricultural by-products along with wood manufacturing by-products as feedstock sources. Organic feedstocks are actively composted for 60 days and then allowed to cure another 30-60 days. RT Soil Sciences uses a water truck for the addition moisture and a compost turner to invert and homogenize the feedstock materials. Homogeneity assures destruction of harmful pathogens at high temperatures. As the compost matures and cures it is then screened using a trommel screen. Material is screen at one half inch for high grade materials and three-quarters of one inch for lesser grades.

CCB fly ash is added by using a front end loader to mix by volume, this is referred to as combing and occurs prior to shipping or packaging of final product. This final product is then marketed to soil brokers, landscape contractors, golf courses, ornamental plant growers and retail accounts.

The following study was conducted by Cedar Creek Nursery, North Carolina Cooperative Extension Service, and RT Soil Sciences to gain knowledge on how Compost/CCB based soils would perform as a standard ornamental growing media. The specific objective was to determine the affect of physical and chemical properties, of four container substrates on the growth of two nursery crops. It also served to support the registration of this product as a potting media in various states.

OUTLINE OF CEDAR CREEK NURSERY STUDY

1. The study was initiated May 22, 1995. Plants, containers, fertilizers and pine bark used in the study were provided by Cedar Creek Nursery. The compost was made up of approximately equal volumes of composted cotton gin stems and hulls: composted hardwood fibers: and fly ash was provided by RT Soil Sciences.

2. Two nursery species: (l) Ilex vomitoria sNana'. Dwarf Yaupon and (2) Ilex comuta 'Burfordi Nana'. Dwarf Burford Hollv were selected for the study. Plants of each species were potted into 3 gallon of four experimental substrates containers and placed in a random complete block design in 10 replicated blocks of each species. All cultural practices including, irrigation, fertilizer, and minor element supplement were those of Cedar Creek Nursery, except for special considerations given for test substrates.

3. Container test substrates were as follows:

   a) CEDAR CREEK Standard: Composed of 2 parts Pine Bark: 2 parts RT Soil Sciences Compost and 1 part Sand (by volume). The Cedar Creek Standard

Substrate was amended with 1.7 lbs (.77 kg.) of $CaSO_4$/ yd^3 and 0.9 lbs./ yd^3 (41kg.) C-Trel. All substrates were ammended with Woodace 20-5-10 Controlled Release Fertilizer was surface applied at the rate of 1.9 oz. product / 3 gallon (10.8 g N) / container.

b) Pine Bark: RT Soil Sciences Compost (50:50 by volume). Substrate was not amended during mixing and potting but .18 oz. (5.6 g) C-Trel was applied to the surface of the container and then feathered into the substrate by hand after placing containers in the test area.

c) Pine bark: RT Soil Sciences Compost (75:25 by volume). Substrate was not amended during mixing and potting but .18 oz. (5.6 g) C-Trel was applied to the surface of the container and then feathered into the substrate by hand after placing containers in the test area.

d) RT Soil Sciences Compost (100% by volume). Substrate was not amended during mixing and potting but .18 oz. (5.6 g) C-Tel was applied to the surface of the container and then feathered into the substrate by hand after placing containers in the test area.

4. Data Collected:

a) Virginia Tech Extraction Method (VTEM) pour through extraction procedure were used to collect leachate samples on 1 cultivar (' Dwarf Burford') from 3 containers in 3 separate blocks (reps) immediately after potting and after 4, 8, 12 and 16 weeks after potting (Approximate dates: May 22, June 19, July 17 and August 14 and September 11). Sample bottles were tabled with the tag number in corresponding pots, dated and the letters CCN (Cedar Creek Nursery) written on the table of each container. In addition to the 3 pour throughs / substrate, 3 irrigation samples were collected for a total of 15 leachate/water samples per collection date and a total of 75 samples for the study. Leachate samples were collected 1-2 hours after irrigation. Samples were refrigerated until analyzed for pH, electrical conductivity (soluble salts) and nutrients.

b) Foliar tissue samples were collected for foliar elemental analysis to determine foliar nutrient levels at the termination of the study.

c) At potting, 5 pots of each substrate were filled and included in the study as fallow pots without plants. At the end of the study, these pots were transported to Raleigh to the Horticultural Substrates Lab and physical property evaluations conducted to determine Total Porosity, Air Space, Container Capacity, Available Water, Unavailable Water Content and bulk density.

d) After the last VTEM collection date, the study was continued and evaluated on September 29 to determine the response of each species to the container substrates.

e) All data collected were analyzed for statistical significance by ANOVA .

SUMMARY OF RESULTS

Physical Properties

Total Porosity, Air Space and Container Capacity were less in the Cedar Creek Standard Mix than other substrates tested, due to the addition of sand (Table 1). Sand increased the Bulk Density but also tended to increase the Available Water Capacity by decreasing Unavailable Water.

Table 1. Physical properties of selected substrates.[a]

Media[b]	Total Porosity	Air Space	Container Capacity	Available Water	Unavailable Water	Bulk Density
	---------------------(%Volume)				-----------------------	(g/cc)
Cedar Creek STANDARD PB:RTC:S (2:2:1)	79.3	13.3	66.0	43.2	22.9	0.43
PB:RTC (1:1)	83.0	12.2	70.8	42.6	28.2	0.31
PB:RTC (3:1)	86.3	18.3	68.0	37.7	30.3	0.26
RTC (100%)	80.6	13.8	67.0	45.3	21.7	0.43
Normal Ranges for Pine Bark : Sand Substrates	50.0-85.0	10.0-30.0	45.0-65.0	25.0-35.0	25.0-35.0	0.19-0.52

[a] All analyses performed using standard aluminum soil sampling cylinders (7.6 cm ID, 7.6 cm h).
Air Space and Container Capacity affected by height of container.
[b] Media (PB = Pine bark; RTC = RT Soil Science Compost; S = Sand : (by volume)
(1) Cedar Creek Standard = 2PB:2RTC:1S ; (2) 1PB: 1RTC ; (3) 3PB:1RTC ;
(4) 100% RTC

Addition of each increment of pine bark increased Total Porosity. Air Space and Container Capacity as well as Available and Unavailable Water were dependent upon the volume of pine bark added.

The 100% RT Soil Sciences Compost had physical properties most similar to the Cedar Creek Standard.

All Substrates were within guidelines for Total Porosity, Air Space and Bulk Density while all exceed normal ranges for pine bark: sand for Container Capacity and Available Water Capacity values. This suggests that in regard to Physical Properties all four substrates could be used to produce plants in containers if irrigation management was tailored to the substrate.

Electrical Conductivity and pH

Initial pH varied considerably among test substrates but was quite variable within each treatment, therefore no statistical differences were established. The 100% RT Compost however, did tend to have the highest pH initially and throughout the study. The Cedar Creek Standard maintained pH within guidelines for good plant response throughout the study. The PB:RTC (1:1) substrates tended to maintain a higher pH than the 3:1 or the Cedar Creek Standard. This may have been due to the high Container Capacity which provided more water for acidity and alkalinity reactions to occur within the substrate. The 3:1 PB:RTC maintained pH levels similar to the Cedar Creek Standard. (Table 2).

Conductivity levels (EC) were initially above the suggested range which is expected in unleached containers immediately after potting. Leachates after 4 weeks indicated adequate conductivity for all substrates and acceptable levels were maintained through 12 weeks after potting.

Table 2. Container leachate pH and electric conductivity (EC) levels from 4 Substrates on 6 sampling dates. Suggested pH 5.2-6.5; EC levels) 0.5 to 2.0 mmhos/cm for VTEM leachates. Each value represents the mean of 3 leachates samples.

Media[a]	0	4	8	12	16	18
	pH					
Cedar Creek STANDARD (2:2:1)	5.9	6.1b[b]	6.4b	6.2b	6.4b	5.9
PB:RTC (1:1)	7.3	6.7a	6.4b	6.9a	6.7ab	6.1
PB:RTC (3:1)	6.0b	6.1b	6.1b	6.2b	6.3b	5.5
RTC (100%)	7.8	7.2b	8.0a	7.1a	7.0a	5.8
	EC (mmhos/cm)					
Cedar Creek STANDARD (2:2:1)	2.6	1.4a	0.9	0.5	0.3	0.2
PB:RTC (1:1)	3.6	0.9a	0.9	0.7	0.2	0.2
PB:RTC (3:1)	3.3	1.6a	1.1	0.6	0.3	0.3
RTC (100%)	3.5	1.1b	0.9	0.7	0.2	0.2

[a] Media (PB = Pine Bark; RTC = RT Soil Science Compost; S = Sand : (by volume)
(1) Cedar Creek Standard = 2PB:2RTC:1S ; (2) 1PB: 1RTC ; (3) 3PB:1RTC ; (4) 100% RTC
[b] Means followed by the same letter are not significantly different.

NO_3-N and P

After 4 weeks, container leachates contained acceptable to high levels of NO_3-N. The 100% RTC had the lowest mean NO_3-N but was within the suggested range. Nitrate Nitrogen levels for all substrates were statistically similar and acceptable at 8 and 12 week, and all were below optimal levels for producing rapid growth at 16 weeks, even though the Cedar Creek Standard held 2 to 3 fold greater NO_3-N. Ammonium Nitrogen levels were consistently below suggested ranges and were not different among substrates, therefore they are not reported. (Table 3).

Phosphorus nutrition is difficult to maintain in organic substrates since soluble phosphate leaches rapidly during irrigation and cannot be fixed by organic components. Recommended levels in leachate solution have been reduced in recent Best Management Practices publications to reflect levels often present when tested. Although P concentration in the 100% RTC substrate were generally below guidelines, the other substrates maintained levels generally higher and within suggested guidelines. In fact the PB:RTC (3:1) and the Cedar Creek Standard maintained surprisingly adequate P throughout 16 weeks of the study.

Table 3. Container leachate NO_3-N and P levels from 4 Substrates on 6 sampling dates.

Media	Sampling Dates (Weeks after potting) 0	4	8	12	16	18
	ppm					
	NO_3-N					
Cedar Creek STANDARD (2:2:1)	28.2	101.3a[a]	81.9	49.4	19.7a	7.9
PB:RTC (1:1)	11.5	73.5ab	89.8	50.7	5.2b	3.0
PB:RTC (3:1)	16.3	107.9a	94.9	57.1	8.7b	4.0
RTC (100%)	12.6	57.7b	79.8	55.5	8.6b	6.7
	P					
Cedar Creek STANDARD (2:2:1)	17.6a	21.1a	8.3a	5.4b	6.3	6.2a
PB:RTC (1:1)	5.7bc	25.1a	8.7a	8.5a	6.1	5.1ab
PB:RTC (3:1)	25.2a	17.9a	11.2a	6.8ab	5.1	2.4b
RTC (100%	2.3c	6.1b	1.8b	2.5c	3.5	1.9b

[a] Means followed by the same letter are not significantly different.

Foliar Nutrition

Eighteen weeks after potting, tissue nitrogen, potassium, magnesium and iron, were within acceptable guidelines for all four substrates. (Table 4)

Table 4. Foliar nutrient levels of Ilex cornuta 'Burfordii Nana holly grown on four substrates at the end of the growing season.

Media	N	P	K	Ca	Mg	Fe	Mn	Zn
	(% tissue dry weight)					(ppm)		
Cedar Creek STANDARD (2:2:1)	2.1bc[a]	0.2	2.0	1.3ab	0.3	172.0a	2432.7a	1319.7a
PB:RTC[b] (1:1)	2.0c	0.2	2.0	1.2b	0.3	122.7b	1484.0bc	773.7b
PB:RTC (3:1)	2.2ab	0.2	2.0	1.3a	0.3	156.5ab	2268.8ab	815.0b
RTC (100%)	2.2ab	0.2	1.9	1.2b	0.3	124.7b	1115.0c	532.7c
sufficiency level	1.8-2.8	0.15-1.00	1.0-2.0	0.2-1.0	0.2-0.8	35.0-250	50.0-200	

[a] Means followed by the same letter are not significantly different.
[b] Media (PB = Pine Bark; RTC = RT Soil Science Compost; S = Sand : (by volume)
(1) Cedar Creek Standard = 2PB:2RTC:1S ; (2) 1PB: 1RTC ; (3) 3PB:1RTC ; (4) 100% RTC

Tissue phosphorus was within guidelines but somewhat low for all test substrates which is somewhat surprising since solution levels were maintained optimal in the PB:RTC (3:1) and the Cedar Creek Standard.

Calcium was evidently available in high levels and therefore was absorbed in large amounts from all four substrates.

Manganese and Zn levels were extremely high in tissue. Hollies are known accumulators of metals particularly Mn and frequently are shown to have levels in the 1 to 2 % tissue range as high as macro-nutrients as is the case in this study. The high Mn and Zn levels do not appear to have created any antagonisms, primarily because high levels of Ca and Fe were also available and absorbed in large amounts. If high Mn and Zn levels were considered to be creating problems, dropping their addition through dropping the micro-nutrient package would seem advisable. The only minor element, which might need to be supplemented when the RT Compost is utilized might be Cu although no tests were run in this study to determine Cu levels or need for other elements.

Growth Measurements

The growth indexes, a measure of plant diameters is calculated using the following formula:

$$\frac{\text{maximum canopy width} + \text{minimum canopy width}}{2} = \text{average canopy width}$$

$$\frac{\text{average canopy width} + \text{maximum plant height}}{2} = \text{growth index}$$

Maximum plus the minimum canopy width is divided by two, which equals the average width, then average width is added to the maximum height from the container lip and the sum of height and average width is divided by two. These measurements were similar for the Cedar Creek Standard, and the Pine Bark: RT Compost substrates both 1:1 and 3:1, however the 100% RT Compost had a lower Growth Index than the PB combination mixes. The same was generally true of the height measurements, except the PB:RTC 1:1 and 100% RTC were the same. Table 5).

Table 5. Growth Index and Plant Height of Ilex Cornuta 'Burfordi Nana' holly at the end of the growing season as affected substrates.

Media	Growth Index[a]	Plant Height (cm)
Cedar Creek STANDARD (2:2:1)	53.4ab[b]	71.0a
PB:RTC (1:1)	59.7a	66.1ab
PB:RTC (3:1)	64.0a	75.6a
RTC (100%)	51.1b	57.1b

[a] Growth Index equals the (Greatest Width plus Least width) divided by two plus the height, then the Sum divided by two. The measurements were made at the termination of the study on September 29, 1995.
[b]Mean separation by Duncan's Multiple Range Test ($p>0.05$). Means followed by the same letter are not significantly different.
[c] Media (PB = Pine Bark; RTC = RT Soil Science Compost; S = Sand : (by volume)
(1) Cedar Creek Standard = 2PB:2RTC:1S ; (2) 1PB: 1RTC ; (3) 3PB:1RTC ; (4) 100% RTC

CONCLUSIONS

Less growth occurred in the 100% RT Compost which generally had less desirable air and water relationships and nutrient capacity characteristics. This substrate benefited from the addition of pine bark and related to plant growth benefited greatest with the addition of 3 parts pine bark to 1 part RT Compost. Although the Cedar Creek Standard potting mix was quite acceptable, removal of the sand might possibly increase growth and reduce handling effort and shipping costs.

Results indicated that Compost/CCB fly ash blend soils can provide a stable growth medium. These soils can increase in value when air space can be increased by adding aged pine bark.

Regulatory Approval

Their are two agencies with in the state of North Carolina that required regulatory approval. One being the North Carolina Department of Environment, Health, and Natural Resources, (NCDEHNR) Division of Solid Waste. NCDERNR require permit for composting of source separated organics. Under current regulations this type of composting falls under Municipal Solid Waste. This posed a problem with finding a suitable site since most municipalities zone this as a Heavy Industrial Area. Most of this type land is scarce and expensive. NCDEHNR, under section .1400 of the Solid Waste Rules, also regulates the type of grade and hence the end use of composted by-products. All of RTSS materials are graded as unrestricted use. This is accomplished by meeting the following parameters:

1. Material size.
2. Percent man made inerts.
3. Total heavy metals.
4. PFRP of material.
5. Soluble salts.

The other agency is the North Carolina Department of Agriculture, Plant Division. This agency regulates fertilizer and lime products. From this agency, it was necessary to gain approval on how this product was to packaged and sold.

Market Acceptance

Market acceptance was gained by experimenting with different mixes. Most of the ornamental growers within the area will use a pine bark and sand mixture when potting. This mixture needs the addition of lime, N-P-K and micro-nutrients to make it an effective media. RTSS has added up to 40% aged pine bark in their ***Grower's Mix*** to increase air space and lighten the media. This along with the benefits of a built in stable pH, N-P-K, and micro-nutrients, has made it an economical choice to the Grower. Most Growers are skeptical in using the ***Grower's Mix*** on all of their stock at first. Many will try a small amount to test and find out that they can use our material in many growing situations. Some Grower's will customize our mix further to suit their particular growing needs.

REFERENCES

Bilderback TE, Everette, Hartzog F & Means R (1990) Use of Fly Ash and Pine Bark as a Potting Medium. Proc. Southern Nurserymen Association Res. Conf. 35:76-79.

Butler SH & Bearce BC (1995) Greenhouse Rose Production in Media Containing Coal Bottom Ash. J. Environ. Hort. 13:160-164.

BLEND BY-PRODUCTS TO MEET CUSTOMER NEEDS - MAKE THESE NEEDS YOUR HIGHEST PRIORITY

KATHY KELLOGG JOHNSON
Kellogg Supply Inc., Carson, California 90745, USA
E-mail kellsup@earthlink.net

Kellogg Supply has, for more than 70 years, been transforming organic feedstocks into commercial products to be used to improve the quality and productivity of soils. These products have been developed both because of the abundance of available materials and the need to fill customer demand. A history of our company as well as descriptions of a series of products that we have developed will illustrate the philosophy of our business.

I will begin by telling you a capitalist tale. This is a story about my grandfather, H. Clay Kellogg a farmer, surveyor, and orange grove farmer. In the early 1920's surveying near the Santa Ana River, he found tremendous lush green growth along the bend of the river. He found the greenery to be buried in a dark sediment some of which he brought home in shoe boxes. He had one orange tree in the front yard and an entire grove in the back. He mulched this one tree generously with the black soil. When it came time to fruit, that orange tree had four times the yield of any other tree in his grove. A lab analysis showed this to be Orange County sewage sludge. It was rich in nitrogen and humus and so he named it Nitrohumus, which has remained our trade mark until today.

Excited by his discovery, he thought he would sell mountains of the stuff, but no one was interested. He was forced to be creative. In the early 1920's, California's Citrus farmers had a gold rush mentality. They farmed a piece of land for a few years and when the yields began to drop, they abandoned it and sought new land. Textbooks of the period validated their belief that farmland, like mines, would only sustain crops for a limited time. My grandfather thought he could change this belief. He scraped together enough money to buy 10 acres of so called "depleted" orange groves, applied Nitrohumus, and when it began to fruit, traded those 10 fertile acres for 20 depleted acres. By the time he acquired 1,000 acres he had a suitable showcase for his Nitrohumus. The farmers found it less taxing to buy Nitrohumus than to trade away their land.

Yesterday, as today, our company faced the challenge of balancing the availability of product and the demand of the customer. In 1925 when the farmers began to use the product, my grandfather needed to go find a suitable supply. He contracted with the Los Angeles County Sanitation District (LASCD), to purchase and remove sewage sludge. As you may know, it's one of the largest Sanitation districts in the world, serving over 78 cities, 4.7 million people, and processing ½ a billion gallons of wastewater each day.

S. Brown, J.S. Angle and L. Jacobs (eds.), Beneficial Co-Utilization of Agricultural, Municipal and Industrial By-products, 315–318.
© 1998 *Kluwer Academic Publishers,*

Kellogg utilizes between 50,000 and 10,000 wet tons per year.

Some people call this sewage sludge. You probably call it biosolids. At Kellogg's we call it "bread and butter". My grandfather saw this as a natural resource. He created a purchase contract with the LACSD, he actually bought this stuff and we still do today.

During World War II Nitrohumus was used in the victory gardens and citrus groves of Southern California and after the war, on the lawns and gardens of the GI's when they settled there in record numbers. As the business grew, Kellogg needed more product. What we found is that our new customer, the homeowner, now demanded Nitrohumus in 100 pound packages. We soon found a 100 LB package with our name on it, built brand recognition faster than a truckload and so Kellogg provided the product the customer wanted. The homeowner now used more Nitrohumus per acre on what used to be farmland than the farmer ever had.

Does anyone here remember what Southern California neighborhoods looked like in the early 1950's? Do you remember the matching tracts of pastel colored houses? What did they mulch with? ROCKS, red rocks, white rocks, green rocks, the shoulder of all the driveways were bordered with DECOROCK. It was at this time that my father H. C. Kellogg, Jr. was visiting a lumbermill in Arizona. He saw a truck being loaded up with the bark waste and hauled away. He asked where it was going. The driver offered to show him. As they crested the hill my father saw an area the size of a football field, 10 feet deep in bark. "What on earth!" "Well the burner is down so we have to do something with this junk" Dad immediately said "I'll take it, I'll take all of it" and went back to Southern California to figure out how he was going to sell it. He thought it was beautiful and that he could market it. He pioneered the Decobark industry, introducing the concept to Southern Californians, incidentally introducing the pincher bug (earwig) at the same time. Nevertheless, today bark is not thought of as a bi-product, nor a waste product. It is a staple in the marketplace...Large, medium, small, pathway bark, shredded bark, and orchid bark. In 1955 my father gets a phone call from Walt Disney who was about to open a theme park in Anaheim. He needed to grow tropical plants for the jungle cruise. The problem was trying to create a tropical paradise in a desert. Jungles require lots of water and acidic fertile soils. Since tropicals had never grown in desert conditions, Disney needed special soil. The answer was Gromulch, a blend of Nitrohumus and the bark from Arizona. Together, it is an acidifying Planter mix which was just what those tropicals needed. Even before the park opened, horticulturist came from around the state to admire plants they had never before seen growing outside of a green house.

In 1969, we absolutely ran out of Nitrohumus. The problem was not the production of the sewage sludge (he had all the contracts in Southern California, Orange County), it was the readiness, the wetness of the material. All the Nitrohumus had been dried in sludge lagoons up until that time. In the Los Angeles County Sanitary District alone there were 300 acres of sludge lagoons all full but not ready to go to market.

The partnership of Kellogg and LACSD pioneered a method to dry out the sewage, faster. This was called windrowing. It was the origin of what we know today as thermophillic windrow composting. We could get the product faster and would soon realize the other benefits that came with composting, like the elimination of the abundant tomato plants that our customer had come to expect from Nitrohumus.

In spite of this solution, the 1969 crisis eventually led to a marketing decision. Kellogg would continue to look for by-products or ingredients that were available in abundance and would be available in a continuous stream. We had been stung from building the market demand for the product then running out, so we diversified. In 1997, less than 15% of Kellogg market is Biosolids. Those that contain biosolids are highlighted in yellow.

In all we sell over 300 different products in different sizes. The diversity of our product line is in direct response to our customers' needs. One customer, Home Depot, asked Kellogg to develop a size of product uniquely for them. Our typical product line sells for about $4 per 1 1/2 cu bag. They asked for a 30 cf bag, the equivalent of 20 of these bags, an $80 value they would price at $45. The consumer is familiar enough with the quality and the cost of the product that paying $45 per bag is a substantial savings to him. We could move the customer away from using the less expensive stuff for larger jobs around his home. Kellogg sells 95% of our production in packages. In March we broke a record, shipping 384,000 units in 5 days from our 4 different locations.

Our customer list includes independent garden retail shops, that recognize the quality and build the credibility of Kellogg products. We also supply chain stores. These are largely responsible for the growth in sales. We are a regional company shipping to customers throughout CA, AZ, NV, HI, OR, WA and most recently ID, MT and CO to support Target's seasonal garden centers. In 1997 we have nearly doubled the units sold by adding several commodities to our line.

Most of Kellogg products contain new and interesting waste materials. Ever wonder what happens to the movie sets from Paramount studios, Twentieth Century and Warner Bros.? Many studios in Southern California build sets and tear them down in a day. That scrap wood was going directly to the landfill. Warner Brothers and Fox studios called me in to determine whether the sets could be used in a beneficial way. They wondered about the nails and the paint and the textures - whether they made the material unusable. It turned out that what they used for the texture was calcium sulfate, known to you and me as Gypsum, which is widely widely in Southern California as an amendment to break up clay soil. It is a beneficial additive. We have been using the wood from the studios as a bulking agent ever since.

Speaking of clay soils, the product you see pictured here in Sunset Magazine was developed as an answer to the clay soils we have in Southern California. We've performed over 30,000 soil tests in Southern California to discover this need. My father was told rice hulls couldn't be composted. They would just come to the surface, float away and stay blond. They are too high in lignin and silica. We compost them, they're black and act as a beautiful fertile wedge in clay soils.

Christmas trees - before 1988, Christmas trees were discarded in landfills until the city of Santa Monica approached Kellogg and asked could we use them if Santa Monica collected them? Absolutely! Christmas trees are a resource of low pH wood materials, which is exactly what rhododendrons, Camellias, and Azaleas need to grow. We now compost them as part of our Azalea/Camellia mix.

There is a lot of turf in Southern California- golf courses, ball fields, lawns. We found a way to improve Nitrohumus as a seedcover. Using crosscut redwood and bark fines composted with Nitrohumus, developed fertile seed cover and top dressing. Another unexpected side benefit was that Topper reduced the incidence of birds scavenging the seeds.

Another example of using an abundant waste resource was began in 1989. In the state of California alone, 76% of our waste stream is Organic or compostable. The California Integrated Waste Management Board (CIWMB) predicts that 10 million tons of Greenwaste will be diverted from landfills annually. Knowing this was coming, Kellogg Supply created a product we called California Compost. We were concerned at the time that Greenwaste would be used by the large waste companies as a product competitive to ours, so we experimented with composting the Greenwaste with biosolids. The quality of the final product was to be weed free, stable, and full of nutrients. We felt that if we once again pioneered a new concept and set the quality standard, our competitors would have to produce a quality product as well.

We set a price point of $1.69, barely enough to cover the cost of ensuring the quality. Home Depot sold 1 million bags the very first spring we introduced Cal Compost. As we suspected, an 8 billion dollar waste company brought out a .49 cent planter mix also made with greenwaste. It was generated from curbside-collected greenwaste. In California everybody thinks that whatever is their yard can be thrown into the yard waste container. Tires, rubber hoses, shovels, empty bags of our product, empty Pepsi bottles, everything gets thrown in, dumped into a grinder, rotted together in a pile, and bagged. The final result was that it contained garbage, glass, shreds of our yellow plastic bags, and the greenwaste. It was of poor quality and the consumer rejected it. Even at .49 cents, the consumer was dissatisfied. Kellogg had set the standard. Our 8 billion dollar competitor was out of our business.

If I could leave you with a final impression, it would be a word on encouragement and admonition. In each of your research projects, focus on the need of your customer, the customer who will ultimately utilize the materials you study and the products you create. In this way your research will result in the development of many successful and well recognized useful products.

SLOW-RELEASE NITROGEN FROM COMPOSTS: THE BULKING AGENT IS MORE THAN JUST FLUFF

D.M. SULLIVAN[1], S.C. FRANSEN[2], A.I. BARY[2] and C.G. COGGER[2]
[1]Department of Crop and Soil Science, Oregon State University, Corvallis, Oregon 97331, USA and [2]Department of Crop and Soil Sciences, Washington State University, Puyallup, Washington 28371, USA
E-mail Dan.Sullivan@orst.edu

ABSTRACT

One of the goals of byproduct co-utilization is to produce products with increased value, and the amount of slow-release nitrogen (N) supplied by composts for plant growth is one component of compost value. To evaluate the effect of bulking agents on the amount of slow-release N derived from composts, we conducted a three-year field trial with a forage-type tall fescue (*Festuca arundinacea* Schreb. 'A.U. Triumph'). Composts were prepared for the field trial from mixtures of food residuals (vegetables, meat, fish, dairy, and bakery by-products) with three bulking agents. Food residuals (33 g N/kg) were bulked with yard trimmings (11 g N/kg), yard trimmings + mixed waste paper (7 g N/kg), and wood chips + sawdust (1 g N/kg). After mixing, the food residual/bulking agent mixtures were composted in a turned windrow supplied with forced air for 70 days, then cured without forced air for 36 days. At the end of curing, total N concentrations in screened compost (< 11 mm) were 17 g N/kg for yard trimmings, 14 g N/kg for yard trimmings + paper, and 8 g N/kg with wood chips + sawdust bulking agent. For the field trial, 155 Mg/ha of compost was incorporated to a depth of 10 cm in a sandy loam soil. Tall fescue was seeded the day after compost application and was harvested 15 times over a three-year period to measure compost effects on grass yield and N uptake. Composts consistently increased yield and grass N uptake in the second and third year after application, demonstrating their slow-release N value. Cumulative apparent N recovery (ANR) over the three-year trial ranged from 7 to 11 % of the compost total N applied. Cumulative ANR was 282 kg N/ha for yard trimmings, 242 kg N/ha for yard trimmings + paper, and 113 kg N/ha for wood chips + sawdust bulking agent. Replacement of wood chips + sawdust bulking agent with yard trimmings more than doubled compost slow-release N value. Thus, yard trimmings are a valuable feedstock when developing compost products with slow-release N value.

S. Brown, J.S. Angle and L. Jacobs (eds.), Beneficial Co-Utilization of Agricultural, Municipal and Industrial By-products, 319–325.
© 1998 *Kluwer Academic Publishers,*

INTRODUCTION

Compost bulking agents are important to the composting process and to the quality of the finished compost as a plant growth media. They are used in compost mixtures to increase porosity, thereby allowing air movement through the compost pile. Bulking agents are needed when composting wet, high N by-products like food residuals, biosolids, or animal by-products. Several properties of bulking agents contribute to their dominance of end-product quality. Compared to wet, high N materials, bulking agents are lower in moisture content, are more resistant to microbial degradation, and are added in larger volumes to typical compost mixtures. Therefore, if the bulking agent is not removed by screening after the completion of composting, it usually becomes a major component of the finished compost.

Basically two philosophies guide bulking agent selection. The first philosophy emphasizes the ease and economics of waste reduction via composting. The bulking agent chosen is low in cost and allows efficient compost production. The second philosophy emphasizes end-product value. The bulking agent chosen adds the most value to the compost product. This paper emphasizes the second philosophy, that of co-utilization of by-products as a way to enhance end-product quality.

Nitrogen is usually the most limiting nutrient for plant growth in both urban and agricultural soils. Composts usually supply some, but not all of the N needed for acceptable plant growth. Most of the N supplied by composts originates from the slow oxidation of compost organic N compounds after application, often called "slow-release N". The amount of slow-release N supplied by composts is one component of compost value.

Slow-release N is especially valuable for perennial agronomic crops that require frequent N fertilizer applications. In western Oregon and Washington, perennial grasses grown for livestock feed require three to six applications of soluble N fertilizers per growing season for maximum production (Hart et al., 1996). The slow-release N provided by composts is much longer lasting than that provided by inorganic N sources such as polymer-coated urea granules, which provide 8 to 16 weeks of slow-release N (Rogers, 1993).

To evaluate the effect of bulking agents on the amount of slow-release N in composts, we composted food residuals with three bulking agents. Slow-release N from composts was evaluated in a three-year field trial with tall fescue.

MATERIALS AND METHODS

Compost Production

The growth trial described here was conducted as part of a pilot food residuals composting project. More detailed descriptions of bulking agents, composting methods, and compost quality has been reported previously (Sullivan et al., 1998; Croteau and Steuteville, 1995).

Food residuals were collected in the Seattle, WA, area over a 10-day period in January

1993. Major components were vegetables, meat, fish, dairy, and bakery residuals. Compost bulking agents chosen for the study were yard trimmings, yard trimmings + mixed waste paper, and wood chips + sawdust.

The food residuals and the bulking agents were mixed with a front-end loader and a Scarab™ windrow turner. Food residuals comprised about 25 % of the initial pile volume. Composting was performed in an unheated shelter. Air was supplied via a perforated pipe to the base of each compost pile, and piles were turned ten times during the active composting period (70 days). This composting method simulated "agitated bay" composting systems where routine mechanical agitation is part of the compost facility design. Temperatures during active composting ranged from 30 to 80 °C, and were greater than 55 °C in all piles for at least 5 days. After 70 days of active composting, piles were cured without aeration for an additional 36 days. Composts were screened to pass a 11 mm (7/16 in.) screen prior to field application.

Representative samples of the bulking agents, food residuals, the initial mixtures of food residuals + bulking agent, and the screened compost were analyzed for total N concentration (Table 1) by a Kjeldahl method (APHA, 1992).

Table 1. Total N concentrations of bulking agents, mixtures of bulking agent + food residuals, and screened compost (dry wt. basis).

Bulking Agent Description	Bulking Agent	Bulking Agent + Food Residuals Calculated[a]	Bulking Agent + Food Residuals Measured[b]	Compost[c]
	g/kg			
Yard trimmings[d]	11.0	19.2	17.0	18.3
Yard trimmings + paper	7.1	13.3	14.7	17.5
Wood chips + sawdust	0.6	8.3	8.7	10.4

[a] Calculated N in mixture based on N concentrations, volume addition, and bulk density of mixture components. Food residuals were added to bulking agent on a volume basis of approximately three parts bulking agent to one part food residuals.
[b] Measured N in mixture based on composite samples obtained immediately after initial pile mixing. Total N in food residuals was 32.6 g/kg.
[c] Screened compost product (< 11mm).
[d] Shredded conifer debris was the major component of the yard trimmings. Yard trimmings were hammer-milled before mixing with food residuals. Initial particle size distribution for the yard trimmings (percentage by weight): greater than 19 mm -- 19 %; 10-19 mm -- 17 %; 5-10 mm -- 12 %; less than 5 mm -- 52 %.

Field Growth Trial

The growth trial was conducted on a Puyallup fine sandy loam soil (coarse-loamy over sandy, mixed, mesic, Fluventic Haploxerolls) located at the WSU Puyallup Research Center in Puyallup, WA. Initial soil test values indicated that only N fertilization was needed for near-maximum perennial grass production (Hart et al., 1996).

Composts were applied 27 May 1993 at a target application rate of 155 Mg per hectare (approximately 7 yd^3 per 1000 ft^2). Actual compost application rates varied from 148 to 172 Mg/ha because of variability in total solids analyses (Table 2).

A cool-season perennial grass (forage-type tall fescue; *Festuca arundinacea* Schreb. 'A.U. Triumph') was seeded the day after compost application. Grass was managed to maintain active growth throughout the growing season. Fertilizer N (ammonium nitrate; 34-0-0) was broadcast-applied after each grass harvest (except the final fall harvest), and about 30 days before the first harvest each year. The rate of fertilizer N (34 kg N/ha per application) was the same for all treatments except the no compost, no fertilizer N control (Table 3).

Table 2. Application rates of compost and compost N in the field trial.

Bulking Agent	Dry Matter[a]	Total N	NH_4-N	NH_4N
	Mg/ha	kg/ha	kg/ha	kg/ha
Yard Trimmings	145	2660	64	< 1
Yard Trimmings + paper	172	3007	90	< 1
Wood chips + sawdust	148	1540	3	< 1

[a] Target application rate was 155 Mg/ha, but actual rate varied due to differences in compost moisture content.

Table 3. Cumulative effects of compost on yield, N uptake, and apparent N recovery (ANR) by grass for the three-year field trial[a].

Compost Applied	Bulking Agent	N[b] Fertilizer	Yield[c]	N Uptake[c]	ANR[d]	Percent ANR[e]
		kg/ha	Mg/ha	kg/ha	kg/ha	% compost N applied
Yes	Yard trimmings	470	33.6 d	822 e	282	10.6
Yes	Yard trimmings + paper	470	33.6 d	782 d	242	8.1
Yes	Wood chips + sawdust	470	29.2 c	653 c	113	7.3
No		470	27.1 b	540 b		
No		0	16.1 a	304		

[a] Cumulative data for 15 grass harvests, 1993-95.
[b] Total of 470 kg N/ha as ammonium nitrate (34-0-0) applied at 34 kg N/ha/harvest. No N fertilizer was applied prior to the final harvest in 1995.
[c] Yield and N uptake values followed by a different letter are significantly different ($P < 0.05$) via a protected LSD.
[d] Apparent N recovery calculated via equation 2.
[e] Percent apparent N recovery calculated via equation 3.

Grass was harvested for yield determination at 30 to 45 day intervals during the growing season. Harvested grass N concentrations were determined via a combustion N gas analyzer.

N recovery equations were used to summarize the grass N uptake data:

Grass N uptake (kg/ha) = A * B/100 [1]

where:

A = grass yield, dry matter basis, kg/ha

B = grass N concentration, dry matter basis, percent

Apparent N recovery (ANR, kg N/ha) = C-D [2]

where:

C = Grass N uptake for compost treatment, kg/ha

D = Grass N uptake for no compost treatment, kg/ha

with both treatments receiving the same rate of fertilizer N (34-0-0)

Percent apparent N recovery (Percent ANR, % of compost N applied) =(C-D)/F *100 [3]

where:

C = Grass N uptake for compost treatment, kg/ha

D = Grass N uptake for no compost treatment, kg/ha

F = compost N applied, kg/ha

with both treatments receiving the same rate of fertilizer N (34-0-0)

RESULTS AND DISCUSSION

Grass yield response to ammonium nitrate and fertilizer N uptake efficiency in our study (calculated from data for no compost treatments in Table 3) was similar to that reported for tall fescue and orchardgrass production under similar management in western Oregon and Washington (Yungen et al., 1977; Turner, 1979). Over the three-year study, an additional 11 Mg per hectare of grass yield was produced with application of 470 kg fertilizer N per hectare. Therefore, yield response to fertilizer N was 23 kg of grass yield per kg of applied N. The N fertilizer applied (470 kg N/ha) increased grass N uptake by 236 kg/ha. Therefore, fertilizer N uptake efficiency was 50 %. Over the three-year trial, the increase in cumulative ANR due to compost application was substantial (113 to 282 kg N/ha). Since fertilizer N uptake efficiency was 50 %, the compost provided the equivalent of 226 to 564 lb of fertilizer N.

Grass yield and N uptake increased with the quantity of compost total N applied. All composts were applied at similar dry matter rates (Table 2). Total N application rates were mainly a function of compost total N concentrations, as determined by differences in bulking agent N concentrations (Table 1). Finished compost total N concentrations were close to the measured and calculated total N concentrations in the initial bulking agent + food residuals mixtures (Table 1), suggesting loss of both C and N during the composting process.

Compost application had only a small effect on ANR in the establishment year (Figure 1). This was probably due to a combination of site factors, including an infestation of annual grasses that reduced measurement precision, and the limited N uptake capacity of the newly-established grass. Visual ratings of grass vigor suggested

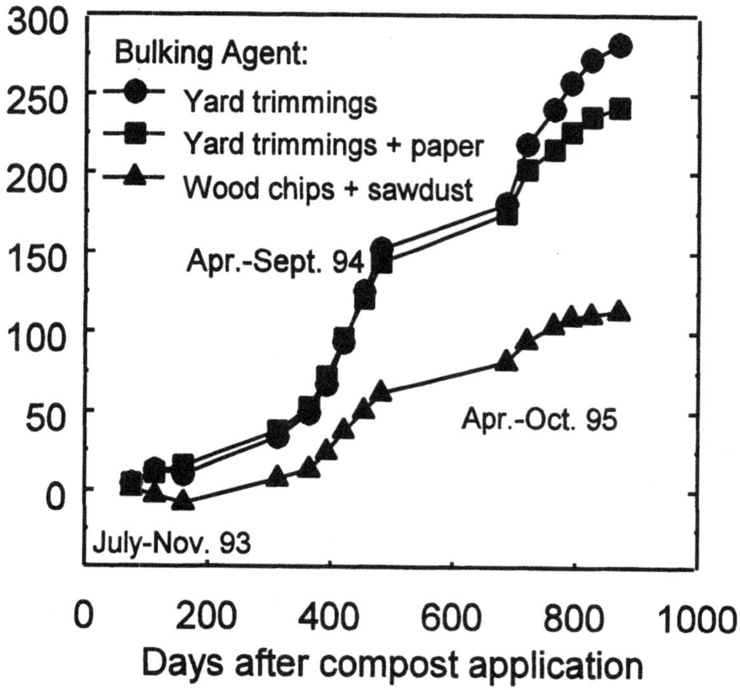

Figure 1. Effect of compost bulking agents on cumulative apparent N recovery by tall fescue. Each data point represents a grass harvest.

that the compost produced with the wood chip + sawdust bulking agent immobilized N during the first 45 days after application, while composts produced with yard trimmings, or yard trimmings + paper bulking agents, provided N for plant uptake. These observations agreed with the compost nitrate-N concentrations at application: 440 mg N/kg for yard trimmings, 523 mg N/kg for yard trimmings + paper, and 20 mg N/kg for wood chips + sawdust bulking agent.

During the second and third year after compost application, compost-treated plots had higher yields and N uptake, resulting in a gradual increase in cumulative ANR (Figure 1). The steady increase in cumulative ANR over time demonstrates the value of compost as a source of slow-release N. Cumulative grass yield, N uptake, and ANR was greater for composts derived from yard trimmings and yard trimmings + paper bulking agent than for the wood chip + sawdust bulking agent (Table 3).

The grass ANR data can be used to roughly estimate the compost N mineralization rate. During the third year of our study, compost increased ANR by 51 to 130 kg N per hectare, which is equivalent to 3 to 5 % of the compost total N applied. Thus, using 50 % N uptake efficiency, we estimate a compost N mineralization rate of approximately 6-10 % for the third year after application.

CONCLUSION

This study demonstrated the long-term benefits of a high rate compost application in providing slow-release N for crop growth. Replacement of wood chips + sawdust bulking agent with yard trimmings more than doubled compost slow-release N value. Yard trimmings are a valuable feedstock when developing compost products with slow-release N value.

ACKNOWLEDGMENTS

Financial support for this project was provided by the Washington State Dept. of Ecology and King County, WA. Cooperators included: Sound Resource Mgmt. Group, Inc.; E & A Environmental Consultants, Inc.; and Herrera Environmental Consultants, Inc.

REFERENCES

American Public Health Association (1992) Nitrogen (ammonia, organic, and nitrate methods). In: Franson MH (Ed) Standard Methods for the Examination of Water and Wastewater (pp. 4-75 to 4-97). APHA, Washington, DC.

Croteau G & Steuteville R (1995) Collecting and composting food residuals. Biocycle 36(8):77-83.

Hart, J, Pirelli G, Cannon L & Fransen S (1996) Fertilizer guide: pastures in western Oregon and western Washington. Oregon State Univ. Ext. and Exp. Sta. Pub. FG-63. Corvallis, OR.

Rogers M (1993) Spruce up turf with slow-release nitrogen. Grounds Maintenance 28(7):12-14, 16, 20, 22, 70.

Sullivan DM, Fransen SC, Bary AI & Cogger CG (1998) Fertilizer nitrogen replacement value of food residuals composted with yard trimmings, paper, or wood wastes. Compost Sci. Util. 6. (In press)

Turner DO (1979) Nitrogen fertilization of orchardgrass pasture as influenced by irrigation. Washington State Univ. College of Agriculture Research Center Bull. 881, Pullman, WA.

Yungen JA, Jackson TL & McGuire WS (1977) Seasonal responses on perennial forage grasses to nitrogen applications. Oregon State Univ. Exp. Sta. Bull. 626. Corvallis, OR.

COMPOSTING OF HAZARDOUS WASTES AND HAZARDOUS SUBSTANCES

K.W. BROWN
Soil and Crop Sciences Department, Texas A&M University, College Station, Texas 77843, USA
E-mail Brown@tamu.edu

ABSTRACT

Composting has typically been used to treat agricultural wastes, yard wastes and sewage sludges which most often contain negligible concentrations of hazardous organic substances. Other wastes including household refuse, industrial wastes and hazardous wastes often contain greater concentrations of hazardous organic substances which could pose a threat to human health or the environment during treatment or after disposal. Some of the organic hazardous substances found in the waste may be amenable to treatment by composting. A series of research projects have been conducted on the fate of volatile organic chemicals found in municipal and hazardous wastes, and the fate of polynuclear aromatics and explosives during composting.

Volatile organic chemicals are rapidly lost during the initial few hours to days of composting both in aerated static pile and in-vessel composting systems. In-vessel composting has been shown to be effective in degrading 65-83% of nitrocellulose propellants in a 6-8 week period. This resulted in first order half lives of 7.7-13.6 days for the nitrocellulose. In-vessel composting of hazardous American Petroleum Institute separator sludge wastes (K051) resulted in the rapid loss of spiked volatile organic chemicals including ethylbenzene, m-xylene, and dimethylnaphthalene. It also resulted in biodegradation which decreased the residual concentrations of anthracene, chrysene and phenanthrene. All of the volatile organic chemicals released in municipal solid waste at the initiation of composting vaporized within a few hours or days at the longest. The results demonstrate that composting can be used to decrease the concentration of hazardous organic substances, but that the more volatile constituents may vaporize, rather than degrade.

INTRODUCTION

Composting has the potential to reduce the toxicity of a variety of wastes by degrading toxic organic substances which are present in many waste streams. Federal laws regulating the disposal of wastes provide a list of hazardous substances (40CFR, Part 261), many of which may be degraded during composting. The hazardous waste disposal regulations

S. Brown, J.S. Angle and L. Jacobs (eds.), Beneficial Co-Utilization of Agricultural, Municipal and Industrial By-products, 327–340.
© 1998 *Kluwer Academic Publishers*,

(40CFR, Part 268) also ban the land disposal of certain hazardous wastes specifically because they contain certain hazardous substances, and thus, if composting is effective in decreasing the concentrations of at least the organic hazardous substances in a hazardous waste, it may serve as an effective pretreatment prior to land disposal. Despite recent improvements in regulation for municipal solid waste landfills (40CFR, Part 258), including requirements for the use of liners and leachate collection and removal systems, the wastes placed in these facilities continue to release hazardous substances to the environment. One possible solution to this problem would be to treat municipal solid waste (MSW) by composting before it is placed in a landfill, as suggested by West et al. (1997). In addition, some specific waste streams, including, for example, pesticide residue, munitions, wood preserving wastes or media contaminated with these and similar products may be amenable to treatment by composting. Caution is needed, however, since not all organic compounds will be fully degraded in a given environment (Irvine, 1989).

Before composting facilities can be designed to decrease the concentrations of hazardous substances in wastes, or to degrade the hazardous substances which result from cleanup activities, information is needed on the fate of such substances during composting. Data are needed on the rates of volatilization, the rate and completeness of biodegradation, and the impact of composting variables which may influence the concentrations of hazardous substances released from or remaining in the compost.

There is very little information available in the literature on the release of volatile hazardous substances to the atmosphere during composting. Information is also needed on the fate of pesticides and semi-volatile hazardous substances including polynuclear aromatics during composting. There have been several studies on the use of composting to degrade explosive wastes and munitions (Williams et al., 1989; Williams and Keehan, 1993; Mohsin, et al., 1993; Brown et al., 1995), but in most instances, while the original compound may have degraded, at least to some extent, complete degradation of the original chemical such as 2,4,6-Trinitrotoluene (TNT) has not been fully achieved and in some instances it appears that toxic metabolites remained at significant concentrations after the composting was complete (Isbister et al., 1984).

COMPOSTING OILY WASTES

Many hazardous waste streams contain hazardous organic substances which may be amenable to treatment by composting. Wastes from the petroleum refining industry are among the more voluminous hazardous wastes which are generated. Among the more toxic of these wastes is the American Petroleum Institute (API) oil water separator sludge, Resource Conservation and Recovery Act number K051. It consists of the oily materials which are denser than water, and settle to the bottom of the oil water separator. API separator sludge contains trace volatile organics including benzene, toluene, ethylbenzene, and xylene, as well as numerous polynuclear aromatic compounds (Donnelly et al., 1985). Presently, such wastes can not be disposed of in landfills or by land treatment and must be incinerated. If the hazardous substances in this waste could be decreased to an acceptable concentration by treatment, it may be possible to dispose of the waste in landfills or by land treatment. Previous studies of composting of oily wastes (Deever and

White, 1978) have demonstrated decreases in the oily fraction of the waste, but they did not present information on the degradation or volatilization of specific waste components.

Studies of the fate of hazardous constituents of API separator sludge were undertaken utilizing in-vessel composters to fill these data needs.

Methods

A series of nine laboratory scale in-vessel composters were designed and constructed from 186 liter portable concrete mixers specially equipped with seals and an air distribution system which provided flow control and the collection of gaseous emissions (Spongberg et al., 1996). The environmental parameters and composition of the initial ingredients were chosen to optimize aerobic biodegradation. Each composter was loaded with 20 kg dry wt of a mixture of coarse sawdust, cow manure, municipal sewage sludge, and 0, 5 or 10% by weight API separator sludge. All ingredients were thoroughly mixed for one hour in the sealed composters prior to the initiation of air flow which commenced the composting. The chemical and physical properties of the compost were as follows: initial C:N = 100:1, porosity = ~30%, bulk density = 0.79 g/cc, O_2 >10%, T_{max} = 50°C, and moisture = 50-60%.

Respiration rate was evaluated using alkali traps to collect the evolved carbon dioxide. An oxygen meter was used to quantify the percent oxygen in the air exiting the composters. Thermocouples were mounted inside the composters to monitor the internal temperature.

The study was subdivided into two phases. Phase I was designed to monitor volatilization loss as a function of waste loading rate and utilized compost matrix with 0, 5 or 10% API waste added. Compounds selected for analysis were chosen based on their prevalence in the API waste and their toxicity and included benzene, toluene, ethylbenzene, xylene and dimethylnaphthalene. Phase II of the study was designed to measure volatilization and degradation of 6 organic compounds which were spiked into the composting matrix containing the above mentioned percentages of API waste. Spiking was achieved by dissolving the appropriate amount of organic chemical in 1 liter of acetone and then mixing it with the waste immediately prior to putting the waste into the composter. Spiked chemicals and concentrations (mg chemical/kg waste) were as follows: 400 and 800 mg/kg concentrations each of ethylbenzene and m-xylene; and 200 and 400 mg/kg concentrations each of dimethylnaphthalene, anthracene, phenanthrene, and chrysene. Each treatment was replicated in three separate composters.

Volatilized organics in the exiting air stream were trapped in a filter containing 25 g of activated carbon inside a 2.5 cm I.D. glass column. Backup filters were used to prevent the loss of volatile compounds when the initial filters might have been saturated. Filters were changed and analyzed at 0.08, 0.17, 0.33, 0.5, 1, 2, 4, 8, 14, 21, 28, and 35 days after the initiation of composting. Volatiles adsorbed onto the carbon were desorbed with carbon disulfide and analyzed using a gas chromatograph. Samples of the compost were collected at 0.33, 1, 4, 8, 14, 21, 28, and 35 days following the initiation of the compost cycle. These samples were analyzed to determine the net loss of extractable hydrocarbons and to quantify residual concentrations of specific compounds. The samples were extracted with methylene chloride using an automated Tecator Soxtec System apparatus and analyzed by gas chromatography.

Results

During the composting of the treatments containing 5-10% API separator sludge waste, 1-2% of the ethylbenzene initially present in the waste was volatilized (Figure 1). The volatilization occurred fairly rapidly and was essentially completed in the first 3-4 days. Volatilization of benzene and m- and p-xylene occurred in a similar fashion but losses only amounted to 0.5-2% of that initially present. Volatile losses of o-xylene and toluene occurred more slowly and extended over an 8 day period and totaled 1-4.5% of that initially present in the waste. The fairly low rates of volatilization of these volatile organic chemicals indicate that they are highly adsorbed to the organic phase of the waste and are unavailable for volatilization.

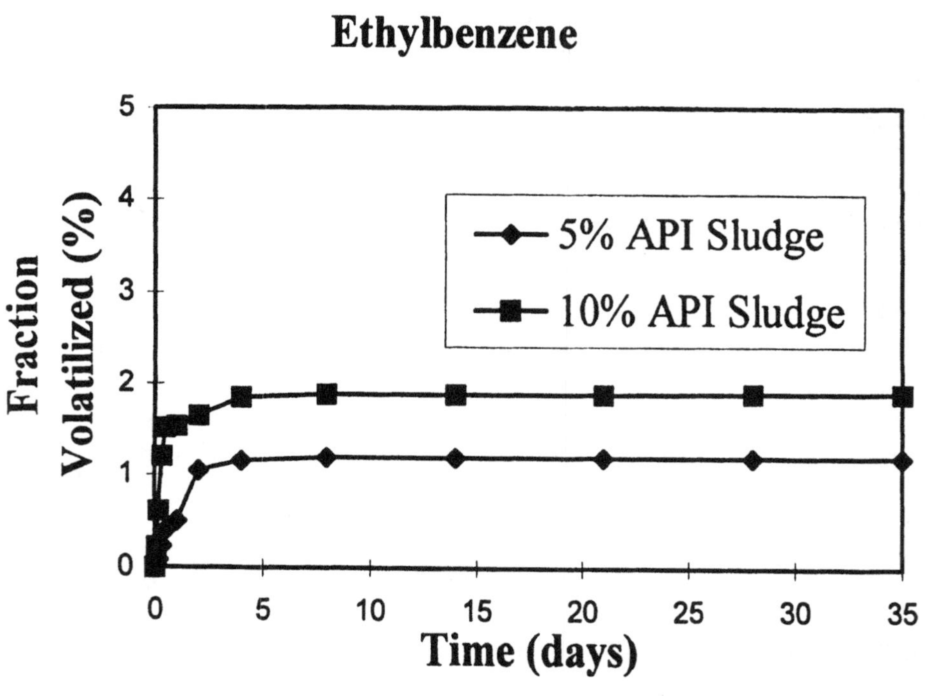

Figure 1. The cumulative volatilized ethylbenzene expressed as a percentage of that initially present from compost containing 5 and 10% by weight API separator sludge waste.

When ethylbenzene was spiked into the waste immediately prior to introduction into the composters, much higher rates of volatilization were observed (Figure 2). From 12-17% of the applied ethylbenzene volatilized from the treatments containing 5% API separator sludge waste spiked with either 400 or 800 mg/kg ethylbenzene or 10% API separator sludge waste spiked with 400 mg/kg. Up to 45% of the applied ethylbenzene

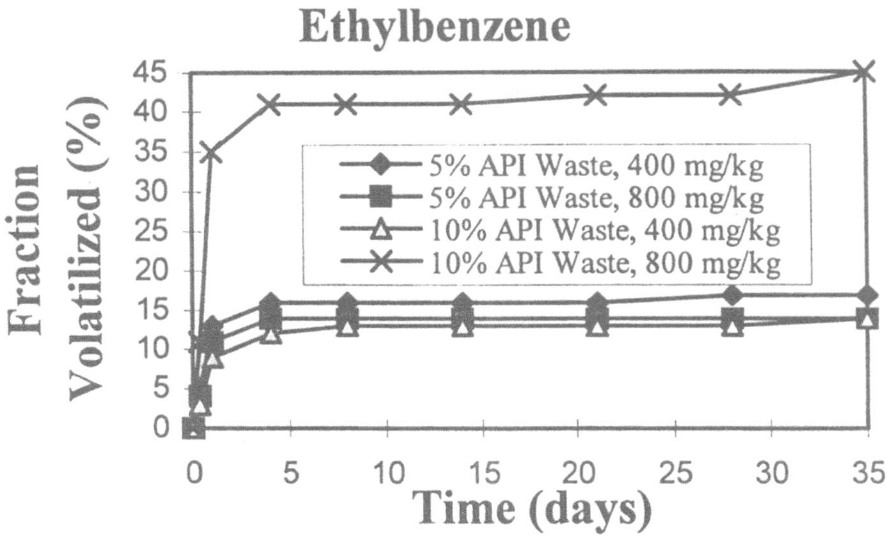

Figure 2. The cumulative volatilized ethylbenzene expressed as a percentage of that added to compost containing 5 and 10% by weight API separator sludge waste spiked with 400 and 800 mg/kg ethylbenzene.

was lost by volatilization from the treatment containing 10% API separator sludge waste spiked with 800 mg/kg ethylbenzene. Almost identical amounts of volatilization were measured for m-xylene. The cause of the higher amount of volatilization from the treatment containing 10% API separator sludge waste spiked with 800 mg/kg ethylbenzene is not known but it is thought that the spiked chemical may not have fully equilibrated with the compost matrix prior to the initiation of composting. Volatilization of dimethylnaphthalene occurred more slowly over the entire 35 day period and the total losses were less. This is likely due to the lower vapor pressure of this chemical as compared to that of xylene and ethylbenzene. These data indicate the need to use caution when using data regarding the volatilization and degradation of organic chemicals in wastes. Chemicals added to the waste appear to be more highly subject to volatilization while those indigenous to the waste are more highly bound to the organic fraction and thus, are less volatile.

Half lives calculated from the rate of disappearance of a volatile organic chemical may be due to both biodegradation and volatilization. Both ethylbenzene and m-xylene had half lives of 0.5-4.2 days for total removal (Table 1). When the effects of volatilization were removed, the half lives based solely on biodegradation ranged from 0.7-5.6 days for m-xylene, 1.3-4.6 days for ethylbenzene and 7.0-11.6 days for dimethylnaphthalene. Although non-volatile in nature, chrysene and phenanthrene had similar half lives which ranged from 0.5-7.0 days and 1.7-11.3 days respectively. Anthracene was more resistant to biodegradation and had the highest half life which ranged from 2.4-24.4 days during

Table 1. Measured half lives for organic compounds in compost (after Spongberg and Brown, 1995).

Organic Compound	Biodegradation	Biodegradation plus Volatilization
Ethylbenzene	1.3 - 4.6 days	0.5 - 4.2 days
Meta Xylene	0.7 - 5.6 days	0.5 - 4.2 days
Dimethylnaphthalene	7.0 - 11.6 days	3.7 - 7.6 days
Anthracene	2.4 - 24.4 days	Non-Volatile
Phenanthrene	1 .7 - 11.3 days	Non-Volatile
Chrysene	0.5 - 7.0 days	Non-Volatile

composting. The large half lives typically occurred in the treatments which received 10% API separator sludge waste.

The mass balance for the treatment receiving 5% API separator sludge waste spiked with 400 mg/kg each of ethylbenzene, m-xylene, dimethylnaphthalene, and anthracene indicated that 17% of the applied ethylbenzene volatilized and the remaining 83% was biodegraded (Table 2). For m-xylene, 12% volatilized and 88% degraded. Only 9% of the dimethylnaphthalene and none of the anthracene volatilized; however, 90 and 98% respectively, biodegraded. The results from the treatment containing 10% API separator sludge waste with the same amounts of spiked chemicals revealed slightly reduced volatilization and increased biodegradation of ethylbenzene and m-xylene. Evidently the increased organic content from the sludge decreased the volatilization of the spiked chemicals. The volatilization rate of dimethylnaphthalene increased to 33% causing a reduction in degradation to only 66%. The degradation rate of anthracene was also lower in the 10% treatment possibly due to less favorable conditions of the compost mixture containing the greater concentration of this hazardous waste.

Table 2. The percentage of spiked chemicals which volatilized and degraded from compost containing 5 and 10% by weight APT separator sludge waste spiked with 400 mg/kg each of ethylbenzene, m-xylene, dimethylnaphthalene, and anthracene over a 35 day period.

Compound	Volatilized (%)	Degraded (%)
Ethylbenzene	17	83
m-Xylene	12	88
Dimethylnapthalene	9	90
Anthracene	0	98

COMPOSTING EXPLOSIVE WASTES

Large volumes of waste propellants are generated each year through the decommissioning of surplus weaponry. Due to the highly flammable and explosive nature of most propellants, disposal options for waste propellants have largely been limited to open burning and detonation; however, these are not desirable due to their adverse environmental effects. Since nitrocellulose can be microbially degraded, and since composting is primarily accelerated microbial degradation of organic materials; composting may serve as an acceptable and environmentally safe treatment option for waste propellants. The process may be further improved through the use of in-vessel composting in which the process and all byproducts are completely contained. The present study was conducted to determine the ability of in-vessel composting to degrade the nitrocellulose content of two commonly used small arms propellants.

Methods

A series of nine laboratory scale in-vessel composters as described by Spongberg et al. (1996) were used to study the degradation of two propellants and a control. Propellants selected for study were WC860 and H5010. The manufacturer's analysis of the propellant compositions are given in Table 3. The primary ingredient of both propellants is nitrocellulose with lesser amounts of nitroglycerin, dibutylphthalate, and calcium carbonate. Three replications of each treatment were run. Each composter was loaded with the equivalent of 10 kg of dry composting matrix consisting of 33% grass clippings and food waste, 25% sewage sludge, 20% cow manure, 11% paper, and 11% wood chips. The moisture content of the mixture was then brought to 50% by weight by the addition of tap water followed by overnight mixing to assure uniformity. A total of 200 g (2% by weight of the initial dry compost) of the propellant to be tested was added to the mixture and the composting cycle was begun. Compost samples were collected weekly for moisture content, ash content, and propellant analysis. Oxygen measurements were taken periodically by inserting an oxygen probe through the sampling door of the composter. Temperature was monitored by means of two thermocouples which were located near the center of the composter. Moisture content was measured by drying the sample to constant weight at 105°C. Ash content was measured as the residue remaining after heating to 550°C.

Table 3. The chemical composition of the two propellants used in the in-vessel composting study (after Brown et al,. 1995).

Constituent	WCS60 (%)	H5010 (%)
Nitrocellulose	73 - 85	87.4 - 92.5
Nitroglycerin	8 - 11	
Dibutylphthalate	6 - 10	
Calcium Carbonate	1.0	
Dinitrololuene	0 - 1	6.5 - 10.0
Diphenylamine	0.75 - 1.5	
Potassium Nitrate	0.1 - 1.5	0.5 - 1.25
Sodium Sulfate	0.5	0.1 - 1.0
Graphite	0.4	
Tin Dioxide	0 - 0.1	0.4

Nitrocellulose concentrations were measured using the spectrophotometric procedures outlined by Weston (1989). Since the initial propellants were used for preparing the standard solutions, the results are expressed in terms of units of propellant rather than units of nitrocellulose. Furthermore; since the organic material in the compost matrix degraded along with the propellants, the results could not be interpreted in terms of concentration in the residual mixture, and we thus normalized it to units of propellant per unit of ash weight.

Results

The cumulative degradation of propellant WC860 expressed as a percentage of that applied per gram of ash, is shown in Figure 3. Except for the first three measurements, the data show a consistent decrease in concentration with time through the first 40 days of composting. The variability in the initial data is due to experimental error associated with sampling a highly heterogenous matrix. Approximately 60% of the applied propellant was degraded within the first 23 days and a total of 72% of the applied propellant was degraded by 48 days after the initiation of composting. The average half life for this propellant under these composting conditions was 13.6 days. The degradation rate did not appear to be highly sensitive to moisture content; however, it was greatly affected by lack of adequate oxygen. Results of the composting of the second propellant, H5010, were similar with an average of 65% of the applied propellant being degraded over a 56 day composting period. Propellant H5010 had a half life of 7.7 days under the composting conditions in these experiments. Since the complete degradation of the propellants was not accomplished in the normal compost cycle, additional work needs to be done to determine the rate of degradation and the length of time needed for these materials to degrade in the curing phase. In addition, no attempt was made to determine the breakdown products of these propellants and it is not known if the nitrocellulose completely degraded to carbon dioxide and water or if there are significant amounts of intermediate breakdown products remaining in the compost.

Williams et al. (1989) and Williams and Keehan (1993) conducted studies of both static pile and in-vessel composting of TNT contaminated soils at the Umatilla Army Depot Activity (UMDA) facility. They found that both static pile and in-vessel composting systems were effective in greatly decreasing the concentrations of TNT, Hexahydro-1,3,5-trinitro-1,3,5-triazine (RDX) and octahydro-1,3,5,7-tetranitro-1,3,5,7-tetraazocine (HMX) in contaminated soil, but that the mechanically agitated systems had superior performance. They reported that the amount of contaminated soil added to the compost was critical to the successful decomposition of these explosives. Additions of over 30-40% contaminated soil in the composting matrix reduced the rate of degradation. Previous work by Isbister et al. (1984) using radiolabeled TNT and RDX indicated that there was no release of radiolabeled carbon dioxide during composting. The explosives degraded to intermediate degradation products and did not fully degrade to carbon dioxide and water. To date no one has fully characterized the degradation pathway for explosives during composting, nor the amounts, persistence and toxicity of the degradation products remaining in the final compost.

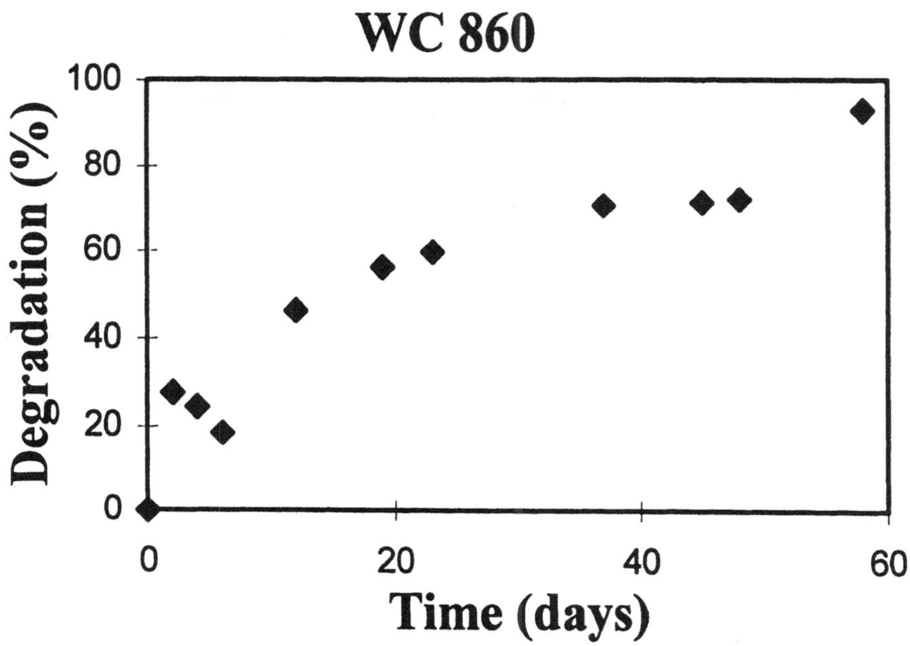

Figure 3. The cumulative degradation of propellant WC860 as a function of time. Degradation is expressed as a percentage of that added to the compost.

COMPOSTING MUNICIPAL SOLID WASTES

Volatile chemicals are components of many products which continue to be used by consumers in their homes, as well as by people working in small commercial establishments (Brown and Nelson, 1994). These products include cosmetics, cleaning products, polishes, waxes, paints, pesticides and auto maintenance products; and contain volatile chemicals including acetone, alcohols, benzene, carbon tetrachloride, cresol, formaldehyde, naphthalene, phenols, toluene, trichloroethylene, xylene, and chlorinated benzenes among others. The volatile constituents disposed of in the municipal solid waste, which includes both household garbage and waste from small commercial activities, would either be dispersed into the waste at the time of collection as would occur from spill clean up activities, or would remain in a container until it is released.

There are several methods by which municipal solid waste may be prepared for composting, and several methods of composting, each of which would influence when a package containing a volatile substance would be ruptured. If the waste is shredded in preparation for composting, volatiles would be released immediately; if bulk waste is introduced into an in-vessel composter, or the bulk waste is put in a windrow, the containers may not be ruptured until the container corrodes or until the waste is turned. Once the volatile contents of a container are released, they may remain dissolved in other components in the product, as xylene may remain dissolved in oil; they may be sorbed onto other components of the MSW; they may dissolve into the water in the waste; they may biodegrade; or they may vaporize. The partitioning between the various components of the waste and rates of release to the atmosphere, will determine if such releases constitute an environmental concern, and if safeguards may be needed to control releases.

Studies were thus undertaken to determine the fate of volatile contaminants from sudden releases which were simulated to occur from a mass of MSW at the initiation of composting in a ventilated static pile.

Methods

A series of nine aerated static pile composters were made from 200 L metal drums. Each was equipped with air entry ports on the top and air and leachate removal ports on the bottom. The composters were covered with insulation and mounted on swivel stands to allow them to be periodically emptied and refilled to simulate mixing of the compost piles. Ambient air was drawn into the top of the composters, passed through the composting matrix, through a condenser to remove the water vapor, through a charcoal trap to collect organic vapors, and finally through a flow meter used to measure the flow rate which ranged from 40-60 standard cubic feet per hour (scfh). The compost matrix was a simulated municipal solid waste composed of 50% shredded paper, 20% food products, 20% yard trimmings, 5% wood products, and 5% plastic (Franklin Associates, Ltd., 1986). Care was taken when collecting these components to obtain only materials which had not been contaminated with any of the compounds to be studied. Appropriate weights of each component were manually mixed. Half the waste was added to the composter and manually tamped to a density of approximately 0.25 g cm^{-3}. The chemicals to be spiked were then added to the composter, followed immediately by the remaining half of the waste which was also compacted. Tap water was added as necessary to increase the moisture content to 50% by weight. The composters were spiked with four commonly occurring (Brown and Nelson, 1994) volatile organic chemicals including benzene, carbon tetrachloride, dichlorobenzene, and xylene. The volatile organic chemicals were mixed together in equal amounts by weight and the mixture was added to the compost to achieve total concentrations of 275, 1375, and 2750 mg/kg which represent 1, 5, and 10 times the 275 mg/kg estimated (Franklin Associates, Ltd., 1986) to be in MSW. The charcoal traps were replaced at 3, 9, 18, 30, 48, 96, 168, 336 hours, 3 weeks and 4 weeks after the start of the composting process. Samples of leachate and waste were collected weekly and were extracted for analysis. The charcoal was extracted with carbon disulfide and all extracts were analyzed by Gas Chromatography.

Results

Benzene, which has a vapor pressure of 75.2 mm Hg at 20°C, was rapidly volatilized from the composting matrix and lost via the exiting air stream. Benzene concentrations in the exiting air stream peaked at 3 hours after the start of composting (Figure 4) and was nearly back to background levels by 9 hours after the start of composting. The maximum air concentration of 12 g/m^3 was far above the air quality standard (Smith, 1996) of 0.00022 mg/m^3 and also well above the threshold limit value (TLV) of 32 g/m^3 (American Conference of Governmental Industrial Hygienists, 1992) and indicates the need to dilute or treat the exhaust prior to release. Volatilization of xylene which has a vapor pressure of 6.16 mm Hg at 20°C was slightly slower and extended over a 30 hour period and the maximum air concentrations were reduced to slightly over 5 g/m^3 which is still far above the air quality standard of 0.73 mg/m^3 and the TLV of 434 mg/m^3. Dichlorobenzene was the least volatile chemical with a vapor pressure of 1 mm Hg at 20°C and required 336 hours (14 days) to completely volatilize; however, the vast majority was volatilized during the first 96 hours (4 days). The maximum air concentration was 1.9 g/m^3 as compared to the air quality standard of 0.150 mg/m^3 and the TLV of 150 mg/m^3.

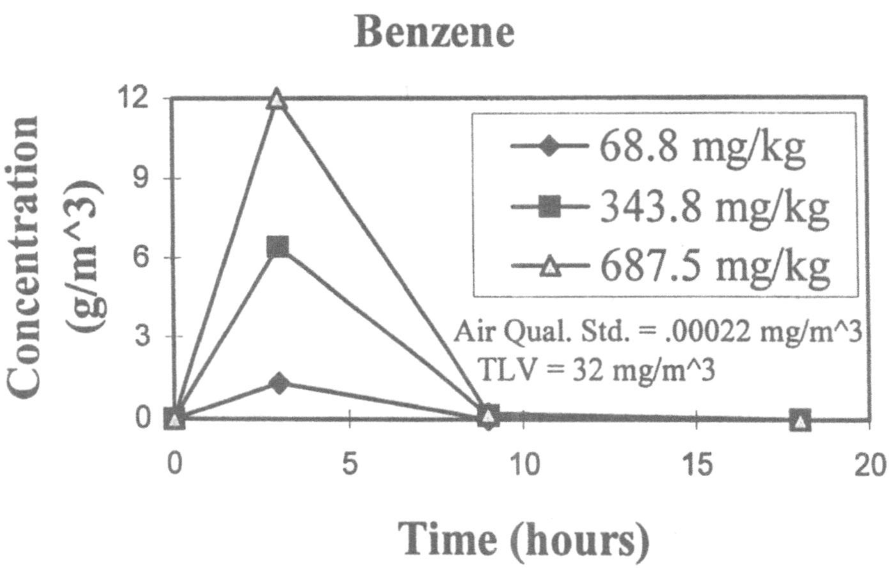

Figure 4. The concentration of benzene in the exhaust air stream as a function of time from ventilated static pile composters containing simulated MSW spiked with 3 concentrations of benzene.

Since the concentrations of all four volatile solvents in the exhaust stream from the ventilated compost with the spiked concentrations tested here greatly exceeded their respective current air quality standards, such exhaust would either need to be controlled or treated for the first few days of composting, or it would need to be released in such a way and at such a location so that it would be mixed with sufficient ambient air to dilute the chemicals prior to reaching a receptor. Treatment of the exhaust air by passing it through a biofilter containing white rot fungus (*Phanerochaete chrysosporium*) has the potential to remove voatile organic chemicals (VOC) including ethylbenzene, xylene, and toluene (Braun-Lullemann et al., 1995).

Plots of the percent of the applied volatile organic chemicals which volatilized as a function of time (Figure 5) show that 100% of the benzene was volatilized in less than 0.25 day. Carbon tetrachloride behaved similarly, while xylene required approximately 0.75 day and dichlorobenzene required 3 days to completely volatilize. In all cases, the cumulative release of the volatile organic chemicals tested at all initial concentrations indicated complete volatilization.

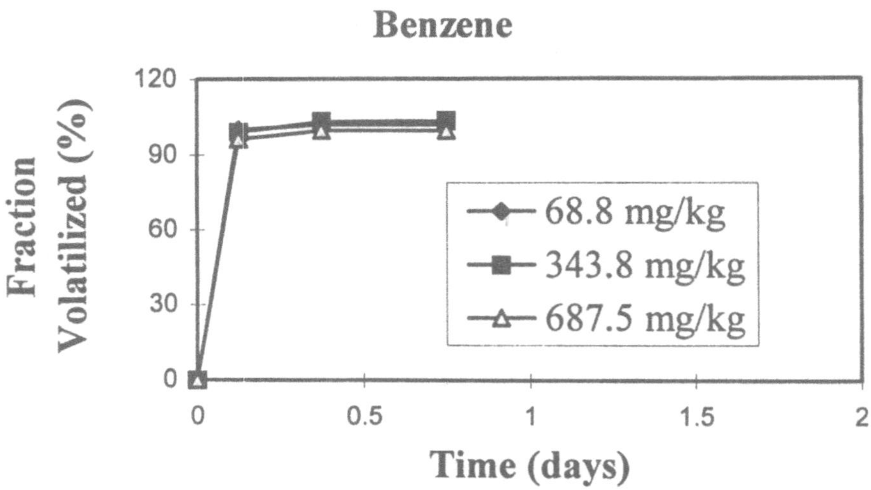

Figure 5. The percent of applied benzene which volatilized as a function of time from ventilated static pile composters containing simulated MSW spiked with 3 concentrations of benzene.

The rapid loss of volatile organic chemicals from composting municipal solid waste is supportive of the data reported by Eitzer (1995) from a field survey of air quality at 8 commercial MSW composting facilities. He found that the highest concentrations of VOCs were found at the tipping floors, the shredders and in areas where the compost first came to operating temperature. During the periods when he made his measurements, the concentrations in the air at these facilities did not exceed the TLV. He did not however, directly measure the exhaust stream from a ventilated compost pile.

The rapid volatilization of the volatile chemicals spiked into the compost minimized the possibility of biodegradation. Furthermore, since all the volatiles added were quickly vaporized, it is not surprising that none were detected in either the leachate or the waste samples collected 1 week after the beginning of composting or thereafter.

Thus, the data indicate that VOCs released into MSW are not highly adsorbed and are rapidly lost by volatilization. Volatile losses will occur mainly when the wastes are initially exposed to air such as during dumping, sorting and shredding operations and when the temperature is elevated as happens during the initial stages of composting. Special precautions, possibly including the dilution or treatment of off-gasses during the first few days of composting of wastes containing freshly released volatile organic chemicals, may be necessary to meet air quality standards and to protect workers health. For composting wastes containing recalcitrant organic chemicals, aging or further treatment may be needed to further decrease their concentrations prior to release.

CONCLUSIONS

- The concentrations of organic hazardous substances including munition propellants, polynuclear aromatics and volatile organic chemicals decrease during composting.
- Munition propellants including nitrocellulose and TNT have half lives on the order of 1 to 2 weeks, but are often not degraded completely during a normal compost cycle. While the original chemicals degrade, significant concentrations of metabolites may remain.
- Polynuclear aromatics in oily waste degrade during composting, with half lives ranging from 0.5-24 days. Again, while the concentrations decrease considerably during a 4 to 6 week compost cycle, these chemicals do not degrade completely.
- Only a very small fraction of volatile organic chemicals indigenous to oily waste are volatilized during composting; however, much larger fractions volatilized from spiked additions of the same chemicals.
- Volatile organic chemicals including benzene, toluene, xylene, and dimethylnaphthalene which were spiked into the waste, to simulate their release from containers in MSW during shredding or during composting were released to the atmosphere typically within hours to a few days of the beginning of composting.

ACKNOWLEDGEMENTS

This manuscript is a contribution of the Texas Agricultural Experiment Station, Texas A&M University System, College Station, TX 77843. The research was supported in part through grants from the Gulf Coast Hazardous Substance Research Center, Lamar University, Beaumont, TX; the Texas Natural Resource Conservation Commission, Austin, TX; and Day and Zimmerman Inc., Lone Star Division, Texarkana, TX 75505

REFERENCES

American Conference of Governmental Industrial Hygienists (1992) 1992-1993 Threshold limit values for chemical substances and physical agents and biological exposure indices; ACGIH: Cincinnati, OH.

Braun-Lullemann A, Johannes C, Majcherczyk A & Huttermann A (1995) The use of White-Rot Fungi as active biofilters. In: Hinchee RE, Sayles GD & Skeen RS (Eds) Biological Unit Processes for Hazardous Waste Treatment (pp. 235-248). Batelle Press, Columbus, OH.

Brown KW & Nelson LD (1994) Sources of hazardous constituents in municipal solid waste and landfill leachate. Paper presented at the Seventh Annual Municipal Solid Waste Management Options for Texas '94 Conference, Jan 19-21,1994, Austin, TX.

Brown KW, Thomas JC & Seago OJ (1995) Degradation of explosive propellants by in-vessel composting. BioCycle 36:56-58.

Deever WR & White RC (1978) Composting petroleum refinery sludges. Paper presented at the Governor's Conference on Economics, Waste, and the Environment, Baton Rouge, Louisiana, Nov 8-10, 1978. Pg. 1-24.

Donnelly KC, Brown KW, Thomas JC, Davol P, Scott BR & Kampbell D (1985) Evaluation of the hazardous characteristics of two petroleum wastes. Hazardous Waste & Hazardous Materials 2:191-208.

Eitzer BD (1995) Emissions of volatile organic chemicals from municipal solid waste composting facilities. Environmental Science & Technology 29:896-902.

Franklin Associates, Ltd. (1986) Characterization of municipal solid waste in the United States, 1960-2000. NTIS No. PB87-178323.

Irvine RL (1989) Overview of biological processes as applied to hazardous waste problems. In: Lewandowski G, Armenante P & Baltzis B (Eds) Biotechnology Applications in Hazardous Waste Treatment (pp. 7-25). Engineering Foundation, New York.

Isbister JD, Anspach GL, Kitchens JF & Doyle RC (1984) Composting for decontamination of soils containing explosives. Microbiologies 7:47-73.

Mohsin MB, Hohnson JH Jr & Wan LW (1993) Bioremediation of 2,4,6-trinitrotoluene contaminated soil via 2,4,6-trinitrobenzoic acid using composting technology. Paper presented at the 25th Mid-Atlantic Industrial and Hazardous Waste Conference, University of Maryland, College Park, MD, July 7-9, 1993.

Smith RL (1996) EPA region III risk-based concentration table. April 19, 1996. 22 pg. Spongberg AL, Thomas JC & Brown KW (1996) Laboratory scale in-vessel composter designed for volatile emissions analysis. Journal of Environmental Quality 25:371-373.

West ME, Brown KW & Thomas JC (1997) Methane production of raw and composted solid waste in simulated landfill cells. Waste Management and Research (In Press).

Weston RR (1989) Field demonstration - composting of propellants contaminated sediments at the Badger Army Ammunition Plant (BAAP). U.S. Army Toxic and Hazardous Materials Agency Report No. CETHA-TE-CR-879061.

Williams RT & Keehan KR (1993) Hazardous and industrial waste composting. In: Hoitink HAJ & Keener HM (Eds) Science and Engineering of Composting: Design, Environmental, Microbiological and Utilization Aspects (pp.363-382). The Ohio State University, Wooster, OH.

Williams RT, Ziegenfuss PS, Mohrman GB & Sisk WE (1989) Composting of explosives contaminated sediments. In: Lewandowski G, Armenante P & Baltzis B (Eds) Biotechnology Applications in Hazardous Waste Treatment (pp. 307-317) Engineering Foundation, New York.

NITROGEN RECOVERY BY BAHIAGRASS FROM PELLETIZED BIOSOLIDS

ROSA M. MUCHOVEJ[1] and J.E. RECHCIGL[2]
University of Florida, [1]Southwest Florida REC, Immokalee, Florida 34142, USA and [2]Range Cattle REC, Ona, FLorida 33865 USA, respectively
E-mail RMM@icon.imok.ufl.edu

ABSTRACT

Nitrogen fertilization has been shown to increase biomass and quality of bahiagrass (*Paspalum notatum* Flugge) dramatically in Florida sandy soils. Nevertheless, with low cattle prices, ranchers are forced to reduce fertilizers inputs. Biosolids contain considerable amounts of N and are inexpensive when compared with commercial fertilizers. Determining application rates that will not exceed crop assimilative capabilities is important to prevent losses to the environment. The objective of this study was to investigate temporal N recovery by bahiagrass pastures from pelletized municipal biosolids. Treatments included 7 rates (0, 0.55, 1.1, 2.2, 4.4, 8.8, and 17.6 Mg/ha) of biosolids, which provided 0, 25, 50, 100, 200, 400, and 800 kg N/ha (organic plus inorganic). Treatments were applied to 6 m x 3 m plots arranged in a randomized complete block design with 4 replications. In addition to the absolute control (0 kg biosolids/ha), control plots were established that received water-soluble N at 180 kg N/ha. Phosphorus and K were applied at the same rate to all plots. Bahiagrass forage was harvested at 35, 69, 104, 151, and 201 days after initial application for yield and total Kjeldahl N determinations. Bahiagrass forage yields increased linearly with increasing biosolids rate in all harvests. Nitrogen accumulation for biosolids treatments was higher than those obtained with the soluble N sources, especially at higher biosolids rates. Nitrogen recovery rates reached values higher than 75% for the lowest biosolids rate. Increasing rates resulted in greater N accumulation in the forage, but comparatively lower recovery rates were found. Recovery rates, in percentage of total N applied, decreased at each subsequent harvest. Most of the N was removed within the first 35 days after application.

INTRODUCTION

Biosolids production has been increasing consistently worldwide in the last few years primarily due to efforts to clean up waterways (i.e. more sewage treatment plants) and due to population increases and urbanization. Land application of wastes is an alternative to disposal methods such as land filling or incineration, and a means to recycle nutrients. Biosolids applied to agricultural land can improve soil physical properties (e.g., water

S. Brown, J.S. Angle and L. Jacobs (eds.), Beneficial Co-Utilization of Agricultural, Municipal and Industrial By-products, 341–347.
© 1998 *Kluwer Academic Publishers*,

retention, infiltration, aggregate stability) as well as chemical characteristics (e.g., cation exchange capacity and plant nutrient availability) (Hue, 1995).

Pastures are a particularly attractive target for biosolids application for several reasons. They often are under-fertilized, they occupy extensive land areas, and high amounts of fertilizer are required for acceptable production. In Florida there are currently more than 5 million hectares of pasture that require fertilization, and nearly 2 million of these are planted with bahiagrass. Most of the grasslands present visible signs of N and Fe deficiencies. Sufficient fertilization is an essential part of pasture management, and N is frequently the most limiting nutrient. However, with low net return for cattle, ranchers are forced to reduce production costs, so pasture fertilization budgets often receive drastic cuts.

Biosolids contain considerable amounts of N, P, S, Ca and Fe, and constitute a relatively inexpensive nutrient source when compared with commercial fertilizers. Additions of both N and S dramatically increased yield and protein content of bahiagrass on Florida Spodosols (Sveda et al., 1992). In Florida, Fe deficiency is also common in pastures in the spring, especially following N fertilization. Biosolids have corrected iron deficiencies of sorghum and barley (McCaslin and O'Connor, 1985; McCaslin et al., 1986).

Determination of the quantity of N recovered by the forage from biosolids is important so that sufficient rates are applied without exceeding crop assimilative capabilities, thereby minimizing losses to the environment. The objective of this study was to determine the effects of varying rates of pelletized municipal biosolids on yield, N accumulation and N recovery by bahiagrass forage established on a Florida sandy soil.

MATERIALS AND METHODS

The experiment was initiated on an established bahiagrass pasture growing on Myakka fine sand (pH 5.0). Granular municipal biosolids [basic composition presented in Table 1], were applied in April at 7 rates (0, 0.55, 1.1, 2.2, 4.4, 8.8, and 17.6 Mg/ha) provided by BioGro Systems, Inc. These rates provided 0, 25, 50, 100, 200, 400, and 800 kg N/ha (organic plus inorganic), respectively. Phosphorus (40 kg/ha) and K (100 kg/ha) were applied to all plots as concentrated superphosphate and KCl, respectively. Treatments were surface-applied to 6 m x 3 m plots arranged in a randomized complete block design with 4 replications. In addition to the absolute control (0 kg biosolids/ha), control plots were established that received inorganic, water-soluble forms of N [$(NH_4)_2SO_4$ or NH_4NO_3] at 180 kg N/ha and 23.5 kg of $FeSO_4$/ha, which provided 2.3 kg of actual Fe/ha. According to the University of Florida / IFAS Standardized Fertilization Recommendations for bahiagrass pastures, three options are provided to the farmer, depending on desired grass forage production: high N (180 kg/ha), medium N (112 kg/ha) and low N (56 kg/ha) (Kidder et al., 1990).

Bahiagrass forage was harvested at 35, 69, 104, 151, and 201 days after initial biosolids application for yield and N determinations. Sub-samples of plant tissue from each harvest were analyzed for total Kjeldahl N. Forage N accumulation was calculated by multiplying yield times N concentration from each plot at each harvest date. The

Table 1. Composition of the municipal biosolids.

Solid 93.5%			
N (TKN)	4.5%	Fe	2600 mg/kg
NH_4	0.25%	Mn	430 mg/kg
NO_3-N	<0.02%	Cu	1200 mg/kg
P	1.75%	Zn	1350 mg/kg
K	0.05%	Cd	11.0 mg/kg
S	1.40%	Ni	75 mg/kg
Ca	1.6%	Pb	350 mg/kg
Mg	0.6%	pH	7.0 mg/kg
Na	0.15%		

amount of N recovered from biosolids in the forage was obtained by subtracting forage N from the control plots (which received no external N) from biosolids-amended plot forage N. The percentage recovery was calculated by comparing N recovered in the forage to the amount of N added with the various amendments (biosolids or soluble fertilizers). Nitrogen recovery rate was calculated by dividing the amount of N recovered in the forage by the number of days elapsed between the previous and the current harvest periods.

RESULTS

Bahiagrass forage yields increased linearly with increasing biosolids rate in all harvests (Figure 1). The overall effect of biosolids on forage production is depicted in the total dry matter yield (Figure 1; sum of forage from individuals harvests). Nitrogen concentration (Figure 2) and accumulation (Figure 3) increased above those obtained with the soluble N sources, especially with the higher rates of biosolids.

Nitrogen recovery rates by the forage ranged from 77.4 to 31.8%, for the lowest to the highest biosolids rates (Figure 4). Increasing rates resulted in greater N concentration and accumulation in the forage, but comparatively lower recovery rates. Recovery rates, in percentage of total N applied, decreased at each subsequent harvest. Most of the N was removed from the biosolids within the first 34 days after application during the first year (Table 2). Nevertheless, with higher biosolids rates, N recovery was still considerable for the period between 104 and 151 days after application.

DISCUSSION

Significant increases in forage yield, and nitrogen accumulation were obtained with biosolids rates higher than 1.1 Mg/ha, when compared with the absolute controls. The 4.4 Mg biosolids/ha rate was expected to supply approximately 200 kg N/ha under Florida conditions, based on their chemical composition (4.5% total N). Dewatered industrial biosolids (8.9% solids) have been applied to bermudagrass grown on a sandy loam and to

Table 2. Nitrogen Recovery Rate by Bahiagrass Forage After Treatment Application at the Various Harvest Dates.

Biosolids Mg/ha	N added kg/ha	June (34 d)[a]	July (69 d)	Sept (104 d)	Oct (151 d)	Dec (201 d)
	----------	----------	---------- g N/Day	----------	----------	----------
0.55	25	132	18.2	13.6	49.9	9.1
1.1	50	145.3	49.9	54.5	45.4	9.1
2.2	100	258.8	72.6	63.6	63.6	13.6
4.4	200	531.2	195.2	304.2	136.2	27.2
8.8	400	767.3	417.7	458.5	172.5	31.8
17.6	800	1085.1	867.1	617.4	286.0	31.8
AN	180	640.1	435.8	376.8	217.9	4.5
AS	180	508.5	199.8	240.6	86.3	9.1
AS + Fe	180	594.7	172.5	231.5	118.0	0

[a] No. of days after application; AN=Ammonium Nitrate; AS=Ammonium Sulfate; AS+Fe=Ammonium Sulfate plus Ferrous Sulfate at 23.5 kg/ha or 2.3 kg Fe/ha.

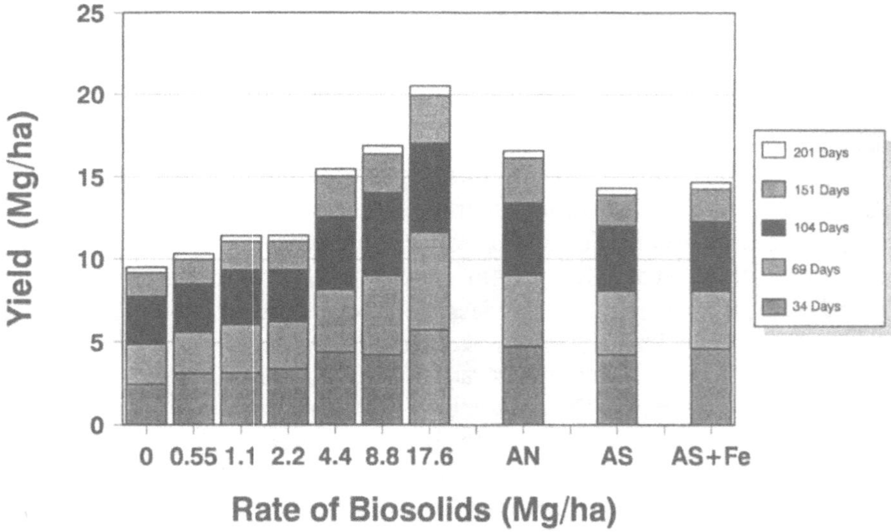

Figure 1. Bahiagrass Dry Matter Yield at various harvest dates: June Harvest, 34 days after application (DAA); July Harvest, 69 DAA; September Harvest, 104 DAA; October Harvest, 151 DAA; December Harvest, 201 DAA; Total dry matter yield of bahiagrass forage (1994) is shown as the sum of all harvest dates. (AN = Ammonium Nitrate; AS = Ammonium sulfate; AS + Fe = Ammonium Sulfate plus Ferrous Sulfate.)

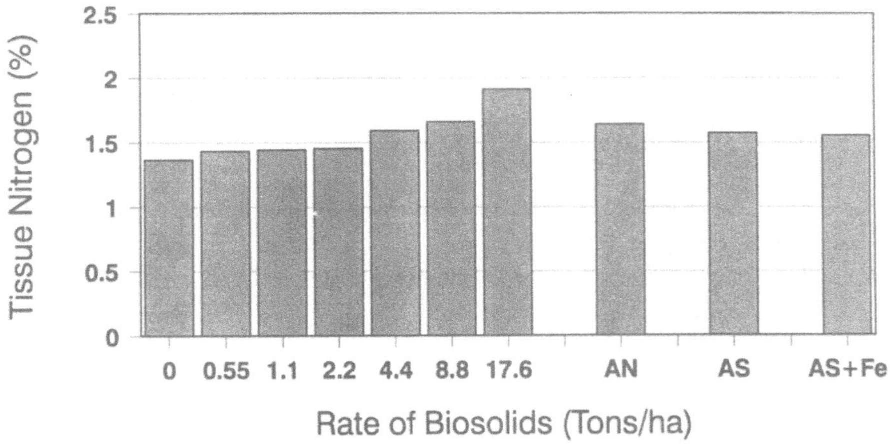

Figure 2. Bahiagrass Tissue Nitrogen Concentration averaged over all harvests. (AN = Ammonium Nitrate; AS = Ammonium sulfate; AS + Fe = Ammonium Sulfate plus Ferrous Sulfate.)

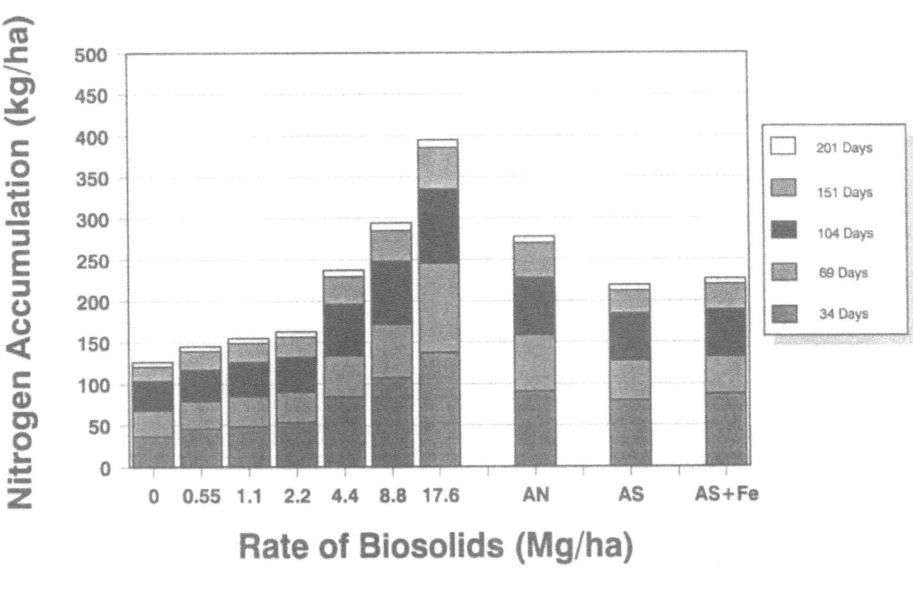

Figure 3. Nitrogen Accumulation by Bahiagrass at various harvest dates: 34 DAA; 69 DAA, 104 DAA; 151 DAA; 201 DAA. (AN = Ammonium Nitrate; AS = Ammonium sulfate; AS + Fe = Ammonium Sulfate plus Ferrous Sulfate.)

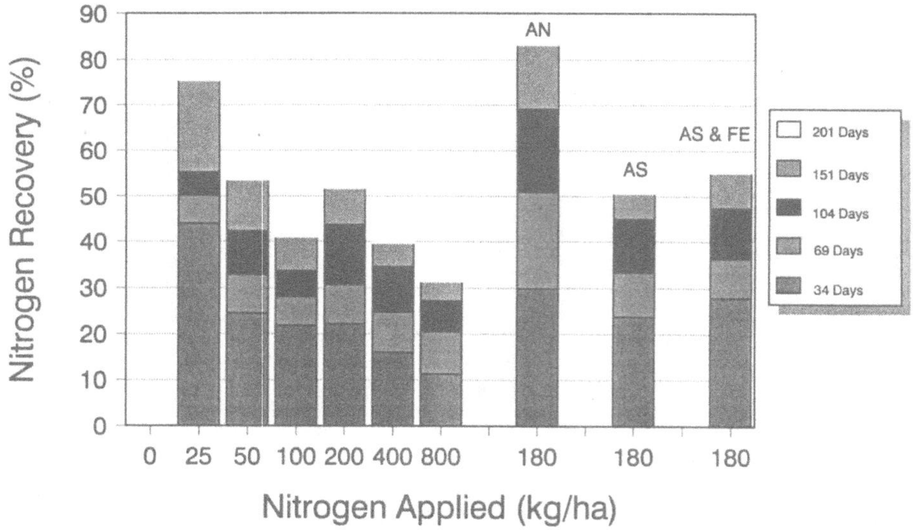

Figure 4. Percentage Nitrogen Recovery in Bahiagrass Forage at various harvest dates: 34 DAA; 69 DAA; 104 DAA; 151 DAA; 201 DAA; total nitrogen recovered by bahiagrass forage is shown as the sum of all harvest dates. (AN = Ammonium Nitrate; AS = Ammonium sulfate; AS + Fe = Ammonium Sulfate plus Ferrous Sulfate.)

wheat and rye grown on a silt loam in Tennessee (Wolt, 1986). The author concluded that application at rates greater than 6 dry Mg biosolids/ha produced bermudagrass yields equivalent to those obtained with high rates of inorganic N application (750 kg/ha).

We had assumed that 40% of the organic N would mineralize during the first year in a warm, humid climate. At the lower biosolids rates, N recovery by the forage alone reached over 70% of the N contained in the biosolids; however, recovery was reduced with increasing rates of biosolids. Most of the N was removed within the first 34 days after application. An estimate of N in the roots was not obtained and might account for some of the "missing N." Since only slight variation on daily uptake was verified for the biosolids treatments and the soluble N sources (Table 2), we assume that a low percentage of the unrecovered N would be still available for plant use during the subsequent year, especially with lower rates of biosolids. Plots were established in the subsequent year and should provide us with some specific information on residual, plant-available N remaining from the biosolids.

It appears that the 4.4 Mg/ha biosolids rate was able to provide as much N as 180 kg N/ha applied from ammonium sulfate soluble sources. Wolt (1986) found that N removal in harvested bermudagrass accounted for 42 to 58% of the total N applied. For wheat and rye managed for haylage, biosolids rate greater than 3 Mg/ha (dry wt.) also produced yields comparable to those achieved with inorganic N sources. Field experiments with corn receiving varying rates of N fertilizer and biosolids indicated mineralization rates of organic N from biosolids to be approximately 55% for the first year of application in

Vermont (Magdoff and Amadon, 1980). In this study, percentage of N recovery in the forage depended on the biosolid application rate.

Biosolids application on pasture lands cannot necessarily replace inorganic fertilization, although many compare favorably with commercial N and Fe fertilizers. However, since pasture grasses are normally under-fertilized, biosolids could provide essential nutrients as well as organic matter for Florida's sandy soils. Slight increases in plant and soil metal contents (data not presented) in this publication were not substantial enough to yield environmental concerns, even with the higher rates of biosolids.

ACKNOWLEDGMENTS

Florida Agricultural Experiment Station Journal Series no. N-01420. The authors gratefully acknowledge Ms. Cindy Holley and Mr. James Platt for help with data collection and analysis, and Ms. Pamela Hunter, assistance with the figures and final typing of this manuscript.

REFERENCES

Hue NV (1995) Sewage Sludge. In: Rechcigl JE (Ed) Soil Amendments and Environmental Quality (pp.199-247). Lewis Publishers, Boca Raton, FL.

Kidder G, Hanlon EA & Chambliss CG (1990) IFAS standardized fertilization recommendations for agronomic crops. Notes in Soil Science. Florida Cooperative Extension Serv. No. 35.

Magdoff FR & Amadon JF (1980) Nitrogen availability from sewage sludge. J. Environ. Qual. 9:451-455.

McCaslin BD & O'Connor GA (1985) Potential fertilizer value of gamma irradiated sewage sludge on calcareous soils. New Mexico Agric. Exp. Stn. Bull. 692.

McCaslin BD, Davis JG, Chacek LC & Schluter LA (1987) Sorghum yield and soil analysis from sludge amended calcareous iron deficient soil. Agron. J. 79:204-209.

Sveda R, Rechcigl JE & Nkedi-Kizza P (1992) Evaluation of various nitrogen sources and rates on nitrogen movement, Pensacola bahiagrass production and water quality. Commun. Soil Sci. Plant Anal. 23:2451-2478.

Wolt JD (1986) Using organic wastes in Tennessee. Biocycle, March, 24-25.

USE OF FRESH AND COMPOSTED DE-INKING SLUDGE IN CABBAGE PRODUCTION

REGIS R. SIMARD[1], J. COULOMBE[2], R. LALANDE[1], B. GAGNON[1] and S. YELLE[2].
[1]Agriculture and Agri-Food Canada, Soils and Crops Research Centre, and [2]Laval University, Horticultural Research Centre, Sainte-Foy, Quebec, Canada
E mail simardr@em.agr.ca

ABSTRACT

Paper mills generate over three hundred thousand tons of de-inking sludge in the province of Quebec, Canada. These residues have C/N and C/P ratios of respectively 200 and 500 and may benefit from mixing with secondary sludge or composting with manure. This would reduce N and P immobilization and related crop deficiencies. The raw de-inking sludge (RDS) corrected for N deficiency was compared with a compost of sludge and dairy manure (SC), a mixture of primary and secondary sludge (MS) and mineral N on winter cabbage (*Brassica oleracea* var. *capitata* L.) yields, soil N and biochemical properties of a Tilly loam (loamy-skeletal, frigid Typic Humaquod). Rates of 0, 15, 30 and 45 Mg ha^{-1} RDS, MS and SC (wet basis) were applied alone or with 0 to 235 kg N ha^{-1} as NH_4NO_3. Samples were taken from the 0-20 cm soil layer five weeks after planting and at harvest. Soil N was significantly increased by MS five weeks after planting but not affected by RDS or SC. Alkaline phosphatase activities were increased by RDS or MS. The highest petiole sap NO_3^- contents were found in the MS treatments. The MS resulted in 20 % larger total and marketable cabbage yields than all other treatments. The RDS performed better than SC without mineral N. The compost had a synergic effect with mineral N on cabbage yields at the highest IN rates. The N and P contents of cabbage heads at harvest were linearly increased by MS addition. The Cu content was increased by RDS. The Zn and Fe tissue contents were increased and Ni decreased by high rates of RDS. The mixed sludges are of particular interest since they resulted in significant increases in cabbage yields and N and P uptake while increasing soil inorganic N and phosphatase activity. The RDS need supplemental N whereas the compost should be viewed more as a soil conditioner than as a true nutrient source.

INTRODUCTION

In the province of Quebec, Canada, the paper industry generates about 160,000 Mg yr^{-1}

S. Brown, J.S. Angle and L. Jacobs (eds.), Beneficial Co-Utilization of Agricultural, Municipal and Industrial By-products, 349–361.
© 1998 *Kluwer Academic Publishers,*

of de-inking sludge and 475,000 Mg yr^{-1} of secondary sludge from the de-inking and paper making processes (Trepanier et al., 1996). Until recently, these residues were landfilled (2/3) or burnt (1/3). Although burning produces energy, more stringent regulations concerning atmospheric emissions increase cost (Cournoyer et al., 1997). Landfilling is expensive and is recognized as a waste of natural resources and as socially unacceptable. It also generates CH_4, a known greenhouse gas (Topp and Pattey, 1997). Alternative uses of these resources are necessary.

Loss of organic matter and compaction are recognized soil degradation problems in horticulture in eastern Canada (Tabi et al., 1990). De-inking sludges are composed of cellulose and lignin fibers, clay fillers and coating agents used in the papermaking processes (Bellamy et al., 1995, Norrie and Gosselin, 1996). Their high C content makes them attractive for soil conditioning (Chong and Cline, 1993; Feagley et al., 1994; Pichtel et al., 1994). There is a growing interest in using the de-inking sludges (raw or composted) as a soil remediation practice in the province of Quebec (Cournoyer et al., 1997).

These residues have very high C/N and C/P ratios and, when added to soils, can result in N and P immobilization by the microbial biomass which may reduce crop growth (Bellamy et al., 1995). A supplement of 3 kg N Mg^{-1} dry sludge as mineral fertilizer was found necessary to limit N immobilization (Trepanier et al., 1996) whereas no information is available about the need of a P supplement. Secondary sludges have C/N ratios as low as 10 and may be mixed with primary sludges to reach an adequate C/N ratio. Also, there is a large manure surplus in many watersheds in the province of Quebec (Simard, 1997). Manure, by providing N and P, can be combined with paper mill sludges which acts as a C source to produce composts with a more balanced composition (Stratton et al., 1995). Land-applied, poultry manure-based compost has been reported to produce a gradual, long-term response in crop yields in warm climates (Altierii et al., 1991). Sikora and Azam (1993) reported higher wheat yields with a fertilizer-compost combination than with mineral fertilizer alone. Compost is a slow-release source of N and P. It may also affect mineral fertilizer nutrient availability by increasing microbial activity or indirectly by its organic acid content (He et al., 1995). However, little is known on the impact of de-inking sludge-manure compost on soil quality, nutrient availability and crop growth on the cool and humic conditions of eastern Canada. The objective of the study was to compare raw de-inking sludge corrected for N with composted sludge, a mixture of primary and secondary sludge and inorganic N on winter cabbage yields, soil N and biochemical properties of a Tilly loam soil.

MATERIALS AND METHODS

The experiment was conducted in 1996 at the Joseph Rhéaume experimental farm of Laval University located in Sainte-Croix (46° 39' latitude north and 72° 06' longitude west), province of Quebec, Canada. The site was a grass ley dominated by timothy *(Phleum pratense L.)* which has been plowed under the preceding fall. The soil is a Tilly loam (loamy skeletal, frigid Typic Humaquod) with a pH of 5.6, 45 g kg^{-1} sand, 31 g kg^{-1} silt, 24 g kg^{-1} clay, 32 mg kg^{-1} Mehlich 3 extractable P and 65 mg kg^{-1} extractable K (Tran and Simard, 1993).

The experiment was a randomized split-plot design in four replicates. The main plots consisted of raw de-inking sludge (RDS) supplemented with 3 kg N Mg^{-1} dry sludge, a mixture of primary and secondary de-inking sludge (MS) and a compost of dairy manure and raw de-inking sludge (SC) applied at 0, 15, 30 and 45 Mg ha^{-1} on a wet basis. These rates of addition were selected for organic amendments to act as soil conditioners. The organic N mineralization of papermill waste and compost was reported to be either negative or of less than 7 % (Sims, 1995; Gagnon et al., 1997). The mixed sludge contained approximately 70 % of primary and 30 % of secondary treatment de-inking paper sludge. The compost was prepared in uncovered windrows, turned frequently with a payloader and was about two years old at the time of spreading. Amounts of 0, 85, 135, 185 and 235 kg N ha^{-1} as NH_4NO_3 (IN) were added as subplots of the 0 and 30 Mg ha^{-1} sludge treatments. Subplot area was 18 m^2.

The sludges were manually applied and immediately incorporated in the first 10-15 cm of soil by a pass of disk harrow. To cover the early nutrient needs of cabbage in the mineral fertilizer treatments only, 85 kg N ha^{-1} as NH_4NO_3, along with 100 kg P_2O_5 ha^{-1} and 70 kg K_2O ha^{-1} were added before planting. Five wk-old plants of winter cabbage (*Brassica oleracea* var. *capitata* L. cv. Bartolo) were transplanted on June 11 at 33,333 plants ha^{-1}. The rest of the N was side-dressed 5 wk after planting. Weed and insect control was applied as described in local growth guides (Conseil des Productions Végétales du Québec, 1993, 1994). Boron was sprayed three times at 0.2 kg ha^{-1}. The heads were harvested from a 7.8 m^2 area on October 28.

Soil samples were taken (10 cores per plot) before the beginning of the experiment, 5 wk after planting (before the N side-dress application), and at harvest from the 0-20, 20-40 and 40-70 cm layers. Tissue samples were taken 1 mo after planting and at harvest. Soil pH was measured at harvest in 0.01 M $CaCl_2$ in a soil/solution ratio of 1:2. Samples were characterized for their 2 N KCl extractable NO_3^- and NH_4^+ contents (Maynard and Kalra, 1993). The profile soil organic N was determined by combining NO_3-N and NH_4-N of the three soil layers, taking in account estimates of their respective soil densities as follows:

$$\text{profile N} = 2.4*\text{N (0-20 cm)} + 2.6*\text{N (20-40 cm)} + 4.8*\text{N (40-70 cm)}.$$

The organic material C content was determined by dry combustion on a LECO CNS 1000 Analyzer (LECO Corp., St. Joseph, MI) on dried samples sieved to 0.5 mm. The acidic and alkaline phosphatase activities of the 0-20 cm soil layer were determined on the 5 wk after planting samples according to Tabatabai and Bremner (1969) and aryl-sulfatase and urease activities as described by Tabatabai (1994). Each value is the mean of two replicated measurements on samples of each plot and is reported on an oven-dry soil basis.

The N, P, K, Ca and Mg contents in the de-inking materials and plant tissues were determined following a digestion in H_2SO_4 and H_2O_2 whereas the content of Mn, Cu, Zn, Fe, Pb, Cr and Cd was determined after dry ashing (Richards, 1993). The soluble NH_4^+ was determined by a 1:2 sludge water (vol:vol) extraction and colorimetry (Nkonge and Ballance, 1982). The cabbage petiole sap NO_3^-content was measured with a Nitracheck 405 reflectometer (Quomed Ltd., England) and Merckoquant test bands (E. Merck, Darmstadt, Germany).

Data transformation was carried out when needed to improve the normality of the distribution of studied variables. The sum of squares for treatments was partitioned according to chosen orthogonal contrasts to show differences amongst mean values. All statistical analyses were performed with the SAS statistical package (Statistical Analysis System Institute, 1990).

RESULTS AND DISCUSSION

De-inking Material Characterization

The RDS had the highest C:N ratio amongst amendments (Table 1); this ratio was slightly lower than reported values (Trepanier et al., 1996). The C:P ratio was also highest in the RDS and exactly twice that of the compost value. The mixed sludge N content was the highest and this is due to the high N content of the secondary sludge. The composted sludge has larger P, K, Ca, Mn, Fe, B, Cd and Pb contents compared to raw materials. The dairy manure added to the RDS at the beginning of the composting process was thesource of these elements enriching the mixture. Lower contents of Mg in the composted sludge may be related to the primary sludge used or to losses by leaching during the composting process since the compost windrows were uncovered. The Mg content of paper sludges is also highly variable (McGovern et al., 1983).

Table 1. Chemical characteristics of the de-inking materials used in the study.

Characteristic		RDS[a]	MS	SC
Water	($g\ kg^{-1}$)	527	602	671
C	($g\ kg^{-1}$)	405	490	382
N	($g\ kg^{-1}$)	24	115	8,2
C/N		169	43	47
NH_4-N	($g\ kg^{-1}$)	ND[b]	1,2	1,2
P	($g\ kg^{-1}$)	215	345	408
C/P		188	142	94
K	($g\ kg^{-1}$)	46	50	660
Ca	($g\ kg^{-1}$)	58	53	169
Mg	($g\ kg^{-1}$)	150	155	23
Cu	($mg\ kg^{-1}$)	103	7,7	4,4
Zn	($mg\ kg^{-1}$)	108	557	19,3
Mn	($mg\ kg^{-1}$)	133	471	58,6
Fe	($mg\ kg^{-1}$)	252	371	570
B	($mg\ kg^{-1}$)	46	7,6	9,8
Cd	($mg\ kg^{-1}$)	110	97	217
Pb	($mg\ kg^{-1}$)	9	6	35
Cr	($mg\ kg^{-1}$)	610	531	ND

[a] RDS = raw de-inking sludge; MS = mixed primary-secondary de-inking sludge and SC = dairy manure - raw de-inking sludge compost.
[b] ND = below detection.

Enzyme Activities

Increases in microbial populations and enzymatic activities with the addition of municipal sludge composts were previously reported (Guidi et al., 1988; Perucci, 1990). The de-inking sludges comprise an array of cellulolitic, nitrogenous and S-containing organic compounds (Trepanier et al., 1996). Phosphomonoesterases are very important in soil organic P mineralization and plant nutrition (Tabatabai, 1994). These enzymes are classified according to their pH range. The acidic phosphatase showed the largest activity amongst studied enzymes (Table 2). There was a definite trend ($p< 0,1$) for acidic phosphatase activity to be increased by all de-inking materials.

Table 2. Soil enzyme activity as influenced by de-inking material additions.

Treatments	Rate (Mg ha^{-1})	Acidic phosphatase (g kg^{-1})	Alkaline phosphatase (mg kg^{-1})	Aryl-sulfatase (mg kg^{-1})	Urease (µg kg^{-1})
Control		102	113	37	94
RDS[a]	15	122	185	51	86
RDS	30	132	171	45	106
RDS	45	1,29	249	66	141
MS	15	108	158	38	79
MS	30	132	205	42	128
MS	45	135	192	53	117
SC	15	135	135	37	101
SC	30	127	131	41	83
SC	45	130	163	40	99
Contrast					
Control vs others		*	*	*	NS
RDS Linear		NS	0	NS	NS
MS Linear		NS	*	NS	NS
SC Linear		NS	NS	NS	NS
RDS vs SC		NS	*	NS	NS
MS vs SC		NS	*	NS	NS

[a] RDS: raw de-inking sludge; MS: a mixture of primary and secondary sludge and SC: de-inking sludge - dairy manure compost.

Alkaline phosphatase is specific to soil microorganisms (Dick et al., 1983). The alkaline phosphatase activity was significantly ($p < 0.05$) increased by RDS and MS applications. A comparable trend was also found for the 45 Mg compost ha^{-1}. The RDS- and MS-treated soils showed larger alkaline phosphatase activity than the SC-soils.

Arylsulfatase is believed to play an important role in the soil organic S mineralization processes (Tabatabai, 1994). There was a trend for larger arylsulfatase and urease activities in the RDS than in all other treatments. The mineral N supplement to the sludges had no significant effect on the phosphatase and arylsulfatase activities but significantly increased urease activity at 235 kg N ha^{-1} (data not shown) The high levels of soil NO_3^- (Figure 1) and SO_4^{2-} (data not shown) may have limited the activity of the N and S enzymes. These results indicate that the addition of RDS or MS will be particularly useful in increasing the soil biological activity as shown by phosphatase measurements.

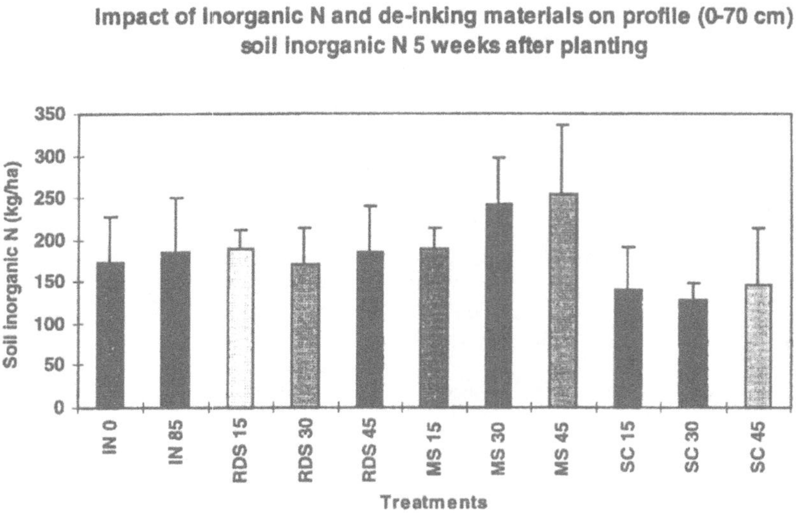

Figure 1. Impact of mineral N and de-inking materials on soil profile inorganic N (0-70 cm) five wk after planting. IN = mineral N only; RDS = raw de-inking sludge; MS = a mixture of primary and secondary de-inking sludge and SC: compost of primary de-inking sludge and dairy manure. The number besides these symbols indicates the rate of application (Mg ha^{-1}) except for IN which is in kg ha^{-1}.

Soil N Content

The soil profile $NO_3^- + NH_4$-N content (0-70 cm) 5 wk after planting was significantly ($p < 0.05$) affected by sludge materials and mineral N addition (Figure 1). The amounts measured were very high in all treatments reflecting the plowing under of the old grass ley the preceding fall. The RDS treatments were not significantly different from the control suggesting that the inorganic N supplement of 3 kg N Mg^{-1} dry sludge was sufficient to counteract the immobilization process. The soil N was linearly increased by MS additions whereas SC at 30 and 45 Mg ha^{-1} resulted in significant N immobilization. These results indicate that, in spite of a C/N ratio of 43, the MS released inorganic N within five weeks of application. Mineral N is added in the secondary process at the de-inking plant and is therefore incorporated in the microbial biomass. The above results suggest that a large part of this microbial biomass N was labile and released in the soil. It is possible that because of the resilient nature of the cellulose and lignin compounds in the MS, soil N was not immobilized by the soil microbes to degrade these compounds (Jansson and Persson, 1982). The possibility that microbial N was released following a microbial population decrease because of the lack of soil labile C deserves further investigation. The impact of the mineral N supplement to the de-inking materials could not be assessed since supplemental N was added immediately after this sampling.

Profile soluble inorganic N content at harvest was not significantly affected by de-inking materials although a trend for higher contents could be observed with the largest rates (Figure 2). These results confirm those of Trepanier et al. (1996) who observed no significant difference in lysimeter water NO_3^- content with RDS additions. Winter cabbage was harvested late and was very active in taking up most of the mineralized N throughout the growing season, leaving little fall residual soil inorganic N. Shorter growing season vegetables may have resulted in increased risk of N leaching. In the absence of an active crop uptake, slow mineralization of previously immobilized N will result in significant N accumulation in late fall. This N may leach in the following spring since uptake will be reduced before significant crop cover may be established. Soil pH at harvest was not affected by the treatments ($F = 1.14$, $p > 0.32$).

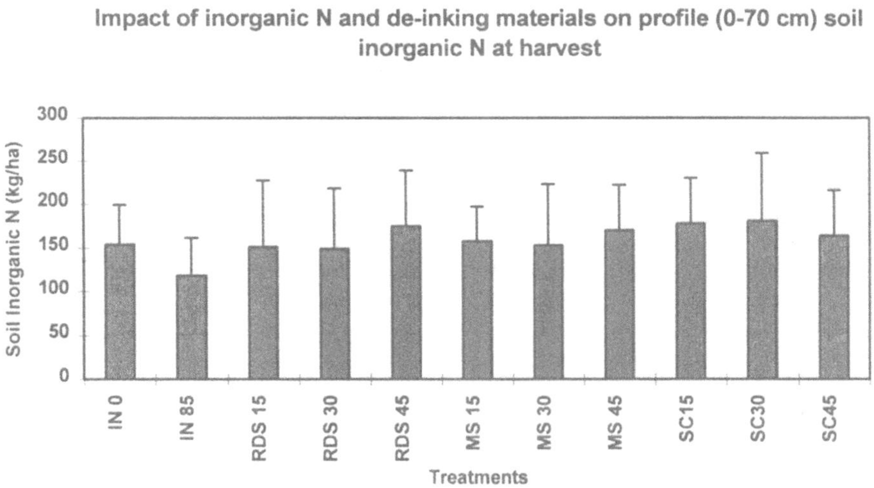

Figure 2. Impact of de-inking materials on soil profile (0-70 cm) mineral N at harvest. RDS= raw de-inking sludge; MS = a mixture of primary and secondary de-inking sludge and SC: compost of primary de-inking sludge and dairy manure. The number besides these symbols indicates the rate of application (Mg ha^{-1}).

Sap Nitrate Content

The sap nitrate content 5 wk after planting was linearly related to the MS rates (data not shown) and was larger than in the other organic treatments at 0 and 85 kg N ha^{-1} (Figure 3). The sap N content of the RDS or SC treatments was not significantly different ($p<0.05$) from the control. This confirms the results of the soil analysis at that time and clearly shows that MS N availability is much larger than from the other organic materials. It was reported that petiole sap N was an efficient way to measure the N status of cruciferae (Prasad and Spiers, 1984, Kubota et al., 1996). The lack of difference between the RDS and the control confirms again that the supplement of 3 kg N Mg^{-1} dry matter was

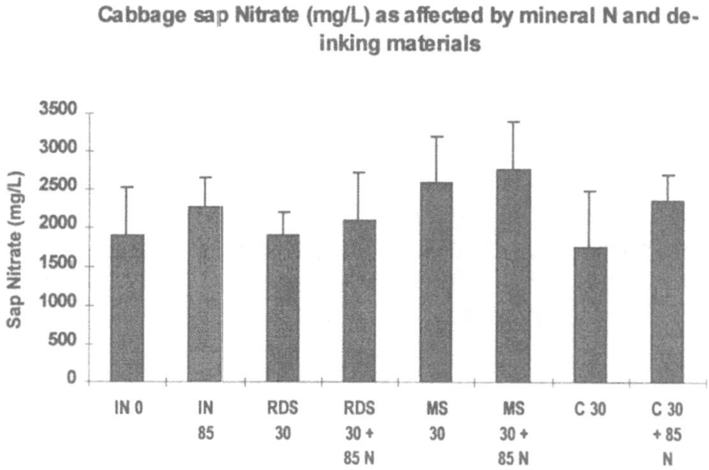

Figure 3. Cabbage sap nitrate content 5 wk after planting as affected by mineral N and de-inking materials. IN = mineral N only; RDS = raw de-inking sludge; MS = a mixture of primary and secondary de-inking sludge and SC: compost of primary de-inking sludge and dairy manure. The number besides these symbols indicates the rate of application (Mg ha^{-1}) except for IN which is in kg ha^{-1}.

sufficient to counteract the potential immobilization of soil N. These results agree with those of reported field studies in the climatic conditions of eastern and central Canada on the need of supplemental N (Bellamy et al., 1995, Trepanier et al., 1996).

Marketable Cabbage Yields

The marketable cabbage yields without supplemental mineral N were in the order MS> RDS> SC> IN (Figure 4). The yields of the MS treatment were at least 20 % larger than RDS, SC or IN at less than 185 kg IN ha^{-1}. There was a synergic effect of the compost and inorganic N at the two highest N rates. Similar results were observed in an experiment involving dry beans (Simard et al., 1996). A synergic effect of compost and mineral fertilizers was reported by Sikora and Azam (1993) and was probably the result of increased compost mineralization (Sikora and Yakovchenko, 1996). This effect happened only late in the season as amounts of soil mineral N were lower for SC than the control 5 wk after planting.

An extrapolation to the yield response obtained with mineral N for the yield obtained with MS without supplemental mineral N (Figure 4) indicates that the N fertilizer replacement value of the 30 Mg MS ha^{-1} is 82 kg NH_4NO_3 ha^{-1}. If we consider a 60.2 % water content in MS, this value is 6.86 kg Mg^{-1} on a dry basis. Considering a total N content of 11.5 kg Mg^{-1} sludge (Table 1), the NH_4NO_3 replacement value of the total N in the MS would be 60%. If the content of inorganic N in the MS is deducted (1.2 kg Mg^{-1}), an apparent mineralization rate of 49 % would be found for the organic N in the

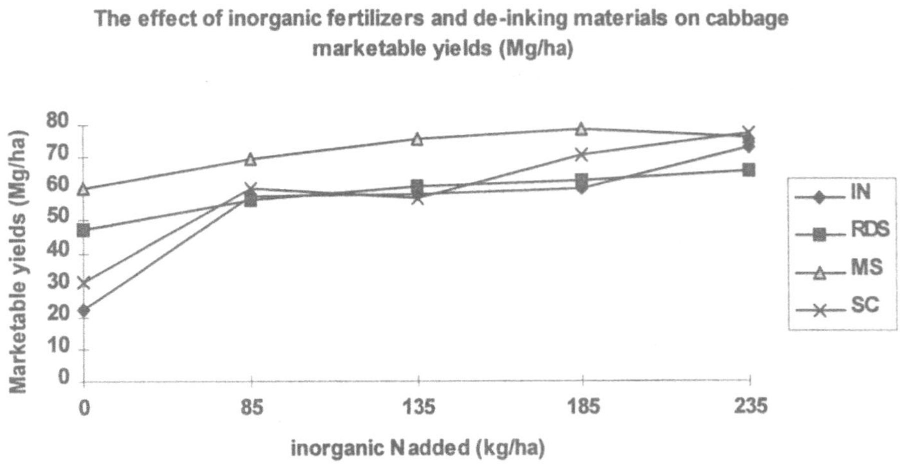

Figure 4. The effect of mineral fertilizer rate and de-inking materials (30 Mg ha^{-1}) on cabbage marketable yields. IN = mineral N only; RDS= raw de-inking sludge; MS = a mixture of primary and secondary de-inking sludge and SC: compost of primary de-inking sludge and dairy manure.

MS. This value is much higher than what was previously reported for other paper mill waste materials (Sims, 1995). The response curves with and without MS are almost parallel and certainly suggests a very large non-N contribution of the mixed sludge to the cabbage crop. Improvements of water availability and of soil physical properties have previously been observed with addition of these materials (Trepanier et al., 1996. The conservation of some spring soil N is also possible since the preceeding crop was forages which are known to leave large amounts of labile N. An average of 127 kg N ha^{-1} was present in the 0-60 cm layer before seeding. Some of this soil N may have been immobilized temporarily with the addition of MS and mineralized later in the season. For instance, the soil inorganic N contents 5 weeks after planting were larger in the 45 Mg MS ha^{-1} than in the 85 kg IN treatment (Figure 1). The mixture of primary and secondary papermill sludge, because of its low C/N ratio, was previously shown to have a positive N mineralization potential (Dolar et al., 1972; Bockheim et al., 1988; Chong and Cline, 1993). The results of the present study indicate that it would also apply to the mixture of primary and secondary treatment de-inking sludge.

The corresponding contribution of the compost was 0.83 kg N Mg^{-1}. This result confirm those of previous studies under the cool and humid conditions of eastern Canada showing the limited nutrient value of composts of manure and wood residues in the first season when spring applied (Gagnon et al., 1997). Manure composted without a resilient C source was previously found to be as effective as fresh manure (Xie and MacKenzie, 1986, Paul and Beauchamp, 1993). Composted papermill sludge with lower C:N ratios than in the RDS used in the present experiment was shown to have significant nutrient value in potting media (Tripepi et al., 1996).

Head Elemental Composition

The N and P contents of cabbage heads at harvest were linearly increased by MS addition (Table 3). These results confirm the large availability of N from the MS source. The increase in P concentration along with the positive yield response indicate an increased soil P availability. A C/P ratio of 200 for the primary sludge and of 142 for the mixed sludge would suggest soil P immobilization and a related decrease in plant P availability. However, the increased microbial activity along with the positive effect of soluble degradation products on P solubility (Ohno and Crannell, 1996) may explain the increased plant tissue content. The K content of the 15 Mg RDS ha^{-1} was significantly larger than in the control. The Ca content of the 15 Mg MS ha^{-1} treatment was significantly larger than in the control and compost treatments (Table 3). The Cu content was increased by all amendments and was particularly high in the 15 Mg RDS ha^{-1} treatment whereas the Zn and Fe tissue contents were increased by all sludges (Table 3). The content of Ni was decreased by all sludges although high Ni and Pb contents were associated with the 45 T MS ha^{-1} treatment. Although some significant effects on metal contents were observed, the contents are in the low range of reported values for edible parts of cabbage (De Pieri et al., 1996). The Mg , Co and Cd contents of the heads were not significantly affected by the treatments (data not shown). These results suggest that, the soil were able to retain the metals in non-extractable forms so their soil Mehlich 3 (data not shown) and cabbages tissue contents were not increased after a single application.

CONCLUSIONS

The results of the present study along with those of related works (Norrie and Gosselin, 1996; Trepanier et al., 1996) indicate that adding de-inking sludge residues improves soil quality by stimulating phosphatase activity and increasing nutrient availability. The combination of primary and secondary sludge is the most efficient source of nutrients. A non-N contribution of the sludge through improved biological and physical soil properties may explain the large cabbage yield with the MS non-supplemented with IN treatments. Also, a synergic effect of compost on mineral fertilizer efficiency was observed. The potential problem of soil P immobilization with the addition of these high C/P materials was not observed; the content of P in plant tissues of the MS and RDS treatments was even increased compared with the control. The gains in microbial activity along with displacement of P from the exchange sites by organic ligands may explain these observations. Conversely, some increases in metal contents were observed with 45 Mg MS ha^{-1} but the levels measured were in the low range of published values. De-inking residues are an excellent source of C for degraded agricultural soils and constitute an excellent opportunity for the horticultural sector. Studies on repeated applications are necessary to determine the long-term impact on soil quality and crop nutritive value.

Table 3. Treatment effects on the elemental composition of the cabbage heads at harvest

Treatment	Rate $Mg\ ha^{-1}$	N	P	K	Ca	Cu	Zn	Fe	Pb	Ni
		%				$mg\ kg^{-1}$				
Control		1.51a[a]	0.225a	1.74a	0.41a	1.10a	7.8a	20.4a	2.47a	4.17c
RDS	15	1.77ab	0.253ab	1.92b	0.44a	1.43b	10.6b	24.4b	2.57a	3.49bc
	30	1.46a	0.222a	1.74a	0.51ab	1.27a	8.7a	23.1a	2.23a	2.42a
	45	1.65ab	0.250ab	1.88ab	0.42a	1.18a	9.6ab	24.5b	2.43a	3.05b
MS	15	1.37a	0.233a	1.77a	0.52b	1.21a	9.0a	24.4b	2.48a	3.16b
	30	1.76ab	0.247ab	1.85ab	0.45a	1.23a	10.6b	23.8ab	2.67b	3.55bc
	45	2.00b	0.277b	1.84ab	0.46a	1.24a	10.6b	25.3b	2.89c	4.91d
SC	15	1.63a	0.240a	1.74a	0.43a	1.13a	9.0a	21.2a	2.63ab	3.77bc
	30	1.43a	0.235a	1.82a	0.39a	1.19a	9.6ab	23.7ab	2.36a	3.47bc
	45	1.52a	0.235a	1.80a	0.43a	1.28a	9.7ab	25.4b	2.38a	3.46bc

[a] Means within a column followed by different letters are significantly different at $p \leq 0.05$ by analysis of variance.

ACKNOWLEDGEMENTS

The authors are grateful to Daishowa Inc., the Composting Council of Canada and the Government of Canada for their financial support. Special thanks are expressed to Alain Larouche, A. D'Avignon and Simon Roy for their technical assistance. The comments of Suzanne Beauchemin and J. Lafond on an early draft of the manuscript are greatly appreciated.

REFERENCES

Alteirii MA, Trujillo J, Astier M, Gersper PP & Bakx W (1991) Low-input technology proves viable for limited-resource farmers in Salinas Valley. Calif. Agric. 45(2):20.

Bellamy KL, Chong C & Cline RA (1995) Paper sludge utilization in agriculture and container nursery culture. J. Environ.Qual. 24:1074-1082.

Bockheim JG, Benzel TC, Lu RL & Thiel DA (1988) Groundwater and soil leachate inorganic nitrogen in a Wisconsin red pine plantation amended with paper industry sludge. J. Environ. Qual. 17:729-734.

Chong C & Cline RA (1993) Response of four ornamental shrubs to container substrate amended with two sources of raw paper mill sludge. HortScience 28:807-809.

Conseil des Productions Végétales du Québec (1994) Répertoire des traitements de lutte anti-parasitaire. AGDEX 250-265. 86 pp.

Conseil des Productions Végétales du Québec (1993) Mauvaises herbes-répression. AGDEX 640. 109 pp.

Cournoyer MS, Granger F & Michon PY (1997) Valorisation des résidus des fabriques de pâtes et papiers: Guide de bonnes pratiques. Urgel Delisle et Associés, 65 pp.

De Pieri LA, Buckley WT & Kowalenko CG (1996) Micronutrient concentrations of commercially grown vegetables and of soils of the Lower Fraser Valley of British Columbia. Can. J. Soil Sci. 76:173-182.

Dick WA, Juma NG & Tabatabai MA (1983) Effects of soils on acid phosphatase and inorganic pyrophosphatase of corn roots. Soil Sci. 136:19-25.

Dolar, SG, Boyle JR & Keeney DR (1972) Paper mill sludge disposal on soils: effects on the yield and mineral nutrition of oats (*Avena sativa* L.). J. Environ. Qual. 1:405-409.

Feagley SE, Valdez MS & Hudnall WH (1994) Bleached, primary papermill sludge effect on bermudagrass grown on a mine soil. Soil Sci. 157:389-397.

Gagnon B, Simard RR, Robitaille R, Goulet M & Rioux R (1997) Effect of composts and inorganic fertilizers on spring wheat growth and nutrient uptake. Can. J. Soil Sci. 77 (In press).

Guidi G, Pera A, Giovanetti M, Poggio G & De Bertoldi M (1988) Variations of soil structure and microbial population in compost amended soil. Plant Soil 106:113-119.

He XT, Logan TJ & Traina SJ (1995) Physical and chemical characteristics of selected U.S. municipal solid waste composts. J. Environ. Qual 24:543-552.

Jansson SL & Persson S (1982) Mineralization and immobilization of soil nitrogen. In: Stevenson FJ (Ed). Nitrogen in Agricultural Soils. Agron. Monogr. 22 (pp. 229-252), ASA, SSSA, Madison, WI.

Kubota A, Thompson TL, Doerge A & Godin RE (1996) A petiole sap nitrate test for cauliflower. HortScience 31:934-937.

Maynard DG & Kalra YP (1993) Nitrate and exchangeable ammonium nitrogen. In: Carter MR (Ed) Soil Sampling and Methods of Soil Analysis (pp. 25-38). Lewis, Boca Raton, FL.

McGovern JN, Berbee JG, Bockheim JG & Baker AJ (1983) Characteristics of combined effluent treatment sludges from several types of pulp and paper mills. Tappi J. 66:115-118.

Nkonge C & Ballance GM (1982) A sensitive colorimetric procedure for nitrogen determination in Micro-Kjeldahl digests. J. Agric. Food Chem. 30:416-420.

Norrie J & Gosselin A (1996) Paper sludge amendments for turfgrass. HortScience 31:957-960.

Ohno T & Crannell BS (1996) Green and animal manure-derived dissolved organic matter effects on phosphorus sorption. J. Environ. Qual. 25:1137-1143.

Paul JW & Beauchamp EG (1993) Nitrogen availability for corn in soils amended with urea, cattle slurry, and solid and composted manure. Can. J. Soil Sci. 73:253-266.

Perucci P (1990) Effect of the addition of municipal solid-waste compost on microbial biomass and enzyme activity in soil. Biol. Fertil. Soils 10:221-226.

Pichtel JR, Dick WP & Sutton P (1994) Comparison of amendments and management practices for long-term reclamation of abandoned mined lands. J. Environ. Qual. 23:766-772.

Prasad M & Spiers TM (1984) Evaluation of a rapid method for plant sap nitrate analysis. Commun. Soil Sci. Plant Anal. 15:673-679.

Richards JE (1993) Chemical characterization of plant tissue. In: Carter MR (Ed) Soil Sampling and Methods of Soil Analysis (pp.115-139). Lewis, Boca Raton, FL.

Sikora LJ & Azam MI (1993) Effect of compost-fertilizer combination on wheat yields. Compost Sci. Land/Util. 1:93-96.

Sikora LJ & Yakovchenko V (1996) Soil organic matter mineralization after compost amendment. Soil Sci. Soc. Am. J. 60:1401-1404.

Simard RR (1997) Manure management in the province of Quebec: the urgent need for solutions. In: Chang C (Ed) Proc. Can. Soc. Soil Sci. Waste Management Workshop (pp. 65-80). Lethbridge, AB, Canada.

Sims JT (1995) Organic wastes as alternative nitrogen sources. In: Bacon PE (Ed) Nitrogen Fertilization in the Environment (pp. 487-535). Marcel Dekker, New York.

Simard RR, Lalande R, Zizka J, Gagnon B, Yelle S & Coulombe J (1996) Effects of composted paper mill sludge on dry bean production. In: Proc. 6th Ann. Conf. Compost. Council of Canada (pp. 39-40). Toronto, ON, Canada.

Statistical Analysis System Institute (1990) SAS/Stat user's guide: Vol 2. GLM-VARCOMP. Version 6. 4th. Ed. SAS Inst., Cary, NC.

Stratton ML, Barker AV & Rechcigl JE (1995) Compost. In: Rechcigl JE (Ed) Soil Amendments and Environmental Quality (pp. 249-310). Lewis, Boca Raton, FL.

Tabatabai MA & Bremner JM (1969) Use of *p*-nitrophenyl phosphate for essay of soil phosphatase activity. Soil Biol.Biochem. 1: 301-307.

Tabatabai MA (1994) Soil enzymes. In: Weaver RW (Ed) Methods of soil analysis Part 2 Microbiological and Biochemical Properties (pp. 775-833). SSSA Book Series, No. 5. Madison, WI.

Tabi M, Tardif L, Carrier D, Laflamme G & Rompré M (1990) Inventaire des problèmes de dégradation des sols agricoles du Québec. Rapport Synthèse. Gouv. Québec, Qc, Can. 127 pp.

Topp EJ & Pattey E (1997) Soils as sources and sinks for atmospheric methane Can. J. Soil Sci. 77:167-178.

Tran TS & Simard RR (1993) Mehlich-3 extractable elements. In: Carter MR (Ed) Soil Sampling and Methods of Soil Analysis (pp. 43-50). Lewis, Boca Raton, FL.

Trepanier L, Caron J, Yelle S, Thériault G, Gallichand J & Beauchamp CJ (1996) Impact of de-inking sludge amendment on agricultural soil quality. TAPPI Int. Env. Conf. 1-9.

Tripepi RR, George MW, Campbell AG & Shafii B (1996) Evaluating pulp and paper sludge as a substitute for peat moss in container media. J. Environ. Hort. 14:91-96.

Xie RJ & MacKenzie AF (1986) Urea and manure effects on soil nitrogen and corn dry matter yields. Soil Sci. Soc. Am. J. 50:1504-1509.

USING COMPOST PRODUCTS IN VEGETABLE PRODUCTION

AZIZ SHIRALIPOUR[1], PHILIP B. LEEGE[2] and KENNETH M. PORTIER[3]
[1]*Center for Biomass Programs, University of Florida, Gainesville, Florida 32611, USA,* [2]*The Procter & Gamble Company, Cincinnati, Ohio 45224, USA and* [3]*Department of Statistics, University of Florida, Gainesville, Florida 32611, USA*
E-mail shirali@gnv.ifas.ufl.edu

ABSTRACT

Six compost products and nitrogen (N) from ammonium nitrate fertilizer were utilized in field experiments to grow lettuce (*Lactuca sativa* var. 'Lollarosa') and broccoli (*Brassica oleracea botrytis*, var. 'Brigadeer') crops to evaluate the effect of various compost products and N application on crop yields. The basic design of the site for each compost type was a replicated split-split plot with compost type on the main plot, three rates of compost application (0, 37, 74 Mg ha^{-1}) on the split plots and three supplemental N rates (0, 75, 150 kg ha^{-1}) applied to split-split plots. Application rates of 37 Mg ha^{-1} and 74 Mg ha^{-1} for all compost types resulted in significant yield increases for lettuce and broccoli. Lettuce yields were significantly different for the different compost types and for combinations of compost rate by nitrogen rate. Broccoli yields were affected by the combination of compost type, compost rate, and N rate applied. In both cases, higher yields were obtained for higher compost rates, with the effect of supplemental N rate less important at these high compost rates.

INTRODUCTION

The major benefit from the application of compost to soil is the improved physical properties related to the increased organic matter content (Gallardo-Lora and Nogales, 1987; Shiralipour et al., 1992). These improved physical properties include, but are not limited to, increased water holding capacity (Mays et al., 1973; Hernando et al., 1989; Shiralipour and Zachary, 1994) and erosion prevention (Watkins and Winch, 1974). In addition, compost application adds nutrients to soil (Sommers and Giordano, 1984; Gallardo-Lora and Nogales, 1987; Shiralipour et al., 1992) and suppresses soil-borne diseases (Hoitink and Fahy, 1986; Hoitink et al., 1991).

The long-term feasibility of composting depends largely upon building a market for the compost by demonstrating the benefits resulting from compost application. To remove

S. Brown, J.S. Angle and L. Jacobs (eds.), Beneficial Co-Utilization of Agricultural, Municipal and Industrial By-products, 363–375.
© 1998 *Kluwer Academic Publishers,*

barriers to compost acceptance, definitive information must be generated for end-users to show the safe use of compost products for public health, safety and the environment. Compost users must distinguish between compost attributes that provide measurable benefits, whereas compost producers need to understand how their choices for compost feedstock and composting process management, determine these desired attributes in the final compost products.

Often end-users (e.g., agricultural, community, landscapers, etc.) are unaware of municipal plans to build compost facilities, while local governments are unaware of the market demands of potential end-users. Field experiments can assist with the development of communication between end-users and local government officials, thus accomplishing the first step in long-term market development.

Studies are needed to measure compost attributes that give measurable benefits under field conditions. To demonstrate the safe use and benefits of compost application in a coastal Southern California climate, greenhouse experiments and field trials were conducted in Santa Barbara County. The main objective of the greenhouse studies was to determine the optimal rates of compost application for the field trial with cash crops. The results of the greenhouse studies were reported earlier (Shiralipour et al., 1996). The objectives of the field studies were to evaluate.(1) the effect of compost application on crop yield and (2) the fate of heavy metals and organic contaminants in plant tissue. This report only explores the benefits of six compost products regarding yield increases of leaf lettuce and broccoli heads under field conditions.

MATERIALS AND METHODS

Plant Types

Leaf lettuce (*Lactuca sativa,* var. 'Lallorosa') and broccoli (*Brassica oleracea botrytis,* var. 'Brigadeer') were chosen for the field experiments, the same cash crops used in the greenhouse studies (Shiralipour et al., 1996). These vegetables were chosen because they represent two of the most important crops in terms of acreage and cash value (personal communication with the Santa Barbara County Agricultural Commissioner).

Soil Type

Soil type was Goleta loam, a well-drained soil on flood plains and alluvial fans. The soil formed in alluvium from sedimentary rock and is one of the most dominant soil types in Santa Barbara County. The chemical analysis results for this soil are shown in Table 1. The soil analyses were conducted by the Soil Control Lab in Watsonville, CA, and the methods they used were as follows:

> Dried samples were stored in a 4°C cold room until ready for analysis, re-dried at 60°C, before grinding by mortar and pestle, then sampled for analysis.

Table 1. Golita loam soil analysis.

Constituents Tested	Unit
(Total)	mg/kg dw
As	5
Cd	1.6
Cr	41
Cu	31
Zn	130
Fe	14,000
Mn	730
Pb	26
B	4
Mo	1.1
Ni	35
P	1,000
Se	<10
Ca	90,000
Mg	3,400
Na	130
K	5,300
Hg	<0.3
Water Soluble (saturated paste)	**% saturation**
Ca	0.8 meq/100 gms
Mg	1.67 meq/100 gms
Na	0.36 meq/100 gms
K	138 meq/100 gms
Cl	0.18 meq/100 gms
NO3	0.57 meq/100 gms
PO4	0.35 meq/100 gms
SO4	0.59 meq/100 gms
Electrical Conductivity (EC10 mmhos/cm)	0.64 meq/1
Exchangeable-cations	
Ca^{++}	5.4 meq/100 gms
Mg^{++}	0.92 meq/100 gms
Na^{+}	0.07 meq/100 gms
K^{+}	0.26 meq/100 gms
Cation Exchange Capacity	16.5 meq/100 gms
Exchangeable Sodium Percentage	0.4 percent
pH on saturated paste	6.7

The elements arsenic (As), cadmium (Cd), chromium (Cr), lead (Pb), molybdenum (Mo), nickel (Ni), selenium (Se), and zinc (Zn) were digested using method EPA 3050 (USEPA) (HNO3 + H2O2) and analyzed using method EPA 6020 (ICP-MS).

The following elements were extracted by EPA Method 3050 and analyzed by AA using the listed EPA procedures: calcium (Ca) - EPA 7140; magnesium (Mg) - EPA 7450; sodium (Na) - EPA 7770; potassium (K) - EPA 7610; manganese (Mn) - EPA 7460; and iron (Fe) - EPA 7380. Total phosphorus (P) was determined using Official Methods of Analysis (A.O.A.C.) 2.020a, which includes digestion in HNO3 + HCl, followed by a spectrophotometric molybdovanadophosphate method of analysis. Boron (B) was extracted using A.O.A.C. method 2.137 (acid soluble), pH corrected, and determined using EPA Method 212.3 (curcumin-colorimetric).

Compost Types

Six compost products were selected for this experiment which included three feedstocks, used alone or in various ratios: urban plant debris (UPD), biosolids/UPD (Bio), and municipal solid waste (MSW). The UPD was composed of grass clippings and local plant leaves and branches. Bio was made from biosolids mixed with UPD at 1:1 ratio (v:v). MSW was an organic fraction of the local municipal solid waste. These three compost products, UPD, MSW and Bio, were obtained from facilities in Santa Barbara, CA, Truman, MN, and Carson, CA, respectively. Six compost types were obtained by mixing UPD, MSW or Bio in specific proportions to produce compost blends with the following radios:

A1 = 100% MSW
A2 = 53% MSW and 47% UPD
A3 = 96% MSW and 4% Bio
A4 = 52% MSW and 46% UPD and 2% Bio
B1 = 96% UPD and 4% Bio
B3 = 100% UPD

These represented the potential compost types expected to be produced at a Santa Barbara facility. Nutrient and trace element contents of the six compost types are shown in Tables 2 and 3. The trace element contents of all compost products were much lower than the Pollutant Concentration Limits (PCL) for sludge in the U.S. (U.S. EPA, 1990).

Compost and Fertilizer N Application Rates

The compost was mixed with upper 15-20 cm of the soil by small garden shovels when used alone, or followed by N application when used in combination. The optimal application rates for broccoli and lettuce were developed based on the results of greenhouse trials (Shiralipour et al., 1996). Rates of 0, 37 and 74 dry Mg ha^{-1} were used in these experiments for all compost products. The N source was ammonium nitrate fertilizer and was applied at N rates of 0, 75 or 150 kg ha^{-1}, either alone or in combination with compost.

Table 2. Nutrient contents (total) of composts[a].

Constituents Tested	Compost Type[b]					
	A1	A2	A3	A4	B1	B3
	% Dry Weight (dw)					
Total N	1.7	1.5	1.7	1.6	1.3	1.2
Organic N	1.6	1.4	1.6	1.4	1.3	1.2
NH3-N	0.002	0.0015	0.001	0.0011	0.0018	0.0016
NO3-N	0.15	0.098	0.16	0.11	0.064	0.041
P2O5	0.65	0.63	0.61	0.61	0.63	0.61
K20	1.1	1.8	1.1	1.6	1.6	2.5
Ca	2.4	2.7	4.0	2.7	2.0	2.4
Mg	0.6	0.78	0.9	0.84	0.6	0.99
SO4-S	0.11	0.083	0.1	0.079	0.079	0.056
Cl	0.37	0.32	0.38	0.35	0.31	0.25
Na	0.29	0.27	0.29	0.25	0.20	0.17

[a] Soil analysis was performed by Soil Control Lab in Watsonville, CA.
[b] Key:
A1 - 100% MSW[c]
A2 - 53% MSW, 47% UPD
A3 - 96% MSW, 4% Bio
A4 - 52% MSW, 46% UPD, 2% Bio
B1 - 96% UPD, 4% Bio
B3 - 100% UPD
[c] MSW = municipal solid waste; UPD = urban plant debris; Bio = biosolids/UPD

Field Site and Design

A one-acre field site was located on an agricultural ranch (Rancho Dos Pueblos) in Goleta, CA, about 18 miles north of Santa Barbara. The site was leveled and disked and then divided into three replicates of nine 1.5m x 1m plots for each compost type and separated by a 0.6 meter-wide alley to provide 27 plots for each compost type, giving a total of 162 plots for each crop (Figure 1). The basic design of the site for each compost type was a replicated split plot with the three rates of compost application as main plots, and the three supplemental N rates as subplots.

Planting, Irrigation and Harvest

From uniformly grown seedling trays, 12 lettuce or 8 broccoli seedlings, approximately five cm in height, were transferred and planted in each plot with the immediate addition of water. Plants were irrigated by installation of drip tapes at the site. Using this method of irrigation, the nitrate movement from one plot to the next was minimized. Crops were harvested by hand at the end of the growing season and placed in paper bags. From each

Table 3. Trace element contents (total) of composts[a].

Constituents Tested	Compost Type[b]					
	A1	A2	A3	A4	B1	B3
	mg/kg (dw)					
As	10	<10	24	22	20	22
Cd	6.6	3	8	5.6	2.2	1.5
Cr	52	26	64	44	22	24
Cu	200	98	260	172	70	58
Zn	800	380	940	680	220	240
Fe	17000	18000	17000	12000	13000	11000
Mn	460	240	500	420	220	280
Pb	240	100	280	200	30	42
B	4	50	32	32	30	50
Mo	4.4	4	5.4	8	13	9
Ni	56	26	74	56	26	28
Se	<10	<10	<10	<10	<10	<10
Hg	0.55	0.45	0.33	0.4	0.61	0.45

[a] Soil analysis was performed by Soil Control Lab in Watsonville, CA.
[b] Key:
A1 - 100% MSW[c]
A2 - 53% MSW, 47% UPD
A3 - 96% MSW, 4% Bio
A4 - 52% MSW, 46% UPD, 2% Bio
B1 - 96% UPD, 4% Bio
B3 - 100% UPD
[c] MSW = municipal solid waste; UPD = urban plant debris; Bio = biosolids/UPD

treatment 9 plants (3 plants from each plot) were randomly harvested. The bags were dried at 70°C for 48 h and weighed as a parameter of crop yield.

Statistical analyses

The field plot layout was a completely randomized design in compost type with compost rate applied as a split of the main compost unit and N fertilizer rate applied as split of the compost rate, thus leading to a split-split plot analysis in a completely randomized design with three replicates. A mixed effect general linear model (Littell et al., 1996) was used to examine main and interaction effects for the three factors (compost type, compost rate, and N rate), as well as estimate main plot, split plot, and split-split plot variance components. Least squares means were computed for appropriate effects and post-hoc means comparisons were performed using the LSD procedure at a comparisonwise Type I error rate of 0.01.

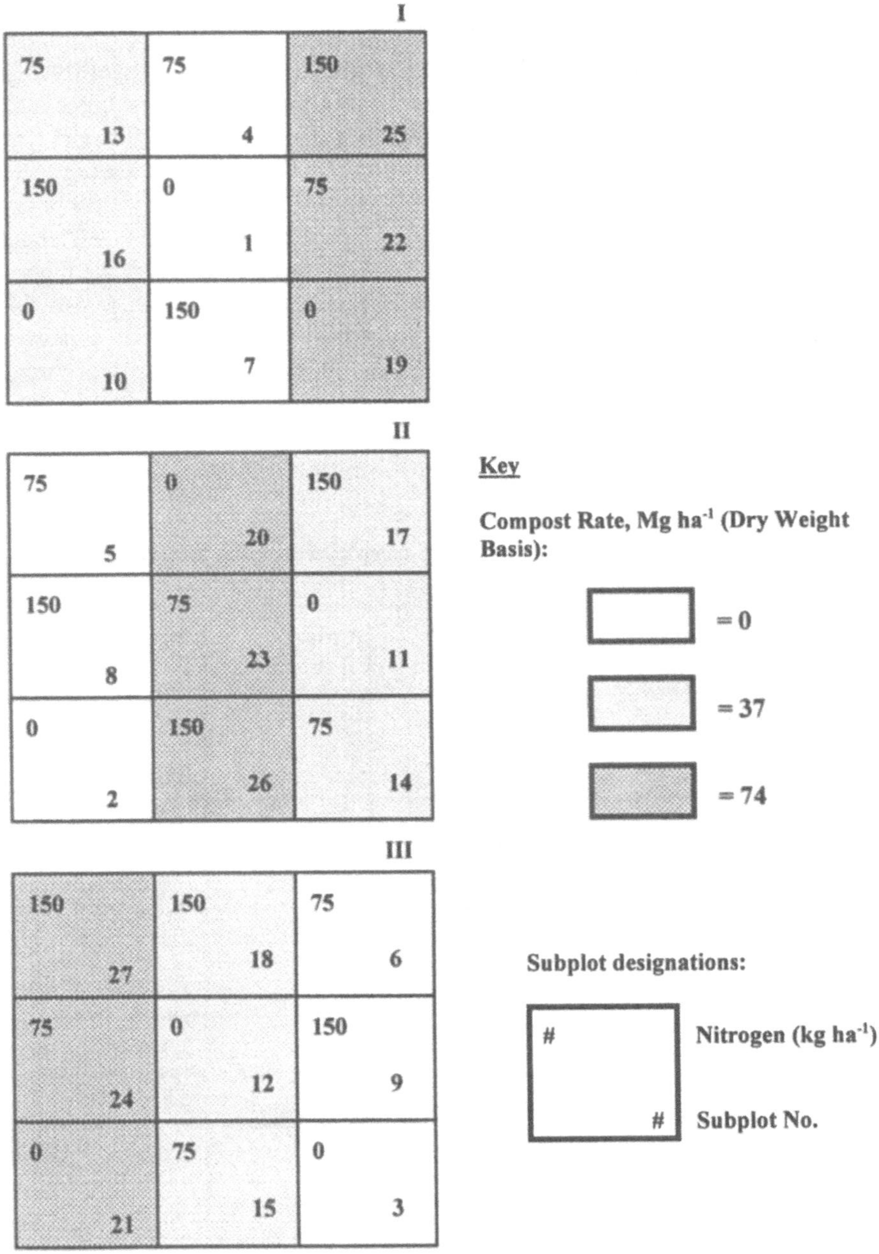

Figure 1. Compost field experiment sample plot design

RESULTS

Compost Effect on Lettuce Yield

The statistical analysis of the field test results on lettuce yields indicated significant ($p < 0.01$) overall compost differences, and compost rate by nitrogen rate interaction effects were also significant ($p < 0.01$) but the rate effects were independent of compost type (Table 4). The relative ranking of the compost types and control treatments are presented in Table 5, along with the estimated lettuce yield for compost without N supplementation. The relative impact of compost rate and nitrogen rate can be seen in the estimated lettuce yields given in Table 6. Note that highest lettuce yields were obtained at the highest compost application rates and that, while there are differences among nitrogen rates at this level, the differences are not great. In general, significant lettuce yield increases were observed at compost application rates of 37 Mg ha^{-1} and 74 Mg ha^{-1} compared to a control with no fertilizer or compost applied (Table 6). Low rates of compost, even when high levels of nitrogen were applied, still resulted in significantly lower yields.

Table 4. Statistical test and variance component estimates for lettuce and broccoli yield data.

Source	NDF[a]	DDF[b]	Lettuce[c] F-statistic	Broccoli[c] F-statistic
Compost Type (Type)	5	12	5.75**	197.34**
Compost Rate (C_Rate)	2	24	158.50**	2531.03**
Type x C_Rate IA	10	24	1.32	84.99**
Nitrogen Rate (N_rate)	2	72	68.21**	595.42**
Type x N_rate IA	10	72	0.86	2.55*
C_rate x N_rate IA	4	72	8.07**	44.83**
Type x C_rate x N_rate IA	20	72	0.62	2.07*
Variance Components	Main Plot		0.72	0.345
	Split Plot		0.00	0.367
	Split-Split Plot		4.84	10.75

[a] Numerator degrees of freedom.
[b] Denominator degrees of freedom.
[c] F-statistic for theoverall testofthe effect.
Significance indicated by *=p<0.05, **=0.05<p<0.01.

Table 5. Average lettuce yields by compost product without N supplementation.

Compost Product	Lettuce Yield[a] *Dry Weight (g/plant)*
(A1) 100% MSW	46.5 a
(A3) 96% MSW and 4%Bio	45.5 a
(A4) 96%YW and 4%Bio	43.5 b
(A2) 52% MSW and 46% UPD and 2%Bio	42.5 b
No Compost / N @ 150 kg ha^{-1}	42.17 b
(B1) 53% MSW and 47% UPD	42.0 b
(B3) 100% UPD	39.0 c
No Compost / N @ 75 kg ha^{-1}	38.3 c
No Compost / N @ 0 kg ha^{-1}	34.0 d

[a] Means with similar letters are not significantly different via LSD at a Type I error of 0.01.

Table 6. Lettuce Yields for Combination of Compost Rate and N Rate.

Composted Rate *(Mg ha^{-1})*	N Rate *(kg ha^{-1})*	Lettuce Yield[a] *Dry Weight (g/plant)*
74	150	46.80 a
37	150	46.17 ab
74	75	45.30 bc
74	0	44.16 cd
37	75	43.83 d
37	0	42.16 e
0	150	42.16 e
0	75	38.83 f
0	0	34.00 g

[a] Means with similar letters are not significantly different via LSD at a Type I error of 0.01.

Compost Effect on Broccoli Yield

The statistical analysis of the field test results on broccoli yields indicated all main, two-way and three-way interaction effects of compost type, compost rate and nitrogen rate were significant ($p < 0.05$) (Table 4). The significant interactions of compost type with the rate factors and concerns that the variability in yield was different for different compost types (Table 7) lead to a separate statistical analysis for each compost type. For each compost type, highest yields were obtained for the combination of high compost rate and high nitrogen rate. The general trend across all composts shows a significant difference in broccoli head yields at compost application rates of 37 Mg ha^{-1} and 74 Mg ha^{-1} compared to control (with no fertilizer or compost added), and between compost application rates of 37 mg ha^{-1} and 74 Mg ha^{-1} (Table 7). High nitrogen rates in the absence of compost still produced lower yields than in the presence of compost. The analysis indicates that broccoli head yields will be increased by a combination of type of compost, amount of compost and amount of N applied.

DISCUSSION

These results clearly indicate that the compost products utilized in these experiments can increase the yields of leaf lettuce and broccoli heads, regardless of whether N supplementation is used. With lettuce, the yield response seems to be to the total available N, regardless of whether the source was added inorganic N or the organic and mineralized N available from the compost (Table 8). Addition of 75 kg ha^{-1} of inorganic N is clearly insufficient to maximize yields. Even when an amount within the range of commercial application rates (115-300 kg ha^{-1} - personal communication with Dr. Ben Faber, University of California at Ventura), in this case 150 kg ha^{-1} N was used, compost produced an additional 10% yield in lettuce, probably due to the combination of improved soil physical condition and added mineralized N.

The broccoli data showed a strong N response, again demonstrating that 150 kg ha^{-1} inorganic N alone is insufficient to maximize head yield (commercial application rates of 150-350 kg ha^{-1} - personal communication with Dr. Ben Faber, University of California at Ventura). For all compost types, the addition of 74 Mg ha^{-1} of compost significantly increased broccoli yields. Those composts with more available N, namely A3 and A1, produced highest overall yields suggesting that the broccoli was responding primarily to the added N.

Urban plant debris, because of its high carbon to nitrogen ratio and lower N mineralization rate, produced the smallest increases in yields. Adding even a small amount of biosolids to yard waste improves the C/N ratio and hence improves yield. When comparing compost products, what seems to be the most important is the amount of available N (Table 8), since all will improve soil physical condition.

An additional benefit of compost over commercial N fertilization is the slow release of the N in compost to the plants through mineralization. Local rate of N mineralization of composted products ranges from 10% to 40% during the first year (personal communication with Dr. Ben Faber and Dr. Ben Crowley, University of California at

Ventura and Riverside, respectively). However, we feel that the rate for this particular area falls at the lower end of the range. This process, in addition to the increased cation exchange capacity of the soil by compost application (Shrilipour et al., 1992), allows the plant roots time to utilize more of the released N, hence minimizing the loss of excess NO_3-N through leaching into the groundwater. This research indicates that at least for some crops, utilization of an appropriate compost type and rate of application may lead to reduction and possible elimination of commercial N supplementation.

Table 7. Broccoli Yields by Compost Type for Combinations of Compost Rate and N Rate

		Compost Product					
		100%MSW[a] (A1)	53%MSW 47%UPD (A2)	96%MSW 4% Bio (A3)	52%MSW 46%UPD 2% Bio (A4)	96%UPD 4% Bio (B1)	100% UPD (B3)
Average Yield *Dry Weight (g/plant)*		**88.1**	**76.9**	**96.4**	**80.3**	**75.0**	**66.7**
C_Rate *Mg ha^{-1}*	N_Rate *kg ha^{-1}*						
0	0	37 a	37 a	40 a	37 a	41 a	37 a
	75	65 b	62 b	58 b	64 b	58 b	59 b
	150	77 c	70 c	65 c	75 c	67 c	69 c
37	0	94 d	72 c	92 d	65 b	62 bc	54 b
	75	99 d	78 d	99 e	80 c	73 c	65 bc
	150	103 d	84 de	102 e	87 d	82 d	76 cd
74	0	100 d	90 ef	129 f	95 e	88 d	72 cd
	75	102 d	95 f	138 g	105 f	98 e	80 de
	150	116 e	104 g	145 h	115 g	106 f	88e
Plot variance		0.0	0.00	1.79	0.56	0.66	0.00
Split plot variance		1.56	2.28	0.00	2.22	0.00	0.00
Split-split plot var		17.22	9.72	6.21	7.44	8.78	10.33

[a] Means in the same column with similar letters are not significantly different via LSD at a Type I error of 0.01.

Table 8. Various Forms of Nitrogen Added to Soil by Different Compost Types and Rates.

Compost Rate	Compost Types	Total N	Organic N	Estimated[a] Mineralized N
	Kg ha^{-1}			
37 Mg ha^{-1}	A1	629	592	59-237
	A2	555	518	52-208
	A3	629	592	59-237
	A4	592	518	52-208
	B1	481	481	48-194
	B3	444	444	44-177
74 Mg ha^{-1}	A1	1258	1184	118-472
	A2	1110	814	104-416
	A3	1258	1184	118-472
	A4	1184	1036	104-416
	B1	962	962	96-385
	B3	888	888	89-355

[a] Based on estimated 10-40% mineralization during the first year of application.

ACKNOWLEDGMENTS

This report is part of a three-year compost markets research project supported by The California Integrated Waste Management Board, the County of Santa Barbara, the Pew Charitable Trusts and The Procter & Gamble Company. The statements and conclusions of this report are those of the authors and not necessarily those of the funding agencies.

REFERENCES

Gallardo-Loro F & Nogales R (1987) Effect of the application of town refuse compost on the soil-plant system: A review. Biol. Waste 19:35-62.

Hernando S, Lobo MC & Polo A (1989) Effect of the application of a municipal refuse compost on the physical and chemical properties of a soil. The Science of the Total Environment. 81/82:589-596.

Hoitink HAJ & Fahy PC (1986) Basis for the control of the soilborne plant pathogens with composts. Ann. Rev. Phytopathol. 24:93-114.

Hoitink HAJ, Inbar Y & Boehm MJ (1991) Status of compost-amended potting mixes naturally suppressive to soilborne diseases of floricultural crops. Plant Disease/September. 869-873.

Littell RC, Milliken GA, Stroup WW & Wolfinger RD (1996) SAS System for Mixed Linear Models. Cary, NC: SAS Institute Inc., 633 pp.

Mays DA, Terman GL & Duggan JC (1973) Municipal compost: effect on crop yields and soil properties. J. Environ. Qual. 2(1):89-92.

Shiralipour A, McConnell DB & Smith WH (1992) Compost: Physical and chemical properties of soils as affected by MSW compost application. Biomass and Bioenergy 3(3-4):261-266.

Shiralipour A & Zachary A (1994) Compost field experiment guide for California Communities. Prepared for the Santa Barbara County Solid Wastè Management Division and the California Integrated Waste Management Board. 89 pp.

Shiralipour A, Faber B & Chrowstowski M (1996) Greenhouse broccoli and lettuce growth using composted biosolids. Compost Science & Utilization 4(3):38-43.

Sommers LE & Giordano PM (1984) Use of nitrogen from agricultural, industrial and municipal wastes. ASA-SSSA, 677 South Segoe Road, Madison, WI 53711, USA.

U.S. Environmental Protection Agency (1990) National sewage sludge survey: Availability of information and data anticipated impacts on proposed regulations. 55FR47210.

Watkins EM & Winch JE (1974) Composted organic waste, sewage sludges, and rock phosphate for the amelioration of acid uranium mine-tailings. In: Tomlinson J (Ed) Proceedings of the International Conference of Land Waste Management. Ottawa, Ontario: Agricultural Institute of Canada.

CO-UTILIZATION OF FLUE-GAS DESULFURIZATION AND ORGANIC BY-PRODUCTS FOR MINE RECLAMATION

RICHARD C. STEHOUWER, JERRY M. BIGHAM and WARREN A. DICK
The Ohio State University, Wooster, Ohio 44691, USA
E-mail stehouwer.1@osu.edu

ABSTRACT

Co-utilization of by-product materials in mine reclamation represents a potential double benefit to society because the materials are diverted from the solid waste stream and utilized to restore degraded landscapes. Several experiments have been conducted or are in progress to investigate minespoil amendment with flue gas desulfurization (FGD) by-products, yard waste compost, and sewage sludge as an alternative to resoiling in abandoned mined land reclamation. We found that interactions between these materials influenced the solubility and mobility of major labile elements in FGD (Ca, Mg, S) and in the spoil (Al, Fe), which in turn affected plant growth.

In most experiments, vegetative growth on acidic spoil was improved by amendment with FGD combined with organic by-products relative to either material alone. In two experiments, however, co-application decreased growth: in one case due to compost-induced N-immobilization, and in another reductions in plant growth may have been due to increased Al uptake.

In all experiments co-application of FGD and organic by-products increased spoil pH and exchangeable Ca, Mg, and S, and decreased Al and Fe in the depth of incorporation. Application of either material alone produced similar effects, but often of a smaller magnitude.

Movement of surface-applied materials into subjacent, unamended spoil produced several interactive effects which were not consistent in all experiments. Application of FGD by-products alone usually increased Ca and Mg and decreased Al and Fe in subjacent spoil due to exchange reactions between these metals. Application of organic by-products alone and with FGD had varied effects on subjacent spoil chemistry. In one greenhouse study, extensive leaching caused surface-applied compost to reduce subjacent exchangeable Al and Fe by 25 to 30% relative to an unamended spoil. With combined compost and FGD, however, subsurface increases in Ca and decreases in subsurface Al were larger than with either material alone. In another greenhouse experiment, however, FGD alone decreased subsurface Al and Fe, but combined FGD and compost increased subsurface Al and Fe. In a field experiment, nine months after application of sewage sludge, exchangeable Al and Fe in the subjacent spoil were one to four times larger than in spoil covered with topsoil. With combined FGD and sewage sludge, however,

S. Brown, J.S. Angle and L. Jacobs (eds.), Beneficial Co-Utilization of Agricultural, Municipal and Industrial By-products, 377–395.
© 1998 *Kluwer Academic Publishers,*

subsurface reductions in Al and increases in Ca were greater than with FGD alone. On severely compacted spoil, however, co-application of FGD and compost caused large increases in soluble Al and Fe just below the depth of amendment incorporation which have persisted for 24 months. These varied results and interactive effects appear to be due to movement and complexation of metals by soluble organic ligands in the compost and sewage sludge. Another experiment showed transport of trace metals from FGD and sewage sludge amended spoil was highly correlated with soluble organic C concentrations. Apparently complexation of metals by these ligands increased their solubility and mobility in the spoil. The extent to which this occurred would be a function of several factors including the amount of soluble organic matter and FGD added, and chemical and hydraulic characteristics of the spoil profile.

Direct revegetation of minespoil through co-application of FGD and organic by-products is a possible alternative to reclamation with borrow topsoil. These experiments have shown that co-application has the potential to enhance amelioration of subsurface toxicity. Elucidation of the mechanisms by which this occurs would enable the development of co-utilization strategies which maximize this beneficial effect.

INTRODUCTION

Our industrial society depends on the extraction of raw materials and fuels from the earth's resources and also produces large volumes of residual materials. Each of these activities can negatively impact environmental quality, on the one hand through drastic disturbance of landscapes, and on the other by improper and concentrated disposal of wastes. Co-utilization of industrial and municipal by-products as amendments for reclamation of drastically disturbed lands therefore represents a potential double benefit to society.

In the eastern United States coal mining region, coal mine spoil and coal cleaning refuse is frequently strongly acidic or acid forming due to oxidation of the pyritic materials they contain (Sutton and Dick, 1987). Reclamation of spoil and refuse is typically accomplished by application of agricultural limestone and covering with at least 20 cm of topsoil. Placement of topsoil can be prohibitively expensive, particularly on abandoned mined lands (AML) and coal refuse piles (operated prior to enactment of the Surface Mining Control and Recovery Act in 1977) where soil material was not conserved and is not available on-site. An alternative to covering with topsoil is direct revegetation of the spoil. Reclamation or revegetation of these materials requires amelioration of phytotoxic conditions, supply of essential plant nutrients, and establishment of a soil-like microbiological community.

Flue gas desulfurization (FGD) by-products, an industrial by-product from scrubbing SO_2 from the emissions of fossil fuel fired boilers, and municipal organic by-products such as composts and sewage sludge each have characteristics which make them beneficial amendments for acidic minespoils. Co-utilization of FGD and organic by-products may improve reclamation success by combining the beneficial aspects of each material.

The numerous type of scrubber processes utilized for SO_2 removal produce several different FGD by-products. In this paper the term "FGD" will be used to refer generically to all types of flue gas desulfurization by-products. When reference to a specific FGD is intended the appropriate name or acronym will be used. One of the most widely utilized FGD processes produces a sludge-like material which will be referred to as "wet FGD" in this paper.

A large body of research reviewed by Sopper (1992) has shown that revegetation of toxic minespoil is improved by the application of organic amendments such as sewage sludge. Reasons for improved vegetative growth from organic amendments include improved soil fertility, reduction in the plant availability of toxic metals, pH buffering, and improved water holding capacity and soil physical properties. Composts produced from organic wastes have not been extensively researched as amendments for mine reclamation, however, they can be expected to provide many of the same benefits as sewage sludge.

Traditionally agricultural limestone (AgLime) has been used to neutralize spoil acidity and to supply Ca. Because FGD by-products have some acid neutralizing potential and also contain large amounts of Ca (primarily as $CaSO_4$ or $CaSO_3$) they can be effective substitutes for AgLime (Stehouwer et al., 1995c; Stehouwer et al., 1996). Application of gypsum to agricultural soils has been found to be more effective than AgLime in reducing phytotoxicity (primarily exchangeable Al) below the zone of application (Sumner, 1970; Pavan et al., 1984; Hammel et al., 1985; Sumner et al., 1986; Farina and Channon, 1988). Several mechanisms are thought to be involved in the reduction of subsurface Al by gypsum, however the primary one appears to be Ca transport into subjacent soil and subsequent exchange with Al. Recent studies with various FGD by-products applied to minespoils has shown they can effectively neutralize acidity in the zone of mixing and that Ca and Mg move rapidly downward into subjacent spoil (Stehouwer et al., 1995a, Stehouwer, 1997, Stehouwer and Mafi, 1997).

Several experiments have been conducted or are in progress at the Ohio State University in which various FGD by-products and either yard-trimmings compost or sewage sludge have been used as amendments for revegetation of acidic minespoil and coal refuse. This paper will review the results of these experiments, focusing on the interactive effects these amendments had on plant growth and amelioration of subsurface phytotoxicity.

MATERIALS AND METHODS

Results from four greenhouse (GH) and four field experiments are reviewed in this paper. The field experiments were conducted either on abandoned mined land (AML) or on a coal refuse pile (CRP). Characteristics of the by-product materials used are given in Table 1. The experiments are listed in Tables 2 and 3 together with the type of material amended and the by-products and application rates used. The citations given in Tables 2 and 3 refer to publications which contain additional detail on the experiments.

The greenhouse experiments were conducted using 15-cm diameter columns. Minespoil and refuse material was collected from field locations, air-dried, and passed through a 10-mm screen before being mixed with air-dry amendments and placed in the columns. In the greenhouse experiments all amendment treatments were replicated three times. In the Fleming GH and Kurtz GH experiment 30-cm columns were used. The columns were filled with spoil material which had been mixed with the amendments and

Table 1. Characterization of by-products used.

Parameter	LIMB[a]	Wet FGD[b]	AFBC[c]	MgGyp[d]	Yard-trimmings Compost	Sewage sludge
pH (1:1, water)	12.5	9.9	12.5	9.7	7.9	6.5
$CaCO_3$ equivalency, %	59	14	39	8	na	na
Major mineral constituents	Lime Anhydrite Calcite Portlandite Fly ash	Hannebachite Calcite Gypsum Fly ash	Anhydrite Lime Fly ash	Gypsum Brucite	na	na
Major elements, %						
Organic C	na	na	na	na	21.4	31.2
Total N	na	na	na	na	1.4	3.8
Ca	36	29	26	24	8.3	5.9
Mg	0.60	na	3.70	2.00	1.5	1.1
S	5.8	17.5	12.3	17.9	na	1.40
K	0.90	na	0.40	0.03	0.40	0.15
P	0.02	na	0.03	<0.01	0.30	1.80
Trace elements (mg kg^{-1})						
B	233	342	418	244	na	na

[a] LIMB, lime injection multi-stage burners; a dry material collected in the fly ash stream.
[b] wet FGD; dewatered slurry of primarily calcium sulfite and calcium sulfate mixed with dry fly ash and lime.
[c] AFBC, atmospheric fluidized bed combustion; a dry material which is a mixture of the bottom and fly ash streams.
[d] MgGyp, $Mg(OH)_2$ enriched gypsum; dewatered gypsum slurry containing some $Mg(OH)_2$.

planted with tall fescue (*Festuca arundinacea*) (Fleming GH) or with orchard grass (*Dactylis glomerata*) (Kurtz GH). In the Rehoboth GH and Zimmer GH experiments 60-cm columns were used. The lower portion of the column (35-40 cm) was filled with unamended material above which a layer of amended material was placed (15-20 cm) and planted with orchard grass (*Dactylis glomerata*) (Zimmer GH) or a mixture of orchard grass (*Dactylis glomerata*), annual ryegrass (*Lolium multiflorum*), birdsfoot trefoil (*Lotus* sp.), and ladino clover (*Trifolium repens* Ladino) (Rehoboth GH). Each experiment lasted approximately 8 mo. Plant growth was harvested monthly beginning three months after planting. In the Fleming GH experiment water was added to meet evapotranspiration loss. Pot leachates were collected at the beginning, middle and end of the experiment by adding enough excess water to produce 100 to 200 ml of leachate. Much more extensive leaching

Table 2. Effect of FGD, compost and sewage sludge on spoil chemistry in the mixing zone of greenhouse experiments.

Experiment name and description	Amendment	Spoil/refuse pH	Ca	Mg	S	Al	Fe
			Leachate concentrations[a]				
			---mg L^{-1}---				
Fleming GH: LIMB (120 g kg^{-1}) and sewage sludge (60 g kg^{-1}) amendment of acidic minespoil Stehouwer et al., 1995a Stehouwer et al., 1995b	no amendment	3.05b	50b	16c	48c	50.2a	1.15b
	sewage sludge	3.95b	82b	35b	170b	3.04b	2.27a
	LIMB	7.80a	696a	11c	648a	<0.05c	<0.01c
	LIMB+sewage sludge	7.83a	770a	58a	631a	<0.05c	<0.01c
			Exchangeable concentrations				
			---g kg^{-1}---			---mg kg^{-1}---	
Kurtz GH: wet FGD (200 g kg^{-1}), AgLime (80 g kg^{-1}), and yard trimmings compost (200 g kg^{-1}) amendment of acidic coal refuse Stehouwer, 1997	no amendment	1.60d	no extractions done of refuse only				
	compost	2.01d	no extractions done of refuse + compost				
	AgLime	6.53a	8.47b	0.04b	4.44b	15.6b	1.46b
	AgLime + compost	6.76a	6.00b	0.09b	3.74b	8.2b	0.71b
	wet FGD	4.16c	19.5a	0.21b	15.0a	162a	35.7a
	wet FGD + compost	5.72b	19.9a	0.81a	14.7a	17.6b	4.79b
Rehoboth GH: wet FGD (500 g kg^{-1}), AgLime (80 g kg^{-1}), and yard trimmings compost (100 g kg^{-1}) amendment of acidic coal refuse Stehouwer and Mafi, 1997	AgLime	6.13b	8.92b	0.41b	5.38b	12.8a	0.52a
	AgLime + compost	5.84b	8.46b	0.60b	5.27b	8.14a	0.63a
	wet FGD	7.04a	18.5a	0.60b	13.9a	2.0a	0.58a
	wet FGD + compost	7.41a	19.1a	1.46a	15.2a	12.3a	1.01a
Zimmer GH: MgGyp (250 g kg^{-1}) and yard trimmings compost (100 g kg^{-1}) amendment of acidic minespoil Yibirin et al., 1996	no amendment	3.38c	0.09c	0.02c	0.05b	480a	158a
	MgGyp	6.15a	17.6a	1.06a	13.1a	14.2cd	23.5c
	compost	4.84b	2.46b	0.46b	0.20b	31.4bc	110b
	MgGyp + compost	6.03a	18.4a	0.94a	13.5a	11.1d	21.9c

[a] Leachates were collected at the conclusion of the eight month experiment.

Table 3. Effect of FGD, compost and sewage sludge on spoil chemistry in the mixing zone of field experiments.

Experiment name and description	Amendment	Spoil/refuse pH	Ca	Mg	S	Al	Fe
			------------- g kg^{-1}	---------		---- mg kg^{-1} ----	
			-------- Water soluble concentrations -------				
Fleming AML: Field experiment of spoil reclamation with topsoil cover (20 cm) and spoil amended with AFBC (280 Mt ha^{-1}) and compost (112 Mt ha^{-1}) Stehouwer et al., 1997	topsoil	6.16b	0.24b	0.06b	0.20b	1.50b	1.42a
	AFBC	7.01a	4.56a	0.59a	4.36a	2.42a	0.33b
	AFBC + compost	5.83b	4.64a	0.66a	4.48a	2.61a	0.26b
Rehoboth CRP: Field experiment of coal refuse reclamation with AgLime (180 Mt ha^{-1}), wet FGD (1120 Mt ha^{-1}), and compost (112 Mt ha^{-1}) Stehouwer and Mafi, 1997	AgLime	4.43bc	5.31b	0.17a	5.08b	98.7a	330a
	AgLime + compost	4.12c	5.69ab	0.25a	5.67ab	35.7a	638a
	wet FGD	5.42ab	5.93a	0.16a	5.74a	93.4a	513a
	wet FGD + compost	5.80a	6.11a	0.29a	5.88a	23.2a	221a
			-------- Exchangeable concentrations -------				
Zimmer AML: Field experiment of spoil reclamation with MgGyp (245 Mt ha^{-1}) and compost (112 Mt ha^{-1})	no amendment	3.30c	0.16c	0.0 7b	0.28b	432a	147a
	MgGyp	5.93a	18.2a	0.37a	13.7a	21.4c	0.86c
	compost	4.30b	2.11b	0.18b	0.68b	164b	48.0b
	MgGyp + compost	5.91a	17.4a	0.35a	12.6a	7.42c	1.23c
USBM AML: Field experiment of spoil reclamation with topsoil cover (20 cm), and spoil amended with wet FGD (670 Mt ha^{-1}), and sewage sludge (100 Mt ha^{-1}).	borrow soil	7.03a	4.27c	0.06a	0.41b	<0.05b	0.66c
	wet FGD	7.35a	17.7a	0.12a	13.4a	0.93b	6.66b
	sewage sludge	4.21b	12.3b	0.11a	0.36b	195a	8.93a
	FGD + sewage sludge	7.09a	17.7a	0.13a	13.6a	1.00b	0.75b

occurred in the Kurtz GH, Rehoboth GH and Zimmer GH experiments where approximately 1000 mm of water was added during the experiment (simulating the average annual rainfall for southeast Ohio). Spoil/refuse samples were collected from various depths in the columns at the end of the experiments. Concentrations of exchangeable cations were determined by extraction with 1 M KCl (1:10, soil:extractant). Extracts were analyzed by inductively coupled plasma emission spectroscopy.

In the field experiments the amendment materials were surface-applied in September or October, and incorporated to a depth of approximately 20 cm with a moldboard plough, chisel plough, and/or roto-tiller. Immediately after amendment application the plots were planted with a mixture of orchard grass (*Dactylis glomerata*), timothy (*Phleum pratense*), ladino clover (*Trifolium repens* Ladino), birdsfoot trefoil (*Lotus* sp.), annual ryegrass (*Lolium multiflorum*), and winter wheat (*Agropyron* sp.). In the field experiments all amendment treatments were replicated four times. Plant growth was measured by cutting, drying, and weighing above ground plant growth. Soil samples were collected by taking two or three, 2.5-cm cores in each plot and combining the appropriate depth segments. Exchangeable and water soluble cations (1:10, soil:water) were determined as described above.

RESULTS AND DISCUSSION

Plant Responses

In most experiments amendment of toxic minespoil and coal refuse with either FGD or organic by-products permitted plant growth to occur (Table 4). Only on the extremely toxic coal refuse material used in the Kurtz GH experiment was plant growth not possible with amendment of compost only. In general, growth with FGD amendments produced growth similar to that with AgLime amendment. Co-application, however, produced varied results. In the Fleming GH, Rehoboth GH, and Rehoboth CRP experiments co-application resulted in the greatest plant growth, while in the Zimmer GH experiment co-application decreased plant growth. Compost-induced yield decreases in the Zimmer GH experiment were due to N-immobilization by the compost. Plant leaves were pale green and tissue N-content was lower than in plants from the non-compost treatments (data not shown). Most of the yield depression occurred in the earlier harvests. Yield depression from compost had largely disappeared by the end of the experiment due to the addition of mineral N fertilizer. In the Kurtz GH experiment compost also depressed yield in the first cutting due to N-immobilization. Addition of mineral fertilizer N, however, increased yields of subsequent harvests and made up for the initial yield reduction.

In the remaining experiments there was no clear advantage or disadvantage to co-application relative to the use of the materials individually. It should be noted that the data for the Zimmer and USBM AML studies are from the first year following treatment application. These experiments are on-going, as is the Fleming AML, and long-term plant growth response could well be much different from what is observed in just the first year. For example in the Fleming AML experiment, most of the large production on the topsoil was from lush growth of winter wheat in the first spring after seeding. In the second year

Table 4. Vegetative yield of minespoil and coal refuse amended with various FGD by-products, compost, and sewage sludge. Amendment rates are given in Table 2.

Experiment	Yield			
	---------------- g pot^{-1} (greenhouse) or kg ha^{-1} (field) ----------------			
Fleming GH	no amendment	sewage sludge	LIMB	LIMB + sewage sludge
	0.0	24.8b	38.1b	62.7a
Kurtz GH (no amendment and compost only gave no growth)	AgLime	AgLime + compost	wet FGD	wet FGD + compost
	38.7b	38.1b	27.9c	48.6a
Rehoboth GH	AgLime	AgLime + compost	FGD	FGD + compost
	46.6a	53.2a	38.9a	47.8a
Zimmer GH	no amendment	compost	MgGyp	MgGyp + compost
	0.0c	30.0b	40.4a	28.5b
Fleming AML	topsoil		AFBC	AFBC + compost
	7331a		3658b	3060b
Rehoboth CRP	AgLime	AgLime + compost	FGD	wet FGD + compost
	299c	410bc	641b	1105a
Zimmer AML	no amendment	compost	MgGyp	MgGyp + compost
	0c	1939b	2249a	2346a
USBM AML	topsoil	sewage sludge	wet FGD	wet FGD + sewage sludge
	208c	2518a	719b	3194a

of this study when wheat was no longer present, the differences among the three treatments were much less than is indicated in Table 4. While the yield totals for Zimmer AML and USBM AML did not show an increase with co-application, there appeared to be a better balance of grass and legume species with co-application than with any other treatment. This may indicate a greater potential for long-term sustainability of the vegetative cover.

Chemistry of the Mixing Zone

The effects of by-product amendment on spoil and refuse pH and Ca, Al, and Fe concentrations in the mixing zone were consistent with their effects on plant growth. In all experiments, application of FGD by-products increased spoil and refuse pH (Tables 2 and 3). These increases in pH also resulted in large decreases in exchangeable and water soluble Al and Fe. Application of FGD materials caused much larger increases in Ca (and S) than did AgLime due to the much larger solubility of gypsum ($CaSO_4 \cdot 2H_2O$) than calcite ($CaCO_3$). The one exception to this was in the Rehoboth CRP experiment. Here the solubility of Ca released in $CaCO_3$ reaction with refuse acidity was apparently controlled by gypsum solubility due to the large background sulfate concentrations in the coal refuse. Application of organic by-products alone caused relatively small pH increases (Fleming GH, Zimmer GH, Zimmer AML), nevertheless significant reductions in exchangeable and soluble Al and Fe were observed. These decreases were likely due to complexation of these metals by organic ligands in the sewage sludge and compost (Stevenson, 1994). These changes in spoil and refuse chemistry represent a significant decrease in phytotoxicity. In none of the experiments, however, did combined application of FGD and organic by-products cause larger decreases in Al or Fe than either FGD or AgLime alone. Co-application did create the situation where there were both high Ca concentrations and low Al and Fe concentrations in the zone of mixing. This may have affected the level of complexation and downward transport of these metals and their effects on spoil/refuse chemistry below the mixing zone.

Subsurface Chemistry

None of the experiments showed evidence of a pH increase below the depth of FGD and/or organic by-product application (data not shown). Surface application of FGD by-products, however, increased subsurface exchangeable and water soluble Ca (and Mg) concentrations in all experiments except Fleming AML (Figs. 1 to 6). These increases are consistent with the literature on gypsum transport in agricultural soils (Sumner, 1995), which indicates much more rapid transport of Ca with gypsum than with AgLime application. Lack of Ca movement in the Fleming AML experiment appeared to be due to the severe compaction of the minespoil at this location. It should also be noted that these experiments gave little evidence that downward transport of Ca (or Mg) was affected by co-application with an organic by-product. Only in the USBM AML experiment was there some indication that downward movement of Ca was increased by co-application of FGD and sewage sludge.

Surface application of FGD by-products alone decreased subsurface exchangeable (or water soluble) Al and Fe in some, but not all, experiments. A FGD induced decrease in subsurface Al and Fe was greatest in the Zimmer GH (Fig. 2), Zimmer AML (Fig. 5), and USBM AML (Fig. 6) experiments where there was also a clear increase in subsurface Ca and/or Mg. Thus the gypsum component of the FGD by-products functioned similarly in these acidic spoils in comparison with what has been observed in agricultural soils (Stehouwer et al., 1995c; Sumner, 1995). In the Fleming AML experiment (Fig. 4), where spoil compaction apparently limited Ca transport, surface application of AFBC had no

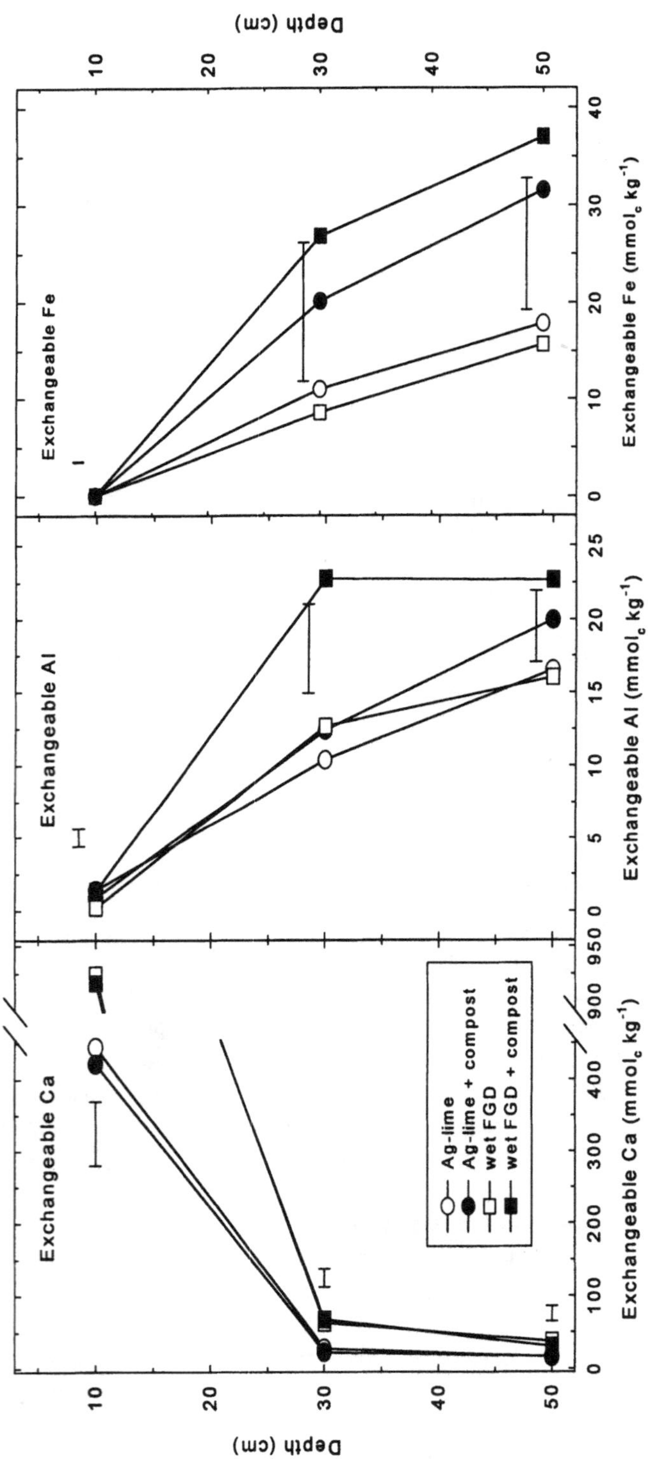

Figure 1. Distribution of exchangeable Ca, Al, and Fe in coal refuse columns at the end of the Rehoboth GH experiment. (Error bars indicate the width of the $LSD_{0.05}$ value for comparison of treatment means at each depth.)

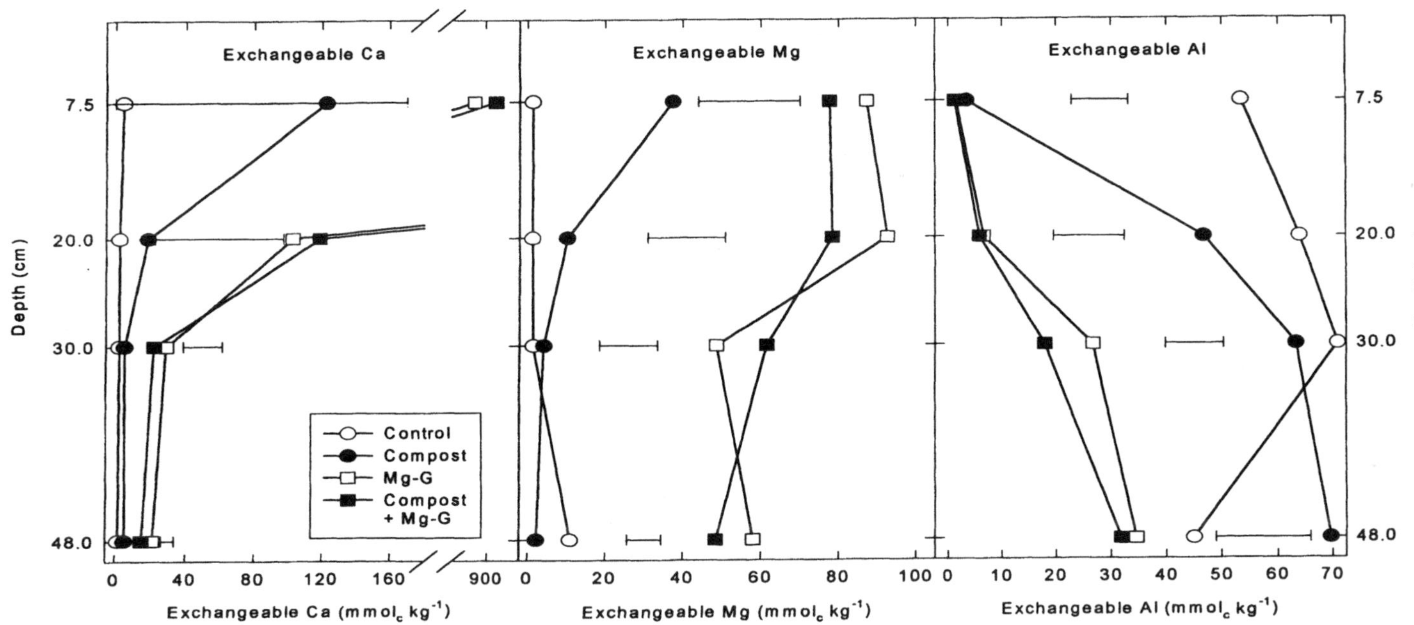

Figure 2. Distribution of exchangeable Ca, Mg, and Al in columns of acidic minespoil at the conclusion of the Zimmer GH experiment. (Error bars indicate the width of the $LSD_{0.05}$ value for comparison of treatment means at each depth.)

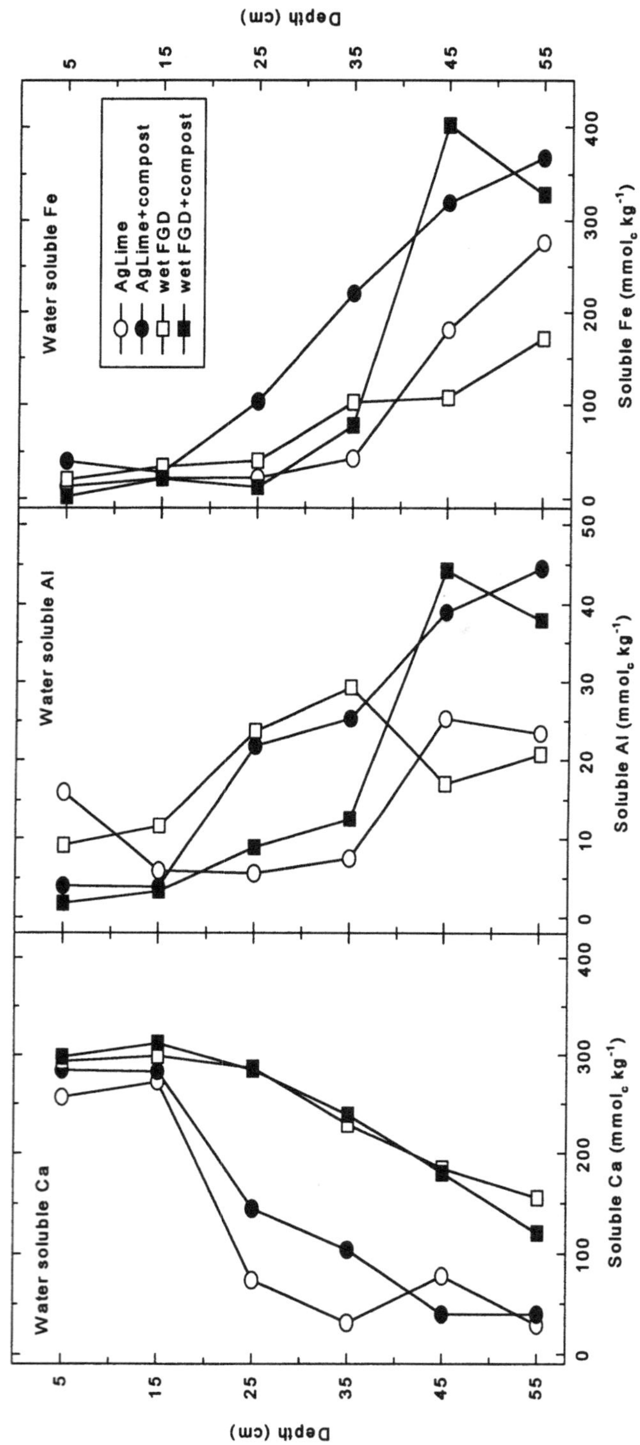

Figure 3. Distribution of water soluble Ca, Al, and Fe concentrations in the acidic coal refuse profile 18 months after treatment application in the Rehoboth CRP experiment. (Treatment means could not be separated using LSD. Analysis of variance with depth as a repeated measure indicated significant treatment differences for the distribution of these metals.)

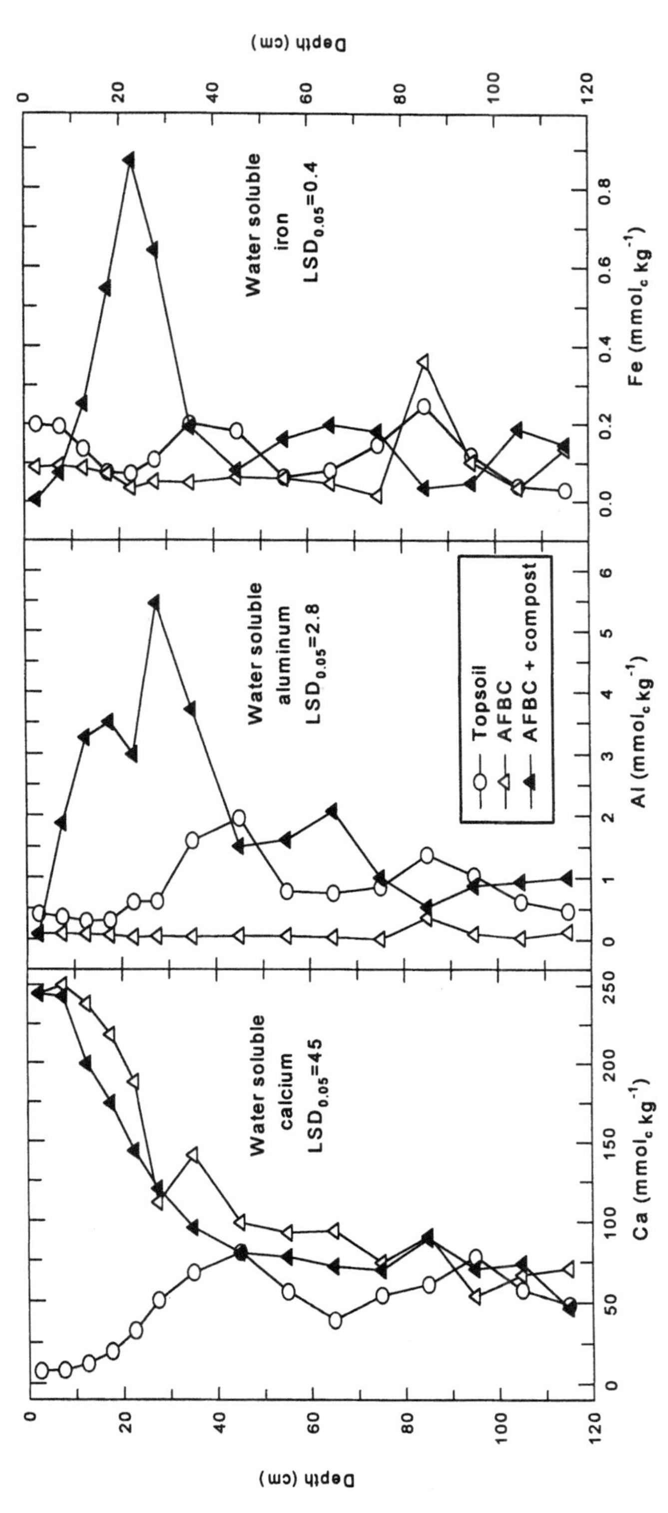

Figure 4. Distribution of water soluble Ca, Al, and Fe in the acidic minespoil profile 24 months after treatment application in the Fleming AML experiment.

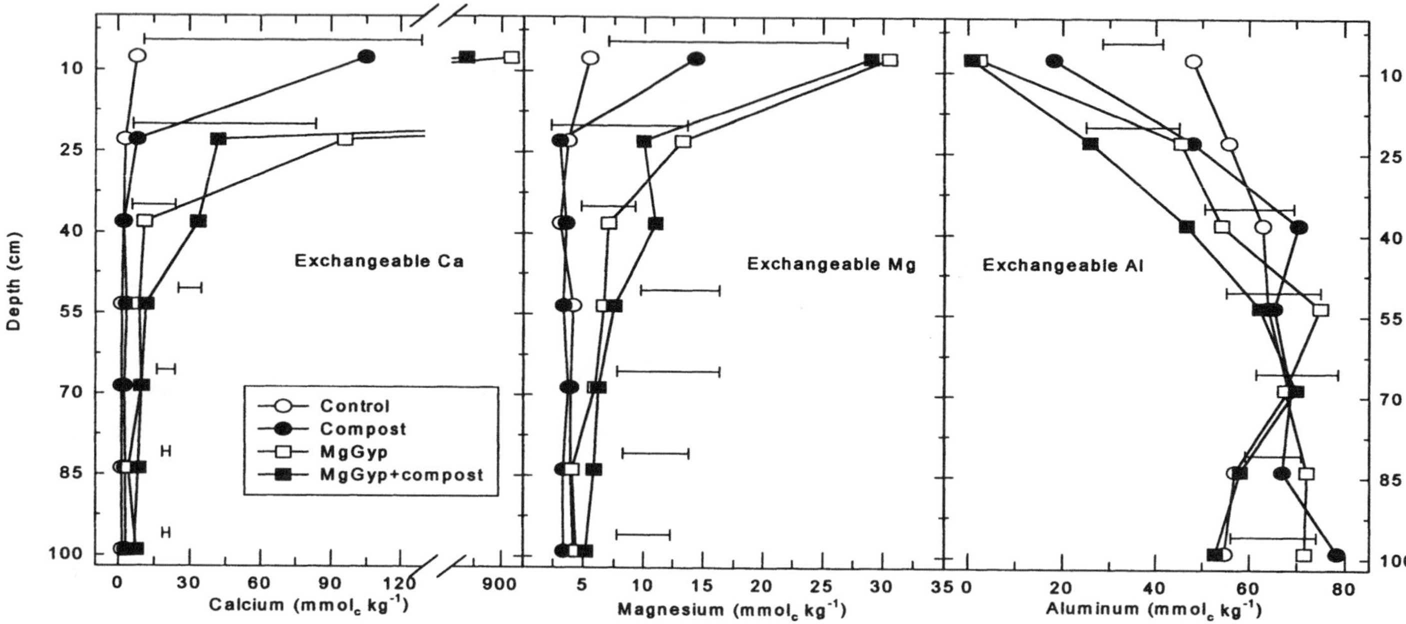

Figure 5. Distribution of exchangeable Ca, Mg, and Al in the acidic minespoil profile 12 months after treatment application in the Zimmer AML experiment. (Error bars indicate the width of the $LSD_{0.1}$ value for comparison of treatment means at each depth.)

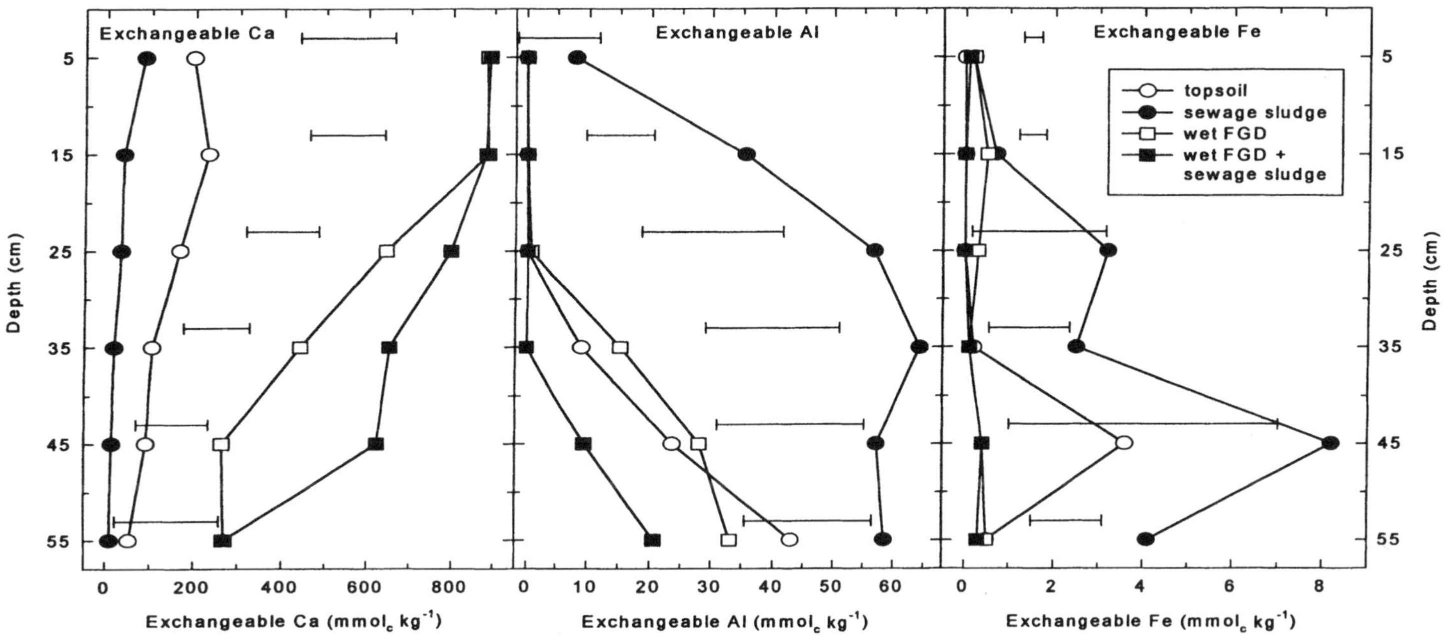

Figure 6. Distribution of exchangeable Ca, Al, and Fe in the acidic minespoil profile nine months after treatment application in the USBM AML experiment. (Error bars indicate the width of the $LSD_{0.1}$ value for comparison of treatment means at each depth.)

effect on subsurface water soluble Al and Fe concentrations compared to the topsoil treatment.

Co-application of FGD and organic by-products had varied effects on subsurface Al and Fe. While amendment with organic materials singly and in combination with FGD decreased Al and Fe in the zone of mixing, in several experiments both exchangeable and water soluble Al and Fe concentrations were increased in subjacent spoil and coal refuse. Small increases in subsurface Al were caused by compost alone in the Zimmer GH and Zimmer AML experiments (Figs. 2 and 5) and large increases in subsurface Al were observed with sewage sludge in the USBM AML experiment (Fig. 6). Application of compost with FGD by-products increased subsurface Al and Fe in the Rehoboth GH (Fig. 1) and Fleming AML (Fig. 4) experiments relative to FGD alone. In the Zimmer GH (Fig. 2), Zimmer AML (Fig. 5), and USBM AML (Fig.6) experiments, however, co-application of FGD and organic by-products appeared to cause larger reductions in subsurface Al than FGD alone. In the Rehoboth coal refuse experiment (Fig. 3) co-application of FGD and compost enhanced Al reduction in the upper part of the profile, while it increased Al in the lower sampling depths relative to FGD alone.

Increases in organic matter, such as produced by the addition of compost or sewage sludge, are generally associated with decreased solubility and availability of Al and Fe (Stevenson, 1994). In these experiments this was true in the zone of mixing as was previously noted. We propose that the observed increases in Al and Fe solubility and availability in subjacent spoil layers is due to downward transport of soluble organic ligands added with the compost or sewage sludge. As these soluble ligands moved into subjacent spoil layers they increased the solubility of Al and Fe by forming soluble organic complexes with Al and Fe. Precipitation of these ligands in the subjacent layers could also increase exchangeable Al and Fe concentrations. Co-application of organic by-products with FGD has been shown to increase concentrations of soluble organic carbon leached from the FGD-organic mixing zone (Stehouwer et al., 1995a). This would increase the potential for complexation of Al and Fe in subjacent spoil layers. In the Fleming GH experiment, leachates from spoil amended with both FGD and sewage sludge had much larger concentrations of organic C than did leachates from spoil with either amendment alone (Table 5). It should be noted that the increase in soluble organic C was associated with increased leachate concentrations of some trace metals. Compost combined with either

Table 5. Effect of LIMB and sewage sludge on pH, dissolved organic carbon and trace metal concentrations of leachates collected at the conclusion of the Fleming GH experiment.

Amendment	pH	DOC	Cr	Cu	Ni
		-------------------	mg L^{-1}	---------------------	
No amendment	2.81	5	0.01	0.07	0.11
Sewage sludge	3.72	14	0.01	0.04	0.24
LIMB	7.86	55	0.11	<0.01	<0.01
LIMB + sewage sludge	8.52	1050	0.15	5.55	0.85
$LSD_{0.05}$	0.55	61	0.04	0.42	0.09

AgLime or wet FGD increased transport of Fe and some trace elements in the Rehoboth GH experiment relative to AgLime or FGD alone (Table 6). Mobilization of trace elements of environmental concern by this mechanism is a potential concern with respect to co-utilization of these by-products and warrants further research.

Co-application also introduces a large amount of soluble Ca whose transport and exchange with Al and Fe could be affected by increased soluble organic ligands. In the USBM AML experiment (Fig. 6) this may account for the greater movement of Ca into, and loss of Al from, subjacent spoil with combined FGD and sewage sludge relative to FGD alone. Calcium transport from the surface mixing zone may have been increased by complexation with organic ligands. Once in the acidic subsurface zone, these Ca-organic ligand complexes may have facilitated Ca-Al exchange due to the high affinity of organic ligands for trivalent metals.

If the increases in subsurface Al and Fe with organic by-product amendment are in fact due to transport and complexation with soluble organic ligands, the extent of the effect

Table 6. Effect of wet FGD, AgLime, and compost on concentrations of metals in leachates collected at the conclusion of the Rehoboth GH experiment.

Amendment	Al	Fe	Cd	Cr	Cu	Ni
	------	------	------ mg L^{-1} ------	------	------	------
AgLime	59	553	0.02	0.05	0.35	0.08
AgLime + compost	94	909	0.03	0.10	0.50	0.14
wet FGD	62	386	0.02	0.05	0.28	0.07
wet FGD + compost	111	1016	0.04	0.10	0.57	0.14
$LSD_{0.05}$	45	565	ns	ns	0.21	0.07

would be dependent both on the amount of soluble organic C and on the extent leaching. Thus in the highly compacted spoil in the Fleming AML experiment (Fig. 4), soluble organic ligands apparently accumulated at the interface between the depth of mixing and the highly compacted subjacent spoil. This would have caused the large increase in Al and Fe in the 20 to 30 cm depth range. This Al and Fe "bulge" was observed in samples collected both 12 and 24 months after treatment application. By contrast, in the Rehoboth CRP experiment (Fig. 3) conducted on coarse, highly permeable refuse material, the Al and Fe increase occurred deeper in the profile 18 months after treatment application.

CONCLUSIONS

The experiments reviewed in this paper indicate that co-utilization of FGD and organic by-products as amendments for revegetation of minespoils is a viable alternative to traditional reclamation by covering with topsoil. In general co-utilization produced as good or better vegetative growth relative to either material alone or compared to amendment with AgLime. The major chemical benefits associated with co-utilization appear to be increased Ca supply, neutralization of acidity, and reduction of Al and Fe by pH increase and complexation with added organic matter. These effects, however, are largely limited to the mixing zone.

Interactions of co-applied FGD and organic by-products resulted in varied effects on subsurface chemistry. These interactions appeared to be associated with downward transport of soluble organic ligands added in the organic by-products. Complexation of these ligands with Ca, Al, or Fe resulted in increased or decreased subsurface concentrations of these metals depending on the extent of leaching and the amount of soluble organic carbon. Evidence for this mechanism in these experiments is highly empirical. More basic studies are needed to further elucidate and substantiate the proposed mechanism. The results of such studies could lead to co-utilization strategies which would enhance the beneficial effects on amelioration of subsurface phytotoxicity.

ACKNOWLEDGMENTS

The experiments reviewed in this paper were conducted with funding from various sources including the Ohio Coal Development Office, U.S. Department of Energy, the Electric Power Research Institute, American Electric Power Co., and Kurtz Brothers Co. Salaries and research support was also provided by State and Federal funds appropriated to the Ohio Agricultural Research and Development Center, the Ohio State University.

REFERENCES

Farina MPW & Channon P (1988) Acid-subsoil amelioration: II. Gypsum effects on growth and subsoil chemical properties. Soil Sci. Soc. Am. J. 52:175-180.

Hammel JE, Sumner ME & Shahandeh H (1985) Effect of physical and chemical profile modification on soybean and corn production. Soil Sci. Soc. Am. J. 49:1508-1512.

Pavan MA, Bingham FT & Pratt PF (1984) Redistribution of exchangeable calcium, magnesium, and aluminum following lime or gypsum applications to a Brazilian Oxisol. Soil Sci. Soc. Am. J. 42:33-38.

Sopper WE (1992) Reclamation of mine land using municipal sewage sludge. Adv. Soil Sci. 17:351-431.

Stehouwer RC (1997) Amendment of acidic coal refuse with yard waste compost and alkaline materials: Effects on leachate quality and plant growth. Compost Science and Utilization, 5(3):81-92.

Stehouwer RC & Mafi S (1997) Efficacy and transport of wet FGD scrubber sludge and compost amendments for revegetation of acidic coal refuse. In: Proceedings: 12th International Symposium on Management and Use of Coal Combustion By-Products (CCBs). American Coal Ash Association, Orlando, FL, January 26-30, 1997.

Stehouwer RC, Dick WA & Lal R (1997) Soil and water quality impacts of a clean coal combustion by-product used for abandoned mined land reclamation. In: Proceedings: 12th International Symposium on Management and Use of Coal Combustion By-Products (CCBs). American Coal Ash Association, Orlando, FL, January 26-30, 1997.

Stehouwer RC, Sutton P & Dick WA (1995a) Minespoil amendment with dry flue gas desulfurization by-products: Element solubility and mobility. J. Environ. Qual. 24:165-174.

Stehouwer RC, Sutton P & Dick WA (1995b) Minespoil amendment with dry flue gas desulfurization by-products: Plant growth. J. Environ. Qual. 24:861-869.

Stehouwer RC, Sutton P & Dick WA (1996) Transport and plant uptake of soil-applied dry flue gas desulfurization by-products. Soil Sci. 161:562-574.

Stehouwer RC, Sutton P & Dick WA (1995c) Use of clean coal technology by-products as agricultural liming materials. p. 1-1 to 1-14. In: Proceedings: 11th International Symposium on Use and Management of Coal Combustion By-Products (CCBs). Orlando, FL, January 15-19, 1995. EPRI TR-104657-V1.

Stevenson FJ (1994) Humus chemistry: genesis, composition, reactions. J. Wiley and Sons, New York.

Sumner ME (1970) Aluminum toxicity - a growth limiting factor in some Natal sands. Proc. Annu. Cong. S. Afr. Sugar Technol. Assoc. 44:1-6.

Sumner ME, Shahandeh H, Bouton J & Hammel JE (1986) Amelioration of an acid soil profile through deep liming and surface application of gypsum. Soil Sci. Soc. Am. J. 50:1254-1258.

Sumner ME (1995) Amelioration of subsoil acidity with minimum disturbance. Adv. Soil Sci. 23:147-185.

Sutton P, & W.A. Dick WA (1987) Reclamation of acidic mined lands in humid areas. Adv. in Agron. 41:377-405.

Yibirin H, Stehouwer RC, Bigham JM & Sutton P (1996) Biomass production and water quality in an acidic spoil amended with Mg(OH)2-enriched by-product gypsum. p. 575-578. In: Proceedings: 211th American Chemical Society National Meetings, Volume 42, No. 2. New Orleans, LA, March 24-28, 1996.

DEHYDRATION OF RESTAURANT FOOD WASTES PRODUCES A NUTRITIOUS FEEDSTUFF FOR USE IN PIG DIETS

DEHYDRATION OF RESTAURANT FOOD WASTES PRODUCES A NUTRITIOUS FEEDSTUFF FOR USE IN PIG DIETS

R. O. MYER, D. D. JOHNSON, K. K. BOSWICK, and J. H. BRENDEMUHL
University of Florida, Marianna, Florida 32446, USA

ABSTRACT

Restaurant food wastes obtained from a resort complex in central Florida were dehydrated. The resulting product, when included in complete, mixed diets, was found to be quite nutritious and well utilized by finishing pigs (55 to 110 kg). In addition, no negative effects on resulting carcass or meat quality characteristics were noted.

INTRODUCTION

Food wastes, mainly food discarded from restaurants, hotels and other food preparation and food service establishments, contain 50 to 85% moisture. However, on a dry basis, food wastes are high in nutrients desirable for pig feeding. Typical analyses (dry matter basis) previously reported include crude protein contents of 15 to 23%, fat (ether extract) of 17 to 24%, and ash (total mineral matter) of 3 to 6% (Kornegay et al., 1970; Pond and Maner, 1984; Ferris et al., 1995; Westendorf et al., 1996).

The use of food wastes as a feed for pigs is not new. For health and safety reasons, many states in the USA have banned the feeding of food wastes to pigs. Those states that allow it, mandate that the food wastes be "cooked" before feeding. The increased regulation, high labor requirements, relatively poor pig performance, and concerns about the meat quality of food waste-fed pigs have resulted in decreased interest in traditional food waste (or garbage) feeding.

However, with other waste disposal options becoming more expensive and scarce (i.e., landfills), recycling these food wastes for livestock feeding is an attractive waste disposal alternative. The technology now exists whereby these food wastes can be converted into a dry, stable product that could be easily incorporated into many pig feeding programs in existence today. The processing and drying costs to obtain such a product are competitive because of high tipping fees at landfills. Costs range from approximately $130 to $250 per ton.

We at the University of Florida have previously conducted two initial studies evaluating dehydrated restaurant food wastes (DFW) as a feedstuff for pigs. The results of these studies were encouraging (Myer et al., 1994; Rivas et al., 1995; Myer et al., 1996).

S. Brown, J.S. Angle and L. Jacobs (eds.), Beneficial Co-Utilization of Agricultural, Municipal and Industrial By-products, 397–403.
© 1998 *Kluwer Academic Publishers,*

Reported here are the results of a larger, follow-up study evaluating a dehydrated restaurant food waste product as a feedstuff for pigs.

EXPERIMENTAL PROCEDURE

Fresh food wastes were obtained from food service operations at a resort complex in central Florida. The food wastes were mostly left over food and plate scrapings, and contained about 60 to 75% water. The food wastes were minced and blended with a dry feedstock (67:33 blend of soy hulls and ground corn) such that the resulting blend was about 40% moisture. The semi-moist blend was then pelleted and dried using the Nutra Feed® system. Drying temperature was maintained at 150 to 200° C in which product temperature reached 110 to 120° C; transit time through the dryer was 4 to 7 min. The blending of fresh minced food wastes with a dry feedstock was done to facilitate the drying process. To increase the concentration of food wastes in the final dried product, initial dried product was blended with additional minced fresh food wastes and dried. The final DFW product utilized in this study contained about 60% dried food wastes with the other 40% being the feedstock. The preparation and dehydration of the DFW product was done at Nutra Feed, Inc. near Clermont, FL.

Representative samples of the DFW product were analyzed for moisture, crude protein, ether extract (crude fat), crude fiber, ash, calcium, phosphorus, sodium, chloride and potassium. Samples of the DFW product along with soybean meal samples were also analyzed for amino acid composition, available lysine using the DNFB (dinitroflurobenzene) procedure, and enzymatic (pepsin) digestibility of the protein. Samples of the DFW product along with samples of a commercially available livestock feed fat (brown grease) were analyzed for fatty acid composition, peroxide value and rancidity (thiobarbituric acid). All analyses were done by a commercial feed analysis laboratory (Woodson-Tenent, Inc., Memphis, TN) using commonly accepted procedures.

The pig feeding trial was conducted to evaluate the DFW product when included as a major component of diets for finishing pigs. The trial utilized 72 crossbred pigs with an average initial weight of 56 kg. Dietary treatments consisted of corn and soybean meal-based diets containing 0% (control), 40%, and 80% DFW product. The diets were formulated following National Research Council (NRC) (1988) guidelines.

At the end of the feeding phase, all pigs were slaughtered at a local state inspected facility. Each carcass was measured for backfat thickness and loin eye (longissimus) area (10th rib), and percentage carcass lean content was calculated (NPPC, 1991). Each carcass was also scored for fat firmness and the exposed loin muscle was scored for lean color, firmness and marbling (NPPC, 1991). A section of the loin (7th to 10th rib area) was obtained and frozen for sensory and shear analyses.

For sensory analysis, the loin sections were divided into chops 2.5 cm thick. The chops were thawed for 18 hr at 2 to 4° C, broiled to an internal temperature of 75° C and evaluated using a trained sensory panel following recommended American Meat Science Association procedures. Loin chops for Warner-Bratzler shear analysis were also thawed for 18 h at 2 to 4° C and broiled to an internal temperature of 75° C.

RESULTS AND DISCUSSION

The DFW product utilized was a pelleted, tan to dark brown colored product that was slightly greasy to the touch. The product had a mild odor that could be characterized as a combination of fish meal, feed fat, ground grain and coffee grounds.

Composition of the major chemical components of the DFW product is given in Table 1. For comparative purposes, corn, soybean meal and feed fat samples were analyzed and results are also reported. The analyses indicated the DFW product to be low in moisture, high in fat, and moderately high in protein, crude fiber and ash (total mineral matter). The chloride and sodium contents, however, were high and equate to a salt content of about 1.5% in the DFW product.

Table 1. Composition and *in vitro* quality assessments of dehydrated restaurant food wastes product (DFW) and of other representative feedstuffs[a].

Item	DFW	Corn	Soybea n meal[b]	Feed fat[c]
Moisture, %	8.4	11.2	11.7	0.2
Crude protein, %	14.4	8.9	47.8	--
Lysine, %	0.64	0.26	2.88	--
Available lysine, %	0.53	--[g]	2.75	--
Threonine, %	0.60	0.30	2.20	--
Isoleucine, %	0.56	0.24	2.02	--
Pepsin indigestible protein, %	3.2	--	4.0	--
Crude fat, %	16.0	3.5	1.2	99.0
Total saturates, %	35.6	--	--	35.1
Total monounsaturates, %	40.1	--	--	48.2
Total polyunsaturates, %	22.9	--	--	15.1
Peroxide value-initial, meq/kg[d]	5.0	--	--	2.6
TBA[e] rancidity, mg/kg[f]	1.6	--	--	<0.25
Crude fiber, %	14.5	2.1	3.6	--
Total mineral matter, %	4.7	1.1	6.4	--
Calcium, %	0.63	0.02	0.30	--
Phosphorus, %	0.38	0.26	0.72	--
Chloride-soluble, %	0.86	0.05	<0.02	--
Potassium, %	0.80	0.32	2.2	--
Sodium, %	0.47	<0.01	<0.01	--

[a] Values, other than for moisture, were adjusted to 88% dry matter basis, except for feed fat which are on an as is basis.

[b] Commercially available 48% product.

[c] Commercial livestock supplemental fat ("brown grease").

[d] Units of peroxide formulation per kg of fat.

[e] Thiobarbituric acid.

[f] mg malonaldehyde per kg of fat.

[g] Not determined.

Extrapolated to 100%, estimated composition of food wastes in the DFW product (dry matter basis) would be 24% crude fat, 20% crude protein, 4% crude fiber, 6% ash, 0.6% calcium, 0.4% phosphorus and 2.0% salt. These levels, including the high salt content, agree with what we previously obtained with dehydrated food wastes (Myer et al., 1994; 1996) and what others analyzing fresh and cooked food wastes (dry matter basis) obtained (Kornegay et al., 1970; Pond and Maner, 1984; Ferris et al., 1995; Westendorf et al., 1996).

Since the DFW product was moderately high in protein, further analyses of the protein were conducted and are also summarized in Table 1. With the possible exception of lysine, the profile of the essential amino acids in the protein of the DFW product was similar to the profile in soybean meal (expressed on g/100 g of total protein basis). Lysine was lower in DFW than in soybean meal but higher than in corn (NRC, 1988; g/100g of total protein). Lysine is usually the most limiting essential amino acid in pig diets.

Two *in vitro* laboratory tests were also done to further assess the quality of the protein in the DFW product for pepsin digestibility and available lysine. Pepsin digestibility of the protein in the DFW product was found to be moderately high, but lower than obtained for soybean meal (80 vs. 92%). The available lysine, as a percent of the total, in the DFW product was also found to be moderately high, but not as high as measured in soybean meal (83 vs. 94%). Soybean meal is the most common supplemental protein source for pig diets in the USA.

Since the DFW product was found to be high in fat, fatty acid composition and quality evaluation of the fat was done. Fat increases the feed value of a feedstuff for pigs. The fatty acid profile of the fat in DFW and results of quality assessment of this fat are also shown in Table 1. The ratio of total unsaturated to saturated fatty acids in fat of the DFW product was similar to that found in the commercial livestock feed fat; however, the percentage of polyunsaturated fatty acids was higher in DFW. The fatty acid profile of the DFW indicates a mixture of fats of animal and vegetable origin. The commercial feed fat product analyzed was a blend of animal and vegetable fats, primarily beef tallow and reclaimed cooking oils from restaurants.

Peroxide and thiobarbituric acid (TBA) numbers obtained for the DFW product indicated some rancidity development (Table 1). By contrast, the peroxide and TBA values obtained for the commercial feed fat were quite low and indicated very little rancidity development. The feed fat product, however, was stabilized by the addition of an antioxidant/preservative.

Growth performance and carcass data from finishing pigs fed diets containing the DFW product are summarized in Tables 2 and 3, respectively. Pig growth rates obtained in the finishing trial were excellent. Average rate of weight gain of the pigs was not affected ($P>0.10$) by the inclusion of the DFW product at 40% or 80% of the diet. Pigs fed the 40% and 80% DFW diets required, on average, 4% and 9% less feed, respectively, per unit of weight gain ($P<0.01$; linear) than pigs fed the control. The better feed-to-gain was likely due to the higher fat content of the DFW diets (7 and 11 vs 3%).

Average estimated carcass lean content was unaffected ($P>0.10$) by the inclusion of the DFW product in the diets of finishing pigs. Lean color, firmness and marbling scores were similar ($P>0.10$) between the DFW and control fed pigs. Carcass fat, however, became softer as the level of DFW increased in the diets ($P<0.01$; linear). The trend

Table 2. Performance of finishing pigs fed diets containing a dehydrated food waste product (DFW)[a].

Item	Dietary treatment (%DFW[b]) 0	40	80	SE[c]
Avg. daily gain, kg	0.91	0.91	0.90	0.016
Avg. daily feed intake[d], kg	2.98	2.88	2.67	0.036
Feed/unit gain[d], kg/kg	3.27	3.17	2.98	0.025

[a] Three pens per treatment with eight pigs per pen. On experiment from 56 to 108 kg average body weight per pig.
[b] Dehydrated food waste blended product (blended prior to dehydration); approximately 60% food wastes (dry) and 40% soy hulls/ground corn blend(67:33).
[c] Standard error of the mean; n = 3.
[d] Means differ (P<0.01; linear).

Table 3. Carcass characteristics of finishing pigs fed diets containing a dehydrated food waste product (DFW)[a].

Item[b]	Dietary treatment(%DFW[c]) 0	40	80	SE[d]
Backfat, cm	2.7	2.5	2.3	0.05
Loin eye area, cm^2	34.8	35.6	34.3	0.42
Carcass lean, %	48.5	49.8	49.7	0.32
Loin color[e]	2.6	2.3	2.4	0.10
Loin firmness[f]	2.5	2.6	2.4	0.06
Loin marbling[g]	2.3	2.3	2.2	0.10
Carcass fat firmness[h]	1.5	2.2	2.6	0.12

[a] Each mean is based on the information from three pens of eight pigs each. Average weight at slaughter, 108 kg per pig.
[b] Means for each measurement do not differ (P>0.10) except for backfat (P<0.05; linear) and carcass fat firmness (P<0.01; linear).
[c] Dehydrated food waste blended product.
[d] Standard error to the mean; n = 4.
[e] Scores of 1 to 5; 2 = gray, 3 = light pink, 4 = reddish pink.
[f] Scores of 1 to 5; 2 = firm, 3 = slightly firm, 4 = slightly soft.
[g] Scores of 1 to 5; 2 = traces, 3 = slight, 4 = modest.
[h] Scores of 1 to 5; 1 = firm, 2 = slightly soft, 3 = soft, 4 = very soft, oily.

Table 4. Sensory evaluations and shear force of broiled loin chops from finishing pigs fed diets containing a dehydrated food waste product (DFW)[a].

	Dietary treatment(%DFW[c])			
Item[b]	0	40	80	SE[d]
Juiciness[e]	5.6	5.6	5.6	0.2
Flavor[e]	5.7	5.7	5.9	0.1
Tenderness[e]	6.3	6.3	6.4	0.2
Off-flavor[f]	5.8	5.8	5.8	0.1
Shear force[g], kg/mm	2.3	2.7	2.6	0.3

[a] Each mean is based on the information from three pens of four pigs each. Average weight at slaughter, 108 kg per pig.
[b] Means do not differ (P>0.10).
[c] Dehydrated food waste blended product.
[d] Standard error to the mean; n = 3.
[e] Scores of 1 to 8 with the trait increasing with an increase in score.
[f] Scores of 1 to 6 with a less intense off-flavor with an increase in value.
[g] Warner-Bratzler shear force values (measure of tenderness).

towards softer fat was probably due to the high level of unsaturated fat contributed to the diets by the DWF product. However, the scores indicated that the carcasses would still be acceptable to the meat packer.

The inclusion of the DFW product in the diets of finishing pigs had no effect (P>0.10) on palatability characteristics of broiled loin chops as determined by a trained sensory panel (Table 4). The values obtained indicated that the chops were acceptable in tenderness and juiciness, and had no detectible off-flavor. The inclusion of the DFW in the diets also had no effect (P>0.10) on shear force values, a measure of tenderness, of the loin chops.

The results of this study indicated the DFW product utilized was a very nutritious feedstuff for inclusion in pig diets. The improvement in feed efficiency, compared to the control, indicated that the DFW product was well digested by the pigs and they utilized the higher fat content contributed by the DFW product in the DFW diets. Furthermore, the DFW product utilized was easy to further process and mix into the meal type diets utilized in the pig finishing trial.

REFERENCES

Ferris DA, Flores RA, Shanklin CW & Whitworth MK (1995) Applied Engr. Agric. 11:567-572.

Kornegay ET, Van der Noot GW, Barth KM, Graber G, MacGrath WS, Gilbreath RL & Bielk FJ (1970) New Jersey Exp. Stn. Bull. No. 829, Rutgers Univ., New Brunswick. 31 p.

Myer RO, DeBusk TA, Brendemuhl JH & Rivas ME (1994) Florida Swine Res. Rep. No. ANS-SW94. Animal Sci. Dept., Univ. of Florida, Gainesville, FL pp. 44-51.

Myer RO, Rivas ME, Johnson DD & Brendemuhl RO (1996) J. Anim. Sci. 74 (suppl. 1):177 (Abstract).

National Pork Producers Council (1991) Procedures to Evaluate Market Hog Performance (3rd Ed.). NPPC, Des Moines, IA.

National Research Council (1988) Nutrient Requirements of Swine (9th Ed.). National Academy Press, Washington, DC.

Pond WG & Maner JH (1984) Swine Production and Nutrition. AVI Publishing Co., Inc., Westport, CT. pp. 336-368.

Rivas ME, Brendemuhl JH, Myer RO & Johnson DD (1995) J. Anim. Sci. 73(suppl. 1):177 (Abstract).

Westendorf ML, Pas, Zirkel EW & Gordon R (1996) The Professional Anim. Scientist 12:129-137.

SOIL REMINERALIZATION FOR SUSTAINABLE CROP PRODUCTION

ALLEN V. BARKER[1], TARA A. O'BRIEN[1] and JOANNA CAMPE[2]
[1]*University of Massachusetts, Amherst Massachusetts 01003, USA*
[2]*Remineralize the Earth, Inc., Northampton, Massachusetts 01060, USA*
E-mail barker@pssci.umass.edu

ABSTRACT

Soil remineralization is the utilization of mineral fines to restore soil fertility through replenishment of plant nutrients removed by agricultural activities or by forces of nature. Basalt dust from a rock quarry or glacial moraine dust from a gravel quarry were evaluated for their effects on nutrient availability in soils and on yields and composition of lettuce (*Lactuca sativa longifolia* Lam.), apple (*Malus sylvestris* L.), or sweet corn (*Zea mays rugosa* Bonaf.). Extractable soil K and Ca were increased for all plots, and some increases in extractable P, Fe, Mn, and Mg were detected in soils treated with basalt rock dusts at rates of 9 to 27 Mg/ha. Soil acidity was decreased by about 0.5 pH unit by the additions of basalt dusts. Availabilities of most micronutrients and nonessential trace elements were not affected by additions of basalt dust. Glacial moraine fines had a lesser effect on available plant nutrients than basalt dusts. Plant tissue composition or yield of produce were not affected by any of the mineral treatments. The principal benefit of soil remineralization noted in this year of research was in the enhancement of soil fertility by elevating levels of certain nutrients in soil.

INTRODUCTION

Soil remineralization is employed to create fertile soils by utilizing natural, fine-textured mixtures of rock and mineral dusts, such as glacial moraine, and single or mixed finely ground rock dusts, which may be naturally occurring or obtained as by-products from manufacture of aggregates (D'Hotman de Villiers, 1961; Korcak, 1996; Leonardos et al., 1987). These mixtures of rocks and minerals contain a broad spectrum of plant macronutrients and trace elements, which, if added to cropland, may enhance soil fertility and crop productivity (Chesworth et al., 1983; Clemens and Singer, 1992). The fine-textured property of these dusts is an important trait in their contribution to soil fertility, for the high surface area of the approximately silt-sized particles ensures a slow but sustained release of plant nutrients. Their fine texture makes the dusts a product that may be superior to more coarsely ground materials, such as granite dust or greensand, which release negligible amounts of nutrients within a growing season (Barker, 1976).

S. Brown, J.S. Angle and L. Jacobs (eds.), Beneficial Co-Utilization of Agricultural, Municipal and Industrial By-products, 405–413.
© 1998 *Kluwer Academic Publishers,*

Many benefits of soil remineralization in agricultural or forestry operations have been reported. These benefits include higher yields and enhanced flavor or quality of produce, faster growth and improved health of trees, and improved resistance to insects and pests (Fragstein, 1993; Leipold, 1993; Sauter and Foerst, 1987). The current project was initiated to evaluate the effects on soil fertility of soil remineralization with local sources of mineral fines.

MATERIALS AND METHODS

The research in 1996 evaluated the benefits of the rocks dusts as amendments to increase soil fertility. Two principal materials were used in this project. One was a basalt rock dust from a traprock aggregate processor in Amherst, MA. This material was used at all test sites. The other was a glacial moraine dust from Westfield, MA. Analyses of these dusts are in Table 1. The analyses of mineral fines were conducted by Elemental Research Inc., Vancouver, BC, Canada. Nitrogen, S, and Cl were not reported by the analytical laboratory. The mineral fines were evaluated for production of lettuce, apples, and sweet corn.

Table 1. Composition of plant nutrients and some nonessential trace elements in mineral fines of mineral fines and food waste compost.

Element	Concentration of element, mg/kg		
	Mineral fines		Compost
	Basalt dust	Glacial moraine	
		Plant Nutrients	
Nitrogen	N/A	N/A	6,100
Phosphorus	700	770	1,900
Potassium	4,490	10,500	1,800
Calcium	20,800	8,410	6,400
Magnesium	7,660	2,920	1,700
Iron	84,300	20,100	4,500
Manganese	1,270	458	274
Nickel	32	12	5
Copper	74	9	20
Zinc	111	43	68
Molybdenum	0.8	0.1	N/A
Boron	<5	53	12
		Nonessential Elements	
Sodium	10,100	14,900	N/A
Cadmium	<0.1	<0.1	0
Lead	13	17	46

N/A, element not determined.
Carbon in the compost was 7.4% giving a C:N ratio of 12:1.

Lettuce Production

This research was conducted at the farm of the Smith Vocational and Agricultural School in Northampton, MA, which is an organic producer of vegetables for a community supported agricultural market. Lettuce was the test crop because it is produced in several seasonal plantings at the farm, allowing for replication of experiments within the regular cropping plans of the farm. Plots were 1.2-m wide by 9-m long with 4 rows of lettuce, one of which was always romaine (*Lactuca sativa longifolia* Lam. 'Parris Island'), the variety on which these results are based. Treatments applied to the soil and tilled in 8-cm deep with a rotary cultivator included:

1. a mixed organic fertilizer (4% N, 2% P, 3% K) applied at 2.25 Mg/ha;
2. composted food wastes (Table 1) applied at 45 Mg/ha (wet basis);
3. the compost at 45 Mg/ha and basalt rock dust (Table 1) at 9 Mg/ha (dry basis).

Four crops were grown with these treatments applied 3 days before planting dates of June 21, July 8, July 23, and August 12, 1996. Time for production of marketable heads of lettuce was about 4 to 5 weeks. Soil samples were taken just after application of materials and at harvest of lettuce. Heads of lettuce were harvested by cutting at the ground and trimming diseased or damaged bottom leaves and determining fresh weights per head. Leaf samples were taken by removing a lower leaf after trimming from 12 plants at each harvest. The leaves were dried and ground for tissue analyses for total element concentrations by the University of Massachusetts Soil and Tissue Testing Laboratory by inductively coupled plasma spectrophotometry. Soils were tested for extractable nutrients and some nonessential elements (Lunt et al., 1950). Harvests and analyses of the multiple plantings were treated as blocks in analysis of variance.

Fruit Production

At a site in Shelburne, MA, rock dusts were evaluated for establishment of newly planted apples (*Malus sylvestris* Mill.). Three treatments were applied to a newly planted orchard.

1. Conventional treatment with no fertilization in the first year;
2. Basalt rock dust (Table 1) applied to surface at 22.5 Mg/ha (dry) and disked in 8 cm at planting;
3. Basalt rock dust mixed into the planting hole at 16 kg/m^3 of soil at planting.

The treatments were arranged in two randomized complete blocks of 'Pioneer McIntosh' and a block of 'Golden Delicious'. Individual plot sizes were 38-m long with 3.7 m between rows of trees. Each row contained 25 trees. Soil samples were taken just after planting (April 19, 1996) and at leaf sampling (August 22, 1996). Soil samples were taken by pooling 8 cores (30-cm deep) from each treatment. Leaf samples were taken on August 16, 1996, a customary time for sampling of apple trees for assessment of their nutrient status. Lengths of the center leader of all trees in a block were measured on August 22 to assess seasonal growth.

Sweet Corn Production

At this site in Deerfield, MA, basalt rock dust and glacial moraine fines were assessed in sweet corn production (*Zea mays rugosa* Bonaf.). This grower used conventional practices of chemical fertilization and pest control in crop production. The treatments included:

1. Basalt rock dust (Table 1) applied at 27 Mg/ha to soil surface and disked in at planting;
2. Glacial moraine fines (Table 1) applied in same amount and manner as the basalt dust;
3. Plots with no mineral fines added.

Chemical fertilization of 100 kg N, 20 kg P, and 80 kg K per ha was broadcast before applications of mineral fines. A sidedressed application of cottonseed meal was applied at variable rates of 0 to 50 kg N/ha according to a soil nitrate test (Magdoff, 1984). Plot size was 4 m x 8 m, with 4 rows of corn at 0.9-m spacings. Experimental design was a 3 x 3 Latin Square. Soil samples from 8-cm deep cores were taken at planting. Yields were taken from 3 meters of a center row in each plot.

RESULTS

Lettuce

The treatments did not have significantly different effects on average yields of heads of lettuce (Table 2), although it appeared that the treatment with rock dust had slightly suppressing effects on head size. Chemical analyses of the leaves indicated that all treatments produced heads with adequate concentrations of macronutrients and micronutrients and with no significant differences among treatments (Table 3). Potassium concentrations seemed to be high into the range of luxury consumption, but all of the other nutrients seemed to be within normal ranges (Mills and Jones, 1996).

In the first harvest of lettuce (planted June 21, 1996), an infection with septoria leaf spot (*Septoria lactucea* Pass.) occurred with the plants in the conventional organic treatment. No infestation occurred with the plants in the soil amended with compost and rock dust. The plots with only compost had a lesser infestation than the plots receiving the conventional treatment. No differences in nutrient compositions were noted between the diseased leaves and the healthy leaves, and the nutrient contents were virtually identical to the means reported in Table 3.

The additions of rock dust to soil enhanced the concentrations of extractable plant nutrients in the soil relative to nutrients present in the soils with the conventional treatment or with the compost addition alone. In many cases, the enhancement was significant (Table 4). No significant differences in extractable nutrients occurred between the dates of planting and the harvest, so the data in Table 4 are pooled across the two sampling dates. Also, in Table 4, some nonessestial elements (Pb, Cd, Al, Cr) of interest are reported. None of the nonessential elements were enhanced in the soil by composts or by

Table 2. Average fresh weights of heads of romaine lettuce from four harvests.

Treatment	Head weight, g/head
Conventional organic	300
Compost	316
Compost and rock dust	272

Table 3. Mean elemental concentrations in lettuce leaves from four harvests.

Element	Treatment		
	Conventional organic	Compost	Compost and rock dust
	-----------------------	mg/kg dry wt	-----------------------
Nitrogen	30,000	29,000	30,000
Phosphorus	2,800	3,500	3,100
Potassium	87,700	80,600	91,300
Calcium	14,800	11,800	13,000
Magnesium	5,800	4,200	4,600
Zinc	21	23	22
Copper	11	13	12
Manganese	48	46	51
Iron	291	244	271
Boron	24	23	24
Molybdenum	3	4	3

composts and rock dust relative to the conventional treatment, indicating that no pollution with heavy metals or no enhancment of soluble aluminum was occurring. Other beneficial effects from additions of composts and rock dusts were slight neutralization of soil acidity. The soil was already at a favorable pH for crop production, but raising of the pH by the rock dusts indicated a potential beneficial effect of rock dust amendments to acidic soils. The beneficial effects occurred in a short period of time, about 4 to 5 weeks. Organic matter appeared to be raised by the additions of composts and rock dusts but not significantly according to analysis of variance.

Apples

Growth of the central leader did not differ among treatments (Table 5). Treatments had no significant effects on nutrients in samples of the middle leaves from the central leaders (Table 6).

Soil samples taken at the time that the trees were planted and at the time that the leaves were samples were tested for available nutrients (Table 7). The soil tests showed no differences in nutrients or nonessential elements concentrations. Differences in soil tests from the time of planting to the time of sampling in August appeared to be normal seasonal variations.

Table 4. Extractable plant nutrients and trace elements, pH, and organic matter from tests of soils treated with conventional organic fertilizer, compost, or compost and basalt rock dust.

Element	Extractable elements, mg/kg		
	Conventional organic	Compost	Compost and rock dust
		Macronutrients	
Phosphorus	17a	39b	50a
Potassium	156a	248b	276b
Calcium	795a	1016b	1510c
Magnesium	209a	242ab	290b
Ammonium-N	9a	4a	5a
Nitrate-N	32a	30a	32a
		Micronutrients	
Boron	0.1a	0.3ab	6b
Zinc	0.7a	1.2b	1.6b
Manganese	3.5a	5.0a	8.8b
Copper	0.2a	0.2a	0.2a
Iron	3.6a	3.9a	4.9b
Nickel	0.08a	0.10a	0.11a
		Nonessential elements	
Lead	0.9a	0.9a	1.0a
Cadmium	0.05a	0.04a	0.09a
Chromium	0.04a	0.08a	0.08a
Aluminum	20	17	16
		Other measurements	
pH	6.09a	6.35b	6.58c
Organic matter(%)	3.9a	4.6a	5.4a

Within rows, means followed by different letters are significantly different by Duncan's Multiple Range Test ($P \leq 0.05$).

Table 5. Length of central leader of young apple trees.

Treatment	Leader length, cm
Rock dust (surface)	43
Rock dust (hole)	35
Unfertilized	47

Corn

Soil tests indicated some differences in plant nutrients just after application of treatments (Table 8). Available Ca, Fe, and Mn were elevated by the application of basalt rock dust relative to these nutrients in soil treated with glacial moraine fines or conventional fertilization. Available Mg was raised by both mineral fines. Nitrate was higher with the rock dust treatment than with the other treatments. Other nutrients appeared not to be affected significantly by treatments.

Table 6. Mean concentrations of nutrients in apple leaves from center of branch of new shoot growth.

Element	Treatment		
	Unfertilized	Rock dust (surface)	Rock dust (hole)
	mg/kg		
Nitrogen	24,800	26,600	25,400
Phosphorus	1,500	1,700	1,700
Potassium	17,600	22,400	21,800
Calcium	6,900	6,700	6,700
Magnesium	3,700	3,100	2,800
Zinc	9	10	8
Copper	8	9	7
Manganese	40	39	36
Iron	56	55	51
Boron	26	26	29
Molybdenum	1	1	1

Table 7. Soil tests results from sampling at time of leaf sampling.

Element	Treatment		
	Unfertilized	Rock dust (surface)	Rock dust (hole)
	mg/kg		
		Nutrients	
Nitrate-N	70	70	76
Ammonium-N	5	5	5
Phosphorus	6	6	7
Potassium	109	102	138
Calcium	1294	1330	1332
Magnesium	174	161	194
Manganese	2	4	2
Zinc	1	1	1
Iron	3	2	2
Boron	0.4	0.4	0.4
Nickel	0.3	0.5	0.3
		Nonessential elements	
Aluminum	19	24	21
Cadmium	0	0.07	0
Lead	12	9	13
Chromium	0.10	0.10	0.07
		Acidity	
pH	6.27	6.73	6.50

Table 8. Soil tests results for available macronutrients and selected micronutients at the planting and sweet corn yields.

Element	Treatment		
	No fines	Basalt rock dust	Glacial moraine fines
		---mg/kg---	
		Soil tests	
Ammonium-N	3	63	
Nitrate-N	28	32	23
Phosphorus	12	8	9
Potassium	104	122	73
Calcium	588	4433	450
Magnesium	109	180	180
Manganese	3	71	4
Iron	4	36	3
		Yields	
Total Mg/ha	11.6	10.0	9.6
Avg g wt/ear	253	220	214

DISCUSSION

The applications of mineral fines resulted in improvements in soil fertility as assessed by analyses of extractable or available plant nutrients. The silt-sized basalt dust appeared to be more effective than the sand-sized and perhaps harder glacial moraine fines in elevating available K, Ca, Fe, and Mn. Both fines gave small increases in available Mg. Phosphorus availability was not increased consistently. The effects on soil fertility did not result in higher crop yields or higher concentrations of nutrients in the plants. With lettuce and corn, this lack of response may have been due to the relative high soil fertility inherent in their plots. With apple, perhaps root development in one growing season was insufficient to access nutrients from the remineralized soil. The basalt dust had value as a liming material, raising the soil pH by about 0.5 units with 9 or 27 Mg/ha applications. No enhancement in extractable nonessential elements or their accumulation in plant tissues was noted with soil applications of either mineral fines. These experiments indicate the potential for improved fertility through increased availability of some plant nutrients, which are constituents of the mineral fines.

REFERENCES

Barker AV (1976) Using tomato plants to assess the release of potassium from fertilizers. J. Agron. Educ. 5:63-65.

Chesworth W, Macias-Vasquez F, Acquaye D & Thompson E (1983) Agricultural alchemy: Stones into bread. Episodes 1:3-7.

Clemens G and Singer A (1992) Ameliorating chlorosis-inducing soils with rock materials of varying porosity and iron content. Soil Sci. Soc. Amer. J. 56:807-813.

D'Hotman de Villiers DO (1961) Soil rejuvenation with crushed basalt in Maritius. Intern. Sugar J. 63:363-364.

Fragstein P von, Pertl W & Vogtmann H (1986) Silicate rock powders: Qualitative and quantitative aspects. In:

Vogtmann H, Boehncke E & Fricke I (Eds) The Importance of Biological Agriculture in a World of Diminishing Resources (pp. 74-84). Proc. 5th IFOAM Intern. Sci. Conf., Univ. Kassel.

Korcak R (1996) Byproduct utilization in agriculture: Past experience and potentials for the future. Mining Engin. 48(8):79-82.

Leipold F (1993) The importance and effects of rock dust in orchards and gardens. In: Campe Joanna (Ed) Soil Remineralization Research Packet. Agriculture (pp. 1-8). Remineralize the Earth, 152 South Street, Northampton, MA 01060.

Leonardos OH, Fyfe WS & Kronberg BI (1987) The use of ground rocks in laterite systems: An improvement to the use of conventional soluble fertilizers? Chem. Geol. 60:361-370

Lunt HA, Jacobson HGM & Swanson CLW (1950) The Morgan soil testing system. Connecticut Agric. Expt. Stn. Bull. 541.

Magdoff FR, Ross D & Amadon J (1984) A soil test for nitrogen availability to corn. Soil Sci. Soc. Amer. J. 48:364-366.

Mills, HA and Jones JB, Jr. (1996) Plant analysis handbook II. MicroMacro Publishing, Athens, GA.

Sauter U and Foerst K (1987) Information for application of silicate rock dusts for amelioration of forest soils (Ger). Forst Holzwirt 42:27-30.

EFFECT OF SURFACE INCORPORATED COAL COMBUSTION BY-PRODUCTS ON EXCHANGEABLE CA AND AL IN SUBSOIL

URSZULA KUKIER[1], WILLIAM P. MILLER[2] and MALCOLM E. SUMNER[2]
[1]*Institute of Soil Science and Plant Cultivation, Pulawy, Poland and*
[2]*University of Georgia, Athens, Georgia 30602, USA*

ABSTRACT

Large amounts of flue gas desulfurization gypsum (FDG) are now available for use as a soil amendment. The efficacy of this material as an acid subsoil amendment was investigated at various rates and in combination with fly ash as a potential contaminant. Flue gas desulfurization gypsum and fly ash-FDG mixtures were applied at rates of 5, 10 and 20 Mg ha^{-1} to an acid Tupelo soil. An increase in exchangeable Ca status combined with a decrease in exchangeable Al in subsoil was a measure of the ameliorative effect. The rate of 10 Mg ha^{-1} FDG was recommended for amelioration of the Tupelo soil. Magnesium losses from the plow layer occurred at all FDG application rates. Application of the mixed fly ash-FDG material partly corrected the gypsum induced Mg losses and increased plant available B in the topsoil.

INTRODUCTION

Gypsum has various beneficial effects on soil physical and chemical properties, including amelioration of subsoil acidity. Acid subsoils usually contain low levels of Ca and high levels of Al which reduce root proliferation. As a result, roots are not able to extract nutrients and water from deeper soil layers and the crop is less resistant to droughts that may occur during the growing season. Incorporation of Ca into subsoil is an effective way of ameliorating subsoil acidity (Sumner, 1993). Traditional liming materials are effective only in the zone of incorporation due to the low solubility of calcium carbonate. In contrast, Ca from surface applied gypsum ($CaSO_4 * 2H_2O$) easily migrates down the soil profile providing an inexpensive method of Ca^{2+} and SO_4^{2-} incorporation into a subsoil. Most published work has dealt with the beneficial effect of mined gypsum and phosphogypsum on subsoil acidity (Sumner, 1993). Recently large quantities of flue gas desulfurization gypsum (FDG), some of which is contaminated with fly ash, have become available. The objective of the investigation was to assess the effectiveness of fly ash and FDG mixtures in ameliorating subsoil acidity. A field experiment on Tupelo soil series

S. Brown, J.S. Angle and L. Jacobs (eds.), Beneficial Co-Utilization of Agricultural, Municipal and Industrial By-products, 415–422.
© 1998 *Kluwer Academic Publishers.*

was conducted to test the effect of different FDG rates on subsoil exchangeable Al and Ca and alfalfa *(Medicago sativa)* yield. Corresponding rates of the 1:1 fly ash-FDG mixture were also applied in order to study any adverse and beneficial effects of agricultural co-utilization of both by-products.

MATERIALS AND METHODS

A field experiment was conducted on a fine mixed thermic Vertic Paleaquult at the University of Georgia experimental station near Calhoun in Northern Georgia. Fly ash and FGD were obtained from a coal-fired power plant employing separate facilities for the precipitation of fly ash and the desulfurizaton of flue gas. The FDG material used in the experiment was a wall-board quality gypsum. High concentration of B was characteristic for the fly ash (Table 1). The FDG had almost neutral reaction. Fly ash was alkaline but had a calcium carbonate equivalent of only 4.4% (Kukier et al., 1994). Deionized water and 0.2 M HNO_3 were used for extraction of Al from fly ash at a solid to solution ratio of 1:5 (m/v). Fly ash was shaken on an end-to-end shaker for 21 hr. Equilibrium pH of the supernatants was measured and dissolved Al was analyzed by inductively coupled plasma emission spectrometry (ICP). The FDG material was applied alone and as a 1:1 mixture with fly ash at the mixture rates of 0, 5, 10 and 20 Mg ha^{-1}. Treatments were broadcast and mixed with 0-20 cm soil layer. Each treatment was replicated three times. Soybeans were planted in the first growing season followed by barley and then alfalfa. Soil cores were collected from the experimental plots to a depth of 80 cm in July of 1994, about 16 months after treatment application. Each experimental plot was represented by 10 cores segmented into 10 cm increments and combined in one composite sample representative of each soil layer. Exchangeable Ca, Mg, K and Na in soil samples were determined in 1M NH_4OAc extracts (Thomas, 1982). Exchangeable Al was extracted in 1 M KCl (Barnhisel and Bertsch, 1982). Cations were determined by atomic absorption spectrometry. Hot water soluble B (Bingham, 1982) was measured only in soil samples

Table 1. The properties of FDG and fly ash used in the field study.

Material	Ca	Al	Mg	B	pH
	----------------g kg^{-1}----------------			mg kg^{-1}	
FDG	192	nd[a]	0.24	160	7.5[b]
Fly ash	22	92	4.40	815	10.2[c]

[a] not detected
[b] pH of saturated extract
[c] fly ash : water ratio 1:2.5

collected from control plots and plots amended with 20 Mg ha^{-1} rate of coal combustion by-products (CCBP) by the azomenthine-H method (Parker and Gardner, 1981). Effective cation exchange capacity (ECEC) of a soil was calculated as the sum of basic exchangeable cations (Ca, Mg, K, Na) and exchangeable Al (Shamshuddin and Ismail, 1995).

RESULTS

The Tupelo soil had an acid pH (Table 2) and Al saturation in the topsoil of 30% increasing with depth. At these levels of Al saturation, a positive crop response to a gypsum application is likely to occur (Sousa et al., 1992). The soil was extremely deficient in plant available B. As soil pH was not affected by any amendment, pH values for CCBP treated plots are not reported. The FDG material applied at rates of 10 and 20 Mg ha^{-1} significantly ($P<0.05$) increased exchangeable Ca levels to a depth of 40 cm with a noticeable effect at 60 cm (Figure 1). In the zone of incorporation, exchangeable Ca was increased by 100 and 150% by application of 10 and 20 Mg ha^{-1} of the FDG, respectively. The increase in exchangeable Ca in the plow layer can be partly accounted for by undissolved gypsum still present in the zone of incorporation. The corresponding rates of fly ash-FDG mixture had a lesser effect on exchangeable Ca than FDG alone, especially at depth, due to a lower $CaSO_4 *2H_2O$ content in the mixed by-product.

Table 2. Chemical properties of the Tupelo soil.

Depth cm	soil pH (1 : 2.5)		Exchangeable cations c mol$_c$ kg^{-1}			ECEC[a] c mol$_c$ kg^{-1}	Al[b] saturation %	HWSB[c] mg kg^{-1}	Clay g kg^{-1}
	water	1*M* KCl	Ca	Mg	Al				
0-20	5.2	4.3	1.16	0.32	0.74	2.5	30	0.13	166
20-30	5.3	4.2	1.66	0.36	0.84	3.1	28	0.08	246
30-40	5.3	4.2	1.92	0.44	0.92	3.5	26	0.07	263
40-50	5.2	4.1	1.76	0.47	1.44	3.8	38	<0.05	294
50-60	5.1	4.0	1.53	0.49	1.95	4.1	47	<0.05	312
60-70	5.1	3.9	1.31	0.47	2.43	4.3	56	<0.05	303
70-80	5.1	3.9	1.06	0.42	2.64	4.3	62	0.12	285

[a] Effective cation exchange capacity

The 5 Mg ha^{-1} application rate of FDG material did not affect exchangeable Al in the soil profile except for the 20-30 cm layer where an increase in exchangeable Al was noted probably due to random variability in soil properties (Figure 1). Higher rates of FDG were equally effective in reducing exchangeable Al. The greatest decrease in exchangeable Al, about 50%, occurred in the zone of incorporation but even at a depth of 50 cm exchangeable Al was decreased by about 20% in comparison to the control. The beneficial effect of FDG application is better expressed when both Ca and Al saturation are displayed simultaneously (Figure 2a). Different response patterns were observed when increasing rates of fly ash-FDG mixture were applied (Figure 1). The lowest application rate of the mixed by-product reduced exchangeable Al levels in the subsoil. In contrast, the higher application rates resulted in exchangeable Al distribution identical to the control profiles.

Application of the CCBP decreased exchangeable Mg levels in the upper zones of soil profiles (Figure 1). The effect was greater in the profiles amended with FDG where a significant decrease was observed not only in the zone of material incorporation but also in the 20-30 cm soil layer. Magnesium leached from the topsoil was arrested on soil exchange sites in subsoil. Mixed fly ash-FDG material served as a Mg source and slightly decreased the leaching effect of FDG alone. The 20 Mg ha^{-1} rate of the mixed material significantly ($P<0.05$) increased plant available B in the incorporation zone. Hot water extractable soil B increased to 0.5 mg kg^{-1}, a concentration that is considered to be sufficient for most crops (Reisenauer et al., 1973).

In the first growing season, the FDG material had no effect on crop yield. Application of high rates of the fly ash-FDG mixture adversely affected soybeans at the early stage of growth. This probably can be attributed to a toxicity of B incorporated with fly ash. Plants fully recovered after some leaching of the plow layer occurred, and no statistically significant yield decrease was recorded (data not shown). Application of the fly ash-FDG mixture in a fall would probably eliminate this adverse effect. There was no statistically significant yield response to CCBP application in the second growing season. Yield increase in the third crop (alfalfa) due to FDG and fly ash application was significant (Table 3). Typically yield increase can be expected after both ions, Ca^{2+} and SO_4^{2-}, move into the subsoil (Sumner, 1993). This process usually requires one to two years. There appears to be some added benefit to the application of the fly ash with FDG probably due to the additional supply of B or other micronutrients.

DISCUSSION

The FDG applied at rates of 10 and 20 Mg ha^{-1} increased the exchangeable Ca status of the subsoil and decreased exchangeable Al levels. This is the typical response of a highly weathered acid soil to gypsum application. The mixed fly ash-FDG material resulted in lower $CaSO_4 *2H_2O$ loads in comparison to corresponding rates of FDG alone but this only partly explains the lack of changes in subsoil exchangeable Al status. An amorphous aluminoslicate glass and crystalline mineral mullite, both very inert, are the predominant forms of the Al in fly ash (Warren and Dudas, 1984). Aluminum release from the fly ash used in the experiment was highly pH dependent as demonstrated by a batch solubility

Table 3. Effect of flue gas desulfurization gypsum (FDG) and fly ash-FDG mixtures on the yield of alfalfa grown three years after application.

Treatment	Rate Mg ha-1	Alfalfa yield kg ha^{-1}
Control	0	10948c[a]
FDG	5	12037bc
FDG	10	12653ab
FDG	20	13111ab
fly ash-FDG	2.5 + 2.5	12767ab
fly ash-FDG	5 + 5	12654ab
fly ash-FDG	10 + 10	13513a

[a] Treatment means followed by the same letter are not significantly different at $P<0.05$ according to Duncan's test.

study. Non-detectable amounts of Al were released from the fly ash in the water extract at pH 11.5 but in 0.2 *M* HNO_3 at pH 3.1 about 2000 mg kg^{-1} of the Al were dissolved. This indicates that relatively small proportion of the total Al can be solubilized under acid environmental conditions. However, at the high application rates used in this experiment, this amount may contribute to soil exchangeable Al. Our results with high application rates of the fly ash-FDG mixture suggest such a possibility but this hypothesis is lacking statistical support due to the high random variability of the field data.

Depletion of exchangeable Mg often observed after gypsum application is caused by Ca displacement of Mg from soil exchange sites. The large quantity of SO_4^{2-} ions enhances mobility of Mg and its downward movement. The fly ash used in this study contained too little Mg to completely offset the leaching effect of gypsum in the plow layer. The application of Mg fertilizer once the gypsum front has moved down the soil profile probably would be a better strategy than simultaneous application of a gypsum and Mg bearing material. The ECEC was markedly increased to a depth of 50 cm by the application of the high rates of FDG and mixed material (Figure 2a). The increase in CEC was explained by Reeve and Sumner (1972) as a result of sorption of the SO_4^{2-} ion by iron and aluminum oxides which generates a negative charge on the oxide surfaces. Shamshuddin and Ismail (1995) provided further experimental support of this research. Increased ECEC enabled simultaneous retention of Ca and Mg in the subsoil. The 10 Mg ha^{-1} rate of the FDG seems to be appropriate for amelioration of subsoil acidity in the Tupelo soil. This rate was almost as effective in changing exchangeable Ca and Al status in subsoil as a 20 Mg ha^{-1} application rate. At this application rate, Mg losses from the soil profile were minimized because the Mg leached from the topsoil was retained in the subsoil. Alleviation of B deficiency and some mitigation of Mg leaching from the topsoil

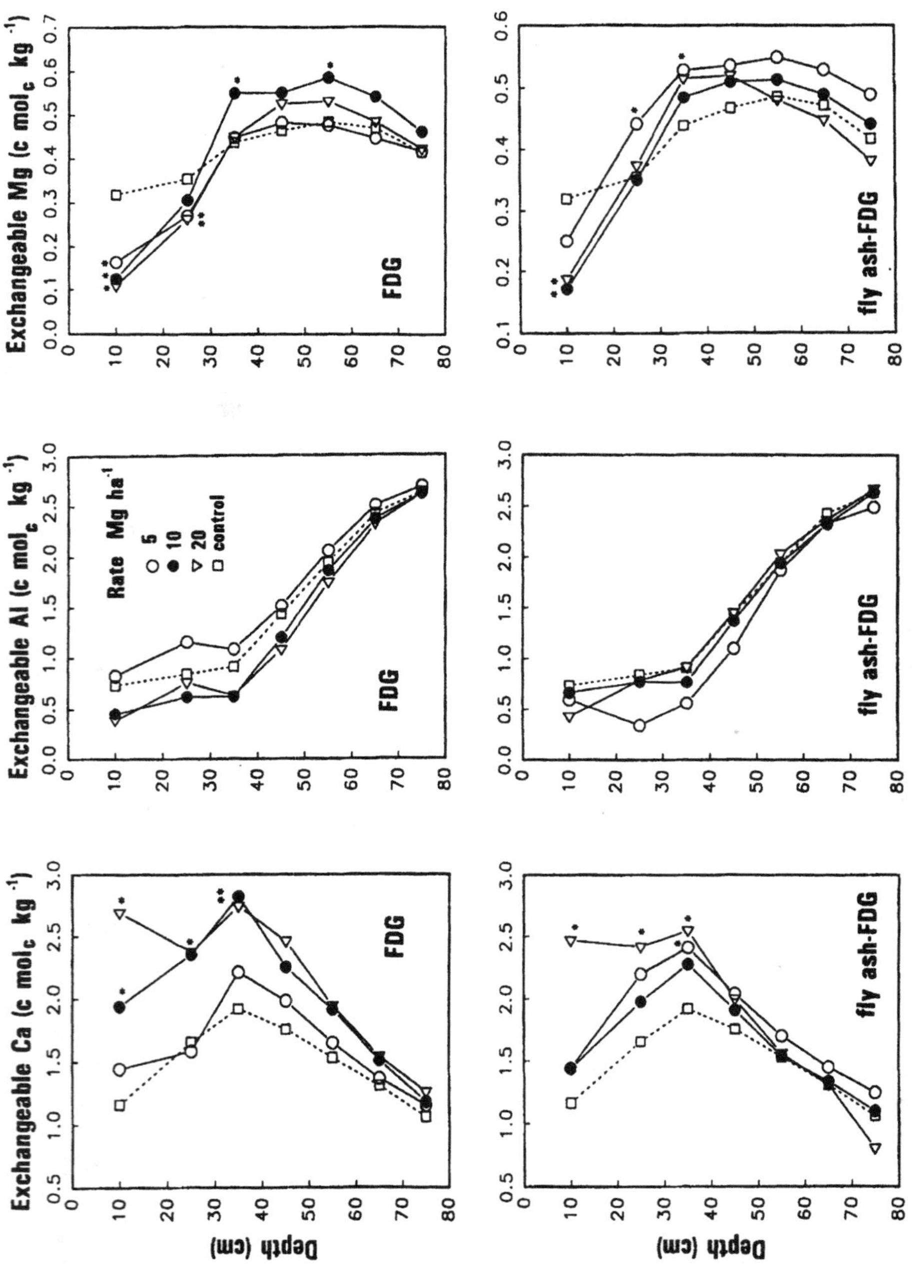

Figure 1. Exchangeable cations in theprofiles of soils amended with different rates of FDG and 1:1 fly ash-FDG mixture. * significantly different from the control (P<0.05)

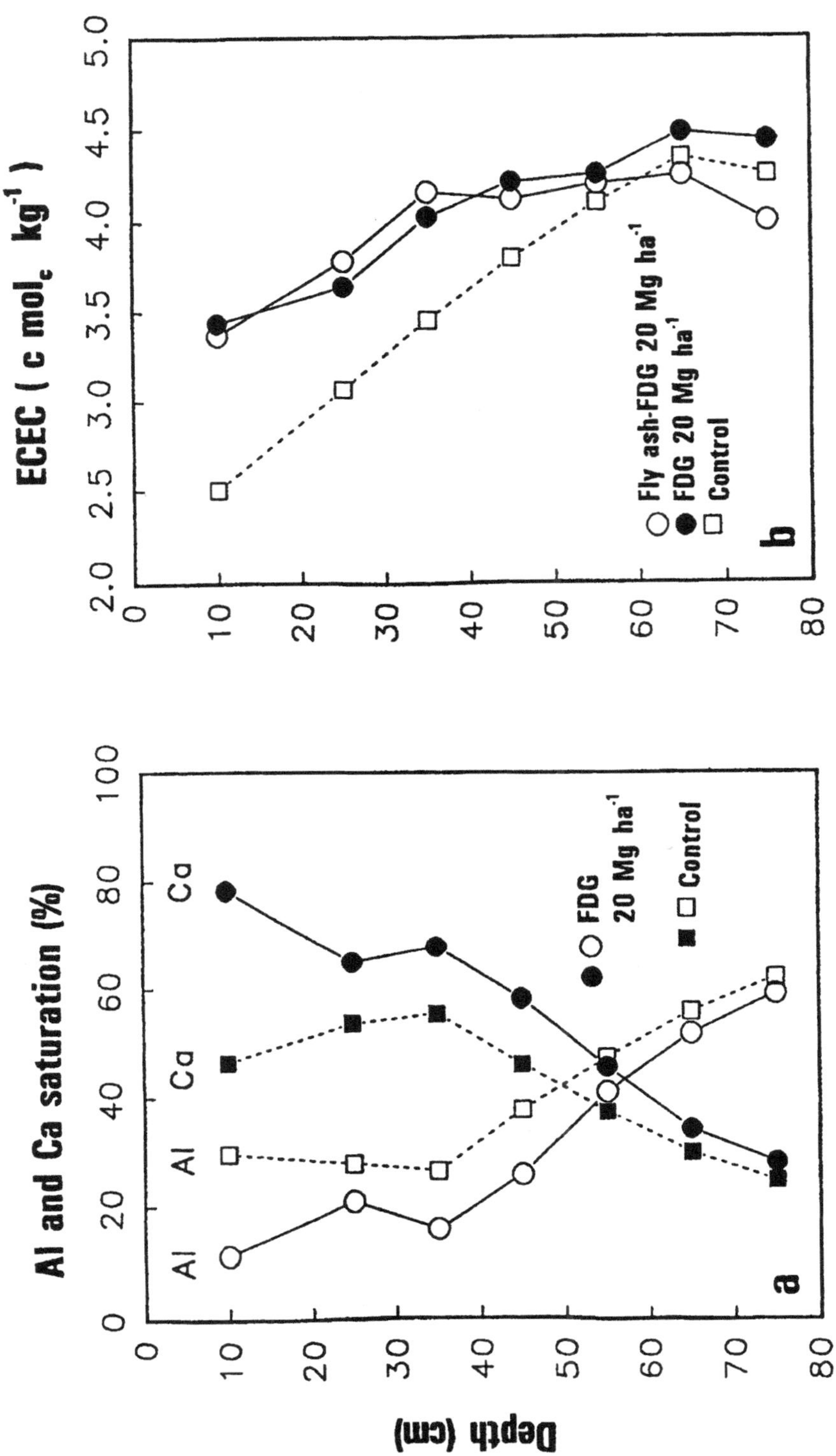

Figure 2. Al and Ca saturation a) and effective cation exchange capacity b) in the profileof soil amended with 20 Mg ha^{-1} rate of CCBP.

were the beneficial effects of the mixed by-product utilization; however economic viability of this strategy is questionable. Further research is needed to establish whether fly ash may offset the ameliorative effect of gypsum in an acid soil.

ACKNOWLEDGMENTS

This research was supported by a grant from Southern Company Services, Inc., under a Clean Technology Project grant from the US Department of Energy.

REFERENCES

Barnhisel & Bertsch PM (1982) Aluminum. In: Page AL, Miller RH & Keeny DR (Eds) Methods of Soil Analysis. Part 2. Agron. Monogr. 9 (pp. 275-300). ASA and SSSA, Madison, WI.

Bingham FT (1982) Boron. In: Page AL, Miller RH & Keeny DR (Eds) Methods of Soil Analysis. Part 2, Agron. Monogr. 9 (pp. 431-447). ASA and SSSA, Madison, WI.

Kukier U, Sumner ME & Miller WP (1994) Boron release from fly ash and its uptake by corn. J. Environ. Qual. 23:596-603.

Parker DR & Gardner EH (1981) The determination of hot-water-soluble boron in some acid Oregon soils using a modified azomethine-H procedure. Commun. Soil Sci. Plant Anal. 12:1311-1322.

Reeve NG & Sumner ME (1972) Amelioration of subsoil acidity in Natal Oxisols by leaching of surface-applied amendments. Agrochemophysica 4:1-6.

Resenauer HM, Walsh LM & RG Hoeft (1973) Testing soils for sulphur, boron, molybdenum and chlorine. In: Walsh LM and Beaton JD (Eds) Soil Testing and Plant Analysis. Revised Edition (pp. 173-200). SSSA Inc., Madison, WI.

Shamshuddin J & Ismail H (1995) Reactions of ground magnesium limestone and gypsum in soils with variable-charge minerals. Soil Sci. Soc. Am. J. 59:106-112.

Sousa DMG, Lobato E, Ritchey KD & Rein TA (1992) Suggestions for diagnosis and recommendations of gypsum application for soils of the Cerrados. In: 2nd Seminar on the Use of Gypsum in Agriculture. Uberaba, 24-26 March (pp. 139-158). IBRAFOS Brazil.

Sumner ME (1993) Gypsum and acid soils: the world scene. Advances in Agronomy. 51:1-32.

Thomas GW (1982) Exchangeable cations. In: Page AL, Miller RH & Keeny DR (Eds) Methods of Soil Analysis. Part 2, Agron. Monogr. 9 (pp. 159-165). ASA and SSSA, Madison, WI.

Warren CJ & MJ Dudas (1984) Weathering processes in relation to leachate properties of alkaline fly ash. J. Environ. Qual., 13:530-538.

AGRO-ENVIRONMENTAL IMPACT OF HIGH RATES OF PHOSPHOGYPSUM APPLIED TO BAHIAGRASS PASTURE ON A FLORIDA SPODOSOL SOIL

I.S. ALCORDO, J.E. RECHCIGL, R.C. LITTELL, A.K. ALVA and C.E. ROESSLER

University of Florida, Gainesville, Florida 33865, USA

E-mail rechcigl@gnv.ifas.ufl.edu

ABSTRACT

Phosphogypsum (PG)is an effective source of S and Ca for bahiagrass (*Paspalum notatum* Fluegge) pasture and is recommended as a source of S and Ca for other crops as well. Earlier studies by the authors showed that rates up to 1.0 Mg ha^{-1} applied annually or 2 or 4 Mg ha^{-1} applied once over a 3-year period increased bahiagrass yields linearly. The present study used 10 and 20 Mg ha^{-1}, with no PG as control, applied to bahiagrass on a spodic soil to help define the potential effects of PG at agronomic rates ($\leq$1.0 Mg ha^{-1})on PG-associated agro-environmental variables. Results showed that 20 Mg PG ha^{-1} reduced forage yield. There was no difference in yield between the control and the 10 Mg ha^{-1} rate. Phosphogypsum reduced soil Mg very strongly, hence, co-utilization of PG with by-products rich in Mg is recommended for crops that need large amounts of Mg in addition to S and Ca. Topsoil and subsurface water pH decreased with PG rates on the short term, however, PG should increase the pH over the long term as H^+ and Al^{3+} are replaced on the exchange complex by Ca^{2+}, brought into the free water phase, and leached into the deeper layers. There were little or no measurable effects on Cd, Cr, Pb, Se, and Hg in soil, groundwater, or bahiagrass forage tissue even at 20 Mg ha^{-1}. Evaluation of the PG used in the study indicated that it is not hazardous as defined by the EPA "toxicity characteristic" criterion for hazardous solid wastes.

INTRODUCTION

The production of phosphoric acid, used in the manufacture of phosphate fertilizers, by acidulation of rock phosphate with sulfuric acid produces phosphogypsum (PG) as by-product. Phosphogypsum is primarily $CaSO_4.2H_2O$. It has high S (19%) and Ca (25%) contents and may be co-utilized with other by-products that require enrichment in S and Ca and with animal wastes or urea fertilizer to reduce loss of NH_3 (Bayrakli, 1990). It can also substitute for the various uses of mined gypsum in agriculture (Alcordo and Rechcigl, 1993). Depending on the process used, 4.3 or 4.9 ton of PG is produced for each ton of P_2O_5 (Kouloheris, 1980). Based on the average PG:P_2O_5 ratio of 4.6 and on the 1991 P_2O_5

S. Brown, J.S. Angle and L. Jacobs (eds.), Beneficial Co-Utilization of Agricultural, Municipal and Industrial By-products, 423–430.

© 1998 *Kluwer Academic Publishers,*

production, estimates of PG productions by region and country are given in Table 1. As Table 1 shows, PG is produced worldwide and ought to be considered an important resource. But in view of some radioactive and/or heavy metal impurities, some producing countries, particularly the U.S., regard PG as a waste whose unregulated use poses a potential hazard to the environment. However, environmental studies showed that: 1) PG samples from 9 Florida PG stockpiles are not hazardous on the basis of the "leaching potentials" of As, Ba, Cd, Cr, Pb, Hg, Se, or Ag under the EPA "toxicity characteristic" criterion (May and Sweeney, 1984); 2) one-time application of 4 Mg PG ha^{-1} showed no measurable effects on the concentrations of some of the above-cited metals and on ^{226}Ra, ^{210}Pb, and ^{210}Po in soils during any year or when averaged over a 3-year period (Rechcigl et al., 1996); and 3) assessment of annual application of PG with about 800 Bq ^{226}Ra kg^{-1} at 0.4 Mg ha^{-1} over a 100-year period, assuming no loss in ^{226}Ra, indicated that the projected additions to ^{226}Ra concentration in soils are less than the existing background in a Florida Spodosol soil and are within the range of variations seen in Florida lands without elevated radioactivity (Rechcigl et al., 1997).

MATERIALS AND METHODS

This paper is part of a larger study which deals with one-time PG application of up to 20 Mg ha^{-1} to help define the effects of PG at low rates on PG-associated agronomic and radiological variables in soil, groundwater, and forage. The field experiment was conducted on an established bahiagrass pasture on a Myakka soil (Sandy, siliceous, hyperthermic Aeric Haplaquods) using a Florida PG (Table 2) applied at 0, 10, and 20 Mg ha^{-1} with 6 replicates. The plots measured 6.5 m x 6.5 m and were fertilized each year with 180 kg N (NH_4NO_3), 45 kg P (TSP), 67.5 kg K (KCl), and 28 kg micronutrient mix per ha. Soil samples were collected at the end of 1993 and 1994 growing seasons, dried, and passed through a 2-mm sieve. Two 10.8 cm ID PVC pipe wells, each provided with a 3-liter capacity collector at the bottom end, were installed at the center of each plot 1.5 m apart to collect subsurface water samples. One had its sampling slits at 45 cm and the other at 90 cm below the surface. A 1.5-liter covered plastic container was placed at the lowest point of a 15-cm deep ditch constructed around each plot to collect runoff. Water samples were collected after each heavy rain. Regrowth forage was sampled every 30-35 days of growth, oven-dried at 60°C, and ground to pass a 0.84 mm sieve. Soil samples were analyzed for pH in water (1:1) and water samples for pH and electrical conductivity (E_c). Plant nutrients and heavy metals in soil samples were extracted or solubilized using Mehlich 3 solution (Mehlich, 1984). Forage samples were ashed at 550°C and the ashes dissolved in 0.3025 M HCl. Except for N and S in forage which were analyzed using a N-S-C analyzer, all nutrients and heavy metals in soil, water, and forage were analyzed using ICP analytical methods. Fluorides in water and in forage were analyzed using F electrode methods. The data were statistically analyzed by ANOVA using LSD for comparison of means. Tests for linear and quadratic trends were done to define or quantify the trends with regression equations.

Table 1. Estimated 1991 phosphogypsum (PG) production (x1000 Mg) by region and country from P_2O_5 productions in "World Fertilizers Plant List & Atlas," 10th Ed (1994). CRU Publishing Ltd, England.

Asia:		**Africa**:		**North America**:	
Afghanistan	28	Algeria	248	Canada	2,765
Bangladesh	230	Egypt	787	United States:	
China, P.R.	20,806	Ivory Coast	9	N. Carolina	5,290
India	11,955	Morocco	4,066	Florida	33,410
Indonesia	2,300	Nigeria	124	Idaho	2,930
Japan	1,684	Senegal	225	Louisiana	6,969
North Korea	630	South Africa	1,147	Mississippi	1,293
South Korea	1,831	Tanzania	14	Texas	1,265
Malaysia	97	Togo	9	Wyoming	1,044
Pakistan	488	Tunisia	2,292	Sub-total	52,201
Philippines	759	Zimbabwe	202		
Sri Lanka	28	**Sub-total**	**9,123**	**South America**	
Taiwan	373			Brazil	5,239
Vietnam	304	**Europe:**		Chile	18
Sub-total	**41,513**	Albania	37	Colombia	129
		Austria	331	Cuba	32
Australia & Oceania:		Belgium/Lux.	1,564	Mexico	212
Australia	1,053	Bulgaria	170	Peru	5
New Zealand	782	Czeck/Slovak		Uruguay	101
Sub-total	**1,835**	Republic	773	Venezuela	253
		Denmark	230	**Sub-total**	**5,989**
Former Soviet Union:		Finland	745		
Belarus	1,191	France	3,947	**Grand Total**	**165,204**
Kazakhstan	460	Germany	994		
Lithuania	805	Greece	833		
Russia	16,270	Hungary	202		
Turmenistan	1,288	Italy	672		
Ukraine	3,165	Netherlands	1,500		
Uzbekistan	4,945	Norway	1,159		
Sub-total	**28,124**	Poland	1,159		
		Portugal	179		
Middle East:		Romania	994		
Iran	736	Spain	1,398		
Iraq	138	Sweden	368		
Israel	865	Switzerland	14		
Jordan	1,270	United			
Lebanon	184	Kingdom	308		
Saudi Arabia	920	Former			
Syria	97	Yugoslavia	1,030		
Turkey	1,725	**Sub-total**	**18,607**		
Sub-total	**5,935**				

Table 2. Total elemental analyses of PG[a] used in the study, as analyzed by two independent laboratories, and analyses of other PG samples from Florida PG stacks as reported by May and Sweeney (1980 and 1984).

Constituents	PG used in the study Lab #1	PG used in the study Lab #2	Average or Range in Florida PG[b]
Major:		mg kg^{-1}	
Calcium (Ca)	245,000	256,000	253,000
Sulfur (S)	198,000	-[c]	202,000
Phosphorus (P)	-[c]	3,000	3,200-23,600
Iron (Fe)	690	460	860-1,000±300-600
Fluorides (F)	4,300	-[c]	2,900-11,690
Aluminum (Al)	-[c]	1,100	2,000±540
Sodium (Na)260	260	520±79	
Potassium (K)	38	110	200-230
Magnesium (Mg)	16	13	<940
Trace:			
Molybdenum (Mo)	-[c]	17.0	2.2-11±1.4-2.2
Boron (B)	-[c]	<10.0	3.0
Zinc (Zn)	6.0	6.2	<340±21
Copper (Cu)	0.6	2.1	<82±<9.6
Manganese (Mn)	-[c]	1.9	25±14
Chloride (Cl)	-[c]	-[c]	<150±<4.7
Nickel (Ni)	-[c]	-[c]	2.0
EPA "TC"[d] metals:			
Barium (Ba)	46.0	45.0	<210±<24
Arsenic (As)	5.0	5.0	0.8-0.9±0.3
Silver (Ag)	2.0	<0.2	<1.3±<0.6
Cadmium (Cd)	0.7	1.1	3.4-4.0
Selenium (Se)	<0.05	1.6	0.7-2.1±0.4-0.7
Mercury (Hg)	-[c]	<0.01	0.3-0.4±0.2-0.3
Lead (Pb)	-[c]	4.0	2.0-13.0
Chromium (Cr)	-[c]	2.9	6.0±1.4

[a] Water-free basis.
[b] May and Sweeney, 1980 and 1984.
[c] Not analyzed.
[d] TC = "toxicity characteristics" (40 CFR, Part 261, 7-1-92).

RESULTS AND DISCUSSION

"Toxicity Characteristic" Determination of PG Used in the Study

A solid waste exhibits "toxicity characteristic" (TC) if the leachate of a sample contains any of the heavy metals As, Ba, Cd, Cr, Pb, Hg, Se, or Ag at a concentration equal to or greater than the maximum concentration level (MCL) established by EPA for that contaminant (EPA, 40 CFR, Part 261, 1992). The "leaching potentials" (LP) of the contaminants are determined using extraction procedure prescribed by EPA.

The determinative study on the LP of the TC metals in Florida PGs was that of May and Sweeney (1984) (Table 3). Using ratio-and-proportion, their data can be used to estimate the LP of any of the TC metals in other PG samples without going through the actual extraction and the analytical determination. Thus, with $C_{US}M$ as the total concentration of metal M in PG used in the study and AVG $C_{SS}M$ the average total concentration and AVG $LP_{SS}M$ the average LP of M in Florida PG stacks samples, then, the LP of M in PG used in the study is estimated by: $LP_{US}M = [(C_{US}M)/(AVG\ C_{SS}M)] \times [AVG\ LP_{SS}M]$. For Ba, $LP_{US}Ba = [(46\ mg\ kg^{-1})/(105\ mg\ kg^{-1})] \times 0.2\ mg\ L^{-1} = 0.09\ mg\ L^{-1}$ (Table 3). Table 3 shows that not a single TC metal in PG used in the study had a LP that equaled or exceeded its MCL. Thus, the PG used in the study is not a hazardous by-product under the EPA "toxicity characteristic" criterion.

Effects on Forage, Soil, and Surficial Groundwater

The beneficial effects of the high rates of PG on bahiagrass are the increase in dry matter, Ca, and S contents in forage (Table 4). The adverse effects are the reduction in yield, Mg, and Zn, and the increase in F in forage. Earlier studies by the authors showed that annual rates of 0.2 to 1.0 Mg ha^{-1} or the higher rates of 2 to 4 Mg ha^{-1} applied once over a 3-year period increased forage yields linearly with PG rates (Rechcigl et al., 1996). Fluoride is a cumulative poison and long-term feeding by cattle on forage with 30 or more mg F kg^{-1} could lead to fluorosis. However, the 2-year mean increase in F in forage ranged only from 0.15, as predicted by the quadratic equation (Table 4), to 0.17 mg F kg^{-1} dry matter per Mg PG ha^{-1}, as predicted by the linear equation which was also significant. Thus, F in PG does not pose a danger to livestock. Phosphogypsum strongly reduced the pH but increased Al concentrations in subsurface waters down to 45 cm and 90 cm. Reduction in pH and increase in Al concentrations were most likely the result of Ca^{2+} replacing H^+ and Al^{3+} from the soil exchange complex and bringing them into the free water phase. The hydrolysis of Al^{3+} could add more H^+ in the free water reducing the pH further. Since bahiagrass requires a soil pH of 5.5 (Kidder et al., 1990) for optimum yield, the low pH of the subsurface water surrounding the roots, and possibly the toxicity of Al^{3+} to the roots, may be responsible for the reduced yield at the high PG rates. Calcium, Mg, and P in soil, and Ca, Mg, F, and E_c in groundwater were also affected by PG (Table 4). The 2-year mean K, Cu, Fe, Na, Ni, Cd, Cr, Pb, Se, and Hg concentrations in soil, groundwater, or forage were not different from those of the control nor were their trends with PG rates significant. The same were true for P and N in forage.

ACKNOWLEDGMENT

The authors wish to thank the Florida Institute of Phosphate Research for the research grant which made the study possible.

Table 3. Maximum concentration level (MCL), mg L^{-1}, for "toxicity characteristic" (TC) determination of solid wastes; "leaching potential" (LP), concentrations, and leachability (%) of toxicity characteristic metals (TCM) in PG stacks samples, and estimated LP of TCM in PG used in the study.

Item	EPA toxicity characteristic" metals (TCM)[a]							
	As	Ba	Cd	Cr	Pb	Hg	Se	Ag
MCL for TC determination of solid wastes:	mg L^{-1}							
	5.0	100	1.0	5.0	5.0	0.2	1.0	5.0
LP in PG stack samples[b]:								
Stack A: 1	0.020	0.2	0.01	0.04	0.01	0.000	0.002	0.08
2	0.019	0.2	0.01	0.05	0.03	0.000	0.003	0.07
3	0.029	0.0	0.02	0.07	0.00	0.000	0.005	0.10
Stack B: 1	0.014	0.0	0.01	0.01	0.01	0.000	0.004	0.01
2	0.011	0.2	0.01	0.01	0.00	0.001	0.003	0.04
Stack C: 1	0.016	0.2	0.02	0.02	0.00	0.001	0.005	0.04
2	0.007	0.4	0.03	0.05	0.00	0.001	0.003	0.10
Stack D	0.009	0.0	0.01	0.11	0.01	0.001	0.003	0.09
Stack E	0.005	0.2	0.03	0.03	0.03	0.001	0.004	0.06
Stack F	0.002	0.1	0.01	0.01	0.04	0.001	0.002	0.05
Stack G	0.005	0.3	0.01	0.01	0.01	0.001	0.002	0.08
Stack H	0.018	0.3	0.01	0.05	0.00	0.001	0.003	0.04
Stack I	0.019	0.2	0.01	0.02	0.01	0.004	0.002	0.07
Average LP (AVG $LP_{SS}M$):	0.013	0.2	0.01	0.04	0.01	0.001	0.003	0.06
Total TCM:	mg kg^{-1}							
In stack samples (AVG $C_{SS}M$)[c]:	0.85	105	0.59	6.00	1.30	0.340	1.400	0.67
Amount leached:	0.24	3.2	0.28	0.80	0.36	0.012	0.060	0.96[d]
In PG used in study ($C_{US}M$)[e]:	5.0	46.0	1.1	2.9	4.0	0.01	1.6	2.0
Estimates of LP in PG used in study ($LP_{US}M$):	mg L^{-1}							
	0.08	0.09	0.02	0.02	0.03	<0.001	0.003	0.18

[a] EPA, 40 CFR (1992)

[b] May and Sweeney (1984)

[c] Average for PG stacks A1, B1, E, F, and H

[d] Indeterminate (quantity leached > total in PG)

[e] See Table 2.

Table 4. Two-year mean effects[1] of high rates PG on soil, groundwater, and bahiagrass forage expressed in regression equations, where 1T = 1 Mg PG ha^{-1} except for dry matter yield where 1T = 10 Mg PG ha^{-1}.

Forage:

Dry matter (DM) yield:	Mg DM ha^{-1} = 5.65 - 0.44T*
DM content:	mg DM kg^{-1} = 322.90 + 0.50T*
Calcium content:	mg Ca kg^{-1} = 4.40 + 0.08T*
Sulfur content:	mg S kg^{-1} = 1.90 + 0.47T - $0.015T^{2*}$
Magnesium content:	mg Mg kg^{-1} = 3.09 - 0.10T + $0.003T^{2*}$
Zn content:	mg Zn kg^{-1} = 19.25 - 0.17T*
Manganese content:	mg Mn kg^{-1} = 33.12 + 1.55T - $0.065T^{2ns}$
Fluoride content:	mg F kg^{-1} = 9.86 + 0.16T - $0.007T^{2*}$

Soil layer: depth (cm)		Groundwater: depth (cm)	
0-5	pH = 5.70 - 0.02T*	Runoff	pH = 5.86 - 0.03T*
15-30	pH = 5.20 - $0.01T^{ns}$	45	pH = 5.61 - 0.05T*
spodic	pH = 5.07 + $0.01T^{ns}$	90	pH = 5.23 - 0.08T*
	----- mg kg^{-1} ------		------- mg L^{-1} -------
0-5	Al = 83.0 - $1.37T^{ns}$	Runoff	Al = 0.28 + 0.04T*
15-30	Al = 41.0 - $0.93T^{ns}$	45	Al = 0.82 + 0.10T*
spodic	Al = 260.4 + $1.71T^{ns}$	90	Al = 1.81 + 0.12T*
0-5	Ca = 1102 + 63T*	Runoff	Ca = 49.0 + 12.9T*
15-30	Ca = 117 + 3T*	45	Ca = 189.6 + 17.5T*
spodic	Ca = 77 + $2T^{ns}$	90	Ca = 52.9 + 6.3T*
0-5	Mg = 101.74 - 4.30T*	Runoff	Mg = 9.54 + $0.13T^{ns}$
15-30	Mg = 9.84 - 0.21T*	45	Mg = 30.86 + 1.49T*
spodic	Mg = 12.53 + 0.16T*	90	Mg = 19.18 + 1.13T*
0-5	P = 22.65 + $0.28T^{ns}$	Runoff	P = 5.65 + $0.05T^{ns}$
15-30	P = 7.42 + 0.11T*	45	P = 3.18 + $0.06T^{ns}$
spodic	P = 37.05 + $0.74T^{ns}$	90	P = 0.67 - $0.02T^{ns}$
		Runoff	F = 0.145 + 0.043T*
		45	F = 0.244 + 0.033T*
		90	F = 0.420 + 0.034T*
			------ mS/m ---------
		Runoff	E_c = 37.8 + 7.2T*
		45	E_c = 110.0 + 10.4T*
		90	E_c = 60.3 + 5.7T*

ns = not significant (P>0.05)

* = significant at $P \leq 0.05$.

REFERENCES

Alcordo IS & Rechcigl JE (1993) Phosphogypsum in agriculture: A review. In: Spark DL (Ed) Advances in Agronomy (49: 55-118). Academic Press, Inc., New York.

Bayrakli F (1990) Ammonia volatilization losses from different fertilizers and effect of several urease inhibitors, $CaCl_2$ and phosphogypsum on losses from urea. Fert. Res. 23:147-150.

EPA, 40 CFR (1992) Part 261 - Identification and listing of hazardous waste (pp. 27-93).

Kidder G, Hanlon EA & Chambliss CG (1990) IFAS standardized fertilization recommendations for agronomic crops. Notes in Soil Science. University of Florida Coop. Ext. Ser. No. 35.

Kouloheris AP (1980) Chemical nature of phosphogypsum as produced by various wet process phosphoric acid processes. In: Borris DP & Boody PW (Eds) Phosphogypsum (pp. 7-33). Florida Institute of Phosphate Research. Bartow, FL.

May A & Sweeney JW (1984) Evaluation of radium and toxic element leaching characteristics of Florida phosphogypsum stockpiles. Report Investigation 8776. U.S. Bureau of Mines.

Mehlich A (1984) Mehlich 3 soil test extractant: A modification of Mehlich 2 extractant. Commun. Soil Sci. Plant Anal. 15(12):1409-1416.

Rechcigl JE, Alcordo IS, Roessler CE & Littell RC (1996) Influence of phosphogypsum on forage yield and quality and on the environment in typical Florida Spodosol soil (Vol I&II). Publ. No. 01-085-127. Florida Institute of Phosphate Research, Bartow, FL.

Rechcigl JE, Alcordo IS, Roessler CE, Alva AK & Littell RC (1997) Impact of high rates of phosphogypsum on radon emissions and on radioactivity and heavy metals in soils, groundwater, and bahiagrass forage (Final Report). Florida Institute of Phosphate Research, Bartow, FL.